THE ROUTLEDGE HANDBOOK OF HOUSING POLICY AND PLANNING

The Routledge Handbook of Housing Policy and Planning provides a comprehensive multidisciplinary overview of contemporary trends in housing studies, housing policies, planning for housing, and housing innovations in the United States, the United Kingdom, and Continental Europe. In 29 chapters, international scholars discuss aspects pertaining to the right to housing, inequality, homeownership, rental housing, social housing, senior housing, gentrification, cities and suburbs, and the future of housing policies.

This book is essential reading for students, policy analysts, policymakers, practitioners, and activists, as well as others interested in housing policy and planning.

Katrin B. Anacker is an Associate Professor at the Schar School of Policy and Government at George Mason University. She co-edited *Introduction to Housing* (2018) and edited *The New American Suburb: Poverty, Race and the Economic Crisis* (2015). Her work has appeared in *Housing Policy Debate*, the *Journal of Urban Affairs*, and *Housing Studies*. Her work has been funded by the National Science Foundation, the U.S. Department of Housing and Urban Development, and the Lincoln Institute of Land Policy.

Mai Thi Nguyen is an Associate Professor of City and Regional Planning at the University of North Carolina at Chapel Hill. Her scholarship has been motivated by a desire to understand how to create a more equitable social and spatial world. Her research focuses on housing policy, resilient communities, and socially vulnerable populations. Her work has been funded by the National Science Foundation, the U.S. Department of Homeland Security, the U.S. Department of Housing and Urban Development, and the John D. and Catherine T. MacArthur Foundation.

David P. Varady, a Professor at the University of Cincinnati, has authored six books, nine book chapters, 76 journal articles, and 89 book reviews on neighborhood development and housing. He has held Visiting Scholar positions at TU Delft in the Netherlands, the City of Helsinki, Rutgers University, University of Glasgow, the National Association of Realtors, and the U.S. Department of Housing and Urban Development.

THE ROUTLEDGE HANDBOOK OF HOUSING POLICY AND PLANNING

Edited by Katrin B. Anacker, Mai Thi Nguyen, and David P. Varady

Routledge
Taylor & Francis Group
NEW YORK AND LONDON

First published 2020
by Routledge
605 Third Avenue, New York, NY 10017

and by Routledge
2 Park Square, Milton Park, Abingdon, Oxon, OX14 4RN

First issued in paperback 2021

Routledge is an imprint of the Taylor & Francis Group, an informa business

Library of Congress Cataloging-in-Publication Data
Names: Anacker, Katrin B., editor. | Nguyen, Mai Thi, editor. | Varady, David P., editor.
Title: The Routledge handbook of housing policy and planning / edited by Katrin B. Anacker, Mai T. Nguyen, and David P. Varady.
Description: New York, NY : Routledge, 2019. | Includes index.
Identifiers: LCCN 2019008927 (print) | LCCN 2019010401 (ebook) | ISBN 9781315642338 (e-book) | ISBN 9781138188433 (hardback)
Subjects: LCSH: Housing policy. | Housing—Planning. | City planning.
Classification: LCC HD7287.3 (ebook) | LCC HD7287.3 .R68 2019 (print) | DDC 363.5/561—dc23
LC record available at https://lccn.loc.gov/2019008927

N 13 978-1-03-224084-8 (pbk)
N 13 978-1-138-18843-3 (hbk)

10.4324/9781315642338

Typeset in Bembo
by Apex CoVantage, LLC

KBA: für Mama und Papa (nein, ihr braucht das Buch nicht zu lesen).

MTN: to Moxy, Layla, and Greg, for making it all worthwhile.

DPV: dedicated to Adrienne Varady who throughout my career has provided me with emotional support and who also has been my most severe (and helpful) critic.

CONTENTS

List of Figures *xi*
List of Tables *xiii*
List of Contributors *xv*

Editors' Introduction 1
Katrin B. Anacker, Mai Thi Nguyen, and David P. Varady

SECTION 1
Right to Housing **9**

1 The Right to Housing: The Goal Versus the Reality 11
W. Dennis Keating

2 A Home Away From Home: Housing Refugees in the Netherlands During the European Refugee Crisis 23
Alfons Fermin and Frank Wassenberg

SECTION 2
Inequality **35**

3 Homeownership and the Racial and Ethnic Wealth Gap in the United States 37
Carolina K. Reid

4 Non-Hispanic White Versus Black Parental Wealth and Wealth Transfers to Enable Homeownership in Five Metropolitan Areas 54
Erin Graves, Ana Patricia Muñoz, Darrick Hamilton, William A. Darity, and Yunju Nam

5 Should Policy Seek to Interfere With Upward Mobility by Deconcentrating Poverty? Reasons for Concern 68
Howard Husock

6 The Affordable Housing Complex: Direct and Exclusionary Displacement in the Lacy and Logan Neighborhoods of Santa Ana, California 76
J. Revel Sims and Carolina S. Sarmiento

7 Neighborhood Centers 92
Emily Talen

SECTION 3
Homeownership **101**

8 Demographic and Policy Factors Affecting Homeownership Across OECD Countries 103
Mark Calabria and Vanessa Brown Calder

9 Declining Homeownership in Liberal, English Speaking Countries 117
Richard Ronald and Christian Lennartz

SECTION 4
Rental Housing **127**

10 Subsidized Rental Housing Programs in the U.S.: A Case of Rising Expectations 129
Kirk McClure

11 Redefining Rental Housing Choice in the Housing Choice Voucher Program 141
Andrew J. Greenlee

12 Accommodating and Accumulating: How the Property Interests of Homeowners and Renters Impact Housing Satisfaction 155
Allison Freeman and Kimberly Manturuk

SECTION 5
Social Housing **171**

13 The U.S. Approach to "Social Housing" 173
Rachel G. Bratt

14 The Privatization of American Public Housing: Leaving the Poorest of the Poor Behind 189
Lawrence J. Vale and Yonah Freemark

15 Public Housing Authorities as Social Enterprises? 207
Rachel Garshick Kleit

16 How the European Commission Affected Social Rental Housing in the Netherlands and Germany 220
Marja G. Elsinga and Marietta E.A. Haffner

SECTION 6
Senior Housing **229**

17 Connectivity as an Indicator of Older People's Housing Quality 231
Stephen M. Golant

18 Housing Models for Aging in Community 243
Sherry Ahrentzen and Ruth L. Steiner

19 Realizing Innovative Senior Housing Practices in the U.S. 256
Deirdre Pfeiffer, Ashlee Tziganuk, Scott Cloutier, Julia Colbert, and Gracie Strasser

20 Designed for All Ages: Multigenerational Housing as a Potential Housing Option in Flanders/Belgium 271
Sebastiaan Gerards, Erik Nuyts, and Jan Vanrie

SECTION 7
Gentrification **283**

21 A Moving Target: The Shifting Genealogy of Gentrification 285
Dennis E. Gale

22 Preventing Gentrification-Induced Displacement in the U.S.: A Review of the Literature and a Call for Evaluation Research 302
Miriam Zuk

23 Urban Restructuring, Demolition, and Displacement in the Netherlands: Uncovering the Janus Head of Forced Residential Relocation 317
Reinout Kleinhans

24 State-Sponsored Gentrification or Social Regeneration? Symbolic Politics and Neighborhood Intervention in an Amsterdam Working-Class Neighborhood 330
Wouter van Gent, Willem Boterman, and Myrte Hoekstra

SECTION 8
Suburbs **345**

25 Housing Policy and the Suburban Metropolis: A Focus on the United States and France 347
Bernadette Hanlon and John Rennie Short

26 A New Generation of "Single-Family" Homes: Multigenerational Homebuilding in the Suburbs of Phoenix, Arizona 357
Willow S. Lung-Amam

SECTION 9
The Future of Housing **373**

27 Addressing Affordability Challenges: The Role and Scope of Local Housing Plans 375
Ingrid Gould Ellen

28 Affordable Housing: Program Financing and Policies in U.S. States 390
Victoria Basolo

29 Prediction Is Difficult: The Future of Housing Policy and Housing Studies 406
Alan Mallach

Index *416*

FIGURES

2.1 First Asylum Applicants in the European Union, September 2013 to August 2017 24
2.2 Number of Asylum Applications in the Netherlands, 1990–2016 26
2.3 Dutch Asylum Procedure 27
2.4 Temporary Refugee Housing Next to Offices in Rijswijk, the Netherlands 30
2.5 Temporary Use of an Old Elementary School in Landgraaf, the Netherlands 30
2.6 The Finch in Leiden, One of Six Winners in the A Home Away From Home Competition 31
2.7 Startblok Riekerhaven, Amsterdam 32
3.1 Trends in the Democratization of Wealth, 1917–2012 38
3.2 Racial and Ethnic Wealth Gap, 1963–2013 38
3.3 Racial and Ethnic Disparities in Credit Scores, 2004–2008 41
3.4 Serious Delinquency Rates by Borrower Race and Ethnicity and Income, 2004–2007 Originations, Observed Through January 2013 42
3.5 Trends in Racial and Ethnic Segregation, 1980–2010 43
4.1 Average Income for Non-Hispanic Whites Versus Blacks for the Five Metropolitan Areas 60
4.2 Homeownership Rates for non-Hispanic Whites versus Blacks for the Five Metropolitan Areas 61
6.1 Aerial View of the Depot at Santiago Project and Other Sites 84
7.1 Hegemann and Peets' Milwaukee Grand Circle, an Effort to "Insert a Pleasant Variation Into an Existing Gridiron" 93
7.2 The Center Is Believed to Provide a Visible, Literal Focal Point, as Well as a Permanent Symbol of the Common Bond That People Living in the Same Neighborhood Share 94
8.1 Homeownership Rate, United States, 1890–2015 104
8.2 Mortgage Debt to GDP Ratio, 2012 112
9.1 Tenure Changes (in Percent) Among 18–34 Year-Olds (2007–2012) in Select Countries in Europe 122
11.1 Housing Choice Voucher Program Structure 142
13.1 Major Components and Approximate Size of the U.S. Assisted Housing Sector, 2016 174

13.2	Components and Approximate Size of the U.S. Social Housing Sector, 2016	175
14.1	Trends in Deep Subsidies for Federally Assisted Rental Housing	192
14.2	Other "Projects" on the Rise in Large U.S. Cities: Privately Developed Public Housing (2015)	197
14.3	The Low-Income Housing Landscape: Three Decades of Dramatic Change	198
14.4	Deeply Subsidized Rental Housing in the United States: A Declining Commitment?	200
15.1	PHAs by Size and Percent of Units for that Size, Vouchers and Public Housing Units Combined	211
17.1	Connectivity Options for the Older Population Aging in Place	233
18.1	Conceptual Framework of Processes through which the NORC Program and Village Models Potentially Influence Aging in Place	247
18.2	Silver Sage Cohousing, a Senior Cohousing Development in Boulder, Colorado	249
18.3	Durham Central Park Cohousing in Downtown Durham, North Carolina	250
20.1	Possible Transformation of Multigenerational Dwelling into Student Housing and Vice Versa	279
21.1	Park Slope, Brooklyn, Experienced Reinvestment in the 1970s	287
21.2	Newcomers to the Lafayette Square Neighborhood of St. Louis Were Attracted to Mansard-Style Houses Such as These	288
21.3	San Francisco Mission District Was Rapidly Gentrifying in the 1980s	289
21.4	Even Structurally Unsound Buildings in Paris' Marais Are Too Highly Desired to Demolish	290
21.5	Example of an Apartment Block in Central Munich	291
22.1	Framework of Affordable Housing Strategies	307
22.2	Publicly Supported Rental Homes With Expiring Affordability Restrictions Within Five Years by Funding Stream	310
26.1	Lennar's NextGen® Evolution Floor Plan for the Layton Lakes Community in Gilbert, AZ	364
26.2	Inside the Evolution NextGen® Suite in Layton Lakes, Gilbert, AZ	365
26.3	Layton Lakes, Gilbert, AZ, Where Multiple Developers Sell New Homes Marketed to Multigenerational Households	365
26.4	In Chandler and Gilbert, Some Homes of the Artesian Ranch Development Include Detached Casitas that May Be Used to Accommodate Multigenerational Households	366
28.1	Number of Housing Units Assisted by Low-Income Housing Tax Credits in the U.S., 1995–2014	394

TABLES

4.1	Percentage of Households that Received Gifts or Loans for Down Payment	62
4.2	Percentage of Respondents Whose Family Members Paid 50 Percent or More of Education Costs	62
6.1	Percent of Eligible Households in Lacy (750.02) and Logan (744.05) Eligible for AMI-Based Income Categories, 2009–2015	82
6.2	Income Limits for Affordable Housing Projects in the Station District	83
8.1	Homeownership Rate in the United States by Age, Race, and Ethnicity, 1994, 2004, and 2014	105
8.2	Homeownership Rates Across OECD Countries	107
11.1	Household-Weighted Census Tract Demographics and Socioeconomics	145
11.2	Household-Weighted Census Tract Demographics and Socioeconomics for Racially Mixed Tracts (Less Than 30 Percent African-American Population)	146
11.3	Household-Weighted Census Tract Demographics and Socioeconomics for Low-Poverty Tracts (Less Than 10 Percent Poverty)	146
12.1	Descriptive Statistics for Sample of Homeowners (n = 1,623)	158
12.2	Descriptive Statistics for Sample of Renters (n = 549)	159
12.3	OLS Regression Model Predicting Housing Satisfaction (1–10 Scale) for Homeowners	162
12.4	OLS Regression Model Predicting Housing Satisfaction (1–10 Scale) for Renters	163
15.1	Variations in Public Housing Authorities, 2015	212
15.2	Variations in Public Housing Authority Enabling Legislation	214
15.3	Subsidies Other than Public Housing for Units Under PHA Ownership	216
20.1	Coefficients of the Variables Predicting the Willingness to Participate in a Multigenerational Dwelling	274
20.2	Coefficients of the Variables Predicting the Preference to Use Different Spaces Privately	276
22.1	Framework for Organizing Neighborhood Stabilization Strategies	304
24.1	Overview of Identified Organizations by Type and Association with the City of Amsterdam and Ymere, and Count of Organizations that (Re-)Produce Middle-Class Symbolic Markers	337
27.1	City Characteristics, 2014	381

27.2	Summary of Local Housing Plans	383
28.1	Private Activity Bond Issuance in the U.S., 2015	393
28.2	Selected State Housing Efforts by Origin of Program and Usage Level	397
28.3	Housing and Socio-Demographics, California and the U.S.	399
28.4	California's Use of Federal Housing Programs/Authorization	400

CONTRIBUTORS

Sherry Ahrentzen, Ph.D., is Shimberg Professor of Housing Studies at the University of Florida. Her research adopts an ecological approach toward building performance for occupant health, ranging from policy to design/construction to human behavior. Her doctorate in Social Ecology from UC Irvine reflects the multidisciplinary, collaborative nature of her work.

Katrin B. Anacker, Ph.D., is an Associate Professor at the Schar School of Policy and Government at George Mason University. She co-edited *Introduction to Housing* (2018) and edited *The New American Suburb: Poverty, Race and the Economic Crisis* (2015). Her work has appeared in *Housing Policy Debate*, the *Journal of Urban Affairs*, and *Housing Studies*. Her work has been funded by the National Science Foundation, the U.S. Department of Housing and Urban Development, and the Lincoln Institute of Land Policy.

Victoria Basolo is a Professor of Urban Planning and Public Policy at the University of California, Irvine. Her research investigates the economic, political, and social forces impacting affordable housing. Her work appears in numerous publications such as the *Journal of Urban Affairs*, *Housing Studies*, *Cityscape*, and *Housing Policy Debate*.

Willem Boterman is at the Department of Geography, Planning, and International Development Studies of the University of Amsterdam. His work centers around spatial reproduction of gender and class in space, focusing on issues of gentrification, segregation, and school choice.

Rachel G. Bratt, Professor Emerita, was on the Tufts University faculty from 1976 to 2014, serving as chair for nearly ten years. She is currently a Senior Research Fellow at the Joint Center for Housing Studies, Harvard University. Her work focuses on affordable housing and community development, particularly the role of nonprofit organizations.

Mark Calabria previously served as the Cato Institute's Director of Financial Regulation Studies. He has also held positions with Harvard University's Joint Center for Housing Studies. Dr. Calabria has held a number of governmental posts, including with the Senate Banking Committee and the U.S. Department of Housing and Urban Development.

Vanessa Brown Calder is a policy analyst at the Cato Institute, where she focuses on housing and urban policy. Previously, Calder was a graduate fellow at The Heritage Foundation, where she analyzed housing subsidies. Calder holds a master's degree in public policy from Harvard's John F. Kennedy School of Government and a B.S. in urban planning.

Scott Cloutier is an Assistant Professor in the School of Sustainability at Arizona State University. He integrates traditional and emerging sustainable community development paradigms and happiness science to promote the move toward a sustainable future.

Julia Colbert is a graduate student in the School of Sustainability at Arizona State University. She focuses on integrating social and environmental justice into sustainability and environmental education. She works primarily with K–12 educators and students to develop sustainability solutions for their schools and communities.

William A. Darity is the Samuel DuBois Cook Professor of Public Policy, African and African-American Studies, and Economics, and the Director of the Samuel DuBois Cook Center on Social Equity at Duke University. His research interests include inequality by race, class, and ethnicity, stratification economics, schooling and the racial achievement gap, North-South theories of trade and development, skin shade and labor market outcomes, the economics of reparations, and the Atlantic slave trade and the Industrial Revolution.

Ingrid Gould Ellen is the Paulette Goddard Professor of Urban Policy and Planning at NYU Wagner and a Faculty Director at the Furman Center for Real Estate and Urban Policy. She has written numerous journal articles and book chapters related to housing policy, community development, and school and neighborhood segregation.

Marja G. Elsinga is a Full Professor of Housing Institutions and Governance at the Faculty of Architecture and the Built Environment at Delft University of Technology in the Netherlands. She is also Visiting Professor in the College of Architecture and Urban Planning at Tongji University in Shanghai/China. She is an expert in teaching, editing, and publishing in housing policy, housing affordability, comparative research, and housing institutions.

Alfons Fermin, Dr., is an independent social scientific researcher at Fermin Research & Consultancy with expertise in migration, inclusion of vulnerable groups, citizen participation, urban development, and multilevel governance. He works on the nexus between policy, practice, and research at the local, national, and European level.

Allison Freeman is a Senior Research Associate at the UNC Center for Community Capital. Her research explores how attributes of place affect lower-income people's ability to move up the economic ladder.

Yonah Freemark is a doctoral candidate in the Department of Urban Studies and Planning at the Massachusetts Institute of Technology. His research lies at the intersection of regional governance, urban history, city design, and local politics.

Dennis E. Gale is a Professor Emeritus at Rutgers University and teaches in the Urban Studies and Public Policy programs at Stanford University. The founding director of the Cornwall Center for Metropolitan Studies at Rutgers, he was a 2013 Fulbright Scholar. He has published widely on gentrification, housing, and neighborhood change.

Sebastiaan Gerards is currently affiliated with Landmarken AG and holds a Ph.D. in Architecture from Hasselt University in Belgium. He has published on housing and multigenerational dwellings in Flanders and in Germany, and is the coauthor of the book *Narrative Architecture* (2017).

Stephen M. Golant, Ph.D., is a Professor in the Department of Geography at the University of Florida (Gainesville). He writes on housing, mobility, transportation, and long-term care needs of the older adult population. His most recent book is *Aging in the Right Place*, published by Health Professions Press.

Erin Graves is a Senior Policy Analyst and Policy Advisor with the Federal Reserve Bank of Boston and an Adjunct Instructor of Sociology at Northeastern University with an interest in low-income housing policy and urban social theory. Erin has a doctorate degree in Sociology and Urban Planning and a master's degree in Urban Planning. She has published work in the *Journal of the American Planning Association*, *City & Community*, *Housing Policy Debate*, and the *Journal of Planning Literature*.

Andrew J. Greenlee is an Assistant Professor in the Department of Urban and Regional Planning at the University of Illinois at Urbana-Champaign. His research focuses on the intersection of neighborhood change, residential mobility, and governance.

Marietta E.A. Haffner is in the Faculty of Architecture and the Built Environment at Delft University of Technology in the Netherlands. An economist by training, Marietta has more than 25 years of experience in conducting European comparative housing studies. She is holding an honorary research appointment at RMIT University in Melbourne in Australia. Marietta is one of the three Managing Editors of *Housing Studies*.

Darrick Hamilton is the Executive Director of the Kirwan Institute for the Study of Race and Ethnicity and a Professor at the John Glenn College of Public Affairs at The Ohio State University. Professor Hamilton is a stratification economist whose work fuses social science methods to examine the causes, consequences, and remedies associated with racial and ethnic inequality.

Bernadette Hanlon is an Associate Professor of City and Regional Planning at The Ohio State University. Dr. Hanlon specializes in the political economy of metropolitan regions, focusing on issues of suburban transformation, housing conditions, and demographic analysis. Her work has been published in a wide range of academic journals and books.

Myrte Hoekstra, Dr., is a postdoctoral researcher at the University of Amsterdam in the Netherlands. Her research focuses on the governance and experience of social differentiation, in particular racial and ethnic diversity. She is currently involved in a research project on the sociocultural dimensions of deindustrialization and population shrinkage.

Howard Husock is Vice President for Research and Publications at the Manhattan Institute, where he is also a contributing editor to its quarterly magazine. He is the author of *Philanthropy Under Fire* (2013) and *The Trillion-Dollar Housing Mistake: The Failure of American Housing Policy* (2003).

W. Dennis Keating is a Professor Emeritus in the College of Urban Affairs at Cleveland State University. He has researched and published widely in the areas of housing policy, neighborhood development, and urban policy. His most recent book was a history of one of Cleveland's oldest neighborhoods—Tremont.

Reinout Kleinhans is an Associate Professor of Urban Regeneration in the Faculty of Architecture and the Built Environment at Delft University of Technology in the Netherlands. His research interests are urban regeneration, self-organization, community entrepreneurship, online/offline citizen engagement, and social capital. He conducts international comparative research on community enterprises, citizen-based co-production, and other forms of area redevelopment.

Rachel Garshick Kleit, Ph.D., is a Professor of City and Regional Planning in the Knowlton School of Architecture and Associate Dean in the College of Engineering at The Ohio State University. Her research focuses on the social impacts of housing, housing mobility and instability, and public housing authorities as developers.

Christian Lennartz, Dr., is a Senior Housing Economist at RaboResearch, a research unit of Rabobank in the Netherlands. Previously he was a postdoctoral fellow at the University of Amsterdam. He has published numerous articles on housing careers and housing systems with a comparative perspective. Most recently he guest edited the special issue "Housing Careers, Intergenerational Support and Family Relations" for *Housing Studies.*

Willow S. Lung-Amam is an Assistant Professor of Urban Studies and Planning at the University of Maryland, College Park. Her research focuses on social inequality and the built environment, including the book *Trespassers? Asian Americans and the Battle for Suburbia* and other projects related to suburban poverty, immigration, gentrification, and neighborhood opportunity.

Alan Mallach is a Senior Fellow at the Center for Community Progress and a Visiting Professor at the Pratt Institute in New York City. His research focuses on the revitalization of older cities and neighborhoods. His latest book is *The Divided City: Poverty and Prosperity in Urban America.*

Kimberly Manturuk is the Head of Evaluation and Learning Experience Research at Duke University. She is also an Instructor in the Department of Sociology at the University of North Carolina at Chapel Hill, where she teaches introductory sociology and criminology. She is an expert on evaluation methodologies and consults with faculty members at both universities to design complex research projects. She has a Ph.D. in Sociology from the University of North Carolina at Chapel Hill.

Kirk McClure is a Professor in the Urban Planning Program at the University of Kansas. He specializes in the areas of housing and real estate development with an emphasis on housing affordability. His work has won multiple national awards and is found in many of the leading journals in planning.

Ana Patricia Muñoz is the Executive Director of Grupo FARO, a think tank based in Quito, Ecuador. She is the former Assistant Vice President at the Federal Reserve Bank of Boston, where she directed the community development research group. Her research focuses on racial and ethnic wealth inequality, Latino entrepreneurship, and urban development.

Yunju Nam is an Associate Professor at the School of Social Work, University at Buffalo, State University of New York. Her scholarly interests include economic inequality and social policy for disadvantaged populations. Her current research focuses on the racial wealth disparity, intergenerational economic mobility, financial capability and asset building, and immigration and refugees.

Mai Thi Nguyen is an Associate Professor of City and Regional Planning at the University of North Carolina at Chapel Hill. Her scholarship has been motivated by a desire to understand how to create a more equitable social and spatial world. Her research focuses on housing policy, resilient communities, and socially vulnerable populations. Her work has been funded by the National Science Foundation, the U.S. Department of Homeland Security, the U.S. Department of Housing and Urban Development, and the John D. and Catherine T. MacArthur Foundation.

Erik Nuyts is a Professor at Hasselt University and University College PXL in Belgium. He has many years of experience in research methodology and statistical analysis in a variety of research domains.

Deirdre Pfeiffer is an Associate Professor in the School of Geographical Sciences and Urban Planning at Arizona State University. She has expertise on housing as a cause and effect of growing social inequality and the role of housing planning in meeting the needs of diverse social groups.

Carolina K. Reid is an Assistant Professor in the Department of City and Regional Planning at the University of California at Berkeley. Carolina specializes in housing and community development, with a specific focus on affordable housing, place-based anti-poverty strategies, and access to credit for low-income and minority households.

Richard Ronald is a Professor of Housing Society and Space at the University of Amsterdam in the Netherlands. Richard also holds an honorary position at the University of Birmingham in the UK. He has published widely on housing in relation to social, economic, and urban transformations in Europe and Pacific Asia. He has also received a number of awards, including JSPS and Japan Foundation Fellowships, and an ERC Frontier grant.

Carolina S. Sarmiento is an Assistant Professor at the University of Wisconsin-Madison in the School of Human Ecology and an Affiliate with the Department of Urban and Regional Planning. Her research focuses on community-based planning in low-income communities of color, addressing inequality and injustice.

John Rennie Short is a Professor in the School of Public Policy at the University of Maryland Baltimore County. Recent books include *Hosting the Olympic Games* (2018) and *The Unequal City* (2018). His work has been translated into Chinese, Czech, Japanese, Korean, and Spanish. His essays have appeared in the *Associated Press, Business Insider, Citiscope, City Metric, Market Watch, Newsweek, PBS Newshour, Quartz, Salon, Slate, Time, US News and World Report, Washington Post,* and *World Economic Forum.*

J. Revel Sims is an Assistant Professor in the Department of Planning and Landscape Architecture at the University of Wisconsin-Madison. Dr. Sims' research focuses on the intersection of housing, urbanization, and power through the concept of "eviction-based displacement." His work includes empirical investigations on eviction patterns as well as theoretical contributions regarding the conflicts that arise around claims to urban land and the production of everyday urban space.

Ruth L. Steiner is a Professor and Director of the Center for Health and the Built Environment in the School of Landscape Architecture and Planning at the University of Florida. Her work focuses on the coordination of land use and transportation to meet the needs of people of all ages.

Gracie Strasser is an undergraduate student at Barrett, the Honors College at Arizona State University. She is interested in industrial design and sustainability, particularly in the design of public space as well as how humans function within it.

Emily Talen is a Professor of Urbanism at the University of Chicago. She has written extensively on the topics of urban design, new urbanism, and social equity. Her latest book is titled *Neighborhood* (Oxford University Press, 2018). She co-founded the *Journal of Urbanism* and is a Guggenheim Fellow.

Ashlee Tziganuk is a doctoral candidate in the School of Geographical Sciences and Urban Planning at Arizona State University. Her dissertation research explores how urban inhabitants in grassroots organizations push back against the planning decisions that affect their communities.

Lawrence J. Vale is Ford Professor of Urban Design and Planning at MIT and Director at MIT's Resilient Cities Housing Initiative (RCHI). He is the author or editor of ten books examining urban design, housing, and planning, including four prize-winning volumes on American public housing.

Wouter van Gent is an Assistant Professor at the Department of Geography, Planning and International Development Studies at the University of Amsterdam in the Netherlands. His work focuses on urban inequalities, particularly urban segregation and gentrification. In addition to analyzing socio-spatial shifts, he addresses how changes are produced by state and housing institutions and in culture.

Jan Vanrie is a Professor of Human Sciences and Research Methodology and the Coordinator of the research group ArcK at the Faculty of Architecture and Arts at Hasselt University in Belgium. His research interests lie at the intersection of environmental psychology, (interior) architecture, and design research.

David P. Varady, a Professor at the University of Cincinnati, has authored six books, nine book chapters, 76 journal articles, and 89 book reviews on neighborhood development and housing. He has held Visiting Scholar positions at TU Delft in the Netherlands, the City of Helsinki, Rutgers University, University of Glasgow, the National Association of Realtors, and the U.S. Department of Housing and Urban Development.

Frank Wassenberg is a Senior Project Officer at Platform 31, where he is responsible for housing, neighborhood, and urban development projects in the Netherlands and abroad. His dissertation was about integrated approaches to improve large housing estates. He is also a visiting researcher at Delft University of Technology.

Miriam Zuk, Ph.D., is the Director of the Urban Displacement Project and the Center for Community Innovation at UC Berkeley. Her research supports the advancement of environmental justice and equitable development. Dr. Zuk holds a Ph.D. from UC Berkeley, an M.S. from MIT, and a B.A. from Barnard College.

EDITORS' INTRODUCTION

Katrin B. Anacker, Mai Thi Nguyen, and David P. Varady

Housing may be understood in many ways. For a homeowner, a home may be a permanent place characterized by stability and a high quality of life; for a renter, an apartment may be a temporary place characterized by household challenges (Edin and Shaefer 2015). For some homeowners, housing may be a solid foundation for building long-term wealth—for others, it may mean foreclosure, eviction, moving in with family or friends, and difficulties rebuilding one's credit history (Bernanke 2015; Dayen 2016; Graham 2017; Immergluck 2015; Martin and Niedt 2015; see also the chapters authored by Carolina Reid, and by Erin Graves, Ana Patricia Muñoz, Darrick Hamilton, William Darity, and Yunju Nam). Neighborhoods and cities may also be understood in multiple ways. Some neighborhoods have a well-constructed and well-maintained housing stock, resulting in property value appreciation and a high quality of life, while other neighborhoods are characterized by vacancies, vandalism, and crime (see also the chapters authored by Andrew J. Greenlee, Alan Mallach, and Kirk McClure). Some cities have booming local economies, resulting in high revenues yet housing affordability challenges, while other cities have depressed economies, resulting in public service challenges (Kinder 2016). Thus, some societies have a large proportion of the housing stock that is safe, affordable, and attractive, while others face many housing challenges (Desmond 2016; see also the chapters authored by Miriam Zuk, Reinhout Kleinhans, and Wouter van Gent, Willem Boterman, and Myrte Hoekstra). In summary, housing can be seen through many lenses.

The housing landscape in the United States has changed dramatically over the past two decades. In 2000 house prices rapidly increased, and then they crashed in the spring of 2006, triggering the much discussed national subprime and foreclosure crises that expanded to the Great Recession from December 2007 to June 2009 and the Global Financial Crisis in 2008 (see also the chapter authored by Richard Ronald and Christian Lennartz). While the economy has largely recovered, many communities and households, and especially communities of color, are still recovering from the Great Recession (see also the chapter authored by J. Revel Sims and Carolina S. Sarmiento). The national homeownership rate has decreased from 69.2 percent in the second quarter of 2004 to 64.2 percent in the first quarter of 2018, a 5 percentage point decrease (U.S. Bureau of the Census, n.d.a). The Black homeownership rate has decreased from 49.7 percent in the second quarter of 2004 to 42.2 percent in the first quarter of 2018, a 7.5 percentage point decrease (U.S. Bureau of the Census, n.d.a). According to the 1970 centennial Census, the national homeownership rate was 62.9 percent, while the Black homeownership rate was 41.6 percent (U.S. Bureau of the Census, n.d.b). In other words, almost all gains made over almost five decades were eradicated (see also the chapters authored by Carolina Reid, and by Erin Graves, Ana Patricia Muñoz, Darrick Hamilton, William Darity, and Yunju Nam).

The housing policy landscape has also shifted dramatically over the past two decades. In the 1990s and 2000s the Clinton and the Bush administrations encouraged homeownership. In the mid- to late 2000s the Bush and Obama administrations intervened for the financial sector in multiple ways. First, through the Emergency Economic Stabilization Act (EESA), which was passed in October 2008 and for which $700 billion was allocated (Anacker 2018). The EESA included the Troubled Assets Relief Program (TARP), for which $445 billion was initially allocated, $377 billion repaid, and $61 billion written off (Lerner et al. 2017). Originally, the government designed TARP to purchase toxic assets, but it ended up being a program that disbursed funds to banks and the auto industry in exchange for dividend-paying preferred stock (Zandi 2012). Due to TARP, the financial system survived the crisis (Zandi 2012).

Second, the government intervened in the economy through the American Recovery and Reinvestment Act (ARRA), one of the largest economic recovery programs in U.S. history, which was signed into law in February 2009 and for which $787 billion was allocated between fiscal years 2009 and 2019 (Reichling et al. 2015). The Congressional Budget Office most recently estimated ARRA's impact to be $836 billion between fiscal years 2009 and 2019 (Reichling et al. 2015). ARRA stimulated the purchase of goods and services by funding construction and other investment activities that may take several years to complete; it provided funds to states and municipalities and increased aid for education and transportation projects; it supported people in need by extending and expanding unemployment and Supplemental Nutrition Assistance Program (SNAP) benefits; and it provided temporary tax relief for individuals and businesses by raising exemptions amounts for the alternative minimum tax, adding a new Making Work Pay tax credit, and creating enhanced deductions for depreciation for business equipment (Reichling et al. 2015).

Third, the government intervened for consumers through the Economic Stimulus Act of 2008, which the George W. Bush administration signed into law in February 2008. This $152 billion stimulus provided tax rebates for low- and middle-income U.S. taxpayers and tax incentives to generate business investment. It also increased limits on conforming mortgages eligible for government insurance and GSE purchase to stimulate the economy (Zandi 2012). In August 2007, the Federal Housing Administration (FHA) established the FHA Secure program, which refinanced mortgages of delinquent borrowers into more affordable loans, resulting in 266 endorsed loans until December 2007 (Immergluck 2013; Jackson 2008). In December 2007, the National Foreclosure Mitigation Counseling (NFMC) program was established, through which local housing counseling organizations provided advice to more than one million borrowers in or at risk of foreclosure (Immergluck 2013).

In July 2008, the Housing and Economic Recovery Act (HERA) introduced the Hope for Homeowners (H4H) Program, through which lenders could write down existing mortgages and refinance borrowers' loans for less than 90 percent of their current property value, resulting in only 340 loans originated through the program (Immergluck 2013). HERA also provided a tax credit for eligible first-time homebuyers (later expanded to include those who had not owned a home in the past three years) of up to $8,000 from 2008 to 2010 (Immergluck 2013; Internal Revenue Service n.d., Pozen 2010). HERA also established the first wave of the Neighborhood Stabilization Program (NSP 1), which allocated $3.9 billion in formula-based funding across both existing Community Development Block Grant (CDBG) entitlement communities and states (Immergluck 2013). ARRA authorized the second wave of the NSP (NSP 2), resulting in $2 billion of competitive funding for communities. Finally, the Dodd-Frank Wall Street Reform and Consumer Protection Act established the third wave of the NSP (NSP 3), which allocated $1 billion to state and local governments on a formula basis (U.S. Department of Housing and Urban Development n.d.).

During the Obama administration, the following foreclosure prevention programs were established: the Home Affordable Modification Program (HAMP), the FHA-HAMP, the Home Affordable

Refinance Program (HARP) in April 2009; the Second Mortgage Program (2MP) in March 2010; the Home Affordable Foreclosure Alternatives (HAFA) in April 2010; the Home Affordable Unemployment Program (UP) in July 2010; the Hardest Hit Fund (HHF) in September 2010; the FHA Short Refinance (FHA Short Refi) in September 2010; the Principal Reduction Alternative (PRA) in October 2010; and the Emergency Homeowners Loan Program in June 2011 (Immergluck 2013). Despite these many efforts, the federal response to the foreclosure crisis may have been "too little, too late, and too timid" (Immergluck 2013, 199). In summary, over the past few decades, U.S. housing finance policy has been fragmented, in addition to the necessity to accomplish more with fewer resources.

Over the past few decades, the quality and accessibility of homes have increased for most residents, while their affordability has decreased for some residents as the number of affordable units has decreased and the gap between housing costs has increased (Watson et al. 2017). The implication for millions of households facing such housing cost burdens are significant. Many households have difficulty affording other basic needs, such as food, transportation to work, health care, and clothing (Joint Center for Housing Studies of Harvard University 2015). Shaun Donovan, former Secretary of the U.S. Department of Housing and Urban Development, has called this trend a "silent crisis" (Donovan 2013). This crisis will most likely worsen in the near and distant future, not only in central cities but also in suburbs (see also the chapters authored by Bernadette Hanlon and John Rennie Short, and by Willow Lung-Amam).

Over the past few decades housing policy solutions have focused on both the demand and supply sides to address the affordable housing crisis. On the demand side, solutions for have included the Mortgage Interest Deduction (MID), the Federal Housing Administration (FHA), and the Federal Housing National Mortgage Association (Fannie Mae). However, resources for renters have fallen far short of the growing need for assistance (Watson et al. 2017). This trend will most likely continue in the near and distant future (see also the chapters authored by Andrew J. Greenlee, Alan Mallach, and Kirk McClure).

On the supply side, solutions can be clustered into three groups. First, policymakers have focused on "brick and mortar" policies, such as public and HOPE VI housing for renters and the Section 8 New Construction and Substantial Rehabilitation program for builders (Schwartz 2015). Second, policymakers, developers, investors, and planners have focused on housing finance policies, including the Low-Income Housing Tax Credit (LIHTC), Community Development Block Grants (CDBG), the HOME Investment Partnership program, below-market interest rate (BMIR) loans, community land trusts, housing trust funds, inclusionary zoning for state and local government, and community development corporations (CDCs) (Schwartz 2015). Third, developers, investors, and planners have focused on manufactured and modular housing, universal design (UD), or accessory dwelling units (ADUs), which will become more common in the near future.

Structure of the Book

This handbook has nine sections and 29 chapters. We start with the section "Right to Housing," an ideal embedded in many international human rights declarations, laws, and treaties, which contains two chapters. W. Dennis Keating compares and contrasts the goal of the right to housing with reality for the United States, France, the United Kingdom, Canada, Australia, and New Zealand. He concludes that the "right" to housing remains aspirational, even where it is legally recognized, and argues that the right nevertheless helps organizations seeking to meet housing needs to better be able to pressure their respective governments to assist people most in need of adequate housing. Alfons Fermin and Frank Wassenberg discuss the current housing situation of refugees in the Netherlands, focusing on processing asylum applications, housing permit holders, and facilitating their integration into society.

The second section contains five chapters on inequality. The first two chapters focus on homeownership and the racial and ethnic wealth gap in the United States. Carolina Reid reviews the historical legacy of past discrimination in the mortgage and housing market and shows how this legacy has persisted in shaping unequal returns to homeownership. She then argues that homeownership has become a riskier investment over time due to the erosion of the social safety net, financial deregulation, and the increasing complexity of financial instruments. Reid concludes that a more ambitious federal homeownership policy is needed to ensure that the benefits and risks of homeownership are more evenly distributed. Erin Graves, Ana Patricia Muñoz, Darrick Hamilton, William Darity, and Yunju Nam focus on non-Hispanic White and Black parental wealth and wealth transfers, which enable homeownership, in five metropolitan areas. They find that a higher proportion of non-Hispanic Whites receive gifts or loans for down payment, compared to Blacks, which partially explains the wealth gap between non-Hispanic Whites and Blacks.

The third chapter in this section, authored by Howard Husock, argues that poverty deconcentration efforts trigger tensions in receiving communities and undermine the perceived status gain by formerly poor households who move to non-poor neighborhoods. J. Revel Sims and Carolina S. Sarmiento study gentrification, direct and exclusionary displacement, and affordable housing development in the Lacy and Logan neighborhoods in Santa Ana in Orange County, California. They move away from the view of gentrification as a localized dualism between gentrifiers on one end and the gentrified on the other, instead suggesting a cultural political economy nested within regime structures where actors do not fit effortlessly into binaries. The authors call for the right to housing as an important part of the struggle for more equitable planning. Finally, Emily Talen traces the history and complications of providing a neighborhood with an identifiable, physical center, which is an important component of neighborhood life, and argues that planners should put more efforts into nurturing neighborhood center definition and designation.

The third section contains two chapters on homeownership. Mark Calabria and Vanessa Brown Calder discuss demographic, economic, and policy contributors to homeownership across OECD countries. They find that demographic characteristics such as age and marriage rates, residential mobility, and urbanization affect homeownership rates and conclude that in some cases government-produced barriers to homeownership, such as transfer and registration fees, could be reduced to increase homeownership rates. Richard Ronald and Christian Lennartz discuss the declining homeownership in liberal, English speaking countries, partially triggered by the Credit Crisis, the Global Financial Crisis, and a new homeownership regime that is characterized by an increasing polarization of housing wealth and diminishing affordability and accessibility, especially for younger households.

Section 4 contains three chapters on rental housing. Kirk McClure discusses the expansion and evolution of two subsidized rental housing programs in the U.S., the Low-Income Housing Tax Credit (LIHTC) and the Housing Choice Voucher (HCV) programs, which provide a subsidy to units for eligible low-income renters. Initially, both programs solved a shortage of units, then addressed housing cost affordability challenges, and currently they provide access to high-opportunity neighborhoods. McClure argues that both programs need to be reformed to achieve their full potential. Andrew J. Greenlee analyzes spatial location data for households subsidized by the HCV program, arguing for an expanded notion of housing choice that more fully acknowledges the role of landlords and local housing authorities in shaping tenant location choices. Allison Freeman and Kimberly Manturuk analyze a survey of low- and moderate-income U.S. homeowners and renters living in similar neighborhoods to determine whether residents have different drivers of housing satisfaction. They find that both owners and renters are affected by the physical characteristics of their homes, although owners seem to be more affected by neighborhood conditions.

The fifth section is about social housing, a concept familiar in most parts of the world although to a lesser degree in the U.S., and contains three chapters on the U.S. and one chapter on the Netherlands and Germany. Rachel Bratt discusses the history and record of two major components of the

U.S. social housing sector, public housing and nonprofit housing organizations. She summarizes their key attributes and current challenges and then concludes with some ideas for how the nonprofit sector could be strengthened through a set of "nonprofit-centric" policies that are explicitly and directly focused on supporting and helping to develop this part of the social housing sector. Lawrence Vale and Yonah Freemark analyze evolving federal housing programs and policies, discussing how the meaning and execution of public housing have shifted and diversified, in tandem with new roles for the private sector. They argue that the 50-year-long trend toward privatization may have reached a peak after 2013 with the burgeoning promulgation of the Rental Assistance Demonstration (RAD) program, which permitted conventional public housing units to be converted to long-term project-based contracts that do not require public ownership or management any longer.

Rachel Kleit focuses on public housing authorities (PHAs), some of which are becoming flexible, hybrid organizations that employ private sector methods and financing to meet their public missions, although there is variation based on their respective local charters and state enabling legislation. Thus, some PHAs may become effective community actors in affordable housing delivery and preservation, supported by a combination of federal, state, and local resources.

Finally, Marja Elsinga and Marietta Haffner discuss the changing role of social housing in the Netherlands and Germany, which are part of the European Union. The European Commission (EC) has promoted competition between those social and private landlords and developers that provide rental housing. While EU legislation is uniform, it has had different impacts on rental housing policies in the Netherlands and Germany. Thus, the authors call for a reexamination of the role of the EC vis-à-vis the different national governments in addressing increasing housing affordability challenges.

Section 6 discusses aspects of senior housing presented in four chapters. Stephen Golant focuses on aging in place, which does not necessarily translate into aging in the right place as homes may become difficult or dangerous to live in and neighborhoods may become inhospitable or unsafe to reside in. The author discusses the challenges seniors have accessing or connecting to their everyday needs and suggests five categories of connectivity options, concluding that there is currently no one-size-fits-all connectivity solution. Sherry Ahrentzen and Ruth Steiner discuss housing models for aging in community, for example, villages, naturally occurring retirement communities, multigenerational and senior co-housing, and accessory dwelling units (ADUs), all of which offer adults opportunities for control and connectivity with community while addressing the challenges of affordability.

Deirdre Pfeiffer, Ashlee Tziganuk, Scott Cloutier, Julia Colbert, and Gracie Strasser discuss two innovative methods for housing seniors, cohousing and ADUs, based on interviews with U.S., Danish, and Canadian housing professionals. They find that innovative senior housing has advantages and disadvantages and conclude with translating a vision of senior housing into implementation, assisted by housing advocates. Finally, Sebastiaan Gerards, Erik Nuyts, and Jan Vanrie investigate attitudes toward multigenerational housing in Flanders/Belgium, based on an assignment given to architecture students who first conducted a convenience survey and then designed a multigenerational home. The authors find that about half of the survey respondents considered multigenerational housing as a future option and then discuss the challenges of creating shared versus private spaces within a home.

The seventh section focuses on community and gentrification and provides four chapters. Dennis Gale discusses the shifting genealogy of gentrification based on five challenges: first, the expansion of the definition of gentrification; second, scholars' disagreement about gentrification's causal roots (demand versus supply side); third, a lack of understanding of the role of public agency; fourth, the reasons for household displacement (gentrification versus neighborhood succession); and fifth, the overestimation of social costs of displacement and the underestimation of social benefits resulting from financial gains. Miriam Zuk reviews the literature on displacement in both gentrifying and non-gentrifying U.S. neighborhoods and concludes that displacement is a process that is separate yet overlaps with gentrification. She suggests increasing discussions about the relative importance of market-rate versus subsidized production on housing affordability and displacement and increasing research on the impacts of the diverse range of anti-displacement strategies.

The last two chapters focus on the Netherlands. Reinout Kleinhans discusses forced residential relocation outcomes for tenants framed in terms of displacement and (disturbed) place attachment, arguing that there is a Janus head of forced relocation, where the process and outcomes show different and sometimes highly contradictory faces. He suggests counseling to mitigate the negative impact on the well-being of relocatees. Wouter van Gent, Willem Boterman, and Myrte Hoekstra discuss state interventions in a low-income former working-class neighborhood in Amsterdam, focusing on "quarter making" by policymakers catering to middle-class preferences through social policies, initiatives, and service provision, in anticipation of in-moving middle-class residents as part of regeneration and gentrification strategies. They argue that state intervention in this neighborhood is a "soft force" approach to state-sponsored gentrification.

Section 8 focuses on suburbs and has two chapters. Bernadette Hanlon and John Rennie Short compare and contrast suburbs and policies in the U.S. and France, where social mixing has become the cornerstone of housing policy in both countries. In the French *banlieues*, social housing has been redeveloped, and market-rate rental housing and ownership options have been introduced, although there are doubts that regenerating the *banlieues* without addressing structural problems will succeed in solving problems for residents. In U.S. suburbs, spatial mobility of low-income people has been encouraged through housing vouchers, although moves to high-opportunity neighborhoods have proven difficult. Thus, the authors argue that the emphasis on social mixing without addressing underlying problems is challenging. Willow Lung-Amam investigates how homebuilders of multigenerational homes in the suburbs of Phoenix, Arizona, have responded to the increasing demand for multigenerational living and reveals planning, design, and policy challenges of this type of home.

Finally, the ninth section discusses the future of housing. Ingrid Gould Ellen analyzes local housing plans that either establish new commitments or increase prior commitments to produce and preserve affordable housing. She discusses ten selected cities in the U.S., most of them high-cost and growing, and pays particular attention to New York City. She concludes that the emergence of these plans is evidence that cities pay attention to the lack of affordable housing. Victoria Basolo provides an overview of common types of affordable housing policies and programs at the U.S. state level and then focuses on housing actions in California. She argues that while states continue to be dependent on federal funding for affordable housing provision, they can supplement federal programs through unique approaches such as housing trust funds. Finally, Alan Mallach opines that the prognosis for housing policy in most developed nations will be problematic as strong policy development depends on political stability and sustained economic growth, which are currently not the case. He foresees retrenchment or at best stabilization of effort in terms of investment in housing and urban revitalization. Thus, housing challenges will increase, resulting in increased housing cost burdens, deteriorating older housing estates, and growing social, spatial, and economic disparities. Mallach predicts incremental change and a possible increase in the focus on housing issues of marginalized populations, shrinking cities, resource allocation under economic constraints, and the role of sub-state actors.

References

Anacker, Katrin B. 2018. "The Great Recession." In *Introduction to Housing*, edited by Katrin B. Anacker, Andrew T. Carswell, Sarah D. Kirby, and Kenneth R. Tremblay, 255–272. Athens, GA: University of Georgia Press.

Bernanke, Ben S. 2015. *The Courage to Act: A Memoir of a Crisis and Its Aftermath*. New York: W. W. Norton.

Dayen, David. 2016. *Chain of Title: How Three Ordinary Americans Uncovered Wall Street's Great Foreclosure Fraud*. New York: The New Press.

Desmond, Matthew. 2016. *Evicted: Poverty and Profit in the American City*. New York: Crown.

Donovan, Shaun. 2013. *Secretary Donovan Highlights Convening on State of America's Rental Housing*. Washington, DC: U.S. Department of Housing and Urban Development.

Edin, Kathryn J., and H. Luke Shaefer. 2015. *$2 a Day: Living on Almost Nothing in America*. Boston: Houghton Mifflin Harcourt.

Graham, Carol. 2017. *Happiness for All? Unequal Hopes and Lives in Pursuit of the American Dream*. Princeton, NJ: Princeton University Press.

Immergluck, Dank. 2013. "Too Little, Too Late and Too Timid: The Federal Response to the Foreclosure Crisis at the Five-Year Mark." *Housing Policy Debate*, 23: 199–232.

Immergluck, Dan. 2015. *Preventing the Next Mortgage Crisis: The Meltdown, the Federal Response, and the Future of Housing in America*. Lanham, MD: Rowman & Littlefield.

Internal Revenue Service. n.d. *First-Time Homebuyer Credit*. Washington, DC: Internal Revenue Service.

Jackson, Paul. 2008. "HUD Kills FHA Secure." *Housing Wire*, December 29, n.p.

Joint Center for Housing Studies of Harvard University. 2015. *State of the Nation's Housing 2015*. Cambridge, MA: Joint Center for Housing Studies of Harvard University.

Kinder, Kimberley. 2016. *DIY Detroit: Making Do in a City without Services*. Minneapolis: University of Minnesota Press.

Lerner, Avi, Elizabeth Cove Delisle, Theresa Gullo, and Jeffrey Holland. 2017. *Report on the Troubled Asset Relief Program: June 2017*. Washington, DC: Congressional Budget Office.

Martin, Isaac W., and Christopher Niedt. 2015. *Foreclosed America*. Stanford, CA: Stanford University Press.

Pozen, Robert. 2010. *Too Big to Save? How to Fix the U.S. Financial System*. Hoboken, NJ: John Wiley & Sons.

Reichling, Felix, Wendy Edelberg, and Benjamin Page. 2015. *Estimated Impact of the American Recovery and Reinvestment Act on Employment and Economic Output in 2014*. Washington, DC: Congressional Budget Office.

Schwartz, Alex F. 2015. *Housing Policy in the United States*. New York: Routledge.

U.S. Bureau of the Census. n.d.a. *Housing Vacancies and Homeownership (CPS/HVS): Table 14: Quarterly Homeownership Rates for the U.S. And Regions: 1964 to Present*. Washington, DC: U.S. Bureau of the Census.

U.S. Bureau of the Census. n.d.b. *Housing Vacancies and Homeownership (CPS/HVS): Table 16: 1994 to Present*. Washingotn, DC: U.S. Bureau of the Census.

U.S. Department of Housing and Urban Development. n.d. *NSP Eligibility Requirements*. Washington, DC: U.S. Department of Housing and Urban Development.

Watson, Nicole, Barry L. Steffen, Marge Martin, and David A. Vandenbroucke. 2017. *Worst Case Housing Needs: 2017 Report to Congress*. Washington, DC: U.S. Department of Housing and Urban Development.

Zandi, Mark. 2012. "Obama Policies Ended Housing Free Fall." *Washington Post*, September 28, n.p.

SECTION 1

Right to Housing

1

THE RIGHT TO HOUSING

The Goal Versus the Reality

W. Dennis Keating

Introduction

Housing encompasses many aspects beyond the basic necessity for adequate shelter. In the physical sense, it should be safe and secure. In the economic sense, it should be affordable. In the social sense, it should not be overcrowded, its location should provide access to services, occupants should have security of tenure, and home seekers should not experience discrimination based on race/ethnicity, disabilities, family structure (i.e., the presence of children), gender, sexual preference, or source of income. While the meaning of the terms "adequate" and "shelter" can be debated, these other characteristics can be measured.

This chapter will identify the extent to which a right to housing has been recognized and whether, formally or not, adequate housing has been provided. While a right to housing has been adopted by a number of countries, no country has yet fully implemented such a right. This chapter will review international declarations of the right to housing and analyze several case studies, including the United States, Europe (including France and the United Kingdom), Canada, Australia, and New Zealand in order to better understand the dilemmas associated with right to housing declarations.

International Declarations

The right to housing has been recognized internationally since the founding of the United Nations (UN). The guarantee of this right can be found in several documents, including Article 25, Universal Declaration of Human Rights (as part of an adequate standard of living) (1948; U.N. Human Rights Council 2015). The International Covenant on Economic, Social, and Cultural Rights (ICESCR) (Article 11[1](1966), the UN Committee on Economic, Social and Cultural Rights (CESCR)) states:

> The States Parties to the Present Covenant recognize the right of everyone to an adequate standard of living for himself and his family, including adequate food, clothing and housing, and to the continuous improvement of living conditions. The States Parties will take appropriate steps to ensure realization of this right, recognizing to this effect the essential importance of international co-operation based on free consent.
>
> *(n.p.)*

Hohmann (2013) notes:

> [I]mplementation of the ICESCR is notoriously underachieved. Even the richest countries regularly fail to provide the minimum core of the Covenant rights, thus "prima facie, failing to discharge their obligations under the Covenant". This is not only a problem of political will, though lack of such is clearly significant. Both theoretical and practical problems lead to legitimate difficulties in the right to housing's implementation and enjoyment.
>
> *(19)*

This quote underscores the reality that a right to housing, if adopted, is only a relative, not an absolute right.

The Office of the UN High Commissioner for Human Rights has provided a Fact Sheet (No. 21/ Rev. 1) (2009) on "The Right to Adequate Housing." Its definition of "adequate" housing encompasses the aforementioned attributes. It specifies three freedoms: (1) protection against forced evictions and arbitrary destruction and demolition of one's home; (2) the right to be free from arbitrary interference with one's home, privacy, and family; and (3) the right to choose one's residence., i.e., to determine where to live and to have freedom of movement. It specifies four entitlements: (1) security of tenure; (2) housing, land, and property restitution (meaning compensation for its seizure); (3) equal and nondiscriminatory access to adequate housing; and (4) participation in housing-related decision-making at the national and community levels.

While the UN labels these attributes as "freedoms" and "entitlements," these are obviously aspirational and not guaranteed by UN agencies involved in housing.

The UN does have a Special Rapporteur on adequate housing as a component of the right to adequate standard of living. This position is honorary, filled by an expert who reports to the Human Rights Council. In her December 2015 report to the UN Human Rights Council, Leilana Farha addressed homelessness as a global human rights crisis in violation of the right to adequate housing as a component of the right to an adequate standard of living and the right to nondiscrimination (Farha 2015). Minimally, she defined it as the absence of "home"—secure shelter—and beyond that as having the absence of discrimination, criminalization, and marginalization. She cited international examples of national and local laws making the homeless into lawbreakers and also cited international examples of marginalized groups of the homeless, including women, children, migrants, and the disabled. Then she proposed a global campaign to eliminate homelessness by 2030.

Selected Case Studies

In addition to these international declarations, some countries have legally established a right to housing. However, such international declarations and national policies do not mean that the right to housing has actually been achieved. In all too many countries there are shortages of housing, physically inadequate shelter, and discrimination. Slum dwellers in informal housing, often lacking basic infrastructure and services, occupy some of the worst housing, typically located on the fringes of mega-cities (Davy and Pellissery 2013). Mostly found in the developing world, this informal housing would generally be considered to be inadequate. In the worst cases, there is homelessness, most notably among refugees. International, national, and local housing and human rights organizations continue to advocate for the right to housing and its implementation, especially for the poorest and most in need. This reflects the failure to achieve adequate housing even where a right to housing is already recognized. It should be noted that while a right to housing may be advocated, often the focus of these organizations is on a particular housing issue. This applies to organizations such as the National Law Center for Homelessness and Poverty in the United States and the International Centre on Housing Rights and Evictions.

United States

Federal housing policy consists of a combination of tax policy and subsidies. The largest federal tax program provides various incentives for homeowners with a majority of the savings going to upper-income homeowners. Beginning in 1986, tax credits have been provided for the development of housing for lower-income tenants. A variety of subsidy programs are aimed at the development of housing for low- and moderate-income tenants and homeowners. Since 1974, poor tenants have received rent subsidies allowing them to live in privately owned rental housing (Hays 2012; Schwartz 2015) (see also the chapter authored by Lawrence Vale and Yonah Freemark).

The reach of federal housing subsidy programs depends upon Congressional funding. For example, the federal standard for "affordable" housing for renters is paying no more than 30 percent of their income for rent. In 2013, the number of cost-burdened renters paying more than this particular rent-to-income maximum ratio was at a new high of 20.8 million (Joint Center for Housing Studies of Harvard University 2015). In that year only about one of four very poor tenants eligible for rent subsidies (known as Housing Choice Vouchers) received them due to limited funding. In his book *Evicted: Poverty and Profit in the American City*, a case study of Milwaukee, Wisconsin, Matthew Desmond argued for a universal entitlement of housing vouchers for poor tenants (Desmond 2016).

Right to Housing

The United States does not have an explicit national right to housing law that would support Desmond's proposal. Housing experts differ on the need for a right to housing. On the one hand, Chester Hartman, then head of the Poverty and Race Research Action Council, has made the case for a right to housing. Citing the housing shortcomings in the United States, Hartman (1998, 224) declared: "I proceed from a normative, philosophical stance that asserts the wisdom and justice of such a right, as well as our society's clear ability to achieve it."

On the other hand, James H. Carr, then with the Fannie Mae Foundation, presented a rebuttal and alternative view.

> [D]eclaring a "right to housing" and pursuing that right as the primary goal for a branch of social policy ignores the underlying causes behind concentrations of shelter poverty among certain populations and residents of certain geographic areas. As such, government programs developed to ensure the "right to housing" would encourage greater concentrations of certain populations in shelter poverty.
>
> *(Carr 1998, 248)*

> Currently, however, a right to housing would merely concentrate minorities and the urban and rural poor, already affected by systematic differential treatment, into a housing situation that perpetuates dependency on the government and vulnerability to shifting political tides.
>
> *(Carr 1998, 256)*

Finally, Carr argued that providing publicly subsidized shelter without the beneficiaries having jobs with livable incomes would be counterproductive because the subsidized housing would undercut work incentives. Desmond, however, disagrees, arguing that the poor want to work to improve their lives beyond having adequate housing (Desmond 2016). These conflicting views represent the competing philosophies regarding governmental intervention in the housing market, a debate between right to housing advocates (e.g., Hartman and Desmond) and right to housing skeptics (e.g., Carr).

In 1972, the U.S. Supreme Court ruled that there is "no constitutional guarantee of access to dwellings of a particular quality" (*Lindsay v. Normet*, 405 U.S. 56). Instead, it pointed to legislation as the appropriate avenue to securing housing rights. In 1949, the U.S. Congress passed a housing act that included the express goal of a "decent home and suitable living environment for every American family" (Hays 2012, 111). However, this act was aspirational only, not a mandate. Federal housing programs are generally limited in term and require periodic renewal and funding by the executive and legislative branches of government. In 1968, the Congressional housing legislation included the ambitious goal of eliminating the shortage of decent housing (including the need for low- and moderate-income housing) within a decade, with annual reports on progress to be issued by the U.S. Department of Housing and Urban Development (HUD; Hays 2012, 111). To achieve an overall goal of 26 million housing units, Congress relied on subsidies for the construction of private housing. Over time, the housing construction programs lagged, and in 1973 HUD declared a moratorium on these construction programs.

Homelessness

Beginning in the 1980s, homelessness became a significant housing problem in the United States. Despite the creation of several federal programs to house the homeless and the expansion of shelters and transitional housing for the homeless beginning in 1987 (Schwartz 2015), homelessness has persisted. In 2015, nearly 600,000 Americans were homeless (National Alliance to End Homelessness 2016). The federal "Housing First" initiative aims at providing enough permanent supportive housing to eliminate homelessness within a decade (Schwartz 2015). This program has made progress in reducing homelessness in many cities, including that of homeless veterans (Janisse 2016). Advocacy organizations such as the National Law Center on Homelessness and Poverty and local housing providers have promoted efforts like these to address the problem and to establish at least a right to shelter.

In some U.S. cities such as San Francisco tent encampments of the homeless have appeared, raising issues of public health and safety (Fuller 2016). Those cities that provide more services for the homeless have found that they have become magnets for many of the homeless and have been unable to cope with their needs, for example, Los Angeles and San Francisco. In 2015, the City of Los Angeles declared a citywide emergency in the face of the escalating number of the homeless totaling 25,000. Including this number in the city, Los Angeles County had 44,000 homeless. In February 2016, the City and County governments proposed complementary plans to spend hundreds of millions to provide expanded housing and housing services for these homeless populations (Lovett 2016). On November 8, voters in the City of Los Angeles voted on a $1 million bond issue for housing for the homeless (Holland 2016).

In his 2017 federal budget President Obama proposed $11 billion over the next decade to eliminate homelessness among families and young children. HUD's 2015 homeless count found that about 64,000 families and 123,000 children were homeless (Stewart 2016).

State and Local Governments

In addition to federal housing programs, much of the responsibility for providing and ensuring that adequate housing is available (including for the homeless) lies with local governments (aided by state and federal housing programs). Local governments regulate housing through licensing landlords, conducting code enforcement, controlling rents (if and where authorized), and enforcing landlord-tenant laws. Many also have public housing authorities, use housing trust funds to subsidize development, and require developers of large-scale residential projects to dedicate a percentage of the units to below-market-priced units (known as "inclusionary housing") (Calavita and Mallach 2010; Schwartz 2015). Despite their best efforts, cities (and their nonprofit partners) have proven incapable of meeting the needs of all who need adequate housing. This has simply been beyond their capacity.

The decline in federal housing support (Schwartz 2015) has limited the ability of these local providers to meet increasing housing needs.

The single most dramatic example of this failure of local governments to meet housing needs is found in New York City, the largest city in the country. The New York City Housing Authority (NYCHA) is the largest in the United States and has been considered one of the most successful in housing the poor (Bloom 2009). In 2014, it housed more than 400,000 residents in its 334 developments. But as of that year its unfunded capital needs totaled $18 billion, reflecting cutbacks in federal spending for public housing (Navarro 2014). In 2015, there were 270,000 individuals and families (totaling more than 500,000 individuals) on NYCHA's waiting list, with only a small number of vacant apartments being filled annually (Navarro 2015).

Despite rent stabilization regulating around one million rental units housing approximately 2.5 million tenants, half of New York City's tenants living in private rental housing covered by rent stabilization pay more than 36 percent of their income for housing (New York City Rent Guidelines Board 2015). In 2015, the Rent Guidelines Board froze rents for one-year leases for the first time in its 46-year life. This freeze reflects the magnitude of the rental housing crisis in one of the most expensive cities in the United States.

New York City itself has funded more housing than any other U.S. city (Schwartz 2015). One of the major initiatives of Mayor Bill DeBlasio is his *Housing New York: A Five-Borough, Ten-Year Plan* (De Blasio 2014). It proposed more than 50 initiatives to build or preserve 200,000 units of high-quality affordable housing by 2024. Despite these attempts to increase the affordable housing supply and regulatory policies to limit housing prices, and the fact that the anti-homeless policy is headed up by one of the city's most well-known homeless advocates, this massive effort is unlikely to meet the need for more and better affordable housing in New York City.

Discrimination and Freedom of Movement

One of the three freedoms cited by the UN High Commissioner for Human Rights is "The right to choose one's residence. To determine where to live and [to have] freedom of movement." Anti-discrimination laws protecting against various types of housing discrimination have greatly reduced overt discrimination based on race, religion, ethnicity, gender, households with children, and disabilities. Litigation has often been required to challenge discrimination under these laws. Yet, much discrimination remains, with it becoming covert rather than overt. And not all discrimination is illegal. For example, most private landlords are free to legally refuse to accept tenants trying to use federal rental housing vouchers to pay part of the rent (Desmond 2016).

Another aspect of housing discrimination in the United States is spatial segregation. There is widespread resistance to the location of subsidized housing for the poor in middle-class and upper-class neighborhoods. This limits the right of choice and access for the lower-income population in general and racial minorities in particular. Sometimes this resistance is loud as in Yonkers, NY (when working-class and middle-class Whites opposed an activist judge's decision to desegregate public housing), and sometimes it is quiet, when suburban governments use land use controls, including zoning, to prevent this type of affordable housing from being built. Under their home rule powers, they are free to exclude this housing as long as racial bias cannot be proven. Residents fearful of a negative impact on housing values often engage in NIMBYism ("Not in My Backyard") (Varady 2005).

Only a few states (e.g., Massachusetts and New Jersey) require most local governments to allow at least some affordable housing to be located in their communities. Even in those states, after decades of controversy many communities have not met their "fair share" housing requirements. In 2015, HUD issued new rules on Affirmatively Furthering Fair Housing that require those communities receiving federal housing and community development grants to make more serious long-term efforts to identify the obstacles to achieving more inclusive communities and then to take specific actions to promote the development of affordable housing. Whether this policy will significantly affect spatial segregation in housing related to income and race will only be known after its implementation (Immergluck and Kao 2016). Special residential mobility programs for tenants with federal rent vouchers (e.g., Gautreaux, Moving to Opportunity) have so far had only a minimal impact in reducing the spatial concentration of the inner city poor (Schwartz 2015) (see also the chapters authored by Howard Husock and Kirk McClure).

Security of Tenure

As for security of tenure, which is one of the four entitlements the UN cites, most low-income U.S. tenants have only short-term leases (many only month-to-month) (Hartman and Robinson 2003). Landlords are under no obligation to renew these leases and can raise the rent at will or convert the housing to other uses unless there is public policy to the contrary. Thus, most poor tenants are forced to rent housing in the worst conditions and locations and are often evicted for failure to pay the rent (Desmond 2016). The exceptions are (1) states and localities that require a "just cause" to evict; in these jurisdictions, as long as the tenant pays the required rent and does not violate the terms of the lease, the tenant is protected against arbitrary eviction; however, if the tenant does either, a right to housing would not prevent their eviction; (2) tenants living in federally subsidized rental housing who are protected by due process, which allows them by right to an administrative hearing to contest eviction prior to a judicial eviction action; and (3) in California, New Jersey, and New York where local rent control exists, where evictions require just cause.

If the government requires residents to move in the face of government projects (e.g., highways), they are entitled to relocation benefits under federal and state laws (for comparative relocation policies in the Netherlands, see also the chapter authored by Reinout Kleinhans). Some local governments also require private owners of housing to provide extended time for relocation and/or relocation benefits to certain categories of tenants (e.g., disabled).

Mortgage Crisis

After the mid-2000s the United States saw one of the worst periods of housing displacement, although this disruption varied by the state and the region. The subprime mortgage crisis resulted from the bursting of a housing bubble that fueled the worst economic recession since the Great Depression of the 1930s. Millions of homeowners saw their houses foreclosed and the equity lost as housing values plummeted (Schwartz 2015). Millions of other borrowers became "under water" (i.e., the amount of their outstanding mortgage loan exceeded the market value of their home), and many simply abandoned their home in a depressed housing market. The efforts of the federal government to assist homeowners to refinance their mortgages to prevent foreclosure achieved limited success, as did counseling programs for homeowners in financial distress (Schwartz 2015). In addition, many tenants were forced to move when their landlords failed to pay their loans and their rental property was foreclosed (Madar and Glashausser 2011).

In the United States there are numerous organizations dedicated to addressing the need for adequate housing. The National Low Income Housing Coalition is the most prominent national advocacy organization for affordable housing. It waged a campaign for a National Housing Trust Fund that finally culminated in its creation in 2008. However, its funding has been delayed due in part to the collapse of its primary funding sources during the subprime mortgage crisis (Schwartz 2015). Now controlled by the federal government after being bailed out in 2008, the Federal National Mortgage Association (FNMA, known as "Fannie Mae") and the Government National Mortgage Association (GNMA, known as "Ginnie Mae"), which are the largest providers of mortgage financing through the secondary market, have an obligation to support housing for lower-income households. With the financial recovery of these two agencies, it is hoped that they will provide significant funding for affordable housing distributed through the states. In early 2016, $174 million was being allocated to build, preserve, and rehabilitate housing for extremely and very low-income households.

Thus, the United States has not only failed to approve a right to housing law, it has also failed to meet existing low-income housing needs. Even if such a legislative policy, a right to housing, existed, that would still not ensure that all unfulfilled housing needs would be met. This has been the case in countries that have adopted a right to housing policy. However, at least it would provide more leverage for housing advocates and stakeholders to pressure governments at all levels to increase their efforts to provide housing for all.

Non-U.S. International Examples

The next sections provide a few examples of countries outside the U.S.—France, the United Kingdom, Canada, and Australia and New Zealand—but without as much detail as in the United States example.

Europe

In 2013, the European Parliament adopted a resolution on social housing in the European Union (EU), recalling that access to housing is a fundamental right but that this right had been under threat due to a shortage of affordable housing, as well as the continuing economic and social crisis. The elected representatives in the European Parliament called on the European Commission, its executive arm, and member states to support and promote innovative exchanges of good practices in upholding the right to housing, especially for particularly vulnerable and marginal groups. Parliament called on the member states to draw up integrated policies for promoting social inclusion, as well as for guaranteeing universal access to decent, healthy, and affordable housing. It pointed out that social housing should be structured to avoid both gentrification and ghettoization. As to the latter issue, it was a warning against the isolation of high-rise public housing, such as the *banlieues*, that are located at the edge of many French cities.

These issues exist despite several European countries guaranteeing a right to housing in their constitutions: Belgium, Finland, Greece, the Netherlands, Spain, and Sweden (Byrne and Culhane 2011). Several European countries have recognized a right to housing through legislation: Austria, France, Germany, Luxembourg, and the United Kingdom. Nevertheless, these and other EU countries have failed to actually provide adequate housing for all of their citizens.

In 2014, according to Eurostat (European Union Eurostat 2014), 11.4 percent of the EU population was confronted with housing cost overburden, meaning that they had to spend more than 40 percent of their disposable income for their accommodations. The overcrowding rate was 17.1 percent. In 2015, Habitat for Humanity's annual *Housing Review* reported:

- More than 10 percent of Europeans had housing costs that exceeded half of their household's income.
- The percentage of young adults aged 18–34 living with their parents was at record highs.
- New housing construction had dropped by 70–90 percent, with social housing running well below 10 percent of current needs.
- There was an increasing gap between poverty and affluence in dynamic urban centers.

These problems have been exacerbated by the flood of refugees from the Middle East and Africa who have been in recent years fleeing wars and poverty. Many are now living in temporary refugee quarters. Recently, there has been a growing backlash against their entry into Europe despite humanitarian concerns (see also the chapter authored by Alfons Fermin and Frank Wassenberg).

France

France is one of the several European countries with an explicit right to housing ("*droit au logement opposable*"; DALO). This legislation was enacted unanimously by Parliament in an emergency session in March 2007. More limited precedents on providing housing short of this broader declaration were proclaimed beginning in 1982. DALO provides French citizens (and non-nationals with residence permission) in need of decent housing the right to seek prioritization from mediation committees established in each Département, which must notify Prefects within several months of successful appeals for housing. Those who successfully appeal then have the right to appeal to the administrative courts if housing is not provided within an established period. The courts can then order Prefects to provide accommodations under financial penalty (which would be paid to a regional and more recently a national fund for social housing construction). DALO was modeled on Scotland's housing legislation, which is judicially enforced, as discussed later (Houard and Lévy-Vroelant 2013; see Olds (2010) for a discussion of the role of the courts in France (and the Netherlands) in supporting a right to housing).

DALO has, however, fallen considerably short of fully implementing the right to housing:

> [F]ive years after its proclamation, the number of people helped by the DALO Act has been lower than expected. By late December 2011, around 250,000 appeals had been lodged (65 percent came from the Île-de-France region). Among these, 80,000 appeals received a favorable ruling, and only 33,000 households were re-housed based on the DALO Act.... The obstacles in implementing the DALO have to be interpreted in broader perspective. . . . The progressive implementation of the enforceable right to housing, from 1990 to 2007, fits well into this landscape formed by an alliance between elite groups composed of lawyers, high-level civil servants, politicians, charities and civil society activists, and fueled by the approbation of public and private actors in the field of urban development and housing construction. However, the system still suffers from fundamental ambiguity created by the mismatch between demand and supply, coupled with an indecisive monitoring of the distribution of public resources.
>
> *(2013, 211)*

This illustrates the dilemma of adopting the right to housing as a national policy; i.e., the resources provided by government usually fall short of the acknowledged need. Nevertheless, Calavita and Mallach (2010) stated that the 2007 adoption of the right to housing was worthwhile. "Precisely what that right will mean remains uncertain, but clearly it will increase the pressure on both the state and the communes to expand the pool of social housing available" (231).

Beginning in 2000 France adopted a national inclusionary housing policy. The "Loi SRU" in Article 55 imposed on all *communes* with more than 1,500 residents in the Île-de-France region or more than 3,500 residents in other urban agglomerations the legal obligation to ensure that a minimum of 20 percent of the housing units be social rental housing. The legislation provides for fines for those *communes* that fail to meet the minimum requirement (Calavita and Mallach 2010). The impact of this policy has not been documented as of this writing.

United Kingdom (UK)

In several legislative acts, the UK Parliament required local authorities to provide "settled" (i.e., permanent) housing to homeless groups that have priority needs (Byrne and Culhane 2011).

However, in the City of London the housing crisis has become so critical due to price inflation and cutbacks in housing benefits by the Tory government that its newly elected Mayor Sadiq Khan has made addressing it his top priority (O'Sullivan 2016). Whether Khan will do any better in addressing the shortage of affordable housing than his counterpart mayor in New York City remains to be seen.

Under its independent powers, in 2003 Scotland adopted a policy requiring local authorities to provide settled accommodations to anyone who becomes homeless through no fault of their own. This goal took effect as of 2012 (King 2015; Tars and Egleson 2009). Graeme Brown, the director of the housing and homeless charity Shelter Scotland, stated in 2013:

> As conceived 10 years ago under the Homelessness etc. (Scotland) Act 2003, reform of the law was to be the locomotive which dragged increasing housing supply and improved services in its wake. That was one of the principal reasons that local government umbrella body Cosla supported the changes. Without the peg of increased duties the argument for more spending on housing would have been so much weaker these last ten years.
>
> *(Brown 2013, n.p.)*

The Homeless Monitor has been conducting a tracking study of homelessness in Scotland in Crisis, a national charity for homeless people. Its report, published in 2015, concluded: "[Homelessness] has been on a marked downward path since 2010/2011, since the introduction of Housing Options [a homeless prevention program]. In 2014/2015 Scottish local authorities logged 35,764 statutory homeless assessments, of which 28,615 resulted in a judgment that the household was 'legally homeless' . . . 37% lower than in 2009/2010" (211).

According to the Scottish government's report on homelessness for 2015–2016, roughly two-thirds of the formerly homeless secured a social rented tenancy or a private rented sector tenancy (The Scottish Government 2016). This demonstrates that with the impetus of the right to housing for the homeless legislation, Scotland has made considerable progress in providing housing for the homeless, as well as reducing their number.

Canada

Canada does not have an explicit right to housing. However, there are rights in the Canadian Charter of Rights and Freedoms that are implicitly related to a right to housing (e.g., the right to security of the person and the right to equality). Also, Canada and its provinces and territories have agreed to the right to housing found in international declarations such as the International Covenant on Economic, Social and Cultural Rights (ICESCR; U.N. Human Rights Council 2015). The Centre for Equality Rights in Accommodations/Social Rights Advocacy (CERA/SRAC) advocates for adequate housing.

In 1973, the housing minister in the Liberal government of Prime Minister Pierre Trudeau described adequate affordable shelter as an "elemental human need." However, in succeeding decades Canadian governments have cut back on funding for social housing. By 2015, Canadian cities such as Toronto and Vancouver have experienced skyrocketing housing rents and prices. Forty percent of Canadian renters spend more than 30 percent of their income on rent and utilities, and about 20 percent spend more than half of their income on rent and utilities (McMahon 2015). In 2015, the newly elected Liberal government of Trudeau's son Justin promised greatly increased spending on social and co-op housing and tax subsidies for new private rental units. It also promised to pursue a national housing action plan to improve housing accommodation options over the coming decade (Raising the Roof 2016).

Australia and New Zealand

Neither Australia (Morris 2010; Bell 2013) nor New Zealand (New Zealand Council of Christian Social Services (NZCCSS) 2015) has a right to housing in their constitutions. While both countries have ratified the various international declarations on the right to housing as part of the right to an adequate standard of living, these are not binding on domestic policies. In the case of Australia, in mid-2009 over 175,000 applicants were on waiting lists for public housing, and 50,000 were on waiting lists for community housing (Bell 2013). Fueled in part by migration, Australia's housing crisis has continued. Population projections for the decade 2012–2022 have predicted a need for 142,060 houses in Sydney and 231,340 in Melbourne in the midst of an affordability and housing shortage crisis (Birrell and McCloskey 2016). Given these housing shortages, the country's ratification of the international declarations of a right to housing has not significantly influenced its attempts to successfully address these housing needs.

Conclusion

Whether in the industrialized countries of Europe, North America, Australia, and New Zealand or in less developed countries such as South Africa, the idea of a right to housing, as declared by international bodies, is an explicit or implicit societal goal. Only in 11 countries in the European Union is it explicitly found in a constitution or national legislation. However, in these examples and those countries that recognize international declarations on the right to housing, the reality is that this ideal is far from being achieved despite the existence of many different housing programs. Even where the right to housing has been adopted as a national policy and can possibly be enforced in the courts as is the case of France, experience indicates that only a fraction of those claiming housing needs have been satisfied after seeking fulfillment administratively and through the courts. As Hohmann states:

> When the right to housing has been interpreted and applied it has not had a radical effect. Even an optimistic reading suggests that it has provided, at most, only partial gains and improvements in the living conditions of some, and by no means always for those who are worst off.
>
> *(2013, 231–232)*

As Alexander (2015) demonstrates in the United States, there are social movements and national, state, and local organizations that are fighting for housing, typically for those most in need; the same is true for other developed countries. While there have been successes in the United States (e.g., the creation and initial funding of the National Housing Trust Fund) and elsewhere (e.g., Scotland's housing of the homeless), clearly the need for adequate housing typically exceeds the available supply.

Even the wealthiest of the industrialized nations such as those included in this chapter have failed to meet the requirements of a right to housing. Low-income tenants in extremely expensive cities such as London, New York, Paris, and San Francisco face the greatest difficulty. Nevertheless, as an aspirational goal having the right to housing embedded in a nation's constitution or national legislation at least encourages efforts to better house its population, most importantly those most in need. This hope has inspired housing advocates to continue to campaign for the general goal of a right to housing while addressing the needs of specific disadvantaged groups.

References

Alexander, Lisa T. 2015. "Occupying the Constitutional Right to Housing." *Nebraska Law Review*, 94: 246–301.

Bell, Kevin. 2013. "Protecting Public Housing Tenants in Australia from Forced Eviction: The Fundamental Importance of the Human Right to Adequate Housing and Home." *Monash University Law Review*, 39: 1–37.

Birrell, Bob, and David S. McCloskey. 2016. *Sydney and Melbourne's Housing Affordability Crisis: No End in Sight.* Victoria: Australian Population Research Institute.

Bloom, Nicholas Dagen. 2009. *Public Housing That Worked: New York in the Twentieth Century.* Philadelphia: University of Pennsylvania Press.

Brown, Graeme. 2013. "Scotland's Law to End Homelessness: A Bold Policy Which Needs More Action." *The Guardian.* February 1.

Byrne, Thomas, and Dennis P. Culhane. 2011. "The Right to Housing: An Effective Means for Addressing Homelessness?" *University of Pennsylvania Journal of Law and Social Change*, 14: 379–390.

Calavita, Nico, and Allan Mallach, eds. 2010. *Inclusionary Housing in International Perspective: Affordable Housing, Social Inclusion, and Land Value Recapture.* Cambridge, MA: Lincoln Institute of Land Policy.

Carr, James H. 1998. "Comment on Chester Hartman's 'The Case for a Right to Housing': The Right to 'Poverty with a Roof'—A Response to Hartman." *Housing Policy Debate*, 9: 247–257.

Davy, Benjamin, and Sony Pellissery. 2013. "The Citizenship Promise: (Un)fulfilled: The Right to Housing in Informal Settings." *International Journal of Social Welfare*, 22: 68–84.

Desmond, Matthew. 2016. *Evicted: Poverty and Profit in the American City.* New York: Crown Publishers.

De Blasio, Bill. 2014. *Housing New York: A Five-Borough, Ten-Year Plan.* New York: The City of New York.

European Union Eurostat. 2014. *Housing statistics: Statistics Explained.* Brussels: European Union Eurostat.

Farha, Leilani. 2015. *Report of the Special Rapporteur on Adequate Housing as a Component of the Right to an Adequate Standard of Living, and on the Right to Non-Discrimination in this Context.* New York: UN Human Rights Council.

Fuller, Thomas. 2016. "San Francisco Wants Homeless to Leave Tent Camp, but Some Vow to Stay." *The New York Times*, February 26.

Hartman, Chester. 1998. "The Case for a Right to Housing." *Housing Policy Debate*, 9: 223–246.

Hartman, Chester, and David Robinson. 2003. "Evictions: The Hidden Housing Problem." *Housing Policy Debate*, 14: 461–501.

Hays, R. Allen. 2012. *The Federal Government and Urban Housing: Ideology and Change in Public Policy.* Albany: State University of New York Press.

Hohmann, Jessie. 2013. *The Right to Housing: Law, Concepts, Possibilities.* Oxford and Portland, OR: Hart Publishing.

Holland, Gale. 2016. "L.A. Officials Launch Campaign for Homeless Housing Bond Measure." *The Los Angeles Times*, September 12.

Houard, Noémie, and Claire Lévy-Vroelant. 2013. "The (Enforceable) Right to Housing: A Paradoxical French Passion." *International Journal of Housing Policy*, 13: 202–214.

Immergluck, Dan, and Mindy Kao. 2016. "The Devils in the Details: Key Issues in Implementing the New AFFH Rule." *Rooflines*, July 5.

Janisse, Flynann. 2016. "From Barracks to Apartments: Serving Vets in Transition." *Rooflines*, August 25.

Joint Center for Housing Studies of Harvard University. 2015. *The State of the Nation's Housing 2015.* Cambridge, MA: Harvard University Press.

King, Fiona. 2015. "Scotland: Delivering a Right to Housing." *Journal of Law and Social Housing*, 24: 155–161.

Lindsay v. Normet, 405 U.S. 56.

Lovett, Ian. 2016. "Los Angeles Approves Plans to Fight Homelessness." *The New York Times*, February 9.

Madar, Josiah, and Allegra Glashausser. 2011. "The Evolving Crisis in Context: Recent Developments for Tenants in the Foreclosure Crisis." In *Forging a New Housing Policy: Opportunity in the Wake of Crisis*, edited by Christopher Niedt and Mark Silver, Hempstead, NY: National Center for Suburban Studies at Hofstra University.

McMahon, Tamsin. 2015. "Affordable Housing Crisis Affects One in Five Renters in Canada: A Study." *The Globe and Mail*, September 10.

Morris, Alan. 2010. "The Lack of a Right to Housing and Its Implications in Australia." *Journal of Australian Political Economy*, 65: 28–57.

National Alliance to End Homelessness. 2016. *The State of Homelessness in America 2016.* Washington, DC: National Alliance to End Homelessness.

Navarro, Mireya. 2014. "New York Public Housing Faces Crisis as Demands and Deficits Grow." *The New York Times*, August 12.

Navarro, Mireya. 2015. "Public Housing Becomes a Lifelong Refuge as Rents in New York Soar." *The New York Times*, August 4.

New Zealand Council of Christian Social Services (NZCCSS). 2015. *Call for More Social Housing and a Right to Housing.* Wellington: New Zealand Council of Christican Social Services.

New York City Rent Guidelines Board. 2015. *2015 Income and Affordability Study.* New York: New York City Rent Guidelines Board.

Olds, Kyra. 2010. "The Role of the Courts in Making the Right to Housing a Reality Throughout Europe: Lessons from France and the Netherlands." *Wisconsin International Law Journal*, 28: 170–199.
O'Sullivan, Feargus. 2016. "London's Housing Crisis and the Inequality Chasm." *Atlantic City Lab*, April 5.
Raising the Roof. 2016. "What Does the Budget 2016 Mean for Housing and Homelessness in Canada" April 1.
Schwartz, Alex. 2015. *Housing Policy in the United States*. New York: Routledge.
The Scottish Government. 2016. *Homelessness in Scotland: 2015–2016*. Edinburgh: The Scottish Government.
Stewart, Nikita. 2016. "Obama Will Seek $11 Billion for Homeless Families." *New York Times*, February 8.
Tars, Eric S., and Caitlin Egleson. 2009. "Great Scot! The Scottish Plan to End Homelessness and Lessons for the Housing Rights Movement in the United States." *Georgetown Journal on Poverty Law & Policy*, 16: 188–216.
U.N. Human Rights Council. 2015. *What Are Human Rights?* New York: United Nations.
U.S. High Commissioner for Human Rights. 2009. *The Right to Adequate Housing*. New York: United Nations.
Varady, David. 2005. *Desegregating the City: Ghettos, Enclaves, and Inequality*. Albany, NY: State University of New York Press.

2

A HOME AWAY FROM HOME

Housing Refugees in the Netherlands During the European Refugee Crisis

Alfons Fermin and Frank Wassenberg

Introduction

Since 2015, Europe has experienced "the greatest mass movement of people since the Second World War" (European Commission 2016, 1). The inflow of refugees traveling across the Mediterranean Sea peaked in the summer of 2015, and has declined sharply since spring 2016. Processing asylum applications and housing permit holders and facilitating their integration into society have been major challenges of countries that have received considerable numbers of asylum seekers. As of this writing, the root causes of the inflow of refugees and the disagreement among EU member states on a common strategy to face the challenges of asylum and migration into the EU have not been solved and thus will remain a major challenge in the near future.

Managing the refugee inflow across the Mediterranean involves formal responsibilities for European, national, local, and in some cases regional governments. Developing a common approach to immigration and asylum and the protection of the EU border are responsibilities shared by the EU and its member states, according to EU Treaties (European Parliament 2017). The national governments are responsible for receiving asylum applicants and establishing and implementing asylum procedures. The local governments are responsible for accommodating recognized refugees and for facilitating their integration into society.[1] Municipalities play an important role in ensuring sufficient decent, affordable housing to cover the needs of refugees and the local population, in promoting community cohesion, and in creating a welcoming social environment (European Union Agency for Fundamental Rights 2016).

This chapter focuses on the responses by the Dutch national and local governments in terms of accommodation for asylum seekers and recognized refugees. It is beyond the intended scope of this chapter to discuss in detail the challenges of integrating immigrants via education, jobs, social contacts, and language training.

Europe's Refugee Crisis

In 2015 and 2016, Europe recorded high numbers of asylum seekers unparalleled since World War II. The largest previous movement was the exodus that resulted from several wars in former Yugoslavia in the early 1990s. About 700,000 people fled to the 15 member states of the EU to apply for asylum between 1989 and 1994 (Ambroso 2006). In 2011 the number of refugees entering

Europe gradually increased due to the start of the civil uprising and wars in Syria and Iraq and due to the intensifying situation of insecurity and war in Afghanistan, peaking suddenly in 2015. This record number was accompanied by disturbing scenes of dangerous sea crossings in unseaworthy and overloaded vessels with the help of smugglers, resulting in 1,308 refugees and migrants drowned or missing in April 2015 (United Nations High Commissioner for Refugees 2015). Large-scale asylum inflows were a new experience for several EU countries (OECD 2015).

The humanitarian crisis was aggravated by a political one as EU leaders were unable to develop a common approach (Grigonis 2017). The refugee crisis has also incited anti-immigrant rhetoric and led to the growth of populist parties in many EU countries (Banulescu-Bogdan and Collett 2015). On March 18, 2016, the EU and Turkey agreed that all new migrants that tried to cross the border between Turkey and Greece without necessary authorization or documents would be returned to Turkey starting March 20. In return, the EU agreed that for every Syrian being returned to Turkey from the Greek islands, another Syrian would be resettled to the EU (European Commission 2016). At that time Hungary, Slovenia, and Croatia, all part of the EU, and along with Serbia and Macedonia, which are not part of the EU, had already closed off the Balkan route. The implementation of the agreement between the EU and Turkey and the closure of the Balkan route resulted in a sharp decrease in the number of refugees and other migrants arriving in Greece from Turkey and, in turn, to a decline in the numbers of asylum seekers entering other EU countries.

Interestingly, the sharp decrease in sea crossings is not immediately reflected in asylum statistics of EU member states as it may take several months before asylum seekers arriving on the European coast submit an asylum application in one of the member states, usually not in the country of first arrival and often not immediately (see Figure 2.1; Connor 2016).

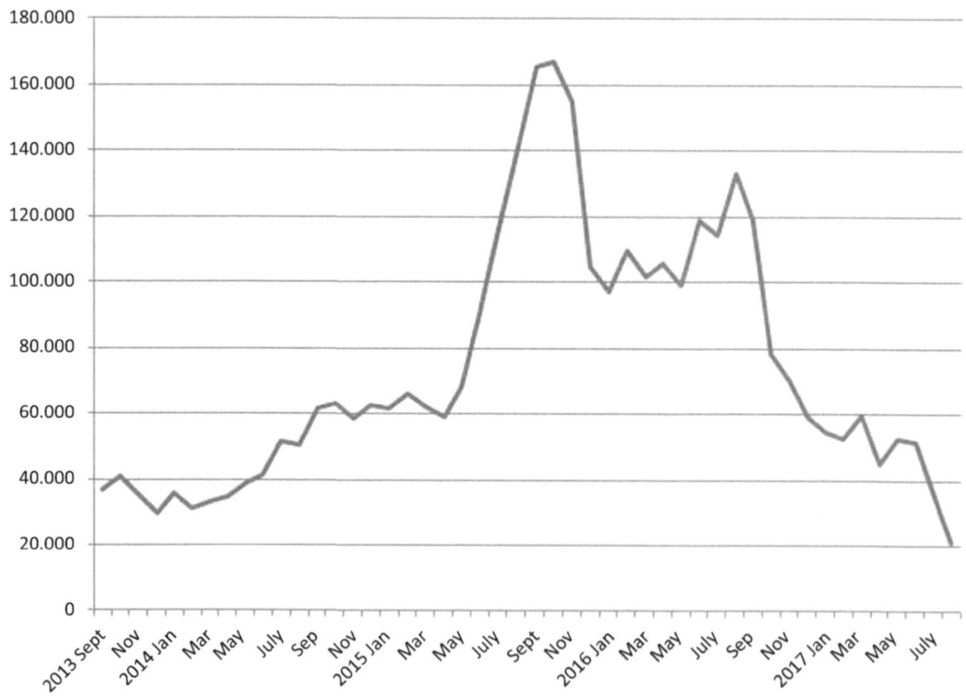

Figure 2.1 First Asylum Applicants in the European Union, September 2013 to August 2017.

Source: Eurostat (2017).

Since the agreement between the EU and Turkey in March 2016, the central Mediterranean route from Libya to Italy has gradually become the main route of undocumented migration to the EU by sea, resulting in about 181,000 arrivals in 2016 and 85,183 arrivals between January 1 and July 3, 2017 (International Organization for Migration 2017a, 2017b). The far longer central Mediterranean route has a higher rate of casualties, resulting in more than 15,150 migrants who attempted to cross losing their life between 2011 and early July 2017 (European Commission 2017; International Organization for Migration 2017b). Most migrants who arrive in Italy are economic migrants from Nigeria, Eritrea, Ivory Coast, Guinea, Senegal, and Gambia (International Organization for Migration 2017c). Since the summer of 2017 there has been a major decrease in the number of undocumented migrants arriving in the EU by sea (United Nations High Commissioner for Refugees n.d.). However, this may be a temporary decrease, as the root causes of migration from Africa and Asia have remained the same.

The influx of migrants is distributed unevenly across the EU. While many migrants enter the EU in Italy and Greece, most migrants attempt to cross the EU border in Southern and Eastern European countries to reach a few popular Western and Northern European countries that are perceived as offering better economic opportunities. In 2015, 90 percent of asylum applicants were registered in nine EU states: Germany, Hungary, Sweden, Austria, Italy, France, the Netherlands, Belgium, and the UK (Eurostat 2017). While the largest number of refugees went to Germany, Sweden received the largest proportion. For example, in 2015 Sweden received about 16,000 asylum seekers per one million inhabitants, Germany about 5,500 per one million, the Netherlands about 2,600 per one million, and the United Kingdom bout 600 per one million inhabitants (OECD 2016).

In 2015, a record number of 58,900 asylum seekers applied for a refugee permit in the Netherlands, about twice the number as in 2014 and 2016 (about 31,600), and about twice the number as in 2013. About 47 percent of the 58,900 asylum applicants in 2015 were from Syria, followed by Eritrea (14 percent), Iraq (6 percent), and several other countries. About 8 percent of the applicants were stateless, typically Palestinians who had formerly lived in Syria (Statistics Netherlands (CBS) 2017).

The Netherlands has had a long history of immigration since World War II, from the former colonies Indonesia and Suriname, the Dutch Caribbean territories that are still part of the Netherlands, and Spain, Italy, Morocco, and Turkey, typically migrant or "guest" workers as well as workers from Eastern EU member states. Currently, refugees from Iraq, Iran, Somalia, Afghanistan, and Syria make up the largest refugee groups. Figure 2.2 shows changes in the number of asylum applicants to the Netherlands over the past 25 years. The results for the Netherlands generally follow the pattern of other Western European countries.

The Dutch Response to the Inflow of Refugees

Unlike several other countries, the Netherlands provided a sufficient number of emergency accommodation for all asylum seekers arriving during the peak period in 2015 and early 2016. Dutch politicians frequently state that in the Netherlands no newly arrived refugee has to sleep on the street, noting that it is a matter of decency. Refugees who complete the asylum procedure and are granted a residence permit have a similar right to housing as any other Dutch citizen. Thus, the demand by new migrants has increased the burden on the already overburdened Dutch housing system, with a particularly strong impact on social housing. Given the shortage of affordable housing, national and local authorities have to balance the rights and needs of the new residence permit holders with those of the resident population.

Asylum applicants arriving in the Netherlands must first report to the Dutch Immigration and Naturalization Service (IND) (see Figure 2.3). Once asylum applicants enter the country, they enter the formal application procedure at an asylum seekers' center, called *azielzoekerscentrum* (AZC), administered by the Central Agency for the Reception of Asylum Seekers (COA).

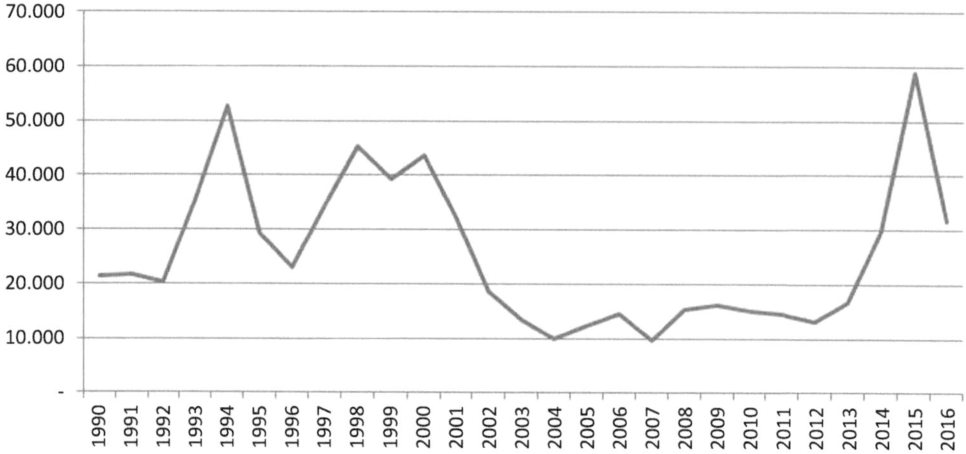

Figure 2.2 Number of Asylum Applications in the Netherlands, 1990–2016.

Note: included: first and subsequent applications as well as applications from family members.

Source: Statistics Netherlands (CBS) 2017

The number of AZCs depends on the number of incoming refugees. At peak, there were about 130 AZCs; as of this writing there are about 65 (Central Agency for the Reception of Asylum Seekers (COA) 2017). They provide basic shelter and accommodate between 300 and 1,000 refugees. AZCs are typically established in vacant schools, monasteries, barracks, or vacation resorts (when not in season). In some cases AZCs may be newly built or expanded units. While some AZCs are located in cities or small towns, others are located in rural areas. Asylum seekers are required to check into their AZC each night and are subject to a curfew.

During their stay at the AZCs, officers from the IND decide whether an asylum seeker is eligible for asylum. Between 2008 and 2013, about 50 percent of applicants were eligible for asylum (Immigration and Naturalization Service (IND) 2017b; Dutch Council for Refugees 2016). This rate peaked to almost 70 percent in 2015 when many Syrians and Eritreans applied for a residence permit, which was typically granted to members of these two groups. However, the number of refugees decreased sharply in April 2016, after the agreement between the EU and Turkey. Since then, more African economic migrants, who typically have low odds of being eligible for asylum, have arrived. In sum, the overall application rate decreased to just over 30 percent from April 2016 to March 2017 (Immigration and Naturalization Service (IND) 2017b). Rejected applicants have to leave the country, although many tend to disappear into illegality.[2]

Asylum applicants who are granted asylum status receive their residence permit and are sent to municipalities in proportion to the municipality's population size. Thus, Amsterdam has had to accommodate the highest number of refugees: 1,377 in 2015 and 2,077 in 2016. At the national level, there were 28,871 permit holders to be housed in 2015 and 43,000 in 2016; the latter is about .25 percent of the Dutch population (Statistics Netherlands (CBS) 2017).

Housing Refugees in the Netherlands

The Right to Housing

The residence permit provides the same right to housing, work, and social welfare benefits as any other Dutch resident. Most recognized refugees receive welfare benefits as they are unable to obtain

Figure 2.3 Dutch Asylum Procedure.

Source: Immigration and Naturalization Service (IND) (2017a).

a job. Thus, they typically receive first priority for a unit in the social housing sector, as is the case for low-income or unemployed Dutch citizens. On average, about 20 to 40 percent of all Dutch applicants for social housing obtain priority, depending on their personal situation, local regulations, and regional pressure on the housing market. Thus, local authorities have to balance the housing needs of refugees with the housing needs of waitlisted Dutch applicants, resulting in a wait time of five to eight years or higher.

Social Housing and Refugees

The Netherlands has a relatively large social housing sector compared with other countries in Northern and Western Europe (Elsinga and Wassenberg 2014; see also the chapters authored by Marja Elsinga and Marietta Haffner, and by Reinout Kleinhans). About one-third of all households rents a social housing unit; most of these are owned by housing associations. Before 2014, most refugees who received a residence permit easily found housing in the social housing sector. For example, a city of, say, 100,000 inhabitants would be able to accommodate about 40 to 60 new refugees each year without significant problems.

However, since 2008 waiting lists in the social housing sector have gradually increased because of a decrease in housing construction due to the national and global financial crisis (Elsinga and Wassenberg 2014) and due to the recent refugee crisis. In December 2015, about 48,000 people lived in AZCs, about 30 percent (or 16,000 people) of whom had already received a residence permit but who were unable to obtain suitable housing from local agencies. These numbers have decreased recently (Central Agency for the Reception of Asylum Seekers (COA) 2017).

Housing needs typically depend on household composition. Between January 2013 and April 2016, 68 percent of all refugee households were single-headed households, while 32 percent were couples with or without children (Mulder 2016). The average household size was about two persons. About 50 percent of households headed by single males wait to be reunited with their families, although the date of reunification is difficult to predict (Central Agency for the Reception of Asylum Seekers (COA) 2017; Immigration and Naturalization Service (IND) 2017b). Interestingly, municipalities found it easier to house families than singles as there is more support for the former, resulting in a relatively high number of young males in AZCs (Central Agency for the Reception of Asylum Seekers (COA) 2017).

Policy Agreements to Support Local Housing Provision for Refugees

In order to accommodate refugees that have received a residency permit, the national government and the Association of Dutch Municipalities (VNG) signed two agreements. The first agreement, signed in November 2015, focused on two housing funding mechanisms: (1) a subsidy to build additional housing units for refugees, and (2) a subsidy to establish temporary housing solutions to address requirements from the COA. Unfortunately, the strict conditions of both funding mechanisms caused local actors to be rather reluctant to use them (Smits van Waesberghe and Razenberg 2016). The second agreement between national and local governments, signed in April 2016, goes beyond housing and includes additional funding for language classes, schooling, and job training for refugees.

Moreover, the Association of Dutch Municipalities established two temporary organizations that assisted municipalities and other stakeholders from 2015 until July 2017. First, *Platform Opnieuw Thuis* (Platform Home Again) supported many local actors that focus on housing refugees, implementing creative solutions, showcasing best practices, advising municipalities, and improving administrative processes. Second, *Ondersteunings Team Asielzoekers en Vergunninghouders* (Support Team for

Asylum Seekers and Residence Permit Holders, OTAV) was a helpdesk for municipalities to answer questions about refugees in regard to housing, education, employment, and social benefits. Moreover, both organizations advised local practitioners about the housing funding mechanisms mentioned earlier.

In addition to these two temporary organizations, *VluchtelingenWerk Nederland* (Dutch Council for Refugees), a non-governmental organization, offers practical support to refugees to facilitate their social integration in Dutch society with the help of about 14,000 volunteers and funding by the national government as well as the private sector.

Unorthodox Ways to Provide Housing

A simple solution for the affordable housing crisis would be to build additional housing units. However, there are many financial and spatial planning constraints, along with a long planning process and insufficient land for constructing housing units. Thus, creativity is needed for finding housing solutions other than new construction. The peak of arrival of the refugees in 2015 stimulated many housing providers to develop temporary or permanent housing alternatives, relieving the shortage of affordable housing by utilizing vacant properties in a more efficient way, i.e., adding temporary or flexible prefabricated housing units, constructing smaller dwellings, and intensifying property usage, as discussed later.

One alternative is to accelerate the transformation of vacant to occupied buildings. In the Netherlands, 19 percent of all offices, 14 percent of all shops, and about 1,000 churches, schools, libraries, monasteries, and other kinds of property were vacant as of 2016, despite the scarcity of land (Bouwstenen voor Sociaal 2016; Statistics Netherlands (CBS) 2016). The vacancy rates are relatively high compared with adjacent countries. From an urban planning and sustainability point of view, reusing vacant properties at closer-in locations is preferable to new construction further out, yet most developers prefer building new structures instead of reusing or transforming the old stock, as building new is easier and more profitable compared to rehabilitating and redeveloping existing buildings.

Another alternative is temporary use or reuse instead of the permanent transformation of vacant buildings, as the former has fewer restrictions in terms of building regulations. Demand for reused office or retail space has declined because some employees work from home or in shared spaces due to the Internet. In 2015, Platform31, a national think tank, ran a pilot study to analyze technical, legal, planning, and financial barriers faced by municipalities and property owners that would like to temporarily reuse vacant buildings via minor basic adjustments in the buildings and temporary leases (Platform31 2015). They found that temporary reuse is easier than most professionals expect, not only technically and legally, but also in terms of planning and finances. With some limited rehabilitation, a vacant property that operates at a loss can be converted into a building that generates some profit in the interim.

Another alternative is to postpone demolition plans. For example, in the town of Vught in the southern part of the country, planners had intended to demolish an old nursing home. As the municipality suddenly had to house a large group of refugees it postponed the demolition and rehabilitated the old building to provide refugee housing. At present, the local housing association houses 32 refugees in this building. The municipality, a housing association, the Dutch Council for Refugees, neighbors, and the refugees themselves helped with rehabilitation activities (see Figure 2.4, Figure 2.5).

Flexible housing units (e.g., prefabricated housing, modular construction, container style housing, or mobile homes) are an attractive alternative because these units are easy to disassemble, to transport, to reconstruct, and to rebuild. Platform Home Again (Platform Opnieuw Thuis), discussed earlier, provides an overview of many types of flexible housing, giving ideas to municipalities and other

Figure 2.4 Temporary Refugee Housing Next to Offices in Rijswijk, the Netherlands.
Photo credit: Frank Wassenberg. Used with permission.

Figure 2.5 Temporary Use of an Old Elementary School in Landgraaf, the Netherlands.
Photo credit: Gemeente Landgraaf. Used with permission.

actors that are suddenly in need of additional housing (Expertisecentrum Flexwonen 2017). In January 2016, the Central Agency for the Reception of Asylum Seekers (COA) and the *Rijksbouwmeester* (the Dutch National Architect, an honorable and thus influential title, although it is only temporary) launched the A Home Away From Home competition for architects, housing designers, developers, and the building industry to develop and provide alternative types of flexible and temporary housing, including housing for refugees (see Figure 2.6). The COA aimed to respond better to the sudden increase of refugees. The solutions had to be easy to transport (for example, tents or caravans), easy to pack and to store (for example, containers), easy to use and reuse (flexible in terms of household size), and low cost. In total, 366 design suggestions were received; 12 were short listed, and six received awards and are being constructed.

Housing people of different backgrounds, for example, refugees, young people who have just started their first job, temporary workers, the homeless, or recently divorced people, within one larger building constitutes a way to address needs quickly and possibly is a way to integrate refugees into society. Platform31 (2016), along with other practitioners, suggested housing these different groups within the same building, called The Magic Mix. While much research on both sides of the Atlantic shows that mixing residents of different backgrounds and incomes at the neighborhood scale may exacerbate tensions, The Magic Mix shows that mixed-income housing may work when there is strict screening of applicants, when tenants are only allowed to stay for a few months or years, and when resident rules are strictly enforced.

Two Magic Mix examples are located in Amsterdam and Utrecht. In Amsterdam, a village of container style housing, Startblok Riekerhaven (2016) has provided housing for 280 refugees and 280 Dutch students who were matched, i.e., one refugee with one Dutch student, and live in a block sharing kitchen facilities and a garden, since 2016, thus combining housing and support for social

Figure 2.6 The Finch in Leiden, One of Six Winners in the A Home Away From Home Competition.

Photo credit: Kees Hummel. Used with permission.

Figure 2.7 Startblok Riekerhaven, Amsterdam.

Photo credit: Frank Wassenberg. Used with permission.

integration (see Figure 2.7). In Utrecht, Place2BU (2017) has offered 490 studios to young people age 23 to 27, a combination of refugees and non-refugees, since May 2017. There are other creative ways to house refugees. For example, Platform31 (2017) issued a report highlighting 45 existing examples of refugee housing.

Housing and Integration

Housing is key to social integration and participation in society. Once refugees receive their permit, they have to rebuild their lives in their new country to become self-reliant by attending language or vocational training classes, learning Dutch, becoming familiar with local customs and institutions, qualifying for and finding a job, and fully participating in society. However, in reality these steps are challenging for most refugees. In December 2015, the influential *Wetenschappelijke Raad voor het Regeringsbeleid* (Scientific Council for Government Policy, WRR) published the policy brief "No Time to Lose: From Reception to Integration of Asylum Migrants," which was a wakeup call to many (Netherlands Scientific Council for Government Policy (WRR) 2015). The authors found that after 15 years, only about one-third of all refugees with a permit had found a job. In December 2016, the Social and Economic Council of the Netherlands (SER), another influential national organization, concluded that although promising small-scale initiatives had been developed in the meantime, they are often too small and fragmented and of too ad hoc a nature to help most refugees who are in need of support (Social and Economic Council of the Netherlands (SER) 2016). Thus, integrating refugees into Dutch society will remain a key challenge for the future.

Conclusion

This chapter has described how the Netherlands, one of the main European destination countries of refugees, has addressed the need for emergency accommodations for asylum seekers and for housing recognized refugees. The Dutch government has responded to the sudden increased influx of asylum seekers in a pragmatic way by organizing emergency shelters and expanding the processing capacity of the Immigration and Naturalization Service and by constructing additional housing for refugees with a residence permit. The national government and municipalities have been well aware of the importance of balancing the right to housing, work, and social benefits of refugees with those of Dutch residents. Refugees as well as Dutch residents have taken advantage of newly constructed and rehabilitated housing units—although in general the national and local governments have emphasized rehabilitation over new construction. The next task for the national government, the municipalities, and other stakeholders is to facilitate the participation and integration of refugees in the labor market, educational systems, and the broader society, as well as to address likely future inflows of refugees.

Notes

1 In addition, there is an intermediate level between state and local governments, i.e., regional, state, or provincial governments, with varying responsibilities to house and integrate refugees. The intermediate governmental level is important in countries like Germany, Austria, and Belgium, but not in a unitary state like the Netherlands.

2 The number of people who disappear into illegality is unknown, but estimates vary from about 20,000 up to 100,000 (*The Post Online* 2015). Numbers typically increase when repeated asylum applications are rejected. Many undocumented refugees do not stay in one country but migrate to another. Some try to increase their odds by discarding their papers. Debates at the local level focus on whether to provide illegal immigrants with basic amenities called "a bed, a bath, and some bread."

References

Ambroso, Guido. 2006. *The Balkans at a Crossroads: Progress and Challenges in Finding Durable Solutions for Refugees and Displaced Persons from the Wars in the Former Yugoslavia.* Geneva, Switzerland: United Nations High Commissioner for Refugees.

Banulescu-Bogdan, Natalia, and Elizabeth Collett. 2015. *Refugee Crisis Deepens Political Polarization in the West.* Washington, DC: Migration Policy Institute.

Bouwstenen voor Sociaal. 2016. *Leegstand.* Amsterdam: Bouwstenen voor Sociaal.

Central Agency for the Reception of Asylum Seekers (COA). 2017. *Bezetting.* Rijswijk, The Netherlands.

Connor, Phillip. 2016. *European Asylum Applications Remained near Record Levels in 2016.* Washington, DC: Pew Research Center.

Elsinga, Marja, and Frank Wassenberg. 2014. "Social Housing in the Netherlands." In *Social Housing in Europe*, edited by K. Scanlon, C. Whitehead, and M. Fernandez Arrigoitia, 25–42. Chichester, UK: Wiley Blackwell.

European Commission. 2016. *The EU and the Refugee Crisis.* Brussels, Belgium: European Commission.

European Commission. 2017. *Irregular Migration via the Central Mediterranean. From Emergency Responses to Systemic Solutions.* Brussels, Belgium: European Commission.

European Parliament. 2017. *Immigration Policy. Fact sheets on the European Union. European Parliament at your Service.* Brussels, Belgium: European Parliament.

European Union Agency for Fundamental Rights. 2016. *Thematic Focus: Impact of the Asylum Crisis on Local Communities.* Vienna, Austria: European Union Agency for Fundamental Rights.

Eurostat. 2017. *Asylum Statistics.* Brussels, Belgium: European Commission/Eurostat.

Expertisecentrum Flexwonen. 2017. *Overview of Temporary Housing Units.* N.1: Expertisecentrum Flexwonen.

Grigonis, Simas. 2017. "EU in the Face of Migrant Crisis: Reasons for Ineffective Human Rights Protection." *International Comparative Jurisprudence*, 2(2): 93–98.

Immigration and Naturalization Service (IND). 2017a. *Infographics Asylum Procedure in the Netherlands.* Den Haag, The Netherlands: Immigration and Naturalization Service.

Immigration and Naturalization Service (IND). 2017b. *Cijfers en Publicaties.* Den Haag, The Netherlands: Immigration and Naturalization Service.

International Organization for Migration. 2017a. *Mediterranean Migrant Arrivals Reach 31,993, Deaths: 664.* Geneva, Switzerland: International Organization for Migration.

International Organization for Migration. 2017b. *Mediterranean Migrant Arrivals Reach 101,210 in 2017; 2,247 Deaths.* Geneva, Switzerland: International Organization for Migration.

International Organization for Migration. 2017c. *Mediterranean Update: Migration Flows Europe: Arrivals and Fatalities.* Geneva, Switzerland: International Organization for Migration.

Mulder, Roxanne. 2016. *Huisvesting van vergunninghouders, succesfactoren en knelpunten bij het realiseren van de gemeentelijke taakstelling.* Utrecht, The Netherlands: Utrecht University.

OECD. 2015. *Is this Humanitarian Migration Crisis Different?* Paris: OECD.

OECD. 2016. *International Migration Outlook 2016.* Paris: OECD.

Place2BU. 2017. *Place2BU, Utrecht.* Utrecht, the Netherlands: Place2BU.

Platform31. 2015. *Temporarily Use of Vacant Buildings.* Den Haag, The Netherlands: Platform31.

Platform31. 2016. *Magic Mix.* Den Haag, The Netherlands: Platform 31.

Platform31. 2017. *Inspiration Book: Housing Refugees with a Permit.* Den Haag, The Netherlands: Platform 31.

The Post Online. 2015, April 17. Illegalen in Nederland: Feiten en Cijfers.

Social and Economic Council of the Netherlands (SER). 2016. *Nieuwe wegen naar een meer succesvolle arbeidsmarktintegratie van vluchtelingen.* Den Haag, The Netherlands: Social and Economic Council of the Netherlands.

Smits van Waesberghe, Eliane, and Inge Razenberg. 2016. *De huisvesting van vergunninghouders.* Utrecht, The Netherlands: Verwey Jonker Instituut and Stedennetwerk G32.

Startblok Riekerhaven. 2016. *Welkom op Startblok.* Amsterdam, The Netherlands: Startblok Riekerhaven.

Statistics Netherlands (CBS). 2017. *Asylumseekers.* The Hague, The Netherlands: Centraal Bureau voor de Statistiek.

United Nations High Commissioner for Refugees. n.d. *Operational Portal Refugee Situations—Mediterranean Situation.* Geneva, Switzerland: United Nations High Commissioner for Refugees.

United Nations High Commissioner for Refugees. 2015, July 1. *Press Release.* Geneva, Switzerland: United Nations High Commissioner for Refugees.

Dutch Council for Refugees. 2016. *Vluchtelingen in getallen 2016.* Amsterdam, The Netherlands: VluchtelingenWerk Nederland.

Netherlands Scientific Council for Government Policy (WRR). 2015. *No Time to Lose: From Reception to Integration of Asylum Migrants.* The Hague, The Netherlands: Netherlands Scientific Council for Government Policy.

SECTION 2

Inequality

3

HOMEOWNERSHIP AND THE RACIAL AND ETHNIC WEALTH GAP IN THE UNITED STATES

Carolina K. Reid

Introduction

In December 2013, President Barack Obama identified income and wealth inequality as "the defining challenge of our time," tying growing inequality directly to the unravelling of the social compact in the United States (2013). The data reveal the magnitude of the divide. In 2013, the top 10 percent of earners took home 50 percent of the nation's income (Saez 2014), while the incomes of households at the bottom end of the distribution continued to decline (Bricker et al. 2016). While much of the public's attention has focused on the income gap and the stagnation of real wages (Mishel et al. 2016), wealth inequalities are even greater: The top 10 percent hold more than 85 percent of the nation's wealth (Saez and Zucman 2016).

The drivers of wealth inequality are complex, but there is no doubt that housing policy has played a significant role in shaping the distribution of wealth in this country (Conley 2009; Oliver and Shapiro 2006; Shapiro et al. 2013). The U.S. government has long promoted homeownership by providing a mortgage interest tax deduction, by guaranteeing mortgages and providing liquidity to the mortgage market, and by leveraging the power of the bully pulpit (Carliner 1998; Carr and Anacker 2015; Collins 2007). Homeownership has thus been the cornerstone of wealth accumulation for the vast majority of households (Turner and Luea 2009; Wolff 2014), and the expansion of mortgage lending after World War II laid the foundation for the growth of the middle class and the democratization of wealth from the 1930s through the late 1970s (Hanchett 2000). Between 1933 and 1985, the share of wealth held by the bottom 90 percent of households more than doubled, from 15 to 36 percent, as shown in Figure 3.1 (Saez and Zucman 2016). But the gains were short-lived; by 2012, this share had dropped to 23 percent, back to levels not seen since the 1940s. Since the housing crisis, there has been no recovery in the wealth of the middle class. The average wealth of the bottom 90 percent of households was equal to $80,000 in 2012—the same level as in 1986. In contrast, the average wealth for the top 1 percent more than tripled between 1980 and 2012 (Saez and Zucman 2016).

The distribution of wealth is also starkly divided along racial and ethnic lines: In 2013, the average non-Hispanic White household held 12.9 times more wealth than the average Black household and 10.3 times more than the average Hispanic household (Kochhar and Fry 2014). The foreclosure crisis is partly to blame: Black and Hispanic households lost their homes to foreclosure at over twice the rate of non-Hispanic White households (Bocian et al. 2011), undermining the homeownership gains achieved during the 1990s and early 2000s (Reid 2014a).[1] But even before the foreclosure crisis, racial and ethnic disparities in wealth were glaring (see Figure 3.2), reflecting the ways in which

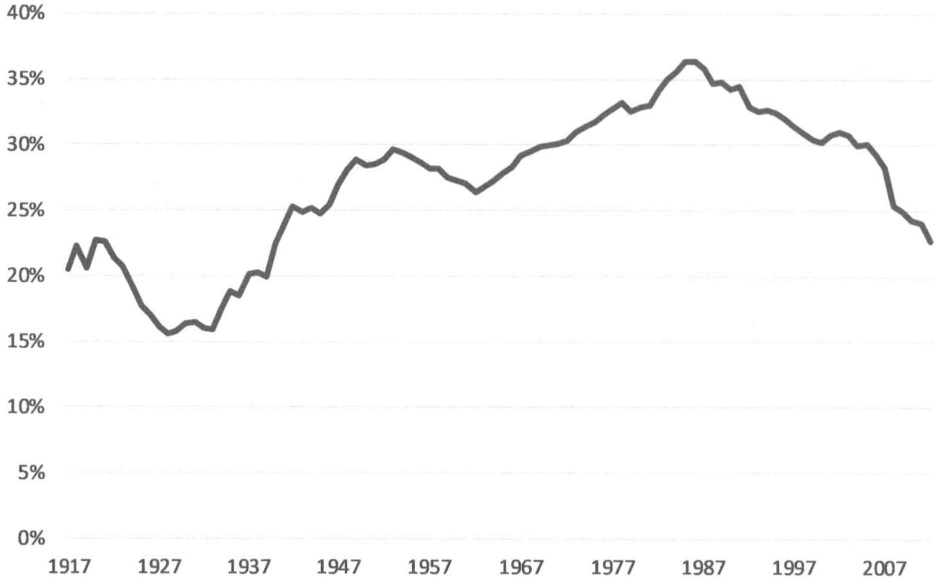

Figure 3.1 Trends in the Democratization of Wealth, 1917–2012.

Note: Graph depicts percent of wealth held by the bottom 90 percent of households.

Source: Saez and Zucman (2016). Used with permission.

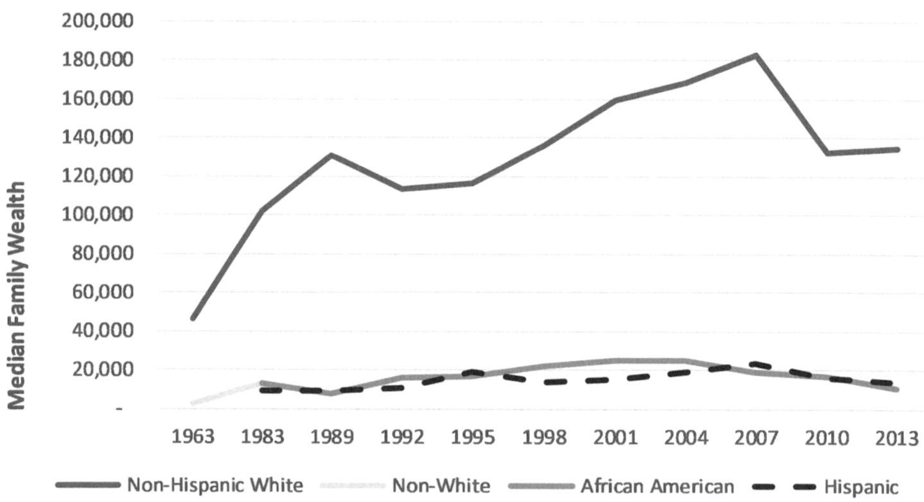

Figure 3.2 Racial and Ethnic Wealth Gap, 1963–2013.

Notes: 2013 dollars. Comparable data between 1963 and 1983 are not available; African American/Hispanic distinction within non-White population available only 1983 and later.

Source: Urban Institute (n.d.). Used with permission.

historical practices of discrimination in housing and mortgage markets have shaped wealth and privilege for generations (Conley 2009; Katznelson 2005; Lipsitz 2006; Massey and Denton 1998; Oliver and Shapiro 2006). Research suggests that one of the largest contributors to the sustained racial and ethnic wealth gap is differences in the number of years of homeownership, eclipsing other factors such as income and educational attainment (Shapiro et al. 2013). While the liberal social democratic reforms of the New Deal Era—including the establishment of the Federal Housing Administration—led to the expansion of wealth in the post-World War II era, they also excluded many Blacks and other racial and ethnic minorities from their benefits (Katznelson 2005).

Access to mortgage loans and homeownership for Black and Hispanic households has certainly improved since the postwar period. On the heels of the Civil Rights Movement, a series of significant federal laws were passed to address racial and ethnic inequalities in mortgage lending. In addition, the 1990s saw an expansion of federal policies to expand access to homeownership for Black and Hispanic households (Reid 2014a), including establishing affordable housing lending targets for Fannie Mae and Freddie Mac and creating funding streams for localized programs such as city-based down payment assistance programs (Reid 2009) and low-income homeownership programs run through community development financial institutions (Quercia et al. 2011).

These policies are justified on the basis of the economic and social benefits attributed to homeownership. Research suggests that homeowners accumulate more assets than their renter counterparts, particularly when homeownership is sustained over time (Boehm and Schlottmann 2004; Di et al. 2007; Grinstein-Weiss et al. 2013; McKernan and Sherraden 2008; Reid 2004; Turner and Luea 2009). Homeownership is particularly important for children: Research has shown that homeownership is associated with improvements in educational attainment, labor market outcomes, and lower rates of teenage pregnancy and other behavioral problems (Aaronson 2000; Green and White 1997; Harkness and Newman 2003; Haurin et al. 2002). Homeownership is also tied to a host of positive social outcomes, including better health and well-being, higher residential and neighborhood satisfaction, and increased civic engagement and political participation (Grinstein-Weiss et al. 2011; Rohe and Lindblad 2014).

However, the strength of association between homeownership and positive outcomes varies, and is often significantly muted when the effects of selection bias are taken into account (Herbert and Belsky 2008). In much of the research prior to 2000, studies on the benefit of homeownership failed to control for many factors that could be underlying the positive outcomes attributed to homeownership, including income, wealth, housing stability, or neighborhood characteristics (Rohe and Lindblad 2014). More recent research has found significant selection effects: For example, Holupka and Newman (2012) focused on the effects of homeownership on children's cognitive achievement, behaviors, and health using statistical matching and instrumental variable analysis and found little evidence of beneficial homeownership effects. Similarly, Engelhardt et al. (2010) found that effectively controlling for bias (with the use of an instrumental variable (IV)) resulted in a null finding for the association between homeownership and political engagement, even though models without the IV showed a positive relationship.

These studies, coupled with the negative spillover effects of the foreclosure crisis, raise important questions about the effectiveness of policies to promote homeownership, particularly among lower-income and lower-wealth borrowers. However, returning to an era when only the wealthiest or those with the most pristine credit scores could buy a home would only serve to calcify entrenched inequalities and create an even more unequal urban landscape in the future, particularly when demographic shifts in the United States mean that the majority of future homebuyers will be non-Whites (Carr et al. 2016). In this chapter, I argue that the response to the foreclosure crisis should not be a retreat from the goals of expanded access to credit, but rather a move toward a more ambitious set of public investments that can create a platform for sustainable homeownership and greater wealth redistribution. To date, federal policy has not gone far enough to undo the legacy of past discrimination in mortgage and housing markets, and until homeownership policy is accompanied with broader efforts to ensure that its benefits are more equally distributed, it will remain an empty promise for closing the racial and ethnic wealth gap. While the expansion of homeownership during the 1950s and 1960s translated into substantial wealth gains for households who were able to access credit over this time period, it is not clear that contemporary first-time homebuyers—particularly Black and Hispanic households—will see the same benefits.

The chapter begins by reviewing the historical legacy of past discrimination in the mortgage and housing market, and shows how this legacy persists today in shaping unequal returns to homeownership. It then argues that the erosion of the social safety net, coupled with financial deregulation and the increasing complexity of financial instruments, has made homeownership a riskier investment than in the past. It concludes with suggestions for how the federal government could help to mediate these inequalities through a comprehensive policy framework that redistributes the risks and benefits of homeownership.

"When Affirmative Action Was White": The Lasting Legacy of Discrimination in the Mortgage Market

As recently as 1965, mortgage discrimination was intentional and overt, a sanctioned practice in both private industry and the public sector (Arrow 1998; Massey and Denton 1998; Stuart 2003). The Home Owners Loan Corporation (HOLC) redlining maps, which laid out which neighborhoods were considered poor insurance bets, were explicitly informed by racialized conceptions of space, simultaneously privileging lending in suburbia while denying credit to borrowers in inner city, predominantly minority communities (Crossney and Bartelt 2005). For example, in St. Louis, houses in African-American neighborhoods were deemed as having "little or no value today, having suffered a tremendous decline in values due to the colored element now controlling the district" (Jackson 1985, 200). The New Deal programs that built a middle class after World War II were similarly designed to exclude Blacks and other racial and ethnic minorities from the benefits of the deal. For instance, the GI Bill was closed off to the majority of Black veterans; fewer than 100 of the 67,000 mortgages insured by the GI Bill in New York and northern New Jersey supported home purchases by non-Whites (Katznelson 2005).

In the 1960s and 1970s, a series of significant federal laws were passed to address racial and ethnic inequalities in mortgage lending, including the Fair Housing Act of 1968, the Equal Credit Opportunity Act of 1974, the Home Mortgage Disclosure Act in 1975, and the Community Reinvestment Act in 1977. While these laws have led to reductions in the prevalence of outright mortgage discrimination, more than four decades later, racial and ethnic disparities in mortgage lending still persist, even as overall access to credit has expanded (Apgar and Calder 2005). Studies have consistently found racial and ethnic disparities in mortgage lending denial rates (Bhutta and Canner 2013; Canner and Passmore 1994; Ross and Yinger 2002), the costs of credit (Bocian et al. 2008; Courchane 2007), and the prevalence of subprime loans or loans with risky product features (Bayer et al. 2014; Bocian et al. 2011; Rugh et al. 2015).[2]

Some of these differences can be attributed to continued racial and ethnic discrimination: unequal treatment based on race and ethnicity. For example, housing discrimination studies based on paired testing show that although overt racial and ethnic discrimination has decreased, minority applicants continue to receive information about fewer units than White applicants (Turner et al. 2013). In addition, recent lawsuits against lenders have revealed the ways in which banks targeted minority borrowers for subprime loans, even when they would have qualified for better loan terms (U.S. Department of Justice 2016). Brokers working in low-income and minority neighborhoods may also have singled out Black and Hispanic borrowers for higher-priced loans: Research has shown that a broker-originated loan was significantly more likely to be a higher-priced loan product or a loan with nontraditional terms (Courchane 2007; Ding et al. 2011; Ernst et al. 2008; LaCour-Little 2009), and that these effects were more marked for Black and Hispanic borrowers (Coulton et al. 2008; Reid and Laderman 2009, 2011; Reid et al. 2017).

However, mortgage lending disparities are more often attributed to racial and ethnic differences in income, assets, or creditworthiness: "race neutral" determinants of borrower risk that justify disparate treatment. Not surprisingly, differences in loan denial rates or loan terms by race and ethnicity are significantly muted when econometric models control for credit scores, loan-to-value ratios, debt-to-income ratios, and neighborhood characteristics, even if they do not go away entirely (Bayer et al. 2014; Bocian et al. 2008; Courchane 2007). The conclusion is then that a disproportionate share of unequal mortgage outcomes for Blacks and Hispanics stems not from discrimination, but from the low credit scores, incomes, or net worth of minority applicants. What is hidden from view is that the amount a household can contribute to a down payment is influenced by not only their income and their ability to save, but also their parents' assets and whether or not they received an inheritance or other financial support from their parents (Myers et al. 2016), which is deeply intertwined with redlining and who was able to buy a house in past generations. Research has shown that parental net worth levels significantly influence racial and ethnic wealth gaps in the next generation, that Blacks are about half as likely as non-Hispanic Whites to have received an inheritance, and that when they do receive an inheritance, the amount is substantially lower (Conley 2001, 2009; Menchik and Jianakoplos 1997; see the chapter authored by Erin Graves, Ana Patricia Muñoz, Darrick Hamilton, William A. Darity, and Yunju Nam).

Contemporary credit scores are also entangled with the history of redlining and the continued uneven landscape of financial service providers. Compared to previous loan allocation regimes in which borrowers were judged by mortgage lenders based on visual cues—for example, how "shifty," "evasive," "seedy," or "flashy" someone looked (Jeacle and Walsh 2002, 743)—credit scores offer a more objective means of assessing risk (Marron 2007). The emergence of risk-based pricing—in which borrowers with lower credit scores and lower down payments are charged more in interest for their loans rather than being denied a loan outright—has also expanded credit allocation for many minority households. As Figure 3.3 shows, Blacks and Hispanics have, on average, lower credit scores than non-Hispanic Whites and Asians. Among borrowers who bought homes between 2004 and 2008, 75 percent of non-Hispanic Whites and 85 percent of Asians had FICO scores above 660—which in general can be seen as representing a "prime" credit score—compared to less than

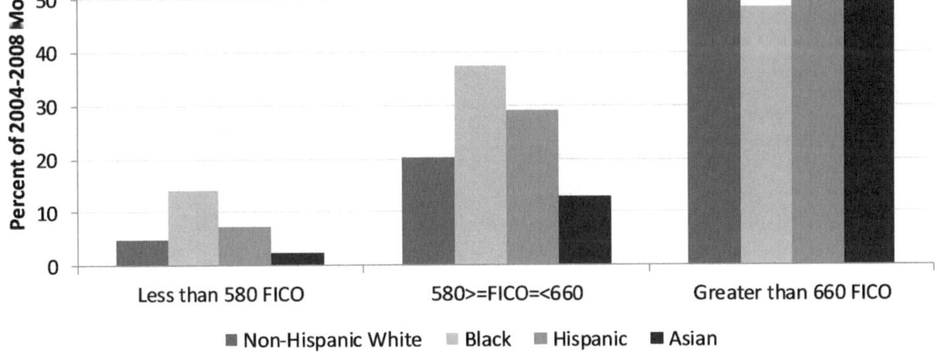

Figure 3.3 Racial and Ethnic Disparities in Credit Scores, 2004–2008.

Source: Author's calculations of HMDA/LPS/Blackbox dataset; for information about the data, see Bocian et al. 2011.

half of Blacks and 64 percent of Hispanics. Lenders thus claim they are justified in charging Black and Hispanic borrowers more, not as a function of "discrimination," but as a business practice warranted by the higher risk these borrowers pose to the lender (Langley 2009b).[3]

However, as with loan-to-value ratios, credit scores disregard past exclusion and make "invisible and hence incontestable the complex array of judgments and decisions that go into a scale and a number" (Rose 2002, 223). A borrower's credit score is dependent on their parental wealth, their labor market position, and their educational attainment, all factors that are in turn influenced by patterns of racial and ethnic discrimination and inequality. In addition, the fact that many alternative financial service providers—e.g., payday lenders, check cashers, tax preparers offering refund anticipation loans, debt settlement companies, and non-bank auto lenders—all concentrate their services in lower-income and certain minority communities compounds the risk that these households take on predatory debt, with attendant impacts on their credit record.

The significance of historical exclusion and social stratification, based on race and ethnicity, in accessing credit and homeownership—what Rugh et al. (2015) refer to as "cumulative disadvantage"—was laid bare during the subprime crisis. Black and Hispanic borrowers were steered into the subprime market channel by lenders who targeted their marketing efforts to households and neighborhoods that had untapped demand, including inner-city minority neighborhoods that had previously been redlined (Hernandez 2009; Mayer and Pence 2009; Reid 2010; Schafran and Wegmann 2012). These borrowers were charged higher interest rates and sold risky mortgage products with no consideration for whether they would be able to afford the home over the long term (Immergluck 2009, 2015). As subsequent research has shown, these loans led to much higher delinquency and foreclosure rates among Black and Hispanic homebuyers over this period, even after controlling for differences in income (Bayer et al. 2016; Bocian et al. 2011; Reid et al. 2017). More than 25 percent of Black homeowners who bought their house between 2004 and 2007 either had lost their home to foreclosure or were at risk of doing so by January 2013, more than double the rate of non-Hispanic White borrowers (see Figure 3.4). While all households saw declines in wealth resulting from the

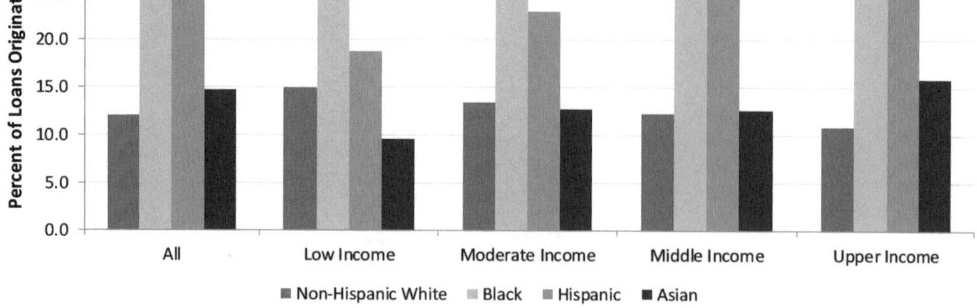

Figure 3.4 Serious Delinquency Rates by Borrower Race and Ethnicity and Income, 2004–2007 Originations, Observed Through January 2013.

Source: Author's calculations of HMDA/LPS/Blackbox dataset; for information about the data, see Bocian et al. 2011. Serious delinquency rates are calculated as a loan ever becoming 90 days delinquent.

housing crisis, the disparate impact of foreclosures resulted in greater declines in median wealth among Black and Hispanic households (Kochhar and Fry 2014).

The legacy of unequal access to credit and discrimination in housing and mortgage markets further plays out through spatial segregation and uneven house price appreciation, meaning that the benefits of homeownership are unequally realized across metropolitan areas. While levels of racial residential segregation have declined over the past 40 years, the United States remains deeply segregated by class and race (De la Roca et al. 2014). The White-Black dissimilarity index for particular metropolitan areas—which represents the share of one group that would need to move in order for the spatial distribution of the two groups to be the same across a metropolitan area—has declined steadily since 1980 (see Figure 3.5). In addition, as the United States has become more culturally, racially, and ethnically diverse, there have been concomitant increases in the number of racially and ethnically integrated neighborhoods (Maly 2005). These improvements notwithstanding, research continues to show that even the highest-earning Black and Hispanic households tend to move to neighborhoods that are lower-income and that have a higher proportion of minority residents than comparable non-Hispanic White households (Pais et al. 2012). Recent research also suggests that subprime lending may have reinforced patterns of racial and ethnic residential segregation. Fischer (2013) found that higher concentrations of subprime lending to Blacks were associated with increases in the segregation of Black homeowners, even as overall racial and ethnic segregation among homeowners declined.

The persistence of racial and ethnic segregation means that many Blacks and Hispanics continue to live in different neighborhoods than their non-Hispanic White counterparts, which contributes directly to differential outcomes in mortgage and housing markets, which in turn shape economic opportunity and wealth inequalities. For example, Rugh and Massey (2010) show that metropolitan racial and ethnic segregation created the opportunity for subprime mortgage targeting, intensifying the consequences of the American housing bubble for Black and Hispanic households. Segregation

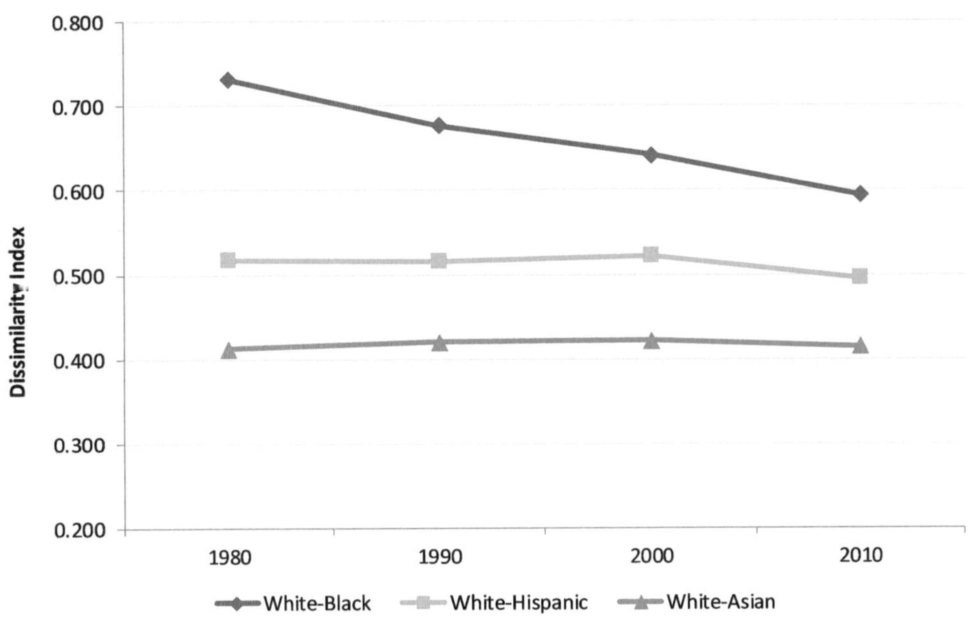

Figure 3.5 Trends in Racial and Ethnic Segregation, 1980–2010.

Source: De la Roca et al. (2014). Used with permission.

also impacts equity accumulation; in metropolitan areas with higher levels of residential segregation, research has shown that the gap between Black and non-Hispanic White home values is wider, even after controlling for household income (Rusk 2001).[4] Black and Hispanic households gain substantially lower returns to homeownership than non-Hispanic White households at similar income and education levels, even during robust economic cycles (Krivo and Kaufman 2004; Newman and Holupka 2016), and even in suburban areas, house price appreciation tends to be lower in tracts that are primarily Black or Hispanic in comparison to those that are primarily non-Hispanic White and Asian (Anacker 2010, 2012). Post-recession, while the housing equity of non-Hispanic White households has started to rebound, evidence suggests that Blacks are being left out of the recovery, suggesting that the foreclosure crisis will likely lead to a widening of the racial wealth gap over time (Burd-Sharps and Rasch 2015; Raymond et al. 2016).

The Privatization of Risk and Its Implications for Mortgage Lending

Layered on top of these historical inequalities and differences in access to credit has been a shift in how the U.S. government protects families and workers from the risks of modern capitalism, which Jacob Hacker refers to as the "great risk shift" (Hacker 2004, 2006). The federal government has long helped to reduce and reallocate risk, from workers' compensation laws (i.e., shifting the risk of economic hardship due to a workplace injury from workers to employers), to federal deposit insurance (i.e., reducing the risk of bank failure for individual depositors), to social security insurance (i.e., ensuring an income for those who can no longer work). Moss (2002) chronicles the evolution of U.S. risk management policy, and shows that in the early twentieth century, the government expanded its risk management policies from those that protected business interests to those that reduced and redistributed risks for workers. However, since the early 1980s, the responsibility for managing economic risks has shifted from government and employers onto individuals and their families (Hacker 2006; Porter and Twomey 2012). This "privatization of risk" can be seen in many social and economic policy domains (Hacker and O'Leary 2012), such as the shift from welfare to workfare, the rise of contingent labor (including an increased reliance on part-time workers and jobs without health or other benefits), and the move away from pensions to individual retirement accounts.

This shift toward the privatization of risk is no less salient in the mortgage market. In the post-World War II era, homeownership was buttressed by strong government oversight and a 30-year, fixed-rate mortgage product (Forrest and Hirayama 2015). As Porter and Twomey (2012) have argued, the term of a mortgage contract not only establishes the affordability of the loan over the long-term, but also allocates the risk associated with the home purchase between the borrower and lender. FHA's role in standardizing the 30-year fixed-rate mortgage—coupled with strong regulatory oversight—substantially reduced the risks of homeownership for families. For example, the fixed rate insulated borrowers from payment shocks due to changes in interest rates, and the amortization schedule removed the need for borrowers to have sufficient funds for large balloon payments.

However, the deregulation of the banking industry in the 1980s weakened consumer protections and spurred product innovation in financial markets. A suite of federal laws, including the 1980 Depository Institutions Deregulation and Monetary Control Act, the 1989 Financial Institutions Reform, Recovery, and Enforcement Act, and the Tax Reform Act of 1986 (Engel and McCoy 2011; Immergluck 2009), paved the way for innovation in mortgage products, including loans with adjustable interest rates, negative amortization, and interest-only payment schedules. The intent of this deregulation was to increase the supply of credit and the profitability of mortgage loans, but it also increased the overall risk of mortgage loans and shifted more of this risk to borrowers (Porter and Twomey 2012). These deregulatory policies emerged shortly after laws passed to expand access to credit for historically excluded groups. Precisely as lenders began to offer a wider array of financial products to more consumers, federal regulators retreated from consumer protections and their oversight of the banking industry, increasing the risk of taking on a mortgage for individuals (see Figure 3.1).

This shift has meant that consumers increasingly need to become sophisticated financial subjects and investors (Langley 2009a), demonstrating individual responsibility and making rational economic decisions that maximize their financial well-being in a world of complex financial products. As with all investments, homeownership has always entailed some degree of risk: For instance, the financial returns to homeownership are influenced significantly by the timing of the purchase and sale. But in the past two decades, the benefits of homeownership have been further influenced by the extent to which a consumer is able to predict the future of interest rate changes and hedge those anticipated changes against a complex array of mortgage products. Borrowers are responsible for ensuring that they are not taking on undue "instrument risk," in which mortgage product features themselves pose an elevated risk of default, even if a borrower's income remains stable (Bostic and Lee 2008). Those with higher levels of wealth, education, and professional social networks are more likely to make decisions that maximize their financial returns, and may also have the largest financial cushion to absorb the costs of bad decisions (Campbell 2013; Campbell and Cocco 2003; Porter and Twomey 2012). While financial disclosures, education, and homeownership counseling can improve borrower decision-making, research in behavioral economics suggests that consumers are susceptible to a wide range of biases in financial decision-making (Essene and Apgar 2007).

Risk-based pricing provides another example of where mortgage contract terms contribute to unequal returns to homeownership. Risk-based pricing relies on measuring, pricing, and allocating risk at the individual level. Yet it also means that those with the least resources pay the most for credit, rather than redistributing the cost of default risk across all borrowers. As noted earlier, risk-based pricing may be a better alternative to the outright denial minorities and other lower-income and lower-wealth borrowers faced previously. For example, if a subprime borrower is able to later refinance into a prime loan, then risk-based pricing could arguably reduce racial and ethnic inequalities in homeownership and wealth by providing access to a mortgage for those who would otherwise be excluded, yet not penalize them with higher interest costs over the long term (Collins et al. 2005). However, research has consistently shown that Black and Hispanic borrowers are less likely to refinance (Agarwal et al. 2013; Deng et al. 2005; Goodstein 2014; Keys et al. 2014; Reid 2014b).[5] For example, among homeowners who bought during the subprime boom and were current on their mortgage in 2013, more than 70 percent of Blacks and Hispanics were paying interest rates above 6 percent, compared to just over half of non-Hispanic Whites and Asians. These borrowers thus continue to pay a subprime "tax," building less equity than a comparable non-Hispanic White or Asian borrower living in the same neighborhood. This "tax" has little to do with their actual default risk: The fact that they have continued to pay their mortgage despite the most severe housing downturn since the Great Depression suggests that they are a "good" lending risk. Although it is not entirely known why refinance rates are so much lower, the fact that Hispanic and Black borrowers have not refinanced into lower priced loans during a period of historically low interest rates has significant implications for their wealth and financial well-being over the long term, and will shape the degree to which homeownership for these families leads to positive spatial, social, and intergenerational outcomes (Landis and McClure 2010).

Finally, the privatization of risk is a useful frame for thinking about how the broader social and economic context in which homeownership occurs has also changed. Those who became homeowners in the postwar era had the benefits of a labor and social insurance system that ensured relative job and income stability over time (Forrest and Hirayama 2015), as well as a tax system that helped to redistribute the benefits of economic growth more broadly across income groups (Chetty et al. 2016). The erosion of the social safety net and an increase in earnings instability increases the risks of homeownership, since sustaining a mortgage is predicated on a steady stream of monthly payments. In addition, to the extent that Black and Hispanic households are overrepresented in lower-skilled, part-time, and contingent work, they will be more vulnerable to job insecurity, earnings instability, and higher housing costs in relation to their incomes, all of which can undermine the sustainability of homeownership.[6] Indeed, during the foreclosure crisis, rates of delinquencies and foreclosures were higher for Black and Hispanic homeowners in counties with higher job losses, suggesting that they may be more vulnerable to changes in employment rates (Bayer et al. 2016). Even prudent financial decisions and "responsible" behaviors will not protect families from foreclosure if they are subject to the vagaries of contingent work, high housing cost burdens, and a tattered safety net.

Policy Implications: Building a Stronger Homeownership Policy

Undoing the legacy of housing and mortgage discrimination and redistributing the risks and rewards of homeownership will require a much stronger role for the federal government going forward, particularly if the goal is to close the wealth gap for Black and Hispanic households. I argue that this policy agenda should entail four components: increasing enforcement of consumer protection and fair lending regulations, creating a stronger role for government policy in undoing the inequalities produced by historical disparities in mortgage lending, reducing the regressive nature of the mortgage interest tax deduction, and supporting sustainable homeownership through broader investments in workforce and community development.

First and foremost, the subprime crisis highlights the importance of consumer protection and regulatory oversight of financial markets in reducing the risks of predatory lending and addressing mortgage market discrimination. The Dodd-Frank Wall Street Reform and Consumer Protection Act (Dodd-Frank) and the establishment of the Consumer Financial Protection Bureau (CFPB) were important steps in this direction. Dodd-Frank and ensuing regulations have largely eliminated the most problematic practices that contributed to the crisis, including the marketing of loans without consideration of a borrower's ability to repay and restricting the origination of loans with the riskiest loan terms. There also needs to be continued efforts to strengthen and enforce fair lending laws, and to fully consider not only disparate treatment, but also disparate impact. The Supreme Court ruling in *Texas Department of Housing and Community Affairs v. Inclusive Communities Projects* in 2015 established that neutral policies and/or practices that lead to unequal outcomes could be subject to action, even if intent to discriminate is not found. Continued exclusion or differential pricing of mortgages for Black and Hispanic borrowers should be tested under this standard, particularly given consistent evidence that pricing differentials do not disappear even when researchers control for borrower differences in standard underwriting criteria. While the CFPB, Department of Justice, and U.S. Department of Housing and Urban Development have all strengthened policies and enforcement of consumer protection, fair lending, and fair housing regulations in recent years, the election of Donald Trump in November 2016 puts these reforms at significant risk of repeal.

Second, efforts to reform the federal housing finance system—including policies related the GSEs and FHA—should incorporate "duty to serve" principles, to ensure that all households have access to responsible lending institutions (Levitin and Ratcliffe 2014). The duty to serve concept is broader than fair lending, and involves taking affirmative steps to reach out to communities traditionally underserved by the housing finance market. The Community Reinvestment Act is an example of such an affirmative obligation, and should be strengthened and reformed to increase its effectiveness (Chakrabarti et al. 2009). Duty to serve obligations, coupled with public funds to subsidize mortgage programs for lower-income and lower-wealth households, could help to overcome the different "starting line" for households with limited assets or lower credit scores. The GSEs have a long history of developing underwriting guidelines and expanding partnerships that have safely expanded credit to underserved communities. For example, the Community Advantage Program, a joint effort with the Ford Foundation and Self-Help Credit Union, leveraged $4.74 billion in financing for low-interest rate mortgages to nearly 52,000 low-income homeowners across the country, using only a $50 million grant as a credit enhancement (Quercia et al. 2011). Given the growth and increased capacity of the community development finance industry, a new federal fund to provide similar credit enhancement capacity to expand this program and bring the number of assisted households to scale could be significant in mediating inequalities in access to credit and homeownership.

There is also an opportunity for the federal government to pilot new loan products that might help to overcome wealth constraints to homeownership. For example, a lease-purchase loan product offered through the FHA could help lower-wealth households stabilize their finances and overcome down payment constraints, while allowing them to benefit sooner from the stability of homeownership (Galante et al. 2017). The federal government could also pilot a mortgage product with an interest rate that automatically adjusts downward when rates decline (Campbell 2013), which could reduce the long-term costs of credit for subprime borrowers who have demonstrated the ability to sustain their mortgage. An automatically refinancing mortgage would further remove the cross-subsidization in the current mortgage finance system, where prime homeowners who use their refinancing option when rates decline are subsidized by subprime households who fail to do so (Keys et al. 2014). While more research is needed to identify which products may be the most beneficial for lower-wealth borrowers, the FHA and/or the GSEs are in a unique position to analyze data and evaluate pilot programs to identify potential responsible, scalable models.

Third, federal tax policy skews the benefits of homeownership to higher-income households, primarily through the mortgage interest tax deduction, which costs approximately $70 billion a year, but the majority—$60 billion—goes to households making more than $100,000. On average, those with incomes under $50,000 receive only $300 in annual deductions, compared to $5,300 for households earning over $200,000 (Galante et al. 2016). There are numerous ways to make the mortgage interest tax deduction more progressive, including replacing it with a credit, limiting its value, targeting it at first-time homeowners, and/or making it refundable (Fichtner and Feldman 2014; Toder et al. 2010). In particular, replacing the deduction with a credit would shift the benefits to lower-income households. While the mortgage interest tax deduction is often considered a third rail of American politics and unlikely to be eliminated outright, there is increasing support along the ideological spectrum for addressing its skewed incentives as part of comprehensive tax reform (see, for example, Desmond 2017; Pethokoukis 2017).

Finally, inequalities in homeownership and its financial benefits are related to a much broader system of disparities linked to race and ethnicity, including labor market outcomes, social service provision, education, and residential segregation and uneven development. Reductions in social welfare programs, labor market shifts, and an increasingly regressive tax policy have all contributed to growing inequalities in income and wealth (Hacker 2016; Katz 2012). Given the significance of these broader forces in shaping not only access to homeownership but also its sustainability and returns, policies to mediate underlying inequalities not only in housing markets but also across other social and economic domains are critical to close the racial wealth gap. For example, improving access to human capital investments—from universal early childhood education to lower college costs—could help close income and employment gaps (Martin et al. 2016). Community development initiatives and place-based investments and support via community development corporations (CDCs) could further help to stabilize property values and reduce intra-metropolitan variation in the returns to homeownership (see the chapter authored by Rachel Bratt).

Conclusion

Racial and ethnic disparities in homeownership and wealth have long been shaped by public policy and planning decisions at the local, state, and federal levels. The long-term repercussions of redlining and overt racial and ethnic discrimination are evident today, from entrenched patterns of racial and ethnic segregation to the unequal "starting line" of savings and credit access that influence who can obtain a mortgage and at what cost. These historical legacies continue to shape the ways in which the gains from homeownership are unevenly realized by different groups. In addition, the deregulation of financial markets coupled with the increasing complexity of financial products requires that at present borrowers have to navigate a much more complicated financial landscape in order to save and build assets. The unequal returns to homeownership for Blacks and Hispanics are further intertwined with inequities in other domains of social and economic life, including disparities in public and private investments in neighborhood infrastructure (e.g., parks, schools, public safety), labor markets, and social welfare policy (Reskin 2012). Addressing these disparities and closing the wealth gap will require a more ambitious set of policy interventions that not only expand access to credit, but also ensure that the conditions of that credit do not serve to further promote stratification on the basis of race or ethnicity. While undoing the legacy of discrimination will not be easy, as Ta-Nehisi Coates (2014) has argued, "as surely as the creation of the wealth gap required the cooperation of every aspect of the society, bridging it will require the same."

Notes

1 Foreclosures in the United States peaked in early 2010, after climbing steadily since 2005 (RealtyTrac 2015). However, in Black and Hispanic communities, evidence of heightened foreclosures (particularly as it related to subprime lending in the refinance market) emerged in the late 1990s (Immergluck and Smith 2005; Li 2011; Li and Morrow-Jones 2010).

2 Some commentators have blamed the growth of subprime lending on government efforts to expand access to credit to underserved groups, including the Community Reinvestment Act (CRA) and the affordable housing goals established for Fannie Mae and Freddie Mac. These arguments have not been substantiated by empirical research. The majority of studies suggest that CRA or other government programs played no role in driving the risky subprime lending of the mid-2000s (Avery and Brevoort 2011; Reid et al. 2013), and that in fact CRA-motivated lending experienced lower default rates than subprime loans made by lending institutions not subject to CRA (Quercia et al. 2011; Reid and Laderman 2011).

3 It is important to note that many Black and Hispanic borrowers with a prime credit score nevertheless received subprime loans with higher interest rates and risky loan terms (Bocian et al. 2008, 2011), suggesting that discrimination operated alongside risk-based pricing to the distinct disadvantage of these minority communities.

4 A recent paper also finds that homeownership in a highly segregated neighborhood may not only influence the direct financial returns to owning, but it may also have a negative impact on intergenerational mobility. For households with incomes below the median, growing up in a neighborhood with higher racial and income segregation reduced the positive impact of homeownership on intergenerational mobility by 35 to 50 percent (Kulkarni and Malmendier 2015).

5 The factors that influence differences in the propensity to refinance are not completely understood. Differences in collateral, income, credit quality, loan terms (e.g., the existence of prepayment penalties), and knowledge all play a role in lower refinance rates for subprime and Black/Hispanic homeowners, but disparities remain (Goodstein 2014; Green and LaCour-Little 1999). Some recent qualitative research suggests that issues of trust and concerns about interactions with mortgage lenders may also play a role for Black and Hispanic borrowers (Reid 2014b).

6 Fergus (2014) has further argued that labor market instability and the retrenchment of social policies like welfare and pensions contributed to the consumer "demand" for financial products, noting that debt has increasingly been used to help cushion the instability of work and volatility of earnings for lower-income households.

References

Aaronson, Daniel. 2000. "A Note on the Benefits of Homeownership." *Journal of Urban Economics*, 47(3): 356–369.

Agarwal, Sumit, Richard J. Rosen, and Vincent Yao. 2013. *Why Do Borrowers Make Mortgage Refinancing Mistakes?* Chicago: Federal Reserve Bank of Chicago.

Anacker, Katrin B. 2010. "Still Paying the Race Tax? Analyzing Property Values in Homogeneous and Mixed-Race Suburbs." *Journal of Urban Affairs*, 32(1): 55–77.

Anacker, Katrin B. 2012. "Shaky Palaces? Analyzing Property Values and Their Appreciation in Minority First Suburbs." *International Journal of Urban and Regional Research*, 36(4): 791–816.

Apgar, William C., and Allegra Calder. 2005. "The Dual Mortgage Market: The Persistence of Discrimination in Mortgage Lending." In *The Geography of Opportunity: Race and Housing Choice in Metropolitan America*, edited by Xavier de Souza Briggs, 101–126. Washington, DC: Brookings Institution Press.

Arrow, Kenneth J. 1998. "What Has Economics to Say about Racial Discrimination?" *The Journal of Economic Perspectives*, 12(2): 91–100.

Avery, Robert B., and Kenneth P. Brevoort. 2011. *The Subprime Crisis: Is Government Housing Policy to Blame?* Washington, DC: Board of Governors of the Federal Reserve System.

Bayer, Patrick, Fernando Ferreira, and Stephen L. Ross. 2014. *Race, Ethnicity and High-Cost Mortgage Lending*. Cambridge, MA: National Bureau of Economic Research.

Bayer, Patrick, Fernando Ferreira, and Stephen L. Ross. 2016. "The Vulnerability of Minority Homeowners in the Housing Boom and Bust." *American Economic Journal: Economic Policy*, 8(1): 1–27.

Bhutta, Neil, and Glenn B. Canner. 2013. "Mortgage Market Conditions and Borrower Outcomes: Evidence from the 2012 HMDA Data and Matched HMDA-Credit Record Data." *Federal Reserve Bulletin*, 99(4): 1–58.

Bocian, Debbie Gruenstein, Keith S. Ernst, and Wei Li. 2008. "Race, Ethnicity and Subprime Home Loan Pricing." *Journal of Economics and Business, Financing Community Reinvestment and Development*, 60(1–2): 110–124.

Bocian, Debbie, Wei Li, Carolina K. Reid, and Robert G. Quercia. 2011. *Lost Ground, 2011: Disparities in Mortgage Lending and Foreclosures*. Durham, NC: Center for Responsible Lending.

Boehm, Thomas P., and Alan Schlottmann. 2004. *Wealth Accumulation and Homeownership: Evidence for Low-Income Households*. Washington, DC: U.S. Department of Housing and Urban Development.

Bostic, Raphael W., and Kwan Ok Lee. 2008. "Mortgages, Risk, and Homeownership among Low- and Moderate-Income Families." *American Economic Review*, 98(2): 310–314.

Bricker, Jesse, Lisa J. Dettling, Alice Henriques, Joanne W. Hsu, Kevin B. Moore, John Sabelhaus, Jeffrey Thompson, and Richard A. Windle. 2016. "Changes in U.S. Family Finances from 2010 to 2013: Evidence from the Survey of Consumer Finances." *Federal Reserve Bulletin*, 100: n.p.

Burd-Sharps, Sarah, and Rebecca Rasch. 2015. *Impact of the US Housing Crisis on the Racial Wealth Gap across Generations*. New York: Social Science Research Council.

Campbell, John Y. 2013. "Mortgage Market Design." *Review of Finance*, 17(1): 1–33.

Campbell, John Y., and João F. Cocco. 2003. "Household Risk Management and Optimal Mortgage Choice." *The Quarterly Journal of Economics*, 118(4): 1449–1494.

Canner, Glenn B., and Wayne Passmore. 1994. "Residential Lending to Low-Income and Minority Families: Evidence from the 1992 HMDA Data." *Federal Reserve Bulletin*, 80: 79–108.

Carliner, Michael S. 1998. "Development of Federal Homeownership 'Policy.'" *Housing Policy Debate*, 9(2): 299–321.

Carr, James H., and Katrin B. Anacker. 2015. "The Past and Current Politics of Housing Finance and the Future of Fannie Mae, Freddie Mac, and Homeownership in the United States." *Banking and Financial Services Policy Report: A Journal of Trends in Regulation and Supervision*, 33(7): 1–10.

Carr, James H., Michela Zonta, Fred McKinney, and Gerald Jaynes. 2016. *State of Housing in Black America*. Lanham, MD: National Association of Real Estate Brokers.

Chakrabarti, Prabal, David Erickson, Ren S. Essene, Ian Galloway, and John Olson, eds. 2009. *Revisiting the CRA: Perspectives on the Future of the Community Reinvestment Act*. Boston, San Francisco: Federal Reserve Banks of Boston and San Francisco.

Chetty, Raj, David Grusky, Maximilian Hell, Nathaniel Hendren, Robert Manduca, and Jimmy Narang. 2016. *The Fading American Dream: Trends in Absolute Income Mobility since 1940*. Stanford, CA: Equality of Opportunity Project.

Coates, Ta-Nehisi. 2014. "The Case for Reparations." *The Atlantic*, June, N.p.

Collins, J. Michael. 2007. "Federal Policies Promoting Affordable Homeownership: Separating the Accidental from the Strategic." In *Chasing the American Dream: New Perspectives on Affordable Homeownership*, edited by William M. Rohe and Harry L. Watson, 69–95. Ithaca, NY: Cornell University Press.

Collins, J. Michael, Eric S. Belsky, and Karl E. Case. 2005. "Exploring the Welfare Effects of Risk-Based Pricing in the Subprime Mortgage Market." In *Refinance and the Accumulation of Home Equity Wealth*, edited by Nicolas P. Retsinas and Eric S. Belsky, 138–151. Washington, DC: Brookings Institution Press.

Conley, Dalton. 2001. "Decomposing the Black-White Wealth Gap: The Role of Parental Resources, Inheritance, and Investment Dynamics." *Sociological Inquiry*, 71(1): 39–66.

Conley, Dalton. 2009. *Being Black, Living in the Red: Race, Wealth, and Social Policy in America*. Berkeley: University of California Press.

Coulton, Claudia, Tsui Chan, Michael Schramm, and Kristen Mikelbank. 2008. *Pathways to Foreclosure: A Longitudinal Study of Mortgage Loans, Cleveland and Cuyahoga County, 2005–2008*. Cleveland, OH: Center on Urban Poverty and Community Development.

Courchane, Marsha J. 2007. "The Pricing of Home Mortgage Loans to Minority Borrowers: How Much of the APR Differential Can We Explain?" *Journal of Real Estate Research*, 29(4): 399–440.

Crossney, Kristen B., and David W. Bartelt. 2005. "The Legacy of the Home Owners' Loan Corporation." *Housing Policy Debate*, 16(3–4): 547–574.

De la Roca, Jorge, Ingrid Gould Ellen, and Katherine M. O'Regan. 2014. "Race and Neighborhoods in the 21st Century: What Does Segregation Mean Today?" *Regional Science and Urban Economics*, SI, 47(July): 138–151.

Deng, Yongheng, Andrey D. Pavlov, and Lihong Yang. 2005. "Spatial Heterogeneity in Mortgage Terminations by Refinance, Sale and Default." *Real Estate Economics*, 33(4): 739–764.

Desmond, Matthew. 2017. "How Homeownership Became the Engine of American Inequality." *The New York Times*, May 9.

Di, Zhu Xiao, Eric Belsky, and Xiaodong Liu. 2007. "Do Homeowners Achieve More Household Wealth in the Long Run?" *Journal of Housing Economics*, 16(3–4): 274–290.

Ding, Lei, Roberto G. Quercia, Wei Li, and Janneke Ratcliffe. 2011. "Risky Borrowers or Risky Mortgages Disaggregating Effects Using Propensity Score Models." *Journal of Real Estate Research*, 33(2): 245–278.

Engel, Kathleen C., and Patricia A. McCoy. 2011. *The Subprime Virus: Reckless Credit, Regulatory Failure, and Next Steps*. New York: Oxford University Press.

Engelhardt, Gary V., Michael D. Eriksen, William G. Gale, and Gregory B. Mills. 2010. "What Are the Social Benefits of Homeownership? Experimental Evidence for Low-Income Households." *Journal of Urban Economics*, 67(3): 249–258.

Ernst, Keith S., Debbie Gruenstein Bocian, and Wei Li. 2008. *Steered Wrong: Brokers, Borrowers, and Subprime Loans*. Durham, NC: Center for Responsible Lending.

Essene, Ren, and William Apgar. 2007. *Understanding Mortgage Market Behavior: Creating Good Mortgage Options for All Americans*. Cambridge: Joint Center for Housing Studies at Harvard University.

Fergus, Devin. 2014. "Financial Fracking in the Land of the Fee, 1980–2008." In *The Assets Perspective: The Rise of Asset Building and Its Impact on Social Policy*, edited by Reid Cramer and Trina R. Williams Shanks, 67–98. New York: Palgrave Macmillan.

Fichtner, Jason J., and Jacob Feldman. 2014. *Reforming the Mortgage Interest Deduction*. Arlington, VA: Mercatus Center at George Mason University.

Fischer, Mary J. 2013. "Black and White Homebuyer, Homeowner, and Household Segregation in the United States, 1990–2010." *Social Science Research*, 42(6): 1726–1736.

Forrest, Ray, and Yosuke Hirayama. 2015. "The Financialisation of the Social Project: Embedded Liberalism, Neoliberalism and Home Ownership." *Urban Studies*, 52(2): 233–244.

Galante, Carol, Carolina K. Reid, and Nathaniel Decker. 2016. *The FAIR Tax Credit: A Proposal for a Federal Assistance in Rental Credit to Support Low-Income Renters*. Berkeley: Terner Center for Housing Innovation Report, University of California, Berkeley.

Galante, Carol, Carolina K. Reid, and Rocio Sanchez-Moyano. 2017. *Expanding Access to Homeownership through Lease-Purchase*. Berkeley: Terner Center for Housing Innovation, University of California, Berkeley.

Goodstein, Ryan M. 2014. "Refinancing Trends among Lower Income and Minority Homeowners during the Housing Boom and Bust." *Real Estate Economics*, 42(3): 690–723.

Green, Richard K., and Michael LaCour-Little. 1999. "Some Truths about Ostriches: Who Doesn't Prepay Their Mortgages and Why They Don't." *Journal of Housing Economics*, 8(3): 233–248.

Green, Richard K., and Michelle J. White. 1997. "Measuring the Benefits of Homeowning: Effects on Children." *Journal of Urban Economics*, 41(3): 441–461.

Grinstein-Weiss, Michal, Clinton Key, Shenyang Guo, Yeong Hun Yeo, and Krista Holub. 2013. "Homeownership and Wealth among Low- and Moderate-Income Households." *Housing Policy Debate*, 23(2): 259–279.

Grinstein-Weiss, Michal, Yeong Yeo, Katrin Anacker, Shannon Van Zandt, Elizabeth B. Freeze, and Roberto G Quercia. 2011. "Homeownership and Neighborhood Satisfaction among Low- and Moderate-Income Households." *Journal of Urban Affairs*, 33(3): 247–265.

Hacker, Jacob S. 2004. "Privatizing Risk without Privatizing the Welfare State: The Hidden Politics of Social Policy Retrenchment in the United States." *American Political Science Review*, 98(2): 243–260.

Hacker, Jacob S. 2006. *The Privatization of Risk and the Growing Economic Insecurity of Americans*. New York: Social Science Research Council.

Hacker, Jacob S. 2016. *The Privatization of Risk and the Growing Economic Insecurity of Americans*. New York: Social Science Research Council.

Hacker, Jacob S., and Ann O'Leary, eds. 2012. *Shared Responsibility, Shared Risk: Government, Markets and Social Policy in the Twenty-First Century*. New York: Oxford University.

Hanchett, Thomas W. 2000. "The Other 'Subsidized Housing': Federal Aid to Suburbanization, 1940s–1960s." In *From Tenements to the Taylor Homes: In Search of an Urban Housing Policy in Twentieth-Century America*, edited by John F. Bauman, Roger Biles, and Kristin M. Szylvian, 163–179. University Park, PA: Penn State University Press.

Harkness, Joseph, and Sandra Newman. 2003. "Differential Effects of Homeownership on Children from Higher- and Lower-Income Families." *Journal of Housing Research*, 14(1): 1–19.

Haurin, Donald R., Toby L. Parcel, and R. Jean Haurin. 2002. "Does Homeownership Affect Child Outcomes?" *Real Estate Economics*, 30(4): 635–666.

Herbert, Christopher E., and Eric S. Belsky. 2008. "The Homeownership Experience of Low-Income and Minority Households: A Review and Synthesis of the Literature." *Cityscape: A Journal of Policy Development and Research*, 10(2): 5–59.

Hernandez, Jesus. 2009. "Redlining Revisited: Mortgage Lending Patterns in Sacramento 1930–2004." *International Journal of Urban and Regional Research*, 33(2): 291–313.

Holupka, Scott, and Sandra J. Newman. 2012. "The Effects of Homeownership on Children's Outcomes: Real Effects or Self-Selection?" *Real Estate Economics*, 40(3): 566–602.

Immergluck, Dan. 2009. *Foreclosed: High-Risk Lending, Deregulation, and the Undermining of America's Mortgage Market*. Ithaca. NY: Cornell University Press.

Immergluck, Dan. 2015. "A Look Back: What We Now Know about the Causes of the US Mortgage Crisis." *International Journal of Urban Science*, 19(3): 269–285.

Immergluck, Dan, and Geoff Smith. 2005. "Measuring the Effect of Subprime Lending on Neighborhood Foreclosures Evidence from Chicago." *Urban Affairs Review*, 40(3): 362–389.

Jackson, Kenneth T. 1985. *Crabgrass Frontier: The Suburbanization of the United States*. New York: Oxford University Press.

Jeacle, Ingrid, and Eamonn J. Walsh. 2002. "From Moral Evaluation to Rationalization: Accounting and the Shifting Technologies of Credit." *Accounting, Organizations and Society*, 27(8): 737–761.

Katz, Michael B. 2012. *Why Don't American Cities Burn? The City in the Twenty-First Century*. Philadelphia: University of Pennsylvania Press.

Katznelson, Ira. 2005. *When Affirmative Action Was White: An Untold History of Racial Inequality in Twentieth-Century America*. New York: Norton.

Keys, Benjamin J., Devin G. Pope, and Jaren C. Pope. 2014. *Failure to Refinance*. Cambridge, MA: National Bureau of Economic Research.

Kochhar, Rakesh, and Richard Fry. 2014. *Wealth Inequality Has Widened along Racial, Ethnic Lines since End of Great Recession*. Washington, DC: Pew Research Center.

Krivo, Lauren J., and Robert L. Kaufman. 2004. "Housing and Wealth Inequality: Racial-Ethnic Differences in Home Equity in the United States." *Demography*, 41(3): 585–605.

Kulkarni, Nirupama, and Ulrike Malmendier. 2015. *Homeownership and the American Dream—An Analysis of Intergenerational Mobility Effects*. Berkeley: University of California, Berkeley.

LaCour-Little, Michael. 2009. "The Pricing of Mortgages by Brokers: An Agency Problem?" *Journal of Real Estate Research*, 31(2): 235–264.

Landis, John D., and Kirk McClure. 2010. "Rethinking Federal Housing Policy." *Journal of the American Planning Association*, 76(3): 319–348.

Langley, Paul. 2009a. *The Everyday Life of Global Finance: Saving and Borrowing in Anglo- America*. Oxford: Oxford University Press.

Langley, Paul. 2009b. "Debt, Discipline, and Government: Foreclosure and Forbearance in the Subprime Mortgage Crisis." *Environment and Planning A*, 41(6): 1404–1419.

Levitin, Adam J., and Janneke Ratcliffe. 2014. "Rethinking Duties to Serve in Housing Finance." In *Homeownership Built to Last: Balancing Access, Affordability, and Risk after the Housing Crisis*, edited by Eric S. Belsky, Christopher E. Herbert, and Jennifer H. Molinsky, 317–350. Washington, DC and Cambridge: Brookings Institution Press and Joint Center for Housing Studies of Harvard University.

Li, Yanmei. 2011. "Geography of Opportunity and Residential Mortgage Foreclosure: A Spatial Analysis of a US Housing Market." *Journal of Urban and Regional Analysis*, 3(2): 195–214.

Li, Yanmei, and Hazel A. Morrow-Jones. 2010. "The Impact of Residential Mortgage Foreclosure on Neighborhood Change and Succession." *Journal of Planning Education and Research*, 30(1): 22–39.

Lipsitz, George. 2006. *The Possessive Investment in Whiteness: How White People Profit from Identity Politics*. Philadelphia: Temple University Press.
Maly, Michael T. 2005. *Beyond Segregation: Multiracial and Multiethnic Neighborhoods in the United States*. Philadelphia, PA: Temple University Press.
Marron, Donncha. 2007. "'Lending by Numbers': Credit Scoring and the Constitution of Risk within American Consumer Credit." *Economy and Society*, 36(1): 103–133.
Martin, Carmel, Andy Green, and Brendan Duke. 2016. *Raising Wages and Rebuilding Wealth*. Washington, DC: Center for American Progress.
Massey, Douglas S., and Nancy A. Denton. 1998. *American Apartheid: Segregation and the Making of the Underclass*. Cambridge: Harvard University Press.
Mayer, Christopher, and Karen Pence. 2009. "Subprime Mortgage: What, Where, and to Whom?" In *Housing Markets and the Economy: Risk, Regulation, and Policy*, edited by Edward L. Glaeser and John M. Quigley, 149–196. Cambridge: Lincoln Institute of Land Policy.
McKernan, Signe-Mary, and Michael Sherraden, eds. 2008. *Asset Building and Low-Income Families*. Washington, DC: Urban Institute Press.
Menchik, Paul L., and Nancy Ammon Jianakoplos. 1997. "Black-White Wealth Inequality: Is Inheritance the Reason?" *Economic Inquiry*, 35(2): 428–442.
Mishel, Lawrence, Elise Gould, and Josh Bivens. 2016. *Wage Stagnation in Nine Charts*. Washington, DC: Economic Policy Institute.
Moss, David A. 2002. *When all Else Fails: Government as the Ultimate Risk Manager*. Cambridge: Harvard University Press.
Myers, Dowell, Gary Painter, and Julie Zissimopoulos. 2016. *The Role of Parental Financial Assistance in the Transition to Homeownership by Young Adults*. Washington DC: Fannie Mae.
Newman, Sandra J., and C. Scott Holupka. 2016. "Is Timing Everything? Race, Homeownership and Net Worth in the Tumultuous 2000s." *Real Estate Economics*, 44(2): 307–354.
Obama, Barack. 2013. "Full Transcript: President Obama's Remarks on the Economy." *Washington Post*, December 4.
Oliver, Melvin, and Thomas M. Shapiro, eds. 2006. *Black Wealth/White Wealth: A New Perspective on Racial Inequality*. New York: Routledge.
Pais, Jeremy, Scott J. South, and Kyle Crowder. 2012. "Metropolitan Heterogeneity and Minority Neighborhood Attainment: Spatial Assimilation or Place Stratification?" *Social Problems*, 59(2): 258–281.
Pethokoukis, James. 2017. *Yes, It's Time to Reform the Mortgage Interest Deduction*. Washington, DC: American Enterprise Institute.
Porter, Katherine, and Tara Twomey. 2012. "Risk Allocation in Home Ownership: Revisiting the Role of Mortgage Contract Terms." In *Shared Responsibility, Shared Risk: Government, Markets and Social Policy in the Twenty-First Century*, edited by Jacob S. Hacker and Ann O'Leary, 142–160. New York: Oxford University Press.
Quercia, Roberto G., Allison Freeman, and Janneke Ratcliffe. 2011. *Regaining the Dream: How to Renew the Promise of Homeownership for America's Working Families*. Washington, DC: Brookings Institution Press.
Raymond, Elora, Kyungsoon Wang, and Dan Immergluck. 2016. "Race and Uneven Recovery: Neighborhood Home Value Trajectories in Atlanta before and after the Housing Crisis." *Housing Studies*, 31(3): 324–339.
RealtyTrac. 2015. *U.S. Foreclosure Activity Increases 5 Percent in January Driven by 15-Month High in Bank Repossessions*. Irvine, CA: RealtyTrac.
Reid, Carolina. 2009. "Sustaining Homeownership: The Experience of City-Based Affordable Homeownership Programs." *Community Investments*, 21(2): 27–30.
Reid, Carolina K. 2004. *Achieving the American Dream? A Longitudinal Analysis of the Homeownership Experiences of Low-Income Households*. Seattle: University of Washington.
Reid, Carolina K. 2010. *Sought or Sold? Social Embeddedness and Consumer Decisions in the Mortgage Market*. San Francisco: Federal Reserve Bank of San Francisco.
Reid, Carolina K. 2014a. "The Promises and Pitfalls of Homeownership." In *The Assets Perspective: The Rise of Asset Building and Its Impact on Social Policy*, edited by Reid Cramer and Trina R. Williams Shanks, 123–149. New York: Palgrave Macmillan.
Reid, Carolina K. 2014b. "The Refinance Gap: The Subprime Crisis and Its Implications for Asset Building among Communities of Color." *Presented at the APPAM Annual Conference*, Albuquerque, NM, November 6.
Reid, Carolina K., Debbie Gruenstein Bocian, Wei Li, and Roberto G. Quercia. 2017. "Revisiting the Subprime Crisis: The Dual Mortgage Market and Mortgage Defaults by Race and Ethnicity." *Journal of Urban Affairs*, 39(4): 469–487.
Reid, Carolina K., and Elizabeth Laderman. 2009. *The Untold Costs of Subprime Lending: Examining the Links among Higher-Priced Lending, Foreclosures and Race in California*. San Francisco: Federal Reserve Bank of San Francisco.

Reid, Carolina K., and Elizabeth Laderman. 2011. "Constructive Credit: Revisiting the Performance of Community Reinvestment Act Lending During the Subprime Crisis." In *The American Mortgage System: Rethink, Recover, Rebuild*, edited by Marvin M. Smith and Susan M. Wachter. Philadelphia: University of Pennsylvania Press.

Reid, Carolina K., Ellen Seidman, Mark Willis, Lei Ding, Josh Silver, and Janneke Ratcliffe. 2013. *Debunking the CRA Myth—Again*. Raleigh, NC: University of North Carolina at Chapel Hill Center for Community Capital.

Reskin, Barbara. 2012. "The Race Discrimination System." *Annual Review of Sociology*, 38(1): 17–35.

Rohe, William M., and Mark R. Lindblad. 2014. "Reexamining the Social Benefits of Homeownership after the Foreclosure Crisis." In *Homeownership Built to Last: Balancing Access, Affordability, and Risk After the Housing Crisis*, edited by Eric S. Belsky, Chris Herbert, and Jennifer H. Molinsky, 99–140. Washington, DC: Brookings Institution Press.

Rose, Nikolas. 2002. "At Risk of Madness." In *Embracing Risk: The Changing Culture of Insurance and Responsibility*, edited by Tom Baker and Jonathan Simon, pp. 209–237. Chicago: University of Chicago Press.

Ross, Stephen L., and John Yinger. 2002. *The Color of Credit: Mortgage Discrimination, Research Methodology, and Fair-Lending Enforcement*. Cambridge: The MIT Press.

Rugh, Jacob S., Len Albright, and Douglas S. Massey. 2015. "Race, Space, and Cumulative Disadvantage: A Case Study of the Subprime Lending Collapse." *Social Problems*, 62(2): 186-218.

Rugh, Jacob S., and Douglas S. Massey. 2010. "Racial Segregation and the American Foreclosure Crisis." *American Sociological Review*, 75(5): 629–651.

Rusk, David. 2001. *The "Segregation Tax": The Cost of Racial Segregation to Black Homeowners*. Washington, DC: The Brookings Institution.

Saez, Emmanuel. 2014. "Income and Wealth Inequality: Evidence and Policy Implications." *Presented at the Neubauer Collegium Lecture*, University of Chicago, October.

Saez, Emmanuel, and Gabriel Zucman. 2016. "Wealth Inequality in the United States since 1913: Evidence from Capitalized Income Tax Data." *The Quarterly Journal of Economics*, 131(2): 519–578.

Schafran, Alex, and Jake Wegmann. 2012. "Restructuring, Race, and Real Estate: Changing Home Values and the New California Metropolis, 1989–2010." *Urban Geography*, 33(5): 630—654.

Shapiro, Thomas, Tatjana Meschede, and Sam Osoro. 2013. *The Roots of the Widening Racial Wealth Gap: Explaining the Black-White Economic Divide*. Waltham, MA: Brandeis University Institute on Assets and Social Policy.

Stuart, Guy. 2003. *Discriminating Risk: The U.S. Mortgage Lending Industry in the Twentieth Century*. Ithaca, NY: Cornell University Press.

Toder, Eric, Margery Austin Turner, Katherine Lim, and Liza Getsinger. 2010. *Reforming the Mortgage Interest Deduction*. Washington, DC: The Urban Institute.

Turner, Margery Austin, Diane K. Levy, Doug Wissoker, Claudia Aranda, Rob Pittingolo, and Rob Santos. 2013. *Housing Discrimination against Racial and Ethnic Minorities 2012*. Washington, DC: U.S. Department of Housing and Urban Development.

Turner, Tracy M., and Heather Luea. 2009. "Homeownership, Wealth Accumulation and Income Status." *Journal of Housing Economics*, 18(2): 104–114.

Urban Institute. n.d. *Nine Charts about Wealth Inequality in America*. Washington, DC: Urban Institute.

U.S. Department of Justice. 2016. *Wells Fargo Bank Agrees to Pay $1.2 Billion for Improper Mortgage Lending Practices*. Washington, CA: U.S. Department of Justice.

Wolff, Edward N. 2014. *Household Wealth Trends in the United States, 1962–2013: What Happened over the Great Recession?* Cambridge, MA: National Bureau of Economic Research.

4

NON-HISPANIC WHITE VERSUS BLACK PARENTAL WEALTH AND WEALTH TRANSFERS TO ENABLE HOMEOWNERSHIP IN FIVE METROPOLITAN AREAS

Erin Graves, Ana Patricia Muñoz, Darrick Hamilton, William A. Darity, and Yunju Nam

Introduction

Helping households become homeowners has been an enduring goal of U.S. housing policy since the mid-twentieth century, ranging from the National Housing Act of 1934 to the current Mortgage Interest Deduction (Shlay 2006). Yet Black homeownership rates have persistently been lower than those of non-Hispanic Whites and have decreased recently (Kuebler and Rugh 2013). Racial and intergenerational wealth disparities have been an important but overlooked factor contributing to the racial homeownership gap.

Despite a recent decrease in the national homeownership rate, homeownership has remained the single largest component of wealth building among most middle- and upper-income households (Wolff 2014). The return on housing as an investment depends on the location, the neighborhood, school quality, the age, size, and condition of the property, and many other factors (Flippen 2001). Mortgages are one of the few mechanisms available to many Americans that allow them to leverage and build wealth (Choudhury and Hartman 2015). Thus, lower homeownership rates among Blacks indicate unequal participation in wealth building (Gordon 2005; Kuebler and Rugh 2013; Piketty 2017).

Homeownership, often framed as an essential component of the American Dream, has also considerable ideological and political appeal (Vale 2007). Homeownership may serve as asset-based welfare, preventing poverty (McKee 2012). Homeownership has been associated with well-being, civic engagement, and positive child outcomes (Bond and Erikson 2016; Green and White 1997) and thus is supported by public policy.

Over the past three decades, the national U.S. homeownership rate peaked at 69 percent in 2004, and then declined to 64.8 percent in the last quarter of 2018. By comparison, the Black homeownership rate stood at 49.7 percent in 2004, and declined to 42.9 percent in the last quarter of 2018, a decline larger than that of any other racial or ethnic group (U.S. Bureau of the Census 2019). The homeownership rate among Blacks may decrease to 40 percent by 2030 (Butrica 2015; Kuebler and Rugh 2013). The White-Black homeownership gap has been persistent over the past five decades. Indeed, the gap between the non-Hispanic White versus the Black homeownership rate increased from 19 percentage points in 1960 to 28 percentage points in 1980 but fell to 22 percentage points at the peak of the housing bubble in 2007 yet increased again to 25 percentage points in 2018 (Collins and Margo 2011; Seah and Fesselmeyer 2017; U.S. Bureau of the Census 2018). Thus, policymakers have tried to increase the Black homeownership rate, employing tools such as homeownership counseling (Bond and Erikson 2016).

The Black homeownership rate varies by region (Kuebler and Rugh 2013). It is highest in the South at 53.8 percent of households, compared to 47.8 percent in the Midwest, 36.9 percent in the West, and 34.8 percent in the Northeast (Leigh and Huff 2007; Kuebler and Rugh 2013). However, home values in the South are lowest compared to other regions (Krivo and Kaufman 2004). These regional variations indicate that place matters for both homeownership rates and wealth accumulation.

Scholars have attempted to explain the racial and ethnic differences in homeownership rates, focusing on differences between non-Hispanic White and Black households (Gabriel and Rosenthal 2005; Gyourko and Linneman 1996). Some scholars attribute the gap to individual differences in human capital (e.g., formal education) or limits on how much a household can borrow (Diaz and Luengo-Prado 2008; Duca and Rosenthal 1994; Gabriel and Rosenthal 2005; Haurin et al. 1996; Haurin and Rosenthal 2007). Other scholars attribute the gap to structural differences that manifest themselves during the home buying process and in mortgage markets (Bond and Erikson 2016; Munnell et al. 1996; Ross and Tootell 2004; Yinger 1998; Zhao et al. 2006). Nevertheless, even after accounting for these differences, the disparity in homeownership rates between non-Hispanic White and Black households has endured (Bond and Erikson 2016), suggesting that additional factors contribute to this gap.

In this chapter we focus on how intergenerational transfers of wealth factor into the homeownership gap and draw a link between homeownership rates, intergenerational wealth building through homeownership, and the wealth gap between non-Hispanic White and Black homeowners. We review the literature on the wealth gap between non-Hispanic White and Black households and its relationship to the racial homeownership gap. We discuss the role of differences in parental wealth on financial transfers to offspring for five metropolitan areas. We analyze the differences between non-Hispanic White and Black households, controlling for many factors. We also discuss some of the impacts of polices on the racial wealth gap and how homeownership is both shaped by intergenerational wealth and shapes subsequent intergenerational wealth.

Wealth and Housing Finance

About two-thirds of U.S. homeowners and 90 percent of first-time homebuyers finance their homes through a mortgage (Lee and Reed 2014). Financing a mortgage requires borrower income—a flow of financial resources, most often received through work, but also through government transfers or investments—and in most cases, a down payment (Killewald et al. 2017). Wealth used for down payment may be based on savings accounts, stocks, bonds, vehicles, gifts, or another home (Chetty et al. 2017; Engelhardt 1996).

In housing finance, a borrower uses his or her income and wealth for different purposes. A portion of a borrower's income is typically used for the monthly mortgage payment. Industry standards state that about one-third or less of the income should be applied toward housing costs, which includes not only the mortgage but also taxes, maintenance, and insurance (Schwartz and Wilson 2008). A portion of a borrower's wealth is typically taken for the down payment. Larger down payments can either lower the monthly mortgage payment or allow a buyer to buy a more expensive home, which may increase the return on the investment and may also enable access to neighborhoods with better public goods, such as high-quality schools, low crime rates, and proximity to public transportation, as well as other desirable amenities (Brueckner 1986).

Mortgages allow households to leverage their assets and build wealth. A homebuyer with a modest down payment (which can actually be 0 percent for a Veteran's Administration (VA) mortgage or as low as 3.5 percent for a mortgage insured by the Federal Housing Administration (FHA) and up to 20 percent for a conventional mortgage) benefits from the gains in the home's value (Case et al. 2005). While the cost of financing may create negative leverage, its risk decreases over time because the value of the house compounds while the debt payment is fixed (Herbert et al. 2013). During the Great Recession, when house prices in most regions and many neighborhoods decreased, households lost significant wealth (Grinstein-Weiss et al. 2015).

For some borrowers, the lack of wealth for the down payment constitutes a serious barrier to homeownership and may affect the level of wealth accumulation through homeownership (Myers et al. 2016; Ortalo-Magne and Rady 1999; Quercia and Wachter 1996). Thus, some buyers use low down payment mortgages, such as VA mortgages and mortgages insured by the FHA, mentioned earlier. Mortgages with a down payment of less than 20 percent require private mortgage insurance, which adds to the monthly payment. Other buyers may delay or forgo a home purchase, which reduces potential wealth accumulation (Bayer et al. 2017; Krivo and Kaufman 2004).

Research has found that intergenerational wealth transfers are an important mechanism for generating down payments (Bond and Erikson 2016; Engelhardt and Mayer 1994). Young adults are more likely to transition into homeownership if their parents are also homeowners partly because parents often draw upon their own housing equity to help their children (McKee 2012; Myers et al. 2016). Thus, the lower homeownership rates in one generation may influence rates in the next.

The Non-Hispanic White Versus Black Gap in Income and Wealth

The income and earnings gap between non-Hispanic Whites and Blacks and its increase in the 2000s and 2010s have been widely recognized (Galster and Sharkey 2017; Mandel and Semyonov 2016; Traub et al. 2017). In terms of income, Chiteji and Hamilton (2010) found that for every dollar that a typical non-Hispanic White person earns, a Black person earns about 60 cents.

The wealth gap between non-Hispanic Whites and Blacks is even more profound than the income and earnings gap and may have serious implications. Studies show that Black households possess about a tenth of the wealth of non-Hispanic White households, although findings vary, depending on the data source (Oliver and Shapiro 2006; Wolff 1998). The Great Recession caused an increase in the racial wealth gap (Shapiro et al. 2013). Shapiro et al. (2013) followed households over 25 years and found that, controlling for inflation, the nominal total wealth gap between non-Hispanic White and Black households nearly tripled, increasing from $85,000 in 1984 to $236,500 in 2009, when the Great Recession technically ended. Even post Great Recession, Blacks' wealth gains were smaller relative to non-Hispanic Whites. Kochhar and Fry (2014) found that in 2010, non-Hispanic White households possessed eight times the median wealth of Black households and that by 2013 they possessed 13 times the median wealth of Black households. Without wealth, households have fewer resources to face hardship or to make investments in education, entrepreneurship, or homeownership (Oliver and Shapiro 2006).

Many factors have contributed to the wealth gap between non-Hispanic Whites and Blacks. These factors include gaps in household income, unemployment rates, and access to and returns from college education. Interestingly, these differences explain a greater share of the wealth gap between non-Hispanic Whites and Blacks at the top of the wealth distribution than at the bottom (Sykes and Maroto 2016; Thompson and Suarez 2015). Educational attainment, income, occupation, and years spent working have been identified as among the most important set of factors explaining the racial wealth gap. These account for one-third to two-fifths of the portion of the wealth gap that can be explained through the analysis of survey data (Thompson and Suarez 2015). However, racial bias can influence these factors. Educational attainment, for example, depends in part on the quality of locally provided education and, due to residential segregation, could differ systematically across racial groups. Incomes are not fully explained by educational attainment, due to biased hiring practices and other forms of racial discrimination (Fryer et al. 2011).

Moreover, the wealth gap is not explained by differences in spending choices between non-Hispanic White and Black households with similar incomes (Charron-Chénier et al. 2017; Traub et al. 2017). Nor is it explained by differences in indebtedness. Using the Survey of Consumer Finances, Thompson and Suarez (2015) show that holding all else constant, the differences in debt between non-Hispanic White and Black families are insignificant and that the difference in mean wealth between these groups is due to assets. Generosity of parents also does not explain the racial wealth gap. Although non-Hispanic White households typically have more wealth to pass on to their children, among Black parents and non-Hispanic White parents of comparable net worth, Black parents provide more monetary support for their children's college education (Nam et al. 2015).

Institutional discriminatory practices and public policies have had a profound effect on the accumulation of wealth for Black households. Institutional practices, such as barriers in workplaces, schools, and health care, have reinforced deeply entrenched patterns of racial and ethnic discrimination (Green 1995; Shapiro et al. 2013). Public policies, such as the Social Security Act and the Fair Labor Standards Act, which established the minimum wage, did not cover farm and domestic workers, who were disproportionately Black, until 1966, leaving them at a disadvantage for wealth accumulation (Katznelson 2013).

Past housing discrimination has also greatly contributed to the wealth gap between non-Hispanic White and Black households. These include previous legal discriminatory practices such as restrictive covenants by municipalities, redlining by lending institutions, and the restrictions by the Home Owners' Loan Corporation (HOLC) that excluded Blacks from purchasing homes in select neighborhoods (Hernandez 2009).

Present housing discrimination is shown by the larger proportion of subprime mortgages and foreclosures among Blacks, which has resulted in a decrease of wealth (Anacker and Carr 2011; Anacker et al. 2012; Hwang et al. 2015; Massey 2016; Pfeffer et al. 2013; Rugh and Massey 2010). Even in 2015, 70 percent of Blacks had non-conventional mortgages, which are more expensive than conventional ones as private mortgage insurance needs to be purchased, compared to 36 percent of non-Hispanic Whites (Bhutta and Ringo 2016).

The wealth gap between non-Hispanic Whites and Blacks is also influenced by regional racial differences in economic mobility, employment rates, household incomes, and intergenerational economic mobility (Ellen and Dastrup 2012; Erikson et al. 1993; Hero and Tolbert 1996; Kline and Moretti 2014; Rugh and Massey 2010). While the Northeast, the Great Plains, and the West have higher intergenerational mobility rates, the Southeast and the Midwest have lower ones due to the variation in regional economic growth and in regional and state policies (Chetty et al. 2014), such as state variations in minimum wages, laws around foreclosure and evictions, early education policy, and tax codes. In regions where there are greater concentrations of Black low-wage workers or higher levels of geographic segregation, wages and housing and education policies can have disproportionate effects on Blacks' wealth accumulation.

Housing-related characteristics further explain some of the wealth gap between non-Hispanic White and Black households. Blacks homeowners enter homeownership later in life, own homes with lower values, tend to own their homes for shorter time periods, and exit homeownership at higher rates compared to non-Hispanic White homeowners (Anacker 2010; Charles and Hurst 2003; Flippen 2001; Keister 2000, 2004; Killewald et al. 2017; Krivo and Kaufman 2004; O'Brien 2012; Sharp and Hall 2014). For example, Shapiro et al. (2013) found that 27 percent of the wealth difference between non-Hispanic Whites and Blacks is explained by the number of years of homeownership.

Finally, racial differences in intergenerational wealth transfer itself may also contribute to the homeownership gap between non-Hispanic Whites and Blacks. Inheritances, bequests, and intra-family transfers account substantially for the racial wealth gap (Blau and Graham 1990; Charles and Hurst 2003; Menchik and Jianakoplos 1997; Thompson and Suarez 2015). A Black person's chances of receiving either an end of life inheritance or a financial transfer from their parents during the parents' lifetime is much lower compared to non-Hispanic Whites (Avery and Rendall 2002; McKernan et al. 2014; Menchik and Jianakoplos 1997). A longitudinal study of Black and non-Hispanic White families found that of those who received more than $50,000 in financial transfers, only 10 percent were Black (Thomas et al. 2014). These gaps in transfers are estimated to explain between 10 and 24 percent of wealth disparities between non-Hispanic White and Black households (Nam et al. 2015; Scholz and Levine 2004). An additional substantial portion of the racial wealth gap can be explained by the fact that for Blacks, a large proportion of relatives and friends in poverty may ask for financial support (Chiteji and Hamilton 2002; Feld 1984; McKernan et al. 2014; Shapiro et al. 2013; Thomas et al. 2014). In all, family financial support is a key factor in wealth building: Households who have received inheritances and other financial support from their families build wealth more easily than those who have not. Non-Hispanic Whites report significantly more of this form of support than Blacks (Thompson and Suarez 2015).

Regional differences in house price changes affect a household's wealth and, paired with differences in homeownership rates, may further contribute to racial wealth gaps by region. An extensive literature analyzes regional house price changes. While nationally house prices rose rapidly in the last quarter of the twentieth century, with a quadrupling of the national house price index between 1975 and the end of 2000, there were substantial regional variations. These ranged from a sevenfold increase in the Pacific region from 1975 to 2001 to a threefold increase in the West South Central region. One consequence of these differences is substantial differences in wealth accumulation by region, and these can have racial implications for wealth accumulation as well (Dietz and Haurin 2003). Because Black households had the lowest homeownership rates in the regions with the greatest house price appreciation, the variation in house price trajectories for different regions may affect household wealth, leaving Black households with less wealth to transfer to their offspring.

The Non-Hispanic White Versus Black Wealth Gap and Homeownership: Parental Financial Contributions

One way parental wealth enables the homeownership of offspring is through in vivo transfers, which are transfers of wealth from parents to children during the parents' lifetime, rather than inheritances granted after parents die. In vivo transfers of parental wealth are typically thought to influence the path to homeownership of the offspring in indirect and direct ways (Bond and Erikson 2016).

In regard to indirect transfers, parents may use their wealth throughout a child's upbringing and support their child's education by locating to a neighborhood with a high-quality school. This move may have implications for homeownership of their offspring as a better school may improve prospects for a better paying job (Shapiro 2004). Some households borrow against their home's value to finance a child's college education, making college enrollment more likely and decreasing the need for a college loan (Greenspan and Kennedy 2005; Myers et al. 2016). While the average transfer for education by non-Hispanic White parents is about $12,000, the average transfer by Black parents is about $4,000 (Addo et al. 2016). Lovenheim (2011) found that an increase of $10,000 in housing wealth increases the likelihood of college attendance by 1.37 percent.

Myers et al. (2016) found that there are racial differences in homeownership outcomes between those who receive parental educational support and those who do not. Controlling for the type of college degree, Black students tend to borrow larger amounts at higher interest rates compared to non-Hispanic White students, possibly because they do not have access to parental wealth (Traub et al. 2017).

Direct transfers directly enable homeownership, and parents may use in vivo transfers to assist their children purchasing homes. Scholars have found that about 25 percent of first-time homebuyers received a gift from a family member or a friend to help with the down payment. This transfer is especially important for those between ages 25 and 44 (Charles and Hurst 2003; Myers et al. 2016). Again, the difference between non-Hispanic White and Black first time buyers who receive gifts is pronounced: 27 percent of non-Hispanic Whites reported receiving down payment assistance, compared to 7 percent of Blacks (Charles and Hurst 2002). Moreover, the difference in homeownership rates is insignificant for Black and non-Hispanic Whites who do receive direct assistance from their parents (Hilber and Liu 2008). This suggests that as a consequence of parental wealth differences, non-Hispanic Whites often have increased access to homeownership than Blacks, and at younger ages, leading to larger gains in equity (Charles and Hurst 2002; Traub et al 2017).

In sum, receiving in vivo transfers, regardless of whether they are indirect or direct, has important implications for homeownership and wealth building (and vice versa). On average, non-Hispanic Whites receive greater parental financial assistance than Blacks, which may explain the racial gap in homeownership rates. In the following text we answer two research questions for our study: First, what is the relationship between financial transfers for education and transfers for down payments, differentiating between non-Hispanic Whites and Blacks for five metropolitan areas? Second, what are the differences between non-Hispanic Whites and Blacks for these metropolitan areas?

Data and Methods

The following analysis uses data from the National Asset Scorecard for Communities of Color (NASCC) survey.[1] Using a systematic approach to ascertain that the metropolitan areas represented different regions across the United States, the survey collected information on households in five targeted metropolitan areas (Boston, Massachusetts; Miami, Florida; Los Angeles, California; Tulsa, Oklahoma; and Washington, DC) about the asset and debt positions of racial and ethnic groups for five metropolitan areas (n = 2,746).[2] The survey contains detailed information within racial and ethnic groups, but the sample size for within groups was too small to make comparisons. Thus, we aggregate the data to include all Black respondents and Non-Hispanic White respondents, the focus of this chapter. The NASCC survey includes questions about intergenerational transfers, including in vivo transfers and inheritances that may shed light on the economic well-being of different racial and ethnic groups for the five metropolitan areas. In the past, other national surveys such as the Survey of Income and Program Participation (SIPP), the Panel Study of Income Dynamics (PSID), and the Survey of Consumer Finance (SCF) have focused on the net worth position of respondents (Killewald et al. 2017) but do not contain enough data to make inferences about specific metropolitan areas or to make comparisons between cities.

Researchers primarily gathered information about a respondent's specific assets, liabilities, financial resources, personal savings, and investment activity at the household level. Additional areas of inquiry included remittance behavior, i.e., sending assets or other resources abroad, or supporting relatives in the United States. The survey also asked for core demographic and socioeconomic characteristics, such as age, sex, educational attainment, household composition, nativity, and household income. Race and ethnic identity were self-identified by the household member family respondent best qualified to discuss household family financial matters. In our analysis we focus on differences between non-Hispanic Whites and Blacks in five metropolitan areas.

Differences in Financial Support for Non-Hispanic Whites Versus Blacks in Five Metropolitan Areas

To assess relationships between wealth and financial transfers for non-Hispanic Whites versus Blacks, we ran basic descriptive statistics of these groups for the entire five area NASCC sample. As seen in Figure 4.1, the average Black income in our sample ranges from a low in the Tulsa metropolitan

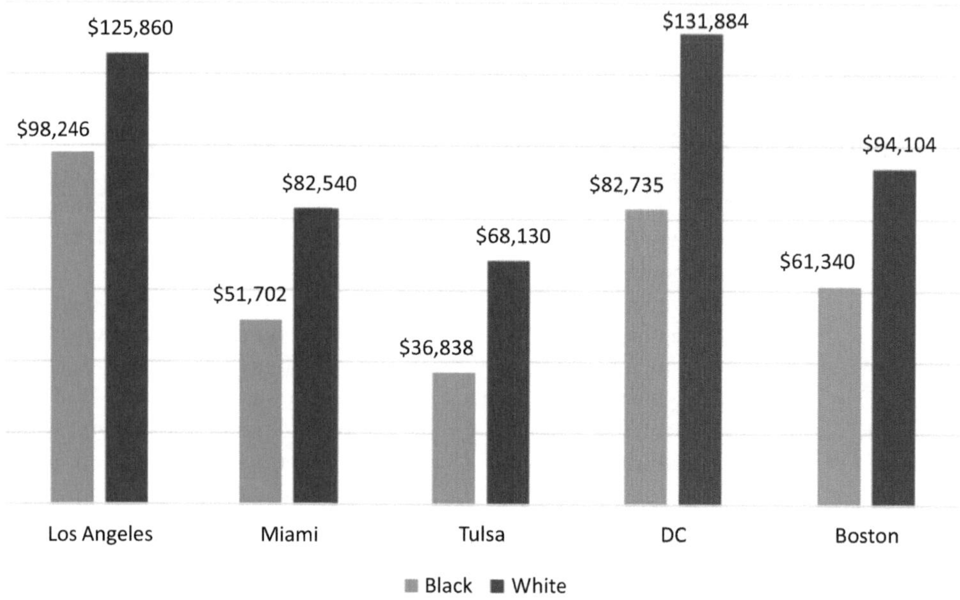

Figure 4.1 Average Income for Non-Hispanic Whites Versus Blacks for the Five Metropolitan Areas.
Source: authors.

area of $36,838 to a high in the Los Angeles metropolitan area of $98,246. Non-Hispanic White income ranges from a low of $68,130 in Tulsa to a high of $131,884 in the Washington, DC, metropolitan area. In all metropolitan areas there is a consistent, large income gap between Blacks and non-Hispanic Whites, with the smallest gap of $27,614 in Los Angeles. The largest gap in income between Blacks and non-Hispanic Whites in the five metropolitan areas is $49,149 in Washington, DC. Thus, as the literature predicts (Florida and Mellander 2016), there is considerable variation in average earnings in all regions and within every region where Blacks on average earn far less than non-Hispanic Whites. Similarly, homeownership rates vary by metropolitan area. Each metropolitan area has a substantial homeownership gap for non-Hispanic Whites versus Blacks, as shown in Figure 4.2. The highest non-Hispanic White homeownership rate is in the Tulsa metropolitan area, and the gap in the homeownership rate of 44 percentage points is nearly the largest of all the five metropolitan areas.

Thus, we find an overall relationship between race and financial transfers, indicating inequality of income and homeownership between non-Hispanic Whites and Blacks. We find that across the five metropolitan areas, 11.3 percent of non-Hispanic White homeowners received direct transfers from families or friends for down payment costs, compared to 3.6 percent of Black homeowners. We also find that non-Hispanic Whites were also more likely to receive financial support to pay for education than Blacks. While educational attainment varies significantly by race, when we limit the analysis to those who attended higher education institutions, we find significant differences for the two analyzed racial groups. Only 15 percent of Blacks that attended higher education institutions received large contributions from family members to pay for costs of education. In addition to those two specific uses of financial support, the NASCC survey inquired whether respondents received large financial gifts from relatives during their lifetime. We found that slightly more than 25 percent of non-Hispanic White households received such gifts versus 14 percent of Blacks and that these differences were significant at the 95 percent level.

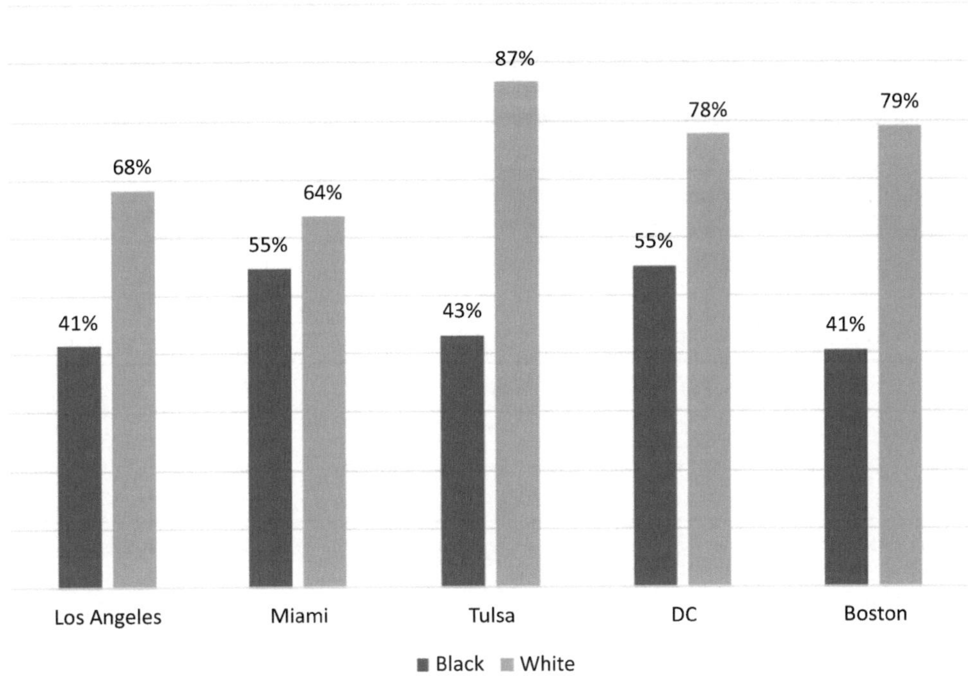

Figure 4.2 Homeownership Rates for Non-Hispanic Whites Versus Blacks for the Five Metropolitan Areas.

Source: authors.

Results from the NASCC survey are consistent with analyses using the PSID, showing that slightly more than 33 percent of non-Hispanic Whites received financial support for higher education compared to 14 percent of Blacks (Nam et al. 2015). Twelve percent of non-Hispanic Whites obtained financial help to buy a home, which includes help with the down payment, versus only 2 percent of Blacks (Nam et al. 2015). Importantly, those differences in transfers are driven by a difference in wealth positions at the outset. Decomposition analyses, which quantify the contribution of each group to the aggregate index, indicate that about 80 percent of the non-Hispanic White-Black gap in parental assistance for higher education and homeownership would disappear if non-Hispanic Whites and Blacks had parents with equivalent levels of income, wealth, and education.

Second, we sought to determine if there were differences in transfers across the five analyzed metropolitan areas. A rich feature of the NASCC survey is that it provides detailed information by race, ethnicity, and geography. Table 4.1 depicts the percentage of households that received a gift or a loan for down payment in the five metropolitan areas. The percentage of non-Hispanic White households that receive such help ranged from 9 percent in Miami to 16 percent in Los Angeles. Close to 5 percent of Blacks received such help across the metropolitan areas studied except for Los Angeles, where 10 percent of Blacks reported having received financial help to pay for down payment. These differences were significant in both Boston and Washington, DC.

The difference in the percentage of non-Hispanic White versus Black households who receive help to pay for education is substantial for the five metropolitan areas. In the Boston, Miami, Tulsa, and Washington, DC, metropolitan areas, the percentage of non-Hispanic Whites that received help to pay for more than 50 percent of their education costs was 30 percentage points higher than the percentage for Blacks (Table 4.2). For example, 40 percent of non-Hispanic Whites in Boston and Miami received financial help for education compared to only 8 percent of Blacks in Miami and

Table 4.1 Percentage of Households that Received Gifts or Loans for Down Payment.

	Boston	*Los Angeles*	*Miami*	*Tulsa*	*Washington, DC*
Non-Hispanic White	0.14	0.16	0.09	0.08	0.14
Black	0.00**	0.10	0.05	0.04	0.04***

Note: The difference in the percentage of non-Whites as compared with the percentage of White households is statistically significant at the *** 99% and ** 95% levels.

Source: NASCC survey, 2014, authors' calculations.

Table 4.2 Percentage of Respondents Whose Family Members Paid 50 Percent or More of Education Costs.

	Boston	*Los Angeles*	*Miami*	*Tulsa*	*Washington, DC*
Non-Hispanic White	0.42	0.37	0.43	0.34	0.50
Black	0.14***	0.26	0.08***	0.06***	0.13
Total	.23	.37	.25	.16	.38

Note: The difference in the percentage of Blacks as compared with the percentage of non-Hispanic White households is statistically significant at the *** 99% and ** 95% levels.

Source: NASCC survey, 2014, authors' calculations.

14 percent in Boston. Interestingly, the differences in Los Angeles do not appear to be as large or significant as in the other metro areas. In Boston, we find that Blacks did not receive loans or gifts for down payment. However, the incidence of gaps in transfers is consistent across the five metropolitan areas: Non-Hispanic Whites consistently transfer more wealth to their children than Blacks. Thus, despite differences in policies and economic trajectories of these five metropolitan areas, we find a similar pattern of differences in financial transfers from parents to children.

Conclusion

The history of housing finance, including the practice of redlining, and the present reality that non-Hispanic White households have more wealth to transfer to their offspring than Black households make race a critical factor in housing finance. Homeownership is often considered an individual achievement, a reflection of a person's ability to generate a steady income and accumulate savings. However, many forces influence an individual's path to homeownership. These include economic conditions, such as whether a region is experiencing economic growth or contraction, and macro-level, structural features, such as high levels of inequality. In this chapter, we focused on how a person's ability to finance a home can be influenced by parental wealth in both indirect and direct ways.

Our focus was on the relationship between homeownership and parental wealth, comparing non-Hispanic Whites and Blacks. We analyzed data about racial differences in indirect and direct transfers as well as information about how these transfers vary for five metropolitan areas. We found that, consistent with earlier research, non-Hispanic Whites receive significantly more direct assistance for homeownership than Black respondents (Hilber and Liu 2008). Nearly four times as many non-Hispanic White respondents reported that they received financial support for a down payment, compared to Blacks. We also found that a higher proportion of non-Hispanic White respondents reported getting indirect assistance in the form of financial aid from their parents for education. Even among those attending higher education institutions, we found significant differences along racial lines, with far greater proportions of non-Hispanic White households receiving financial help from their families. Offspring who receive financial assistance from their parents finish school with less debt and can therefore begin saving earlier for a down payment. While not every college graduate becomes a homeowner, parents who use their wealth to help with education are indirectly helping their children move along this path.

Economic and housing policies vary across regions in the United States, and these could have implications for the racial wealth gap and parental wealth as well. We found differences in wealth and wealth transfers by race for the five metropolitan areas. In all five metropolitan areas we surveyed, non-Hispanic Whites receive more assistance with down payments for homeownership and with higher education compared to Blacks. This finding suggests a pattern where non-Hispanic White parents are able to aid their children on the path to financing a home and subsequently generating wealth and equity. It was beyond the intended scope of this chapter to explain these inter-metropolitan differences with respect to receiving assistance. Future research on this subject is needed.

The question of whether the fundamental source of homeownership is wealth or whether homeownership is the origin of wealth is certainly difficult to answer, as homeownership is both a cause and an effect of wealth. In this chapter, rather than resolving this issue, we simply want to emphasize that the relationship between wealth and homeownership has a certain "chicken or egg" quality. Regardless of whether one starts with the chicken or the egg, the process that ensues results in racially uneven economic outcomes.

Scholars and policymakers have suggested a variety of policy responses, from mainstream to somewhat radical. The first one is to further reduce the down payment requirement imposed by FHA loans. To date, there are only a few rigorous evaluations on the loan performance for low down payment products. It is possible that even lower down payment requirements may result in higher defaults (Campbell and Cocco 2015). The second policy response is to increase public funding for programs that provide down payment loans. Massachusetts has one such program, called The One. Approved buyers make a down payment as little as 1.5 percent and receive a subsidized down payment loan (Smith 1999). The third policy response is to provide down payment grants, for example, the loans provided by the U.S. Department of Housing and Urban Development to qualifying buyers as part of the now-ended HOPE VI program (Graves and Curley 2013).

Additional measures would be more politically difficult. Some have proposed down payment savings products, which facilitate savings for a down payment. Others have introduced Baby Bonds, a publicly funded program that provides low-income children with a sizable trust, about $20,000 at birth (Hamilton and Darity 2010). The United Kingdom instituted a much more modest program until 2011, The Child Trust Fund, which granted children around $400 at birth (Butrica 2015). It is unlikely that such a small grant would have much of an impact if applied in the U.S. Finally, some scholars and activists have proposed a publicly funded reparations program to people of African-American descent as the only just solution to racial wealth inequality (Coates 2014; Feagin 2004), although the chances of this strategy being implemented would require a level of political will currently absent in the U.S.

Notes

1 The NASCC project is funded by the Ford Foundation (for any data besides the Boston data) and the Federal Reserve Bank of Boston (for the Boston data). William Darity (Duke University) and Darrick Hamilton (The New School) serve as primary investigators. The NASCC-Boston project manager was Ana Patricia Muñoz. The NASCC survey was administered by the University of Virginia's Center for Survey Research from January to March 2014. Surveyors collected 2,746 complete surveys across the five metropolitan areas.

2 For example, the survey asked the country of origin of Black, Asian, and Latino respondents. More specifically, the NASCC dataset includes data for respondents of Korean, Vietnamese, Chinese, Filipino, Puerto Rican, Dominican, Mexican, Cuban, and African descent. While detailed questions were included on ethnicity, the sample sizes were too small to draw statistical inferences.

References

Addo, Fenaba R., Jason N. Houle, and Daniel Simon. 2016. "Young, Black, and (Still) in the Red: Parental Wealth, Race, and Student Loan Debt." *Race and Social Problems*, 8(1): 64–76.

Anacker, Katrin B. 2010. "Still Paying the Race Tax? Analyzing Property Values in Homogeneous and Mixed-Race Suburbs." *Journal of Urban Affairs*, 32(1): 55–77.

Anacker, Katrin B., and James H. Carr. 2011. "Analyzing Determinants of Foreclosure among High-Income African-American and Hispanic Borrowers in the Washington, DC Metropolitan Area." *International Journal of Housing Policy*, 11(2): 195–220.

Anacker, Katrin B., James H. Carr, and Archana Pradhan. 2012. "Analyzing Foreclosures among High-Income Black/African American and Hispanic/Latino borrowers in Prince George's County, Maryland." *Housing and Society*, 39(1): 1–28.

Avery, Robert B., and Michael S. Rendall. 2002. "Lifetime Inheritances of Three Generations of Whites and Blacks." *American Journal of Sociology*, 107(5): 1300–1346.

Bayer, Patrick, Fernando Ferreira, and Stephen L. Ross. 2017. "The Vulnerability of Minority Homeowners in the Housing Boom and Bust: Corrigendum." *American Economic Journal*, 9(1): 344–345.

Bhutta, Neil, and D. Ringo. 2016. "Residential Mortgage Lending from 2004 to 2015: Evidence from the Home Mortgage Disclosure Act Data." *Federal Reserve Bulletin*, 102(6): 1–26.

Blau, Francine D., and John W. Graham. 1990. "Black-White Differences in Wealth and Asset Composition." *The Quarterly Journal of Economics*, 105(2): 321–339.

Bond, Shaun A., and Michael D. Erikson. 2016. *The Role of Parents on the Home Ownership Experience of Their Children: Evidence from the Health and Retirement Study*. Cincinnati, OH: University of Cincinnati Lindner College of Business.

Brueckner, Jan K. 1986. "The Downpayment Constraint and Housing Tenure Choice: A Simplified Exposition." *Regional Science and Urban Economics*, 16(4): 519–525.

Butrica, Barbara A. 2015. *A Review of Children's Savings Accounts*. Washington, DC: Urban Institute.

Campbell, John Y., and Joao F. Cocco. 2015. "A Model of Mortgage Default." *The Journal of Finance*, 70(4): 1495–1554.

Case, Karl E., John M. Quigley, and Robert J. Shiller. 2005. "Comparing Wealth Effects: The Stock Market versus the Housing Market." *Advances in Macroeconomics*, 5(1), n.p.

Charles, Kerwin Kofi, and Erik Hurst. 2002. "The Transition to Home Ownership and the Black-White Wealth Gap." *Review of Economics and Statistics*, 84(2): 281–297.

Charles, Kerwin Kofi, and Erik Hurst. 2003. "The Correlation of Wealth across Generations." *Journal of Political Economy*, 111(6): 1155–1182.

Charron-Chénier, Raphaël, Joshua J. Fink, and Lisa A. Keister. 2017. "Race and Consumption: Black and White Disparities in Household Spending." *Sociology of Race and Ethnicity*, 3(1): 50–67.

Chetty, Raj, Nathaniel Hendren, Patrick Kline, and Emmanuel Saez. 2014. "Where is the Land of Opportunity? The Geography of Intergenerational Mobility in the United States." *The Quarterly Journal of Economics*, 129(4): 1553–1623.

Chetty, Raj, László Sándor, and Adam Szeidl. 2017. "The Effect of Housing on Portfolio Choice." *The Journal of Finance*, 72(3): 1171–1212.

Chiteji, Ngina S., and Darrick Hamilton. 2002. "Family Connections and the Black-White Wealth Gap among Middle-Class Families." *The Review of Black Political Economy*, 30(1): 9–28.

Chiteji, Ngina S., and Darrick Hamilton. 2010. "Wealth in the Extended Family: An American Dilemma." *Du Bois Review: Social Science Research on Race*, 7(2): 357–379.

Choudhury, Askar H., and Nathan S. Hartman. 2015. "Regional Differential Wealth Effect on Home Value: A Cross-Sectional Analysis." *Journal of Economics and Economic Education Research*, 16(1): 273–290.

Coates, Ta-Nehisi. 2014. "The Case for Reparations." *The Atlantic*, June. n.p.

Collins, William J., and Robert A. Margo. 2011. "Race and Home Ownership from the End of the Civil War to the Present." *The American Economic Review*, 101(3): 355–359.

Díaz, Antonia, and María José Luengo-Prado. 2008. "On the User Cost and Homeownership." *Review of Economic Dynamics*, 11(3): 584–613.

Dietz, Robert, and Donald Haurin. 2003. "The Social and Private Micro-Level Consequences of Homeownership." *Journal of Urban Economics*, 54(3): 401–450.

Duca, John V., and Stuart S. Rosenthal. 1994. "Do Mortgage Rates Vary Based on Household Default Characteristics? Evidence on Rate Sorting and Credit Rationing." *The Journal of Real Estate Finance and Economics*, 8(2): 99–113.

Ellen, Ingrid Gould, and Samuel Dastrup. 2012. *Housing and the Great Recession*. Stanford, CA: Stanford Center on Poverty and Inequality.

Engelhardt, Gary V. 1996. "Consumption, Down Payments, and Liquidity Constraints." *Journal of Money, Credit and Banking*, 28(2): 255–271.

Engelhardt, Gary V., and Christopher J. Mayer. 1994. "Gifts for Home Purchase and Housing Market Behavior." *New England Economic Review*, May, 47–53.

Erikson, Robert S., Gerald C. Wright, and John P. McIver. 1993. *Statehouse Democracy: Public Opinion and Policy in the American States*. Palo Alto, CA: Cambridge University Press.

Feagin, Joe R. 2004. "Documenting the Costs of Slavery, Segregation, and Contemporary Racism: Why Reparations are in Order for African Americans." *Harvard Black Letter Law Journal*, 20: 49–81.

Feld, Scott L. 1984. "The Structured Use of Personal Associates." *Social Forces*, 62(3): 640–652.

Flippen, Chenoa A. 2001. "Racial and Ethnic Inequality in Homeownership and Housing Equity." *The Sociological Quarterly*, 42(2): 121–149.

Florida, Richard, and Charlotta Mellander. 2016. "The Geography of Inequality: Difference and Determinants of Wage and Income Inequality across US Metros." *Regional Studies*, 50(1): 79–92.

Fryer, Roland, Jörg Spenkuch, and Devah Pager. 2011. *Racial Disparities in Job Finding and Offered Wages*. NBER Working Paper 17462. Cambridge, MA: National Bureau of Economic Research.

Gabriel, Stuart A., and Stuart S. Rosenthal. 2005. "Homeownership in the 1980s and 1990s: Aggregate Trends and Racial Gaps." *Journal of Urban Economics*, 57(1): 101–127.

Galster, George, and Patrick Sharkey. 2017. "Spatial Foundations of Inequality: A Conceptual Model and Empirical Overview." *Russell Sage Foundation Journal of the Social Sciences*, 3(2): 1–33.

Gordon, Adam. 2005. "The Creation of Homeownership: How New Deal Changes in Banking Regulation Simultaneously Made Homeownership Accessible to Whites and Out of Reach for Blacks." *The Yale Law Journal*, 1(5): 186–226.

Graves, Erin M., and Alexandra Curley. 2013. "Homeownership among Low-Income Families." In *Urban Ills: Twenty-First-Century Complexities of Urban Living in Global Contexts*, edited by Carol Camp Yeakey, Vetta L. Sanders Thompson, and Anjanette Wells, 101–136. Lanham, MD: Lexington Books.

Green, Richard K., and Michelle J. White. 1997. "Measuring the Benefits of Homeowning: Effects on Children." *Journal of Urban Economics*, 41(3): 441–461.

Green, Rodney D. 1995. "Review of *Black Wealth, White Wealth: A New Perspective on Racial Inequality* by Melvin L. Oliver and Thomas M. Shapiro." *The Journal of Negro Education*, 4(4): 477–479.

Greenspan, Alan, and James E. Kennedy. 2005. *Estimates of Home Mortgage Originations, Repayments, and Debt on One-to-Four-Family Residences*. Washington, DC: Federal Reserve Board.

Grinstein-Weiss, Michal, Clinton Key, and Shannon Carrillo. 2015. "Homeownership, the Great Recession, and Wealth: Evidence from the Survey of Consumer Finances." *Housing Policy Debate*, 25(3): 419–445.

Gyourko, Joseph, and Peter Linneman. 1996. "Analysis of the Changing Influences on Traditional Households' Ownership Patterns." *Journal of Urban Economics*, 39(3): 318–341.

Hamilton, Darrick, and William Darity. 2010. "Can 'Baby Bonds' Eliminate the Racial Wealth Gap in Putative Post-Racial America?" *The Review of Black Political Economy*, 37(3–4): 207–216.

Haurin, Donald R., Patric H. Hendershott, and Susan M. Wachter. 1996. *Borrowing Constraints and the Tenure Choice of Young Households*. NBER Working Paper 5360. Cambridge, MA: National Bureau of Economic Research.

Haurin, Donald R., and Stuart S. Rosenthal. 2007. "The Influence of Household Formation on Homeownership Rates across Time and Race." *Real Estate Economics*, 35(4): 411–450.

Herbert, Christopher E., Daniel T. McCue, and Rocio Sanchez-Moyano. 2013. "Is Homeownership Still an Effective Means of Building Wealth for Low-Income and Minority Households? (Was It Ever?)." *Joint Center For Housing Studies Working Paper*. Cambridge, MA: Harvard University Press.

Hernandez, Jesus. 2009. "Redlining Revisited: Mortgage Lending Patterns in Sacramento 1930–2004." *International Journal of Urban and Regional Research*, 33(2): 291–313.

Hero, Rodney E., and Caroline J. Tolbert. 1996. "A Racial/Ethnic Diversity Interpretation of Politics and Policy in the States of the US." *American Journal of Political Science*, 851–871.

Hilber, Christian A.L., and Yingchun Liu. 2008. "Explaining the Black—White Homeownership Gap: The Role of Own Wealth, Parental Externalities and Locational Preferences." *Journal of Housing Economics*, 17(2): 152–174.

Hwang, Jackelyn. 2015. *Racialized Recovery: Post-Foreclosure Pathways in Distressed Neighborhoods in Boston*. Cambridge, MA: Joint Center for Housing Studies, Harvard University.

Katznelson, Ira. 2013. *Fear itself: The New Deal and the Origins of Our Time*. New York: W.W. Norton & Company.

Keister, Lisa A. 2000. "Race and Wealth Inequality: The Impact of Racial Differences in Asset Ownership on the Distribution of Household Wealth." *Social Science Research*, 29(4): 477–502.

Keister, Lisa A. 2004. "Race, Family Structure, and Wealth: The Effect of Childhood Family on Adult Asset Ownership." *Sociological Perspectives*, 47(2): 161–187.
Killewald, Alexandra, Fabian T. Pfeffer, and Jared N. Schachner. 2017. "Wealth Inequality and Accumulation." *Annual Review of Sociology*, 43: 379–404.
Kline, Patrick, and Enrico Moretti. 2014. "People, Places, and Public Policy: Some Simple Welfare Economics of Local Economic Development Programs." *Annual Review of Economics*, 6(1): 629–662.
Kochhar, Rakesh, and Richard Fry. 2014. *Wealth Inequality has Widened along Racial, Ethnic Lines Since End of Great Recession*. Washington, DC: Pew Research Center.
Krivo, Lauren J., and Robert L. Kaufman. 2004. "Housing and Wealth Inequality: Racial-Ethnic Differences in Home Equity in the United States." *Demography*, 41(3): 585–605.
Kuebler, Meghan, and Jacob S. Rugh. 2013. "New Evidence on Racial and Ethnic Disparities in Homeownership in the United States From 2001 to 2010." *Social Science Research*, 42(5): 1357–1374.
Lee, Chyi Lin, and Richard G. Reed. 2014. "The Relationship between Housing Market Intervention for First-Time Buyers and House Price Volatility." *Housing Studies*, 29(8): 1073–1095.
Leigh, Wilhelmina A., and Danielle Huff. 2007. *African Americans and Homeownership: The Subprime Lending Experience, 1995 to 2007*. Washington, DC: The Joint Center for Political and Economic Studies.
Lovenheim, Michael F. 2011. "The Effect of Liquid Housing Wealth on College Enrollment." *Journal of Labor Economics*, 29(4): 741–771.
Mandel, Hadas, and Moshe Semyonov. 2016. "Going Back in Time? Gender Differences in Trends and Sources of the Racial Pay Gap, 1970 to 2010." *American Sociological Review*, 81(5): 1039–1068.
Massey, Douglas S. 2016. "Residential Segregation is the Linchpin of Racial Stratification." *City & Community*, 15(1): 4–7.
McKee, Kim. 2012. "Young People, Homeownership and Future Welfare." *Housing Studies*, 27(6): 853–862.
McKernan, Signe-Mary, Caroline Ratcliffe, Margaret Simms, and Sisi Zhang. 2014. "Do Racial Disparities in Private Transfers Help Explain the Racial Wealth Gap? New Evidence from Longitudinal Data." *Demography*, 51(3): 949–974.
Menchik, Paul L., and Nancy Ammon Jianakoplos. 1997. "Black-White Wealth Inequality: Is Inheritance the Reason?" *Economic Inquiry*, 35(2): 428–442.
Munnell, Alicia H., Geoffrey MB Tootell, Lynn E. Browne, and James McEneaney. 1996. "Mortgage Lending in Boston: Interpreting HMDA Data." *The American Economic Review*, 86(1): 25–53.
Myers, Dowell, Gary Painter, and Julie Zissimopoulos. 2016. *The Role of Parental Financial Assistance in the Transition to Homeownership by Young Adults*. Washington, DC: Fannie Mae.
Nam, Yunju, Darrick Hamilton, William Darity, and Anne E. Price. 2015. *Bootstraps are for Black Kids: Race, Wealth, and the Impact of Intergenerational Transfers on Adult Outcomes*. Oakland, CA: Insight Center for Community Economic Development.
O'Brien, Rourke L. 2012. "Depleting Capital? Race, Wealth and Informal Financial Assistance." *Social Forces*, 91(2): 375–396.
Oliver, Melvin L., and Thomas M. Shapiro. 2006. *Black Wealth, White Wealth: A New Perspective on Racial Inequality*. New York: Taylor & Francis.
Ortalo-Magne, Francois, and Sven Rady. 1999. "Boom in, Bust out: Young Households and the Housing Price Cycle." *European Economic Review*, 43(4): 755–766.
Pager, Devah, and Hana Shepherd. 2008. "The Sociology of Discrimination: Racial Discrimination in Employment, Housing, Credit, and Consumer Markets." *Annual Review of Sociology*, 34: 181–209.
Pfeffer, Fabian T., Sheldon Danziger, and Robert F. Schoeni. 2013. "Wealth Disparities Before and After the Great Recession." *The Annals of the American Academy of Political and Social Science*, 650(1): 98–123.
Piketty, Thomas. 2017. *Capital in the Twenty-First Century*. Cambridge, MA: Harvard University Press.
Quercia, Roberto, and Susan Wachter. 1996. "Homeownership Counseling Performance: How Can It Be Measured?" *Housing Policy Debate*, 7(1): 175–187.
Ross, Stephen L., and Geoffrey MB Tootell. 2004. "Redlining, the Community Reinvestment Act, and Private Mortgage Insurance." *Journal of Urban Economics*, 55(2): 278–297.
Rugh, Jacob S., and Douglas S. Massey. 2010. "Racial Segregation and the American Foreclosure Crisis." *American Sociological Review*, 75(5): 629–651.
Scholz Karl John, and Kara Levine. 2004. "U.S. Black-White Wealth Inequality." In *Social Inequality*, edited by Kathryn M. Neckerman, 895–929. New York: Russell Sage.
Seah, Kiat Ying, Eric Fesselmeyer, and Kien Le. 2017. "Estimating and Decomposing Changes in the White—Black Homeownership Gap from 2005 to 2011." *Urban Studies*, 54(1): 119–136.
Shapiro, Thomas M. 2004. *The Hidden Cost of Being African American: How Wealth Perpetuates Inequality*. New York: Oxford University Press.

Shapiro, Thomas, Tatjana Meschede, and Sam Osoro. 2013. *The Roots of the Widening Racial Wealth Gap: Explaining the Black-White Economic Divide*. Waltham, MA: Institute on Assets and Social Policy.

Sharp, Gregory, and Matthew Hall. 2014. "Emerging Forms of Racial Inequality in Homeownership Exit, 1968–2009." *Social Problems*, 61(3): 427–447.

Shlay, Anne B. 2006. "Low-Income Homeownership: American Dream or Delusion?" *Urban Studies*, 43(3): 511–531.

Smith, David A. 1999. "Mark-To-Market: A Fundamental Shift in Affordable Housing Policy." *Housing Policy Debate*, 10(1): 143–182.

Sykes, Bryan L., and Michelle Maroto. 2016. "A Wealth of Inequalities: Mass Incarceration, Employment, and Racial Disparities in US Household Wealth, 1996 to 2011." *Russell Sage Foundation Journal of the Social Sciences*, 2(6): 129–152.

Thomas, Hannah, Tatjana Meschede, Alexis Mann, Janet Boguslaw, and Thomas Shapiro. 2014. *The Web of Wealth: Resiliency and Opportunity or Driver of Inequality?* Waltham, MA: Institute on Assets and Social Policy.

Thompson, Jeffrey P., and Gustavo A. Suarez. 2015. "Exploring the Racial Wealth Gap Using the Survey of Consumer Finances." *Finance and Economics Discussion Series 2015–76*. Washington: Board of Governors of the Federal Reserve System.

Traub, Amy, Laura Sullivan, Tatjana Meschede, and Tom Shapiro. 2017. *The Asset Value of Whiteness: Understanding the Racial Wealth Gap*. New York: Demos.

U.S. Census Bureau. 2019. (February 28. 2019). *Quarterly Residential Vacancies and Homeownership, Fourth Quarter 2018* (Report No. CB19-10). Retrieved from https://www.census.gov/housing/hvs/files/currenthvspress.pdf

Vale, Lawrence J. 2007. The Ideological Origins of Affordable Homeownership Efforts. In *Chasing the American Dream: New Perspectives on Affordable Homeownership*, edited by William L. Rohe and Harry L. Watson, 15–40. Ithaca, NY: Cornell University Press.

Wolff, Edward N. 1998. "Recent Trends in the Size Distribution of Household Wealth." *The Journal of Economic Perspectives*, 12(3): 131–150.

Wolff, Edward N. 2014. *Household Wealth Trends in the United States, 1962–2013: What Happened over the Great Recession?* NBER Working Paper 20733. Cambridge, MA: National Bureau of Economic Research.

Yinger, John. 1998. "Housing Discrimination is Still Worth Worrying About." *Housing Policy Debate*, 9(4): 893–927.

Zhao, Bo, Jan Andrich, and John Yinger. 2006. "Why Do Real Estate Brokers Continue to Discriminate? Evidence from the 2000 Housing Discrimination Study." *Journal of Urban Economics*, 59(3): 394–419.

5

SHOULD POLICY SEEK TO INTERFERE WITH UPWARD MOBILITY BY DECONCENTRATING POVERTY?

Reasons for Concern

Howard Husock

Introduction

The vast majority of scholars and policymakers agrees with the statement that public policies should deconcentrate poverty. It is not hard to see why. The term "concentration of poverty," widely credited to sociologist William Julius Wilson (1987), conjures the image of disadvantaged low-income people residing in challenged neighborhoods. Indeed, a wide range of scholars has not only identified an increased extent of concentrated poverty in the U.S., including poverty areas dominated by Blacks and/or Hispanics, but discussed its negative consequences as well. For example, Jargowsky has documented an increase in the number of high-poverty Census tracts between 2000 and 2011 and expressed the view that these Census tracts are detrimental to the prospects of residents:

> For any given number or percentage of poor families in a society, a more concentrated residential pattern of the poor will result in more poor adults living in dangerous neighborhoods with less access to information about jobs. More poor children will grow up with fewer employed role models and attend schools that, on average, function at far lower levels than those of the middle class. Physical and mental health of the poor will also suffer. While the exact extent of these effects is debated, few would dispute that there are costs to the poor of living in economically devastated ghetto or barrio neighborhoods, rather than middle-class or better neighborhoods with good schools, good connections to the labor market, and other public amenities.
>
> *(Jargowsky 2014, 2)*

A similar view is expressed by Glaeser, who reaches a similar conclusion in an analysis based on the changing fortunes of regions when high-poverty concentrations develop. "There is an externality when low human capital people move into (a region) that they create by lowering the probability that the next generation in that area will be high human capital" (Glaeser 2008, 231).

Such analyses of the negative effects of concentrated poverty have been accompanied by policies and proposals that aim to deconcentrate low-income people by dispersing them to higher-income neighborhoods. It is assumed that these dispersal efforts will lead to an improved quality of life and better life outcomes. Historic and present examples of these policies and proposals are the Housing Choice Voucher (HCV) program, established in the 1970s, which encourages moves from high-poverty to low-poverty neighborhoods, the Gautreaux demonstration, which emphasized minority deconcentration, the Moving to Opportunity (MTO) pilot program, an experimental randomized housing mobility demonstration program, conducted between 1994 and 2004 in five cities, Baltimore, Boston, Chicago, Los Angeles, and New York (Kling et al. 2007), and the currently proposed Small Area Fair Market Rents (Small Area FMRs) program. In regard to the latter, the U.S. Department of Housing and Urban Development (HUD) proposed to adjust rent subsidies provided by the voucher program.

> Known as Small Area Fair Market Rents (Small Area FMRs), the policy replaces a single voucher subsidy ceiling which applies across all neighborhoods in a metro area with a policy tying the subsidy maximum to the typical rent within each zip code. [...] This allows the voucher to be large enough to pay for more expensive rents in higher rent zip codes while cutting the maximum subsidy in low-rent communities.
>
> *(Collinson 2016, 1)*

Recent research on mobility by Chetty and colleagues shows that the MTO program has been relatively successful. For example, Chetty et al. (2016) focused on the life prospects of children raised in low-income families that participated in the MTO program and found that children who moved to lower-poverty neighborhoods before the age of 13 have higher college attendance rates and earnings, and lower single parenthood rates. As Chetty et al. (2016) state: "The findings imply that offering families with young children living in high-poverty housing projects vouchers to move to lower-poverty neighborhoods may reduce the intergenerational persistence of poverty and ultimately generate positive returns for taxpayers" (1). Or, as Chetty et al. (2016) state in a related paper about poor households who choose to move to places where general upward mobility rates are higher: "Low-income children are most likely to succeed in counties that have less concentrated poverty, less income inequality, better schools, a larger share of two-parent families, and lower crime rates" (862).

Indeed, *The New York Times* applauded the findings by Chetty and colleagues with the following headline: "Why the New Research on Mobility Matters" (Wolfers 2015). The research conducted by Chetty et al. (2015, 2016) implies not only that MTO has been successful but that additional or expanded deconcentration programs may be successful as well. Indeed, Chetty's work has fueled already existing support for expanding the extent of housing vouchers and dispersing eligible low-income households to so-called "higher opportunity neighborhoods." For example, Julian Castro, when he served as Secretary of the U.S. Department of Housing and Urban Development in the Obama administration, stated, "We're interested in, in terms of the Housing Choice Voucher program, how do we enhance mobility to get families who choose to live in higher opportunity areas into those higher opportunity areas?" (Axel-Lute 2015–2016, n.p.). Similarly, Sard and Rice (2016) focused on reforming the housing voucher program, encouraging voucher holders to locate to low-poverty areas. Furthermore, Hertz (2016) goes so far as to advocate that housing vouchers—i.e., individual household rent subsidies—should be made an entitlement.

Poverty Deconcentration: Reasons for Skepticism

There are, however, other ways to look at the goal and methods of poverty deconcentration that should give policymakers pause. First, there are practical political and fiscal barriers to broad implementation of a deconcentration policy. Second, it is crucial for those who would shape such policies to recognize that they do more than move a specific population to a specific new geography. Indeed, they represent an intervention in the system of norms along which American neighborhood residential patterns are organized. In other words, Americans have looked to climb and occupy what can be termed a housing ladder in order to live in communities in which they generally share the socioeconomic status of their neighbors—and believe that, in so doing, they have earned that status.

Policy-based efforts aimed at poverty deconcentration risk undercutting the sense of achievement through self-sacrifice of those low-income households that have successfully climbed the economic ladder. First, establishing such a policy is impractical due to current political and fiscal realities. Second, poverty deconcentration would undercut the self-sacrifice and the sense of achievement of those low-income households that have successfully climbed the economic ladder. Third, deconcentrating poverty and achieving upward mobility can only be accomplished through improving the quality of public goods provided in low-income neighborhoods. Finally, deconcentrating poverty risks diverting resources from, and shifting policy focus away from, laying the foundation for broad upward mobility through the improvement of the quality of public goods provided in low-income neighborhoods.

I now turn to potential complications stemming from poverty/minority deconcentration policy adoption and implementation. The positive results in the Chetty et al. (2016) study attracted such wide attention in part because they cast a new light on the MTO program, whose very name reflects the core assumption of the deconcentration goal. Previous work had been sanguine in its assessment. Kling et al. (2004) stated "no significant overall effects on adult employment, earnings or public assistance receipt" (1). Two years later, Sanbonmatsu et al. (2006) "did not find evidence of improvements in reading scores, math scores, behavior or social problems, or school engagement, overall for any age group" (682). Certainly, Chetty's new work provides additional and positive findings. But it is important to note that positive effects were limited to those families who moved when their children were young—and were less positive the older children were at the time of the move. This matters much in terms of any federal housing program, whether one that would use higher-value vouchers (i.e., those providing higher rent subsidies) to allow the poor to pay the higher rents of more affluent communities (and perhaps require landlords to accept them), or one that would use financing tools such as the federal Low-Income Housing Tax Credit (LIHTC) subsidies to support the construction of subsidized housing in non-poor communities where land costs are higher and thus subsidies would produce fewer units. Indeed, in practice participation in federal housing programs is based not on motivation or demonstrated benefit for select groups but on eligibility based on the income level, among other factors. It is difficult to imagine—even as a legal matter—limiting participation to those households whose children are pre-adolescents, those for whom the new research by Chetty et al. (2016) specifically finds to benefit.

Additional Reasons for Skepticism About Poverty Deconcentration

Additional reasons to be cautious about policy interventions aimed at jumpstarting upward mobility are provided by history. At least twice before, federal policy has put its thumb on the scales of upward social mobility—neither time with acclaimed results. Although they differ from that which is proposed currently by advocates of government-directed poverty deconcentration, they should give us pause about the collateral damage to which initially attractive social and housing polices can lead. First, urban renewal cleared neighborhoods with a high proportion of homeowners. Second, changes in public housing policy reduced the effectiveness of public housing as a mechanism to promote mobility, if, indeed, it was ever likely to have such capacity.

Consider public housing, which, beginning with the National Housing Act of 1937, sought to provide a safe and sanitary environment with community facilities that would further self-improvement and upward mobility. Von Hoffman (1996) labeled these assumptions "environmental determinism," a belief that a high-quality residential environment will improve the quality of life and thus enable self-improvement and subsequent upward mobility. This is much the same guiding principle as that implied by the contemporary goal of relocating low-income households to higher-income neighborhoods (Von Hoffman 1996, 2). Notably, the initial public housing such an environment proved difficult for public authorities to maintain—a challenge made greater by the 1968 amendment to federal housing law, which limited rent payments to 30 percent of tenant income. Not only did the change limit the revenue stream received by local housing authorities, but, because it meant that increased income would lead to increased rent, it sent a deleterious message that such habits as self-sacrifice, work, education, marriage, and thrift were implicitly discouraged, as public housing projects became redoubts of the very poor.

Interestingly, as is shown by the following St. Louis example, the values associated with the enunciated goals of public housing could be found in neighborhoods that were demolished to make way for public housing, specifically the well-known Pruitt-Igoe development built in 1954 in a neighborhood that had a relatively high proportion of homeowning Blacks. Such neighborhoods would have qualified as areas of concentrated poverty, yet had significant positive aspects, as manifested notably by homeownership rates. In 1950, St. Louis was a city with 856,796 people, 702,348 of whom were White (about 82 percent) and 153,755 of whom were Black (about 18 percent), and a median income of $2,718 in 1950 (about $25,000 in 2011 dollars) (U.S. Bureau of the Census 1950).

An examination of two nearly all-Black neighborhoods based on 1950 U.S. Census data disproves the claim that absentee slumlords controlled much of the real estate and bilked tenants who had no choice of where to live because of racial segregation. The proportions of owner occupiers for both neighborhoods were below the citywide average—34 percent owner occupiers and 41 percent of structures with an owner present—although not by much. For instance, Census tract 11-A, which I call Northside A, was 98 percent Black, and 21 percent of all housing units (828 of 3,696) were owned by non-Whites, most likely Blacks, and at least a quarter of the units had an owner occupier. Interestingly, the proportion of owner occupants (21 percent) in this particular Census tract exceeds that of an adjacent, predominantly White Census tract (18 percent). In contrast, Census tract 21-B, which I call Northside B, was 96 percent Black, and most of the units were owned by non-Whites, 95.8 percent (460 out of 480), and at least 23 percent of all housing units had an owner occupier.

In Northside A, Northside B, and the predominantly White Census tract adjacent to Northside B, rents ($27.71, $21.85, and $17.53 per month in 1950 dollars, respectively) were lower than the citywide median ($28.55) and the predominantly White Census tract adjacent to Northside A ($33). Not surprisingly, the housing quality in Northside A and B was low. For example, in Northside A, Northside B, in the predominantly White Census tract adjacent to Northside B, 30, 56, and 44 percent of the housing stock, respectively, did not have a private bath or were deemed dilapidated, compared with 28 and 30 percent of the housing stock citywide and in the predominantly White Census tract adjacent to Northside A, respectively. In sum, there were poor Black as well as White Census tracts in St. Louis in 1950. The preceding refutes the conventional tale that, historically, low-income housing was chiefly provided by slumlords, i.e., absentee landlords who took advantage of the limited housing opportunities available to the poor to reap outsized profits.

One could argue that public housing generally, and Pruitt-Igoe specifically, was needed to upgrade so-called slum conditions. However, that argument rests on faulty assumptions—that poor physical conditions should be the main gauge of the health of a community, and that demolition and redevelopment are the only options available to promote social mobility. In reality, building Pruitt-Igoe eliminated the option of transforming neighborhood residents into homeowners, the traditional mechanism for providing economic improvement for Americans. In other words, we must be cautious in assuming that policy can be designed to definitively increase the likelihood of household upward mobility and that we can ascertain what the social and environmental preconditions for upward mobility actually are.

Let us next consider both the practical and normative dimensions of the policy levers that might be employed to use housing policy as an expanded means to deconcentrate poverty. The means of choice would be the HCV program, a HUD program administered through local housing authorities that provides income-eligible households subsidies for use in the private housing market such that their out-of-pocket housing cost would be, as for public housing, limited to 30 percent of household income. The current annual HUD budget includes about $19 billion of funding for the HCV program, which currently supports 2.38 million voucher households, composed of 5.3 million individuals (U.S. Department of Housing and Urban Development n.d.c.). However, only a small proportion of eligible low-income renters actually apply for Housing Choice Vouchers. Expanding the program to serve all eligible low-income renters would result in an additional 17 million households, comprising 40 million individuals who would need to be served. Thus, about $138 billion of funding would be needed. Indeed, Congress has consistently declined to support relatively small budget increases for the program proposed by the Obama administration; the most recent proposal was to increase its budget to $21.2 billion. (Let us see if the new administration will bring any increases to the budget.)

Deconcentration inevitably brings with it higher costs, as reflected in the complications of a proposed HUD initiative known as the Small Area FMR program, which focuses on increasing rental subsidies for some eligible renters, such that tenants can afford rents in higher-cost "better" neighborhoods. This modification is going to result in an increase in rental subsidies for some eligible recipients in select ZIP code areas that have higher income levels and higher rents. Assuming fixed funding, this modification would result in an overall decrease in the total number of HCV households served (U.S. Department of Housing and Urban Development n.d.a, n.d.b). Thus, the Small Area FMRs initiative is unrealistic, given current political and fiscal realities—if poverty deconcentration aimed at reaching a larger, income-eligible population was to be a paramount federal housing policy goal.

Deconcentration, whether through the means of the Small Area FMR or other means, faces a more serious, normative problem, however. It risks sparking neighborhood tensions based on race, ethnicity, and class. Indeed, government-led deconcentration will be at odds with the unwritten yet powerful norm that residential patterns are largely shaped by socioeconomic status and resident upward mobility, resulting in an urban spatial structure with varied housing styles and types (based on class) in different neighborhoods.

This powerful, but informal, system has long been recognized by scholars starting with the Chicago School of Sociology in the 1920s. Rees (1979) observed that "socioeconomic status is a universal sorting principal in American cities" (47). He found that education, income, and occupation strongly correlated with residence. Similarly, White (1987) noted that residence and social status are interrelated, as the following quote illustrates:

> Time and again it has been argued that socioeconomic status is the principal differentiator in the metropolis. Poor neighborhoods with residents who work in lower status occupations and have lower levels of education are found to be separated from the high status neighborhoods in the urban mosaic. Educational attainment and occupation are good overall indicators of a neighborhood's status. The level of income, correlated with these two characteristics, is the most direct indicator of a household's ability to "purchase" status, or at least purchase a residence in what is regarded as a high status neighborhood.
>
> *(White 1987, 27)*

These findings have stood the test of time, as evidenced by findings of Hardman and Ioannides (2004):

> For the vast majority of US households, neighbors' incomes and other characteristics are the market-driven outcome of individual choices. Households' tastes for housing space, quality and access to jobs and amenities, together with their incomes and assets, define

> demand for housing types and locations. Prices set in the housing market determine what housing units and neighborhoods households can afford.
>
> *(Hardman and Ioannides 2004, 4–5)*

In the 1960s, during a time when race-based discrimination was common, Gans observed that rival ethnic groups such as the Irish, Italians, Jews, and Poles (some of whom were not even considered White by nineteenth century nativists) made common cause in Levittown, New York, on Long Island. Most Levittowners understood that there was a virtuous conspiracy of self-sacrifice, shared values, and smart personal decisions in terms of work, education, marriage, and thrift, resulting in desirable communities characterized by high-quality public amenities, such as a low crime rate, high-quality public schools and parks, and clean streets, among others.

Gans' observations have remained true to this day: Forcing residents with a different socioeconomic status to live in the same neighborhood is a recipe for tension. As Gans writes, "If one group is threatened by another's demands, intolerance may [...] increase" (Gans 1967, 171). Forcing people to live together raises the questions of whether new subsidized residents deserve to be in a community that houses those who moved there based on self-improvement, self-sacrifice, and upward mobility, not government subsidies. The federal court decision ordering affluent suburban Westchester County, New York, to place subsidized housing in high-income neighborhoods provides an excellent example. As a self-identified Black resident of high-income Larchmont, located in Westchester County, observed:

> As an African-American who happily resides in one of the aforementioned towns targeted for this deplorable lower income housing, I am appalled at this decision to reward those individuals who ... chose the easy way out instead of dedicating oneself to hard work.... My wife and I worked hard to be able to purchase a home and PAY TAXES in one of these towns, just like everyone else who resides in them.
>
> *(Comment posted in response to Roberts (2009))*

Another reader made a similar point:

> As an African-American, I am tired of the practice of placing government housing in otherwise middle class and affluent neighborhoods.... All it does is reinforce a stereotype that all African-Americans are laggards when it comes to educating ourselves, rising socially and advancing economically. It is not as though there aren't African-Americans already living in these neighborhoods in Westchester.
>
> *(Comment posted in response to Roberts (2009))*

These quotes reflect the current U.S. housing norm that residential patterns are largely shaped by socioeconomic status and resident upward mobility. Interestingly, Blacks comprise 1.5 percent of the population in the Village of Scarsdale, which has a median income of $260,000 and is also located in Westchester County, whose population is about 16 percent Black (U.S. Bureau of the Census n.d.). I suspect levels of discrimination have dropped to the point that middle-class Blacks who want to live in predominantly White middle-class communities can do so.

The two commentators seem to be yearning for a post-racial America, where Blacks are no longer a special concern of public policy and where their own accomplishments may serve as their ticket to prosperity. They understand that desirable communities are neither inevitable nor inherited. Rather, they must be built and maintained day by day, meeting by meeting. Virtuous conspiracy, however, is predicated on a sense of commonality. As Putnam (2007) found, "[n]ew evidence from the US suggests that in ethnically diverse neighborhoods residents of all races tend to 'hunker down.' Trust (even of one's own race) is lower, altruism and community cooperation rare, friends fewer" (Putnam 2007, 137). Putnam expresses hope of overcoming "such fragmentation by creating new, crosscutting forms of social solidarity and more encompassing identities" (Putnam 2007, 137). Indeed, U.S. society has created these new, crosscutting forms of social solidarity and more encompassing identities over the centuries.

I believe that the proposed Small Area FMRs initiative is going to create social tensions in receiving neighborhoods. Integration by fiat will undercut the spontaneous ways in which residents form tolerant, encompassing communities. The Small Area FMRs initiative is also going to undermine self-improvement, self-sacrifice, and the sense of achievement by those low-income households that have successfully climbed the economic ladder unassisted by government subsidies. They will not feel it necessary to work hard to be able to live in the most desirable middle-class communities.

Dispersal programs, such as MTO, Gautreaux, and the proposed Small Area FMRs, divert funds from the historic role of government, i.e., providing high-quality public amenities (including public education). Therefore, an enhanced dispersal effort draws funds away from education and other public amenities, thereby undermining the mobility prospects for low-income people who have moved up and into higher-income communities through their own initiative. Public policy should support the self-sacrifice and resident upward mobility and not interfere with these values and efforts. The proposed Small Area FMRs initiative is going to alter, by fiat, the spontaneous ways in which Americans form tolerant, encompassing communities.

Conclusion

Glaeser is indeed correct to say that it is unwise for government to subsidize housing in such a way that "poor people stay in a [neighborhood] that is filled with other poor people" (Glaeser 2008, 231). If we are not, however, to subsidize the poor to move to non-poor areas, the thesis of this paper, what should our approach to encouraging upward mobility be?

It certainly cannot be to send a government-approved message that predominantly poor (or minority) neighborhoods are, inevitably, bad neighborhoods. Indeed, one wonders whether, at some point, minority public officials—the election of whom has been a key goal of the Voting Rights Act—will object to that implicit assumption. Put another way, public policy simply cannot give up on the possibility that poor neighborhoods can become good neighborhoods. That does not mean investing only or mainly in their physical improvement. To do so would simply mean making these areas more comfortable places to be poor (see also the chapter authored by Kirk McClure).

Rather, policy must follow the time-honored path of supporting high-quality public goods that help lay the foundation for those of low income to improve their socioeconomic status. Effective public education must top the list of such public goods—and it must include good neighborhood libraries, clean parks and recreation facilities, and, crucially, public safety. Ultimately, there is no practical substitute for supporting a culture of self-improvement and upward mobility.

Put another way, is concentration of poverty *per se* the most important problem that the poor, notably the African-American poor, face—and is deconcentration of poverty the only way many American poor, regardless of race or ethnicity, will improve their prospects? It is both more practical and more in keeping with history and the character of housing markets for the concentrated poor to deconcentrate through private sales and individual decisions, and organically by adopting the norms and habits of upward mobility. This is an approach that makes sense only in the context of a growing economy—but that caveat applies broadly to the prospects of all Americans. It is worth keeping in mind that Americans now look back on the concentrated poverty of the nineteenth and early twentieth century and both celebrate and romanticize it—as with New York's Tenement House Museum, which, in its literature, refers to itself as the "urban log cabin"(Thirteen.org n.d., n.p.). At least in retrospect, concentrated poverty—in the context of effective public education, a growing economy, and a culture of striving—is a stage in development, not a condition that itself must be directly remedied.

References

Axel-Lute, Miriam. 2015–2016. "Shelterforce Exclusive: Interview with HUD Secretary Julián Castro." *Shelterforce*, Winter, n.p.

Chetty, Raj, Nathanel Hendren, and Lawrence F. Katz. 2015. *The Effects of Exposure to Better Neighborhoods on Children: New Evidence from the Moving to Opportunity Experiment*. Cambridge, MA: National Bureau of Economic Research.

Collinson, Rob. 2016. *Expanding Neighborhood Choices for Voucher Tenants using Small Area Fair Market Rents*. New York: New York University Furman Center.

Comments posted in response to S. Roberts. 2009. "Westchester Adds Housing to Desegregation Pact." *The New York Times*, August 10.

Gans, Herbert J. 1967. *The Levittowners*. New York: Pantheon Books.

Glaeser, Edward L. 2008. *Cities, Agglomeration, and Spatial Equilibrium*. Oxford: Oxford University Press.

Hardman, Anna, and Yannis M. Ioannides. 2004. "Neighbors Income Distribution: Economic Segregation and Mixing in Urban Neighborhoods." *Journal of Housing Economics*, 13: 368–382.

Hertz, Daniel. 2016. "Make Housing Vouchers an Entitlement—We Can Afford it." *City Commentary*, January 5, n.p.

Jargowsky, Paul. 2014. *Concentration of Poverty in the New Millennium: Changes in Prevalence, Composition, and Location of High Poverty Neighborhoods*. New York and Camden, NJ: Century Foundation and Center for Urban Research and Education, Rutgers University.

Kling, Jeffrey R., Jeffrey B. Liebman, Lawrence F. Katz, and Lisa Sanbonmatsu. 2004. *Moving to Opportunity and Tranquility: Neighborhood Effects on Adult Economic Self-Sufficiency and Health from a Randomized Housing Voucher Experiment*. Cambridge, MA: Harvard University Press.

Kling, Jeffrey R., Jeffrey B. Liebman, and Lawrence F. Katz. 2007. "Experimental Analysis of Neighborhood Effects." *Econometrica*, 75: 83–119.

Putnam, Robert D. 2007. "*E Pluribus Unum*: Diversity and Community in the Twenty-First Century: The 2006 Johan Skytte Prize Lecture." *Scandinavian Political Studies*, 30: 137–174.

Rees, Philip H. 1979. *Residential Patterns in American Cities*. Research Paper 189. Chicago, IL: University of Chicago, Department of Geography.

Sanbonmatsu, Lisa, Jeffrey R. Kling, Greg J. Duncan, and Jeanne Brooks-Gunn. 2006. "Neighborhoods and Academic Achievement: Results from the Moving to Opportunity Experiment." *The Journal of Human Resources*, 41(4): 649–691.

Sard, Barbara, and Douglas Rice. 2016. *Realizing the Housing Voucher Program's Potential to Enable Families to Move to Better Neighborhoods*. Washington, DC: Center on Budget and Policy Priorities.

WNET, Thirteen.org. "The Urban Log Cabin: Over Ten Thousand People Lived at 97 Orchard Street Between the Years 1870 and 1915." https://www.thirteen.org/tenement/logcabin.html.

U.S. Bureau of the Census. n.d. *State and County Quickfacts*. Washington, DC: U.S. Bureau of the Census.

U.S. Bureau of the Census. 1950. *Census of Housing, Volume 1: General Characteristics*. Washington, DC: U.S. Bureau of the Census.

U.S. Department of Housing and Urban Development. n.d.a. *Budget Outlays by Program: Comparative Summary, Fiscal Years 2014–2016*. Washington, DC: U.S. Department of Housing and Urban Development.

U.S. Department of Housing and Urban Development. n.d.b. *Public and Indian Housing Tenant-Based Rental Assistance 2017: Summary Statement and Initiatives*. Washington, DC: U.S. Department of Housing and Urban Development.

U.S. Department of Housing and Urban Development. n.d.c. *Picture of Subsidized Households*. Washington, DC: U.S. Department of Housing and Urban Development.

Von Hoffman, Alexander. 1996. "High Ambitions: The Past and Future of American Low-Income Housing Policy." *Housing Policy Debate*, 7(3): 423–446.

White, Michael J. 1987. *American Neighborhoods and Residential Differentiation*. New York: Russell Sage Foundation.

Wilson, William Julius. 1987. *The Truly Disadvantaged: The Inner City, the Underclass, and Public Policy*. Chicago, IL: University of Chicago Press.

Wolfers, Justin. 2015. "Why the New Research on Mobility Matters: An Economist's View." *The New York Times*, May 4.

6

THE AFFORDABLE HOUSING COMPLEX

Direct and Exclusionary Displacement in the Lacy and Logan Neighborhoods of Santa Ana, California

J. Revel Sims and Carolina S. Sarmiento

Introduction

The production of subsidized affordable housing is often seen as one of the key pillars of anti-gentrification and anti-displacement mitigation strategies. We suggest that this view is hegemonic in that it has become a "common sense" approach shared by a range of actors including community activists, scholars, policymakers, and planners. While perspectives on the application of subsidized affordable housing differ, most proponents ascribe to what economists refer to as "market failure," where housing markets are unable to distribute housing efficiently. In this case, housing that is affordable for low-income households tends to be under-produced and unavailable—both of which are magnified by gentrification. Such so-called inefficiencies have been described in econometric terms as a "less than perfect" elasticity (Solomon 1974, 63) as well as more frank assessments that emphasize the fundamental inability of the market to provide quality housing for low-income populations (Schwartz 2006, 36).[1] Under a broad framework encompassing this range, the production of subsidized affordable housing is thought to help address two aspects of housing inefficiency that are exacerbated by gentrification and that negatively impact vulnerable tenants: rising rents and spatial exclusion, or what Marcuse referred to as "exclusionary" displacement (Marcuse 1985).

As an entry point, our analysis begins with the debate about the causes of gentrification that is sometimes referred to as the "production" versus "consumption" origins of gentrification (Lees et al. 2008). The general assumption of consumption theorists is that preferences diverge along a number of socioeconomic and demographic axes and those people with greatest locational choice have more power to manifest their desires in urban space. This perspective on urban change has been criticized for confusing symptoms for causality and ignoring important structural shifts in capitalist development (Harvey 1973; Smith 1979; Storper and Scott 2009). For our purposes specifically, we argue that the consumption approach is problematic for two additional reasons. First, the central assumption that wealthier or more privileged groups such as Florida's "creative class" (2003)—a term we will employ for gentrifiers generally because of its broad, albeit imprecise, applicability—drive urban change at the neighborhood level effectively marginalizes the preferences of non-White, low-income communities, and ultimately devalues their own agency within the production of space. Second, emphasizing the power of the consumption preferences of gentrifiers redirects attention away from the important role property owners, developers, and financial institutions play in shaping urban change and gentrification processes. As Smith (1979) argued—we believe correctly—gentrification is primarily a "spatial fix" to uneven capitalist urbanization, and thus research that reduces analysis to a narrow dualism between gentrifiers and the gentrified often misses crucial, sometimes definitive, actors in housing and labor markets. Failure to conceptually incorporate these subjects runs the risk of also reproducing the neoclassicist notion that preferences and amenities alone drive housing markets and urban development (Glaeser 2005).

Finally, our analysis builds off of a framework that not only emphasizes the role that actors and ideology play in the production of gentrification (Rose 1984) but also sees gentrification as involving "both a change in the social composition of an area and its residents and a change in the nature of the housing stock" (Hamnett 1991, 176). This approach is based on the notion of an affordable housing complex (AHC) that functions hegemonically in order to bring together actors and provide a space for the development of housing strategies (Sarmiento and Sims 2015). Together, the notion of an AHC and our enlarged view of gentrification allow us to examine how the production of subsidized affordable housing can form part of a gentrification and displacement process at the neighborhood scale in Santa Ana, a city in Orange County, California, with a high proportion of low-income Latinx[2] residents. This approach is significant in that it extends the notion of exclusionary displacement beyond the methodological preference for income measures alone toward other considerations such as cultural representation and documentation status that are important to the context of Santa Ana specifically. In sum, we find that through the AHC, subsidized affordable housing production can result in negative outcomes that range from direct displacement to a complex form of exclusionary displacement, thus facilitating a larger gentrification process.

Theoretical Framework: The AHC and Displacement

The first component of our theoretical framework is the AHC—i.e., the use of a complex to describe subsidized housing production. President Eisenhower's now famous warning regarding the rise of a post-World War II "military industrial complex" (MIC) has been useful for problematizing the motivations, practices, and trajectories of industrial sectors. According to this view, a *complex* formed around "an interwoven constellation of interests that tied together the growth of large-scale bureaucracies such as the Defense Department with particular firms that profited from military procurement" (Cox 2014, 1). One of the most important features of complexes such as the MIC is the role of ideology and how it is employed to redirect resources toward individuals, firms, and even entire regions. Similar conceptualizations drawing attention to the systems of interests behind complexes have since been applied to other sectors such as the "prison industrial complex" (Davis 1998), the "nonprofit industrial complex" (Incite! Women of Color Against Violence 2007), and most recently the "affordable housing complex" (Sarmiento and Sims 2015). Sarmiento and Sims' (2015) articulation of the AHC is furthermore distinctive in that it is a specifically urban, local scale utilization that emphasizes the spatial distribution of Low-Income Housing Tax Credit (LIHTC) developments and their relationship to gentrification processes.

A second component of our theoretical framework draws from the regulationist approach. Regulationist approaches begin from the Marxian notion that capitalist accumulation is an inherently unstable process that requires a certain amount of consent or agreement among political actors in order to be reproduced (Aglietta 1979; Boyer 1990; Lipietz 1987). Due to inherent crisis tendencies, regulationists argue that regimes form to construct consent around a Gramscian "common sense" (1971) in order to bind a wide variety of social actors to strategic objectives. Two historic examples that are commonly used to illustrate this process are (1) the rise of a post-Great Depression Keynesian Welfare State (KWS) that promoted demand-side development models and instituted a process for managing conflict between industrial labor and capital, and (2) a post-1970s neoliberal turn that emphasized privatization, de-unionization, and a devolution of certain state practices toward urban entrepreneurialism (Jessop 2002). In this case, we believe that both the process of gentrification generally and the AHC support regulationist interpretations. On one hand, it has been persuasively argued that gentrification is essentially a revanchist, global urban *strategy* to restructure urban space (Smith 1996, 2002). On the other hand, the AHC is operationalized as a type of sub-regime that includes financial institutions, state agencies, politicians, nonprofit and for-profit housing providers, community-based organizations (CBOs), community development corporations (CDCs), and other intermediaries such as the Local Initiatives Support Corporation (LISC) and the Enterprise Foundation. Furthermore, we theorize that one of the main roles of the AHC is to provide consent around the production of subsidized affordable housing as an anti-displacement, anti-gentrification mitigation strategy even when it may actually function as a gentrifying force through different forms of displacement.

In order to evaluate the role of the AHC as an anti-gentrification/anti-displacement strategy, it is critical to understand the different forms that displacement may take. Following the publication of a series of studies on displacement in New York by Lance Freeman (Freeman 2005; Freeman and Braconi 2004) that minimized the magnitude of displacement and argued that many long-term residents in Harlem benefited from gentrification, the question of displacement returned as an important topic of gentrification scholarship (Davidson 2009; Newman and Wyly 2006; Slater 2009). One of the central pieces missing from this body of research, as pointed out by Slater's (2009) widely recognized contribution, was a critical engagement with the concept of displacement itself. In particular, Marcuse's pioneering work (1985) that set forth a typology of displacement and broadened the understanding of the process was missing. Most notions of displacement have tended to reduce displacement to egregious actions within a strict landlord-tenant relationship such as is the case with Grier's (1978) early and influential definition.[3] Within this limited scheme, voluntary moves by tenants who leave a community that is increasingly becoming unfavorable to them due to gentrification pressure from a variety of economic and social sources are typically not considered as displacement. To fill this gap, Marcuse therefore introduced the notion that displacement, in addition to being direct, could also be exclusionary; that is, rent and other housing changes could make a neighborhood that was previously economically, socially, and culturally accessible to a particular income demographic inhospitable and inaccessible.

Methods and Context

In order to better understand the process through which the AHC both builds consensus among actors and also foments displacement pressures, this chapter analyzes both quantitative and qualitiative data including formal and informal interviews with current and displaced residents, participant observation at city sponsored meetings, archival research of documents such as city minutes and planning documents, and a community-based collaborative housing survey of the Lacy neighborhood carried out by the authors along with community reesidents and organizers. This chapter highlights the newest affordable housing development project, the Depot at Santiago, to demonstrate the process of building consensus among different community actors that helped reproduce the AHC as well as direct and exclusionary displacement in two neighborhoods in Santa Ana.

This research takes place in two neighborhoods, Lacy and Logan, that have been vulnerable to gentrification processes due to their demographic and socioeconomic composition and location. The adjacent neighborhoods are located in the northeast corner of the City of Santa Ana—a "majority-minority" city where 78.2 percent of the population identify as Hispanic or Latino (U.S. Bureau of the Census 2010). Lacy and Logan, by comparison, are almost entirely Latino. Within the two Census tracts that encompass the neighborhoods—750.02 and 744.05—Latinos represent between 90.5 and 89.4 percent of the respective populations (U.S. Bureau of the Census 2014a). More than half of the residents are foreign born—i.e., 50.8 and 52 percent, respectively (U.S. Bureau of the Census 2014d). The neighborhoods are also low income as anywhere from to two-fifths (41.7 percent) to a third (33.2 percent) of the tracts' populations are below the poverty line (U.S. Bureau of the Census 2014e).

Relatively low incomes, however, do not correlate with the availability of low-cost housing as residents in these neighborhoods face high housing cost burdens. A traditional measure of housing cost and affordability is the proportion of a household's rent relative to its income. Typically, households paying more than 30 percent of their income toward rent are considered rent burdened, while those paying more than 50 percent are considered severely rent burdened. The more a household pays toward rent, relative to their income, the less a household can afford to spend on other non-discretionary costs such as food, health care, and transportation as well as important discretionary expenses that are important to the quality of life such as recreation. In the two Census tracts, 65.1 and 54.7 percent of renter households paid more than 30 percent of their income toward rent in 2014, respectively (U.S. Bureau of the Census 2014f).

Relatively high rents and low incomes in a city characterized by high degrees of spatial and social exclusion have resulted in severe overcrowding in these neighborhoods. According to recent estimates, anywhere between 40.6 and 50.1 percent of households had more than one person per bedroom—which by conventional standards means they are considered overcrowded (U.S. Bureau of the Census 2014g). In both tracts, *severe* overcrowding is common, with 30.3 and 23.1 percent of households with more than 1.5 persons per room, respectively. A recent study based on interviews with 401 households in the Lacy neighborhood specifically demonstrates that these Census figures most likely underestimate the degree of overcrowding (Sarmiento and Sims 2016). Among the 401 households surveyed for this research, there were an average of 2.5 families per household. High numbers of multifamily households suggests a particular type of overcrowding involving possible sub-letting of space within individual housing units that may not be captured in the Census. The majority of households were also severely overcrowded, with 61 percent of respondents reporting two or more persons per bedroom—a standard measure of overcrowding (U.S. Department of Housing and Urban Development 2007)—as well as an additional 18 percent reporting rates from 1.51 to 2 (Sarmiento and Sims 2016). In other words, four out of every five renter households could be considered overcrowded.

Lacy and Logan fall within the bouandaries of plans to spatially restructure Santa Ana's urban core toward new residential and transportation patterns. The close proximity of these neighborhoods to the Santa Ana Regional Transportation Center (SARTC), which includes access to Amtrak rail trains, regional Metrolink commuter trains, and Orange County Transportation Authority (OCTA) buses, combined with the area's strategic location within the Station District Redevelopment Plan make these neighborhoods especially vulnerable to displacement pressures. The Station District is part of a complex history that includes the larger Santa Ana Renaissance Specific Plan (City of Santa Ana 2009), which was described by one *L.A. Times* writer as "instant gentrification along a 421-acre corridor stretching from the county's civic center to the city's train station" (Delson 2008, n.p.). The development of the 7-acre Station District was consequently significant in that it represented a piecemeal version of the Transit Zoning Code (TZC) that triggered the rezoning of a much larger area that included downtown, the First Street corridor, Civic Center, and several other nearby neighborhoods such as French Park in 2010.

Stressing the intended direction of this effort, city officials have also publically referred to these neighborhoods as the "gateway to the city" and a gentrifying downtown district. Recent development efforts have attempted to make this connection explicit by linking the SARTC to the downtown core through a combination of housing and transit-oriented development (TOD) projects. Planning efforts have involved both market-rate and affordable housing such as the Santiago Street Lofts built in 2006 and the Terraces at Santiago built in 2013 (C&C Development n.d., Caumiant n.d., Delson 2006). On the one hand, the Santiago Street Lofts, located directly across from the

Depot at Santiago, are live-work condominiums built to attract the so-called "creative class" to Santa Ana. A celebratory website that lists condominium developments boasts the live-work appeal of the Santiago units, highlighting that they come complete with industrial, roll-up doors and open floor plans that were created to help "the neighborhood gain the needed critical mass of retail to attract gallery goers and shoppers," and which are "Home to many artists and creative professionals" (Orange County Lofts n.d., n.p.). On the other hand, the area has also included subsidized affordable projects as part of the restructuring effort. Following the approval of the TZC, a number of new affordable housing development projects have been built. These include the Triada, Terraces at Santiago, and the Depot at Santiago. However, the affordability levels of these projects were immediately recognized as important points of struggle and reflected a significant amount of community mobilization, participation, and organizing. Affordable housing development projects in the Station District have increased the percentage of affordable units for extremely low-income households beyond what was originally proposed by developers (Beard and Sarmiento 2014; Gonzalez et al. 2012; Sarmiento and Sims 2015).

Establishing Housing Affordability in Lacy and Logan

This section analyzes and critiques the housing affordability achieved by the LIHTC subsidy program in both the Lacy and Logan neighborhoods. By calculating the increase in rent, the rising rent burden, the area median income (AMI), and the breakdown of units in recent LIHTC-subsidized projects, the results demonstrate both the serious housing situation in Logan and Lacy and how new housing, either subsidized affordable or market-rate, may actually function in these specific neighborhood contexts.

Increasing Housing Costs, Rising Rent Burdens, and Density

Over the past few years both the Lacy and Logan neighborhoods have experienced an increase in housing costs. Between 2009 and 2014 the proportion of one-bedroom rental housing units with gross rents of $1,000 or more increased significantly in the Lacy neighborhood. In this neighborhood, the proportion of rental units with rents at $1,000 or more increased 47.6 percent. In the Logan neighborhood, they increased even more, 76.3 percent. During the same time frame, the proportion of rental units with gross rents at $1,000 or above only increased 11 percent in Orange County overall (U.S. Bureau of the Census 2014b). These increases in housing costs are likely due to a number of factors including scarcity—especially at the low end of the market—as well as gentrification pressures from new housing and commercial developments directly to the west of the Lacy and Logan neighborhoods toward the downtown core.

In addition to experiencing some of the highest overcrowding rates in the country (Reyes and Menezes 2014), renters in the Lacy and Logan neighborhoods have experienced disproportionately increasing housing costs. In 2009, roughly half of all renter households in both neighborhoods were rent burdened—i.e., 51.7 percent in Lacy and 49.3 percent in Logan (U.S. Bureau of the Census 2014c). However, not only has the overall proportion of rent burdened households increased, but the proportion of severely rent burdened households has risen as well. Thus, while the total proportion of rent burdened households in Orange County increased only 6 percent between 2009 and 2014, in Lacy and Logan, this proportion increased 26 and 11 percent, respectively. When looking at the increase in those households paying more than 50 percent of their income toward rent, the different trajectories between Orange County and Lacy and Logan are magnified even further. In Lacy, it increased an astonishing 188.7 percent—meaning that by 2014, more than a third (36.7 percent) of all renter households were paying more than 50 percent of their income toward rent. In Logan, the proportion of severely rent burdened households increased 34.9 percent. By comparison, in Orange

County the proportion of severely rent burdened households increased 14.7 percent during the same time frame (U.S. Bureau of the Census 2014c). The increase in rents was supported by our survey data from the Lacy neighborhood that showed that more than 43 percent of respondents reported rent increases in the previous year (Sarmiento and Sims 2016).

Area Median Income (AMI) and Affordability

In 1986, the Internal Revenue Service (IRS) introduced a tax code to authorize the use of tax credits to promote affordable housing development through the LIHTC program. Tax credits are distributed to state agencies by the IRS based on population and generally require syndication—i.e., developers who have been awarded credits sell them to an investor in exchange for equity in the project. The LIHTC program is distinct from other housing subsidies such as Section 8 vouchers in that the subsidy is tied to the unit but not the occupants (see the chapters authored by Rachel Bratt; Lawrence Vale and Yonah Freemark; Howard Husock; Kirk McClure; and Rachel Garshick Kleit). In order for the program to work, the U.S. Department of Housing and Urban Development (HUD) establishes maximum allowable rents for units based on the AMI that HUD determines for all U.S. counties and metropolitan areas (U.S. Department of Housing and Urban Development n.d.). A rent-restricted, affordable housing unit subsidized through the LIHTC program is open only to eligible households earning a certain specified percentage of the AMI. These low-income categories are defined accordingly: Households earning between 51 and 80 percent of the AMI are designated "low income" (LI); those earning between 31 and 50 percent of the AMI are designated "very low income" (VLI); and those earning at or below 30 percent of the AMI are defined as "extremely low income" (ELI). A subsidized unit's maximum rent is set so that a household with an income equal to 30, 50, or 80 percent of the AMI would pay only 30 percent of its income toward rent. For example, if an area's AMI is $60,000 and a unit was set at 50 percent of the AMI (or very low income), then the maximum allowable rent for the unit would be 30 percent of $30,000, or $9,000—i.e., $750 per month.

Due to the comparatively severe housing conditions in Lacy and Logan and the way in which housing affordability is determined through HUD, subsidized affordable housing development in extremely low-income neighborhoods can potentially exacerbate housing needs. For example, in order to approximate how many households would fall within the income limits for subsidized LIHTC affordable housing in the Lacy or Logan neighborhoods, we find that from 2009 and 2015, anywhere from 81 percent to 92 percent of renter households in Lacy and Logan would be eligible[4] for low income restricted affordable housing—i.e., households earning below 80 percent of the AMI (see Table 6.1).[5] Breaking these proportions down further by the AMI-based income restrictions illustrates the growing severity of the housing needs for low-income renters in the two neighborhoods. In both neighborhoods, the proportion of eligible households at the LI and VLI levels declined from 2009 to 2015. In Logan, the proportion of eligible households within the LI range decreased 29.5 percent from 23 in 2009 to 17 percent in 2015. A similar 22.9 percent decrease was also seen among VLI-eligible households in Logan. In Lacy, there was a greater decline of LI-eligible households—35.5 percent—than in the VLI category, which declined 3.3 percent. However, while these declines suggest a reduction in affordable housing needs, when we compare these figures to ELI-eligible households we see both the magnitude as well as the growing severity of housing needs. Not only are there larger relative numbers of eligible households at the ELI level than either the VLI or LI levels—for example, from a high of 57 percent in Lacy in 2015 to a high of 47 percent in Logan in 2014—but importantly, there has been considerable growth at the extremely low end of the income scale. Accordingly, we see an 85.8 percent increase in ELI-eligible households in Logan from 24 to 45 percent between 2009 and 2015. In Lacy, the increase was comparatively less but nevertheless noticeable due to the relatively high numbers of ELI-eligible households, increasing 17.6 percent from 48 to 57 percent between 2009 and 2015.

Table 6.1 Percent of Eligible Households in Lacy (750.02) and Logan (744.05) Eligible for AMI-Based Income Categories, 2009–2015.

	2009		*2010*		*2011*		*2012*		*2013*		*2014*		*2015*	
	Logan	*Lacy*	*Logan*	*Lacy*	*Logan*	*Lacy*	*Logan*	*Lacy*	*Logan*	*Lacy*	*Logan*	*Lacy*	*Logan*	*Lacy*
Extremely low income (ELI)	24%	48%	25%	42%	25%	38%	35%	40%	38%	44%	47%	53%	45%	57%
Very low income (VLI)	33%	22%	31%	22%	26%	24%	26%	24%	24%	25%	27%	25%	26%	21%
Low income (LI)	23%	22%	25%	26%	31%	27%	22%	25%	20%	22%	16%	16%	17%	14%
Median income	7%	3%	6%	3%	5%	3%	4%	2%	5%	3%	3%	2%	2%	3%

Source: authors' calculations based on HUD 4-person household AMI figures and ACS household income ranges for all renter-occupied households from table B25118 (U.S. Bureau of the Census 2014h).

Table 6.2 Income Limits for Affordable Housing Projects in the Station District.

	30% AMI		*40% AMI*		*50% AMI*		*60% AMI*	
Triada	20.5%	(23)	20.5%	(23)	58.9%	(68)	---------	
Terraces	11.4%	(4)	---------		88.6%	(31)	---------	
Depot	10.1%	(7)	20.3%	(14)	40.6%	(28)	29.0%	(20)

Source: authors.

Using the actual incomes of renters in these neighborhoods as a basis for determining affordability levels for newly built affordable housing units offers a different perspective on what affordability might mean for existing residents. For example, in the Logan Census tract anywhere from 24 percent of project units in 2009 to 45 percent of units in 2015 would need to be reserved for ELI residents to meet existing needs. In the Lacy tract, the affordability need is even greater as anywhere from 38 percent of project units in 2011 to 57 percent of units in 2015 would need to be set aside to meet the affordability needs of the neighborhood's large proportion of ELI residents. While the effort of current projects to meet the housing needs of households with incomes around the VLI range is commendable—and partly due to community organizing and engagement efforts—it is important that in a context of severe overcrowding and housing cost burdens among the lowest-income households, none of the current LIHTC-subsidized affordable housing projects demonstrate adequate affordability by this measure (see Table 6.2). Thus, we argue that "unaffordable-affordable housing" can function as an exclusionary displacement mechanism, effectively raising the rent levels in the neighborhood and preventing the lowest-income households from accessing future housing in the neighborhood, despite the addition of a certain number of income restricted housing units at 60, 50, and even 40 percent of the AMI.

The Depot at Santiago: From Direct to Exclusionary Displacement

The Depot at Santiago, which bridges the Lacy and Logan neighborhoods, is the most recent affordable housing project in the Station District area. This approximately 2-acre development is located directly across from the Santiago Street Lofts and the SARTC. Prior to being demolished for affordable housing, the industrial site included multiple individual units, several small outbuildings (a smaller detached building), and about 61,000 square feet of yard space. The developers who purchased the land for the Depot at Santiago in 2013 were the same developers of the Terraces at Santiago, an adjacent affordable housing project. Once constructed, the Depot at Santiago will be a four-story building with 69 units of income-restricted affordable housing (15 one-bedroom, 24 two-bedroom, and 31 three-bedroom units) as well as one unit set aside for a building manager and space for commercial uses on the ground floor. Current plans for the commercial space call for between 8,500 and 10,900 square feet for retail, 4,400 square feet of "community space"—which plans date-stamped on the url and delivered to the city on February 21, 2014, identify as a 1,927-square-foot "community room" and a 1,504-square-foot "leasing/community room"—as well as parking for both residents and customers (City of Santa Ana 2014a; City of Santa Ana n.d.b).

The location of the Depot is also important due to its proximity to the proposed OC Streetcar light rail project, which is envisioned to go from the SARTC through Santa Ana and into the adjacent city of Garden Grove in long-term plans that position the Depot at Santiago as part of Santa Ana's TOD and environmental planning initiatives (see Figure 6.1). In addition to a $3 million Community Development Block Grant (CDBG), the Depot at Santiago received a state grant of $3.9 million in cap-and-trade funding from the Affordable Housing and Sustainable Communities

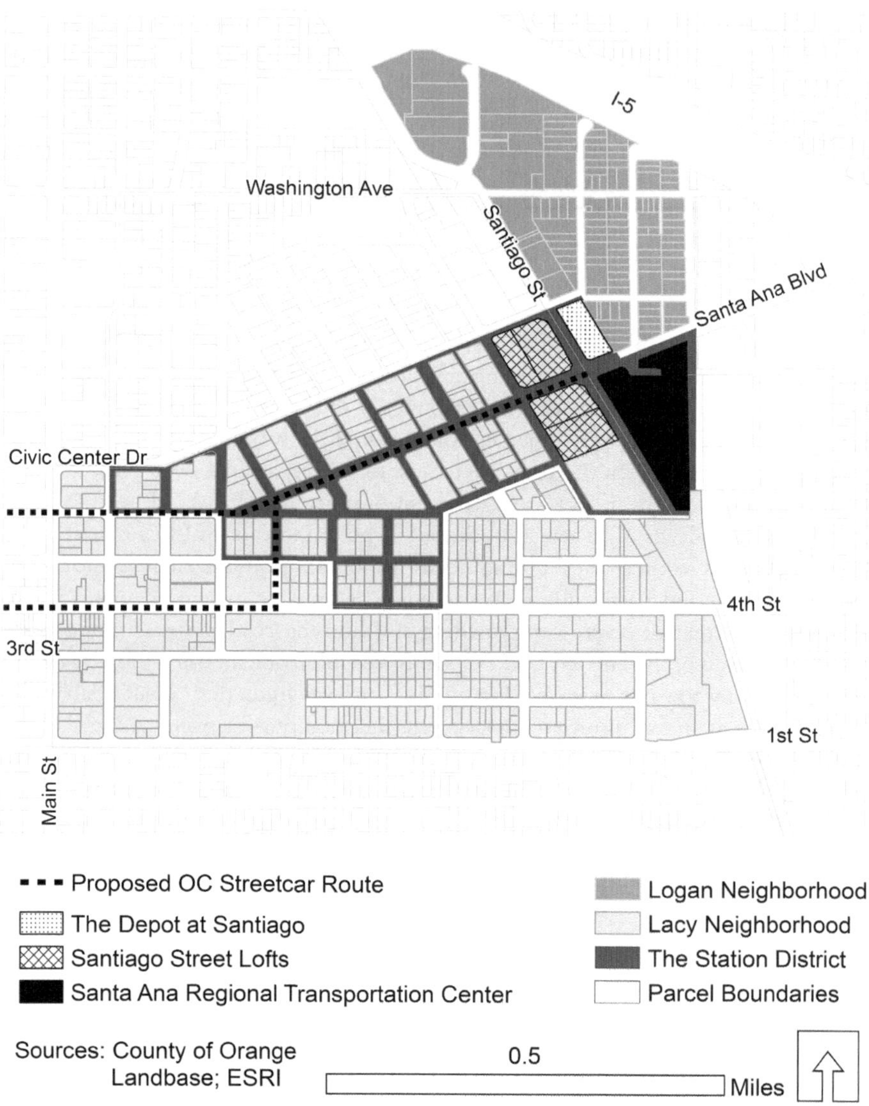

Figure 6.1 Aerial View of the Depot at Santiago Project and Other Sites.

Source: authors.

(AHSC) program, which "provides grants and affordable housing loans for compact transit-oriented development and related infrastructure and programs that reduce greenhouse gas emissions" (California Strategic Growth Council 2015, n.p.). The current development plan is to provide units for households with incomes that range from 30 to 60 percent of AMI. While this income mix represents a significant effort toward greater affordability relative to other affordable housing projects (see, for example, Sarmiento and Sims 2016), it nevertheless falls short of the greater need for housing that is affordable to the large and growing extremely low-income households in Lacy and Logan. The difference between what would actually be affordable for most people in the Lacy and Logan neighborhoods and what subsidized housing is legitimized and produced ultimately exposes an important distinction that is crucial for critically evaluating the mobilization of LIHTC-subsidized housing within the AHC.

Building Consensus Around Housing Affordability

The Depot at Santiago case study demonstrates how different actors, including developers, government officials, and community organizations, converge through a common sense understanding of LIHTC affordable housing production that claims to meet the housing needs of low-income residents while also providing neighborhood amenities and design elements that align with the assumed preferences of the so-called "creative class." Despite a range of perspectives and interests, the AHC allows different actors to unite around the mobilization of resources and interests of the city, developers, artists, and arts-based organizations.

Prior to the Depot at Santiago's approval, the city council had already articulated a vision for the area that focused on a particular kind of TOD. At city council meetings, the mayor stated that affordable housing development and transportation were part of a larger effort to "lift up the entire city." Other city council members also highlighted the needs and preferences of certain amenities for live-work residents, including retail and coffee shops. In May 2013, the city council approved a $3 million loan agreement for construction—a 55-year agreement at 3 percent interest that included $365,434 in CDBG funds (Gonzalez 2013). The council also approved up to eight project-based vouchers tied to particular units. At the city council meeting, city staff supported the project by affirming the quality of the developers, citing the successful track record of the Orange Housing Development Corporation (OHDC), Magis Realty, and C&C Development in building affordable housing. The project also brought together state and federal support for the development. In addition to the $3.9 million state grant through AHSC to reduce greenhouse gas emissions that C&C Development received, the project was also allotted about $2.6 million in federal HOME program funds that target development of affordable rental housing.

Building consensus among the community was key to the project's approval as similar research in the United States on the framing of affordable housing has shown (Goetz 2008; Nguyen et al. 2013). While many different community actors and organized interests came together around affordable housing, in the case of the Depot at Santiago, local residents raised concerns about the development that highlight a peculiar process. On one hand, the development demonstrates the initial juxtaposition of interests, perceived or otherwise, between an AHC and adjacent residents who reside in the Santiago Street Lofts. Certainly, conflict over urban space is not unique. However, in our case, it exposes the strategic in-road that was pursued by C&C Development and the city between art-led housing development and "creative class" gentrification in the form of the Santiago live-work condominums such that a bridge between groups that were initially opposed could be formed, thus further strengthening the AHC. For example, during the final planning commission meeting on May 12, 2014, there was a divide between adjacent loft residents and the Santa Ana Collaborative for Responsible Development (SACReD), a coalition made up of community-based organizations and residents. Unlike the city council, the planning commissioners were not entirely supportive and voted 5–2 in favor of the project. The disagreement on the commission was largely due to the concern that all units in the project were intended to be income restricted and available to only low-income households. Many of the loft owners who were present at the planning commission meeting feared that because of the low-income requirements the new development would not be an appropriate project for the "gateway to the city." The Santiago Street Loft residents critiqued the affordable housing project as potentially negatively impacting their property values and alluded to the uplifting vision originally presented by the city when they had purchased their condos. In contrast, SACReD pointed out the importance of community benefits for low-income residents, such as greater affordability and open space.

Following this initial confrontation, the development was explicitly placed within an arts framework as C&C Development sought participation from a local art institution to fill the "community room" space. The intentional participation of artists and arts-based organizations in the Depot project reveals a level of strategic complexity and an effort to assuage the fears of adjacent, live-work condominium residents and the formation of consensus around low-income affordable housing. Both the city council and the developer specifically worked to engage artists and arts-based organizations in the planning for the Depot. During a city council meeting, one council member requested that staff reach out to artists as potential tenants. At the same time, C&C Development mentioned that it had already promised space to The Wooden Floor, a long-time arts organization in Santa Ana that provides underserved youth with dance classes. The Wooden Floor subsequently participated in community meetings. The mobilizing idea became that in addition to affordable housing, the development would also provide the "community room" for a dance space, as well as an education and family resource center, where "The Wooden Floor will provide social services" (City of Santa Ana n.d.a.). According to The Wooden Floor's press release dated April 25, 2016, this new space would allow it "to serve 100 more youth and their families annually" (The Wooden Floor 2016, n.p.). Moreover, C&C Development and OHDC selected this arts organization in part through a previously established relationship through the Senior Vice President and OC Market Manager of the Enterprise Business and Community Engagement unit at Bank of America. As one of the Neighborhood Builders Award winners in 2009—one of the bank's top leadership awards for nonprofit excellence—the arts organization had already established a long-term partnership with the bank. Bank of America was also the primary lender on the project, providing $23,434,785 in construction financing and $4,403,840 in permanent financing (California State Treasurer 2015).[6]

The mobilization of arts-based organizations as part of this mixed-use, low-income affordable housing project demonstrates how actors within the AHC oriented the development toward the production of a larger "creative class" vision rather than one that saw affordable housing as a means to alleviate the actually existing housing and cultural needs of the neighborhoods' residents. For example, according to The Wooden Floor, all the major actors, including the City of Santa Ana, C&C Development, and OHDC, "share a mutual commitment . . . to improve the lives of children and families by providing a foundation of excellence in their homes, at school, and after school" (The Wooden Floor 2016, n.p.). Upon further questioning, however, it became clear that not all children currently living in the area, nor those living in the future development, would actually have access to the services provided by The Wooden Floor. The organization's representative conveyed to one of the authors that not only would participation be restricted to children who passed a qualifying audition, but also there would not be any specialized programs or admittance process for local children from low-income households. For example, through personal communications the respondent stated, "I am not aware of any advantages [for residents of the Depot]"; all children will have "an equal chance of acceptance," and there would not be any "separate audition process" for neighborhood residents; and families living in the complex would only be able to participate "in select community workshops" (Anonymous email communication July 12 2016).

An important part of this argument is that the Depot at Santiago not only directly displaced artists living at the location as well as other poor tenants occupying different units but also denied them just relocation assistance.[7] During the relocation efforts, only ten paying tenants were officially identified from a larger group of people who were both living and working in various units. One unit was occupied by a number of residents who used the space for various informal income-generating activities such as trash collection, recycling, and "DIY" auto repair services. In addition to regular tenants who occupied the unit along with the leaseholder, this unit was also continually occupied by a number of partially homeless residents who periodically lived and worked there for decades with the full knowledge of the prior owner. Two of the units were also occupied by self-identifying artists. One used his space as a residence, visual art gallery, and convening location for other artists; and the other was an important Mexican-Chicanx art institution, colloquially known as "La Bodeguita," or "little warehouse." From 2009 to 2015, La Bodeguita served as a residence and a cultural space for Mexican immigrant "artivists"—both documented and undocumented—from Santa Ana who used the space to host cultural events, as a location for affiliated organizations to use and meet, and as a space for temporary lodging for domestic and international artists performing and holding workshops. Local immigrant rights, community media, and cultural organizations also used La Bodeguita's space to organize, hold meetings, and offer a wide variety of events. At the time of the relocation, all of those living at La Bodeguita were low-income Mexican immigrants. In comparison to the arts-based organization that is intended to replace these tenants in the new building, neither the artivist organization nor the artists occupying units at the site were formally established or incorporated entities.

Just before demolishing the space to start construction on the Depot at Santiago, the city declared the lot "currently a *vacant* [emphasis added] 1.47-acre site across from the Santa Ana Regional Transportation Center" despite the fact that the council was informed in a public meeting of the presence of a number of existing residential and live-work users (City of Santa Ana 2014b, n.p.). The developer hired various consultants to undergo the relocation processes. According to semi-structured informal interviews with some of the tenants, the entire process included at least three different relocation agents and a number of entities. Thus, despite the fact that initial relocation documents given to tenants from C&C Development notified tenants of its plans to build affordable housing units soon after it had purchased the property in 2013, each interaction regarding relocation involved a different relocation agent who would consequently restart the entire process all over again, making the relocation process difficult and unnecessarily confusing for residents and tenants. Ultimately, according to La Bodeguita tenants, the developer provided a little over $1,000 in moving fees and a promise of a second round of reimbursement for reestablishment if residents found a new location within a year. Relocation documents not only reiterated a denial of two of the three residential users—"There are eight business tenants and one individual who stores personal property on-site . . . [and] one residential tenant" (City of Santa Ana 2015, 55A-11).—but also articulated an assurance that "the remaining businesses should easily be able to relocate into alternate commercial or light industrial spaces in Santa Ana" (City of Santa Ana 2015, 55A-13). However, even with the nonresidential relocation assistance, the self-described "artivist" La Bodeguita organization, which was defined as a "artist, urban planning/cultural events firm," did not have sufficient funds available to rent a comparable space at a similar location due to scarcities and rising costs (City of Santa Ana 2015, 55A-12). It is unknown what happened to the poor, marginally homeless tenants in the other unit after they left.

Finally, a lack of various types of formality complicated the relocation process and reinforced a particular intersectionality, which contributes to further marginalization and exclusion. Not only were tenants actually residing in the space despite its nonresidential industrial designation, but they also lacked formal business licenses and necessary documentation. In the case of La Bodeguita, the cultural work being done in the space, institutional documentation of the costs and expenses of cultural work, and lease agreements for all individuals inhabiting the unit were unattainable. With regard to the poor tenants occupying one of the units and using it for various informal economic subsistence activities, their severe poverty and informality allowed the relocation firm to deny them just compensation entirely, making meaningful reestablishment unlikely. Ultimately, a number of tenants residing at La Bodeguita were denied just relocation assistance due to a combination of federal legislation[8] that prohibits any benefits to individuals not lawfully present in the country as well as a disregard for the actual Relocation Plan submitted to the city, which stipulated that "The Developer will pay relocation benefits to otherwise eligible non-lawfully present persons with non-federal funds" (City of Santa Ana 2015, 55A-27). Furthermore, none of the residential tenants we spoke with were offered housing in the new development. In sum, this process underscores the relationship between direct and indirect exclusionary displacement that resulted from the Depot project—i.e., a process that displaces through the production of unaffordable income restrictions, cultural dislocation, and exclusionary documentation regimes that (re)produce marginality.

Conclusion

As Marcuse (1985) articulated, displacement operates through a number of forms; some of these are direct, while others are more complex and less immediately observable. The Depot at Santiago represents at least two forms of displacement. The first is direct. The Depot at Santiago led to the direct displacement of at least ten, and possibly more, residents and eight businesses. In addition to a local artist and two students who had converted their light industrial units to live-work spaces, a revolving number of poor residents had occupied an additional unit. None of these residents were considered residential tenants, and various mechanisms were used to deny appropriate relocation assistance. The relocation process furthermore demonstrates how displacement negatively affects those on the margins of structures of formality—either undocumented residents or informal businesses—excluding them from relocation assistance as well as a chance to be re-housed in the new development. At the same time, this case shows how the Depot at Santiago also furthers a form of Marcuseian exclusionary displacement where new housing units are no longer accessible to existing socioeconomic demographic groups. Not only will the new subsidized affordable housing project effectively raise rent levels in the neighborhood, making it more difficult for members of the severely overcrowded and rent-burdened low-income community to afford to live in the neighborhood, but the explicit "revitalization" effort that the Depot at Santiago falls within is intended to transform the neighborhood from its previous association with immigrant households from Mexico to a mixed-use regional TOD site more fully connected to adjacent live-work artist lofts, a civic center government employment district, and a rapidly gentrifying downtown. Thus, as the doors are opened for the new "gateway," fewer opportunities remain for those outside of this vision such as undocumented Mexicans, the lowest-income households, and the forms of cultural and familial reproduction that residents in these neighborhoods create.

Urban planners have the opportunity to set higher standards for subsidized affordable housing projects that may contribute to displacement, especially as gentrification increases the need for low-income housing in advantageous, centrally located neighborhoods. Raising standards for affordable housing development projects means increasing the affordability depth to meet the actual needs of both resident populations as well as those being displaced. However, at the same time, deepening affordability levels alone does not sufficiently address the question of displacement, as the production of affordable housing can also operate socio-spatially as a gentrifying force when connected to larger revitalization efforts. Linking affordable housing projects to visions that inherently dislocate the most vulnerable populations normalizes the displacement of individuals from the community and reaffirms exclusion. Understanding how ideologies help build consensus in order to mediate housing-based community struggles is critical for enabling planners to recognize the presence of and potential negative outcomes produced by an AHC. For example, many of the ideologies used to legitimate the Depot can be consdiered "common sense" planning objectives and include a predilection for concepts such as environmentalism, dense transportation-oriented development, and affordable housing that together supplant concerns over potential displacement. In addition to local measures of affordability that more accurately reflect income distributions, planners should be creative in applying relocation assistance so that the goals of fairness and equity are upheld. Meaningful resolutions affirming the right to housing and just cause eviction protections for all residential tenants should also be considered by planners and policymakers interested in alleviating some of the more harmful aspects of displacement (see also the chapter authored by W. Dennis Keating).

Notes

1 It is important to note that radical critiques of econometric and neoclassical understandings of the capitalist housing market emphasize not only the systematic failure to supply quality housing at the lowest income levels, but also the role of state intervention in order to return or facilitate the profitability of housing (Castells 1977, 162–169; Harvey 1982).
2 The convention of adding the "x" to the end of Spanish-language words in place of the masculine "o" or feminine "a" has evolved from LGTBQ struggles to escape binaries and introduce a level of gender consciousness into discourse on populations and social movements.
3 Grier's (1978) definition emerged as one of the earliest and most widely adopted articulations of the displacement process. In this work, the cause of displacement is reduced to a narrow landlord-tenant relationship that excludes factors outside these actors such as neighborhood changes and/or policy.
4 We excluded owner occupiers from all eligibile households for two reasons. First, these tracts are overwhelmingly renter-occupied. Second, we considered it highly unlikely that owner occupiers (even low-income ones) would vacate their housing and give up their tenure protections for subsidized rental housing.
5 Estimates of the percent of eligible renter households are approximations based on data available from the U.S. Bureau of the Census on income ranges for renter-occupied housing units (Table B25118) in the two Census tracts and AMI-based income limits published yearly by HUD. The tables from the U.S. Bureau of the Census provide data on the number of renter-occupied housing units within certain income ranges using $5,000, $10,000, and $25,000 increments. While these ranges approximate the AMI-based income limits, they do not correspond exactly. We calculated percentages of eligible renter households in each tract by taking the difference between U.S. Census income ranges and HUD AMI income limits and converting this difference into a percentage, which was then applied to the number of households. A certain amount of error is therefore assumed in these estimations.
6 Reinforcing other key relationships within the AHC, the California State Treasurer document also shows that tax credits for the Depot at Santiago were sold—through syndication—to the National Equity Fund, an affiliate of LISC.
7 California Relocation Assistance Law requires that relocation assistance be provided to all households or businesses that are displaced due to the action of any public entity in order to promote fair and uniform treatment.
8 Pursuant to U.S. Public Law 105–117, undocumented people in the United States are not eligible for relocation assistance.

References

Aglietta, Michel. 1979. *A Theory of Capitalist Regulation: The US Experience*. London: New Left Books.

Anonymous. 2016, July 12. Email communication.

Beard, Victoria A., and Carolina S. Sarmiento. 2014. "Planning, Public Participation, and Money Politics in Santa Ana (CA)." *Journal of the American Planning Association*, 80(2): 168–181.

Boyer, Robert. 1990 *The Regulation School: A Critical Introduction*. New York: Columbia University Press.C&C Development. n.d. "Terraces at Santiago."

California State Treasurer. 2015. Project Staff Report.

California Strategic Growth Council. 2015. Affordable Housing and Sustainable Communities (AHSC) Program.

Castells, Manuel. 1977. *The Urban Question: A Marxist Approach*. Cambridge: MIT Press.

Caumiant, Carly. n.d. "Award-Winning Affordability with Sustainable Design. Builder and Developer."

City of Santa Ana. 2009, October. *City of Santa Ana General Plan: Housing Element 2006–2014*. Santa Ana, CA: City of Santa Ana Planning Division.

City of Santa Ana. 2014a. "Community Meeting Presentation."

City of Santa Ana. 2014b. "Depot at Santiago."

City of Santa Ana. 2015. Council Resolution to Approve the Relocation Plan for the Depot at Santiago Project (Strategic Plan No. 5, 3C), November 3.

City of Santa Ana. n.d.a. "Community Meeting Presentation."

City of Santa Ana. n.d.b. "Depot at Santiago Fact Sheet."

Cox, Ronald W. 2014. "The Military-Industrial Complex and US Military Spending After 9/11." *Class, Race and Corporate Power*, 2(2): 1–20.

Davidson, Mark. 2009. "Displacement, Space/Place and Dwelling: Placing Gentrification Debate." *Ethics, Place and Environment*, 12(2): 219–234.

Davis, Angela Y. 1998. "Masked Racism: Reflections on the Prison Industrial Complex." *Colorlines Magazine*, 4(27): 4–7.

Delson, Jennifer. 2006. "New Homes Help Old in Changing Santa Ana Enclave." *Los Angeles Times*, January 1.

Delson, Jennifer. 2008. "Plan Seeks to Make Santa Ana the Heart of Orange County." *Los Angeles Times*, January 20.

Florida, Richard. 2003. *The Rise of the Creative Class*. New York: Basic Books.

Freeman, Lance. 2005. "Displacement or Succession? Residential Mobility in Gentrifying Neighborhoods." *Urban Affairs Review*, 40(2): 463–491.

Freeman, Lance, and Frank Braconi. 2004. Gentrification and Displacement in New York City in the 1990s." *Journal of the American Planning Association*, 70(1): 39–52.

Glaeser, Edward. 2005. *Smart Growth: Education, Skilled Workers and the Future of Cold-Weather Cities*. Cambridge, MA: Harvard University Press, Kennedy School, Public Policy Brief PB-2005-1.

Goetz, Edward G. 2008. "Words Matter: The Importance of Issue Framing and the Case of Affordable Housing." *Journal of the American Planning Association*, 74(2): 222–229.

Gonzalez, Erualdo, Carolina S. Sarmiento, Ana Siria Urzua, and Susan C. Luévano. 2012. "The Grassroots and New Urbanism: A Case from a Southern California Latino Community." *Journal of Urbanism*, 5(2–3): 219–239.

Gonzalez, Ron. 2013. "New Development Planned for Station District." *Orange County Register*, May 21.

Gramsci, Antonio. 1971. *Selections from the Prison Notebooks*. London: Lawrence & Wishart.

Grier, George W., and Eunice S. Grier. 1978. *Urban Displacement: A Reconnaissance*. Washington, DC: U.S. Department of Housing and Urban Development.

Hamnett, Chris. 1991. "The Blind Men and the Elephant: The Explanation of Gentrification." *Transactions of the Institute of British Geographers*, 16(2): 173–189.

Harvey, David. 1973. *Social Justice and the City*. London: Edward Arnold.

Harvey, David. 1982. *The Limits to Capital*. Chicago: University of Chicago Press.

Incite! Women of Color Against Violence. 2007. *The Revolution will not be Funded: Beyond the Non-Profit Industrial Complex*. Cambridge, MA: South End Press.

Jessop, Bob. 2002. "Liberalism, Neoliberalism, and Urban Governance: A State-Theoretical Perspective." *Antipode*, 34(3): 452–472.

Lees, Loretta, Tom Slater, Elvin Wyly. 2008. *Gentrification*. New Yok: Routledge.

Lipietz, Alain. 1987. *Mirages and Miracles: The Crisis in Global Fordism*. London: Verso.

Marcuse, Peter. 1985. "Gentrification, Abandonment, and Displacement: Connections, Causes, and Policy Responses in New York City." *Journal of Urban and Contemporary Law*, 28(1/4): 195–240.

Newman, Kathe, and Elvin Wyly. 2006. "The Right to Stay Put, Revisited: Gentrification and Resistance to Displacement in New York City." *Urban Studies*, 43(1): 23–57.

Nguyen, Mai Thi, Victoria Basolo, and Abhishek Tiwari. 2013. "Opposition to Affordable Housing in the USA: Debate Framing and the Responses of Local Actors." *Theory and Society*, 30(2): 107–130.

Orange County Lofts. n.d. "Santiago Street Lofts/Santa Ana/CA."

Reyes, Alpert E., and Ryan Menezes. 2014. "L.A. and Orange Counties are an Epicenter of Overcrowded Housing." *Los Angeles Times*, March 7.

Rose, Damiris. 1984. "Rethinking Gentrification: Beyond the Uneven Development of Marxist Urban Theory." *Environment and Planning D Society and Space*, 2(1): 47–74.

Sarmiento, Carolina S., and J. Revel Sims. 2015. "Façades of Equitable Development: Santa Ana and the Affordable Housing Complex" *Journal of Planning Education and Research*, 35(3): 323–336.

Sarmiento, Carolina S., and J. Revel Sims. 2016. *Lacy in Crisis and in Action: A Report on Housing and Tenant Conditions in the Lacy Neighborhood in Santa Ana, California*. Santa Ana, CA: People's Data Project.

Schwartz, Alex F. 2006. *Housing Policy in the United States: An Introduction*. New York: Routledge.

Slater, Tom. 2009. "Missing Marcuse: On Gentrification and Displacement." *City*, 13(2): 292–311.

Smith, Neil. 1979. "Toward a Thery of Gentrification a Back to the City Movement by Capital, not People." *Journal of the American Planning Association*, 45(4): 538–548.

Smith, Neil. 1996. *The Urban Frontier: Gentrification and the Revanchist City*. New York: Routledge.

Smith, Neil. 2002. "New Globalism, New Urbanism: Gentrification as Global Urban Strategy." *Antipode*, 34(3): 427–450.

Solomon, Arthur P. 1974. *Housing the Urban Poor: A Critical Evaluation of Federal Housing Policy*. Cambridge, MA: MIT Press.

Storper, Michael, and Allen J. Scott. 2009. "Rethinking Human Capital, Creativity and Urban Growth." *Journal of Economic Geography*, 9: 147–167.

U.S. Bureau of the Census. 2010. 2010 Census Summary File 1. "Hispanic or Latino by Type." Table QT-P10.

U.S. Bureau of the Census. 2014a. 2010–2014 American Community Survey 5-Year Estimates. "ACS Demographic and Housing Estimates." Table DP05.

U.S. Bureau of the Census. 2014b. 2010–2014 American Community Survey 5-Year Estimates. "Bedrooms by Gross Rent." Table B25068.

U.S. Bureau of the Census. 2014c. 2010–2014 American Community Survey 5-Year Estimates. "Gross Rent as a Percentage of Household Income in the Past 12 Months." Table B25070.

U.S. Bureau of the Census. 2014d. 2010–2014 American Community Survey 5-Year Estimates. "Place of Birth by Nativity and Citizenship Status." Table B05002.

U.S. Bureau of the Census. 2014e. 2010–2014 American Community Survey 5-Year Estimates. "Poverty Status in the Past 12 Months." Table S1701.

U.S. Bureau of the Census. 2014f. 2010–2014 American Community Survey 5-Year Estimates. "Tenure by Housing Costs as a Percentage of Household Income in the Past 12 Months." Table B25106.

U.S. Bureau of the Census. 2014g. 2010–2014 American Community Survey 5-Year Estimates. "Tenure by Occupants per Room." Table B25014.

U.S. Bureau of the Census. 2014h. "2010–2014 American Community Survey 5-Year Estimates." Table B25118.

U.S. Department of Housing and Urban Development. 2007. *Measuring Overcrowding in Housing*. Washington, DC: U.S. Department of Housing and Urban Development.

U.S. Department of Housing and Urban Development. n.d. *Income Limits*. Washington, DC: U.S. Department of Housing and Urban Development.

The Wooden Floor. 2016. "Revered OC Arts-Based Youth Development Nonprofit to Serve 100 More Youth in 2017."

7
NEIGHBORHOOD CENTERS

Emily Talen

It has often been thought that residential areas can be improved if they are given an identifiable center. A center can be used to transform housing from being merely residential land to being a bona fide neighborhood. This chapter traces the history and complications of that quest.

Planning for centers in new neighborhoods engenders a slightly different set of concerns than planning for centers in already existing neighborhoods as part of a neighborhood revitalization process. In the former case, planning for the neighborhood center is a matter of prescribing the right location, size, and functionality. In the latter case, it might be a matter of repurposing an existing space to create a neighborhood center where none currently exists. But whether planned in advance or fashioned retroactively, the purpose of a neighborhood center is essentially the same. On a social level, centers are thought to provide a means of connecting people—to one another, and to some larger common, public purpose. In this way they can be thought of as a physical articulation of "community"—a tangible, permanent symbol of the common bond that people living in the same neighborhood share. The center provides a place for spontaneous interaction, which, potentially, fosters a sense of social connection. In addition, centers may, over time, promote a sense of shared responsibility.

Centers are often thought of as being the actual geographic heart of a neighborhood, but that is not a strict requirement. However, if they can be positioned in a geographically central way—that is, in the center of a population—this is considered an efficient arrangement, because it maximizes the number of people with access to them. The twentieth century sociologist Clarence Perry, who was famous for articulating the theoretical basis of the neighborhood unit, maintained that the center should be equidistant from all residents, because that is the best way to maximize access (Patricios 2002; Perry 1929).

It is difficult to have a neighborhood center function as a center if it is excessively large, like a baseball stadium or a high school surrounded by acres of open space. The exception is where a large open space is surrounded by buildings. Ironically, one critique of Radburn, NJ, the iconic 1929 planned community that was based on Perry's neighborhood unit, was that it lacked centrality. Serge Chermayeff and Christopher Alexander (1963, 187) wrote that Radburn's centrality was ambiguous, because "the generous provision of communal park space raises new problems of ownership and responsibility."

In car-dominated cities, if large land uses make up what is generally regarded as the neighborhood's center, access to it may be low, because of all the surrounding land area required. Lower walking access requires the addition of parking lots, which further increases the amount of land area needed. For urban planners, large, publicly owned vacant parcels are often avoided as possible centers, because they lack good accessibility and are usually positioned not as centers, but as peripheral uses. Although centers can consist of any combination of land use, normative planning often contends that centers should provide daily life services for surrounding residents, ideally within walking distance (although this depends on whether the surrounding density is sufficient to support these services). In addition, there is often a public component. The public space incorporated in a center—intended explicitly for public use and publicly owned—can be formally designed as a green, a plaza, a square, or other formal space.

Historians have documented that centralized civic spaces have long played an essential role in neighborhood life. In an example from nineteenth century Sicily, neighborhoods traditionally consisted of households grouped around a partially enclosed courtyard, called a *cortile*, that functioned as an outdoor shared room. The *cortile* was the center of social life, the location of friendships, and the basis of the Sicilian concept of neighborhood. These were not family compounds. One's extended family lived outside the *cortile*, not within, giving meaning to the Sicilian proverb "your neighbor is your real relative," and providing a counter-narrative to the idea that kinship is all that mattered in southern Italy. Once immigrants landed in the dense tenement districts of New York City, however, absent the socializing force of the *cortile*, reliance on kin grew (Gabaccia 1982, 56).

In Western cities, interjecting a public space that could function as a center was seen as a way of countering what Peter Marcuse has called the "open grid," a system of land parcelization that became the basis of nineteenth century expansion and was focused solely on land speculation and consumption. The nineteenth century grid lacked a socialized notion of land value, where for centuries land became valuable only after a building was placed on it (Marcuse 1987). The center was an attempt to address that, and in the early twentieth century, city planners tried to create neighborhood identity in an otherwise endless grid system by inserting elaborate, centrally positioned civic spaces. Figure 7.1 shows Hegemann and Peets' Milwaukee Grand Circle, a center that was an effort to "insert a pleasant variation into an existing gridiron" (Hegemann and Peets 1922, 280).

A neighborhood center mitigates argumentation over boundaries by shifting the focus to the "heart" of the neighborhood rather than its external delimitation. Rather than being exclusionary, as

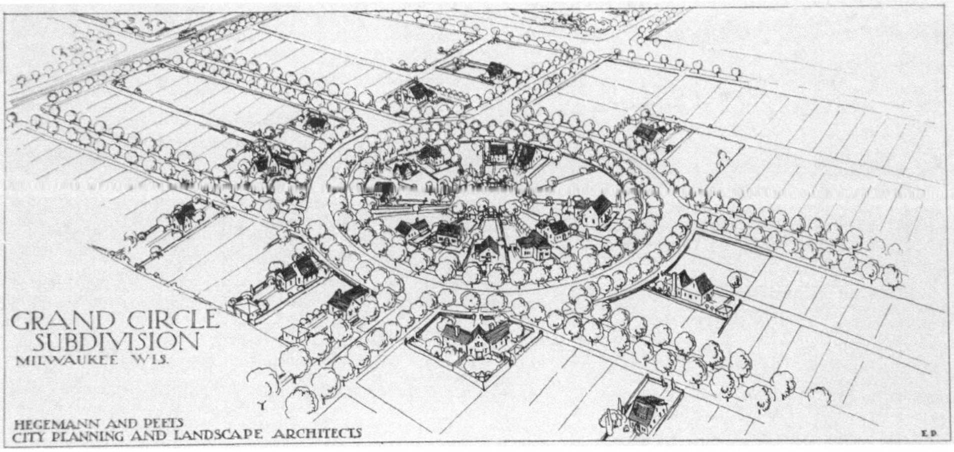

Figure 7.1 Hegemann and Peets' Milwaukee Grand Circle, an Effort to "Insert a Pleasant Variation Into an Existing Gridiron." In the early twentieth century, city planners tried to create neighborhood identity in an otherwise endless grid system by inserting elaborate, centrally positioned civic spaces.

Source: Hegemann and Peets 1922, 280. Public domain.

boundaries can be, centers are centripetal, providing a common, centrally located destination for surrounding residents. Clarence Perry had stressed that center and boundary were mutually reinforcing. Australian architect Gilbert Herbert implored planners to "worry less about the size and boundaries of neighborhood units and concentrate instead on the institutions and amenities which constitute the focal points of the area," stressing that design was essential for this to succeed. He lauded Walter Burley Griffin's plan for Canberra, because he believed it was able to work out a solution involving vague boundaries and strong centers recognizable for their repetitive pattern rather than their functionality (Herbert 1963, 184).

Planners later sought to elevate the neighborhood center over its boundary as one of two deviations from Perry (the other being that boundaries could become more accentuated by using green belts to surround and isolate neighborhoods, as in the case of the British New Towns; Herbert 1963). Then as now, the center is believed to provide a visible, literal focal point, as well as a permanent symbol of the common bond that people living in the same neighborhood share (Larice 2005). As Leon Krier put it graphically in Figure 7.2, in the urban quarter, a "beautiful boundary is a luxury"; a "beautiful center is a necessity" (Krier 1984, 71 2011). Resident surveys have given this center-over-boundary idea some legitimacy. For example, a survey of 322 people in a hilly area

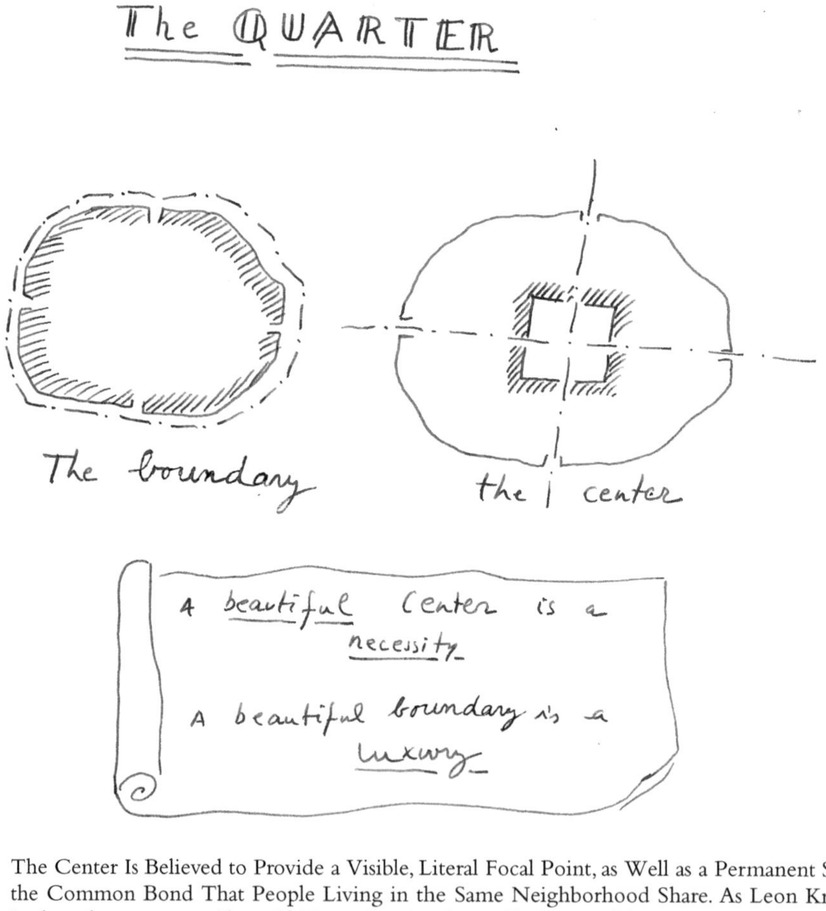

Figure 7.2 The Center Is Believed to Provide a Visible, Literal Focal Point, as Well as a Permanent Symbol of the Common Bond That People Living in the Same Neighborhood Share. As Leon Krier put it, in the urban quarter, a "beautiful boundary is a luxury"; a "beautiful center is a necessity."

Source: Leon Krier. Used with permission.

with curvilinear streets in Brisbane, Australia, asked residents about their neighborhood boundaries. Results showed weak agreement on boundaries, but strong agreement on the neighborhood's core (Minnery et al. 2009).

Of course, centers can sometimes function as boundaries. In the 1970s, it was suggested that neighborhoods could be defined by interviewing a few adolescent boys and understanding their perception of barriers, since it was an age and gender group that represented "a free ranging, diurnally tied, family member with little managerial, sexual or ceremonial restraint or responsibility." The method showed that schools were not really neighborhood centers at all, but were instead neighborhood boundaries (Bowden 1972, 228; Green 1956). In the adult world, enclaves can be formed from a core location outward, a magnet that attracts by way of goods and services, and that has the same exclusionary outcome as a hard boundary (Abrahamson 1996).

Physically, the historical precedent for a neighborhood center was the central square. Spanish colonial town planning required them, William Penn established them for Philadelphia, James Oglethorpe for Savannah, and hundreds of courthouse squares embodied the physical expression of community life. In a slightly loosened interpretation, Lewis Mumford argued that neighborhoods needed "some point of focus" outside the home, which could be a health center, a settlement house, a school, a community center—these were "all worthy attempts." Patrick Geddes suggested that neighborhoods should set aside a "handsome historic house" that families could rent for parties, while in England a room attached to a pub was considered an appropriate focal point for the neighborhood (Mumford 1949, 12). French historians pointed to the influence of the "public monument" on defining a neighborhood's layout, life, and "aspect," and public buildings were conceived as the "factor of movement" that helped neighborhoods form around them (Bardet 1951, 249).

In the U.S., centers associated with the Settlement House neighborhood took the form of a building or group of buildings. Jane Addams's settlement in Chicago was known as the Chicago Commons Buildings, serving, for example, as neighborhood meeting place, health clinic, school, and communal kitchen. In a manner, Jane Jacobs would have agreed; it was also a place where the neighborhood connected with ward aldermen/alderwomen and other city politicians in an effort to strengthen its political impact. Above all, the Chicago Commons was "a center for a higher civic and social life" in a diverse industrial district. The Commons was meant to bridge diversity and provide a way "to understand, to interpret and to cooperate" among "strangers in a strange land" (Chicago Commons Association 1920, 8, 31). Later in the 1960s, planners believed that neighborhood centers would be strongest in areas with "concentrations of minority groups," for example, "a Buddhist Church in an oriental neighborhood, a bocce ball court in an Italian area." The "elementary school-common green-community center complex" was an attempt to replicate the magnetism of ethnically specific center functions with only mixed success (American Society of Planning Officials 1960, 14).

According to historian Howard Gillette, the neighborhood center as a deliberate planning idea was launched in St. Louis' plan of 1907. Prior to that, under the reign of the City Beautiful movement, centers were larger, serving the needs of an entire city. The St. Louis plan was the first to decentralize, providing a "neighborhood approach [to] civic concerns," advanced by placing a half-dozen centers throughout the city (Gillette 1983, 423). Of note is Henry Wright, who together with Clarence Stein planned Radburn, was instrumental in developing the St. Louis neighborhood civic center concept. The center was supposed to be a neighborhood space that took on "the characteristics of a centrally located magnet that exerts its influence radially" (American Society of Planning Officials 1960, 14). But in a recent turn toward pragmatism, as well as a shift from institutionally based to retail-based centers, some planners now advocate putting the neighborhood center along a busy thoroughfare so that it has a better chance of surviving.

In the 1920s, neighborhood centers formed around playgrounds or recreation buildings, famously in Chicago and Washington, DC, where such venues went beyond the mere provision of equipment and were explicitly intended to "promote neighborhood consciousness" (Peets 1931, 230). But it was the school as center that resonated most. Even Ebenezer Howard located a school at the center of each ward in his 1898 garden city diagram. Perry was under the influence of the social center movement, which advocated in the early nineteenth century that schools should function as neighborhood centers complete with adult education and town forums. Edward Ward originated the concept in 1907 in Rochester, New York, and was committed to the idea of using neighborhood schools for multiple community functions, an idea also advocated by Jacob Riis. Ward wrote in 1914: "The whole United States is divided into two units of neighborhood. One is the voting precinct, the other is the public school district" (Ward 1914, 519). The movement sought to merge the two, such that voting would take place exclusively in public schools.

These Progressive Era reforms marked the beginning of a long tradition in the U.S. of trying to use schools as centers. The school was part of the community center at the heart of Clarence Perry's 1929 neighborhood unit idea. Perry's school-centered neighborhood was meant to counter the "centerless tendency" (Warner 1962, 158) of urban expansion prior to the 1920s, and as such it was an attempt to reverse what was mostly a transportation based method of city-building. The U.S Department of Education made some effort to strengthen the school-as-center tradition and issued a publication in 2000 called *Schools as Centers of Community: A Citizens' Guide for Planning and Design*, summarizing the benefits and laying out policy directives.

In his 1914 book *The School as a Factor in Neighborhood Development*, Perry laid out all the ways that schools were essential for civic life. There was solid evidence, he wrote, that "throughout the country there is a growing tendency to resort to school buildings for deliberation upon matters of civic and political import" (Perry 1914, 3). Perry detailed the ways in which the "school plant" was being physically altered to accommodate voting booths, museums, employment bureaus, branch libraries, art galleries, and theatrical productions (as Perry wrote, the novelty of movable chairs was making it easy to transform the school into something else). In some cases, these school transformations corresponded to neighborhood identity, but if school administration was centralized rather than school-based, the connection to the neighborhood seemed more ad hoc (Ward 1914). On the other hand, national organizations were supportive: The American Public Health Association's Committee on the Hygiene of Housing published *Planning the Neighborhood: Standards for Healthful Housing* in 1940 to argue that an elementary school-based neighborhood was the source of physical, social, and emotional health (American Public Health Association 1948).

The centrality of the school in neighborhood identity building was even more pronounced in rural settings. One 1923 study of rural neighborhoods in New York State delineated areas of "human association" by working with rural school teachers. Teachers had their students take a questionnaire home, and each family was asked to identify what their "country neighborhood" was. The teachers then transferred these descriptions and names to a map. Where there was sufficient overlap in the 150 returned questionnaires, a neighborhood was indicated. Using a definition of neighborhood as a "geographic group" beyond the family that "is conscious of some local unity," neighborhood identity seemed mostly based on the school; one-half of the neighborhoods were identical in name and geographic area to the school district. On the other hand, social functioning was mostly a matter of helping out with farm projects (harvesting) or exchanging tools, as opposed to connection via organized social events. The authors of the study concluded: "In general, the rural neighborhood [is] ceasing to function as a social unit except where its life is centered in some local institution" (Sanderson and Thompson 1923).

But rural schools, abandoned via school consolidations, were increasingly unable to serve that function (Holt 1940). Proponents of rural neighborhoods seemed to understand that neighborhoods needed to have centers and that "centers need accommodations, buildings, which even the local farmer can feel are his very own" (Kolb 1921, 105). A Wisconsin plan called for establishing cooperative agricultural communities called "neighborhoods," and similarly argued that neighborhoods needed to have a physical focal point in the form of a neighborhood center (Williams 1912). County employees in rural 1940s North Carolina were specifically charged with identifying neighborhood centers where meetings could take place (Holt 1940).

In more urban areas, the historical emphasis on schools as neighborhood congealers now has to be weighed against the reality that schools are not the social integrators they once were. A serious problem in the U.S. is the association between school quality and wealth, where the "good school" attracts affluence and virtually defines neighborhood segregation. Added to this, the ability of a school to function as a center is compromised, because many public schools are more school board-based than neighborhood-based, thus losing their incentive for neighborhood connection (Sanchez-Jankowski 2008). Schools are a far cry from their position as centers of "neighborhood life" that an earlier generation viewed as fundamental to a child's education. Planners had previously focused on the "school-neighborhood nucleus," arguing that "every neighborhood planner is an educator," and that "the best education is the result of a well-conceived neighborhood plan in which the school has been created as an integral part of the daily life of all the people who reside in the community" (Engelhardt 1943, 89). Now, the inter-relation of school and neighborhood is spatially and programmatically ambiguous. School consolidation trends as well as modern site design requirements (having to do with parking and the size of playing fields, for example) have made the connection especially problematic (Norris 2015). This is not just a U.S. problem. India ran into problems when it tried to implement a Right to Education (RTE) Act that required every child aged 6 to 14 be provided education "in an age-appropriate classroom in the vicinity of his/her neighborhood." Turmoil ensued when it was realized that India had no definition of neighborhood (*The Hindu* 2011).

If the neighborhood is diverse in terms of housing types and land uses, a neighborhood center is believed to be a good way to link and integrate the mix, which might otherwise lack any deliberate, meaningful connection in terms of physical design. One idea is that if diverse housing types front the neighborhood center equally, they may be more likely to be equally acknowledged as important, even if only symbolically (Duany 2013). Related to this are studies documenting the special importance of neighborhood centers—shops, schools, bars, streets, and health centers—in multicultural and multilingual urban neighborhoods (Blommaert et al. 2005). In heterogeneous neighborhoods social connection via homogeneity cannot be relied on, so the emphasis shifts to public space rather than social relationships as a way of countering alienation. In the Netherlands, sense of neighborhood was found to be especially important for establishing social contacts in unfamiliar terrain, prompting a policy recommendation to create more "meeting places in public spaces" (Bloem et al. 2013, 44).

Neighborhood space as central meeting ground is used to purposefully draw divergent populations together. A project called Barriers Not Included to transform a park in Charlotte, NC, between two very different neighborhoods had the purpose of creating a common area that would use light, sound, and play "to stimulate conversation." The project's designers believed that the transformed park, which sits at the border of the two neighborhoods, would be able to "build relationships and connections" (Knight Foundation 2015, n.p.).

In high-crime neighborhoods, the neighborhood center takes on the added role of being a safe haven. Neighborhood schools, churches, or other forms of center are protected places that have been shown to be significant factors in a child's ability to cope with violence, reduce anxiety, and increase resiliency (Osofsky 1999). Churches play a special role as neighborhood centers, and their form matters. One study showed that storefront churches, as opposed to freestanding churches, were valuable in poor neighborhoods, because, positioned near commercial uses, they offered "a level of continuity and favorable visibility" that was able to provide "material benefit" to other adjacent properties" (Kinney and Winter 2006, 348).

There is a certain stability that accompanies the idea that a place of worship is a neighborhood center, especially if the spatial connection between church and church-goer can be maintained. One study contrasting "resident" versus "commuter" churches in Philadelphia found that the former tend to be within neighborhoods that are more stable in terms of household income, homeownership rates, change of residence, population density, and racial make-up. Member proximity to their place of worship, in other words, is related to neighborhood stability. The obvious challenge is to avoid the situation where over-protection of the church-based neighborhood leads to exclusion (Sinha et al. 2007).

Another update of the neighborhood center involves connecting multiple centers to create a series of linked focal points. So in addition to drawing residents in, centers facilitate movement through a neighborhood, and potentially enhance the connections between them. One proposal argued for the installation of "artifacts" along "pathways," providing meaningful destinations that not only motivate walking, but enhance neighborhood identity (Hess et al. 2013, 52). Focal points can also be a kind of territorial marker that informally regulates and enhances interaction. Conceptually, this parallels the decorations on housing units that promote social encounter—like flower boxes or porch decorations (Greenbaum and Greenbaum 1981). A downscaled version of the neighborhood center might be similarly small, like gardens or renovated houses, or perhaps the result of an urban homesteading program in which vacant property is transferred to communal ownership (I Grow Chicago 2017). Neighborhoods that lack such memorable places are not benign—they can exert negative social effects, like alienation (Brown et al. 2003; Ward 1979).

An idea from an earlier generation (particularly Edward Ward's community center movement) that has never died out is that centers—or more simply, a neighborhood's primary facilities—should serve more than one function. A neighborhood center should support multifunctionality, because multi-use is neighborhood-supporting, broadening the reach of a single building or place. Examples are multifunctional schools, multifunctional parks and libraries (that service the recreational or library needs of schools as well as the general public), senior centers connected to elementary schools or nursery schools, or school auditoriums functioning as community theaters. The relationship is reciprocal: Neighborhood consciousness makes mixed services possible; mixed services elevate neighborhood consciousness.

Yet mostly, proposals for what the neighborhood center should consist of have expanded in recent years, such that the center might be conceived more often as a full-blown mixed-use retail destination than a neighborhood school. The urban planning firm Dover Kohl suggested that the neighborhood center should be not only mixed use but relatively high density (although ideally limited to four stories outside of a metropolitan core), and take the form of a square, plaza, or simply the "four corners" of a prominent street intersection. The firm of Duany, Plater-Zyberk suggested that the neighborhood center should consist of neighborhood shops, institutions, and a bus stop, with the elementary school transferred to the edge to make it accessible to adjoining neighborhoods. Architect Doug Farr provided a further update by calling for a "higher intensity transit mode" (BRT, trolley, light rail) at the center (Farr 2008, 126; Plater-Zyberk 2000).

Recent research has revealed that even in what might be considered exemplary neighborhoods—i.e., those that are compact, diverse, and walkable—neighborhood centers, if specifically defined, tend to have a commercial as opposed to an institutional basis. One study of 79 award winning neighborhoods in the U.S. classified neighborhood centers as either commercial, open space, institutional, or a mix of the three (Talen et al. 2015). The study found that only about half of the neighborhoods had readily identifiable centers, and that the overwhelming majority of those (75 percent) were classified as commercial centers. Only 11 percent had institutional elements. America's best neighborhoods, the study concluded, either do not have clear centers or have centers that are mostly commercial and not oriented around communal or institutional uses.

In this light, more effort could be put into nurturing neighborhood center definition and designation. This necessarily starts with identifying the resources at hand: What uses, public or private, currently function as a neighborhood's center? If there are no obvious choices, a starting point could be to identify the various definitions of neighborhood that might exist, each of which might give a clue about what the center might be: the post office definition of neighborhood, the police precinct definition of neighborhood, the school board definition of neighborhood. As a practical matter, an initial charge for residents might be to catalog where and in what condition all publicly owned space (land and buildings) in the neighborhood is located. Once this is known, residents could evaluate what the neighborhood's requirements for public space are and whether those needs are being met. Is there a public space that acts as a center? Are the requirements for children's play, senior activities, or family recreation understood and accommodated? From there, neighborhood residents might also develop an understanding of all public actions affecting their neighborhood centers. This includes zoning and other regulations, street "improvements" and closings, infrastructure investments, and policies like tax increment districts. This understanding is needed if centers are to be adapted to resident needs (Sorlien 2015).

These practical needs can be juxtaposed against a more fundamental idea about centers: that they are necessary for a collective identity, potentially nurturing a sense of belonging needed to address the "crisis of connection" modern life faces (New York University 2015). It might even be possible to activate the social diversity implications of a neighborhood center that Lewis Mumford recognized: "the spotting and inter-relationship of schools, libraries, theaters, and community centers is the first task in defining the urban neighborhood and laying down the outlines of an integrated city" (Mumford 1937, 187).

References

Abrahamson, Mark. 1996. *Urban Enclaves: Identity and Place in America. Contemporary Social Issues*. New York: St. Martin's Press.

American Public Health Association. 1948. *Planning the Neighborhood: Standards for Healthful Housing*. Chicago: American Public Health Association.

American Society of Planning Officials. 1960. *Neighborhood Boundaries*. Chicago: American Society of Planning Officials.

Bardet, Gaston. 1951. "Social Topography: An Analytico-Synthetic Understanding of the Urban Texture." *The Town Planning Review*, 22(3): 237–260.

Bloem, Brigitte A., Theo G. Van Tilburg, and Fleur ThomÉse. 2013. "Starting Relationships with Neighbors after a Move Later in Life: An Exploratory Study." *Journal of Housing for the Elderly*, 27(1/2): 28–47.

Blommaert, Jan, James Collins, and Stef Slembrouck. 2005. "Polycentricity and Interactional Regimes in 'Global Neighborhoods.'" *Ethnography*, 6(2): 205–235.

Bowden, Leonard W. 1972. "How to Define Neighborhood." *The Professional Geographer*, 24(3): 227–228.

Brown, B., Perkins, D., and Brown, G. 2003. "Place Attachment in a Revitalizing Neighborhood: Individual and Block Levels of Analysis." *Journal of Environmental Psychology*, 23(3): 259–271.

Chermayeff, Serge, and Christopher Alexander. 1963. *Community and Privacy: Toward a New Architecture of Humanism*. Garden City, NY: Doubleday.

Chicago Commons Association. 1920. *Chicago Commons. Twenty-Fifth Year, 1894–1919*. Chicago: Chicago Commons Association.

Duany, Andres. 2013. "Commentary." In *Charter of the New Urbanism*, edited by Emily Talen, 231–238. New York: McGraw-Hill Professional Publishing.

Engelhardt, Nickolaus L. 1943. "The School-Neighborhood Nucleus." *Architectural Forum*, 10(4): 88–90.

Farr, Douglas. 2008. *Sustainable Urbanism: Urban Design with Nature*. Hoboken, NJ: Wiley.

Gabaccia, Donna R. 1982. "Sicilians in Space: Environmental Change and Family Geography." *Journal of Social History*, 16(2): 53–66.

Gillette, Howard. 1983. "The Evolution of Neighborhood Planning from the Progressive Era to the 1949 Housing Act." *Journal of Urban History*, 9(4): 421–444.

Green, Norman E. 1956. "Aerial Photographic Analysis of Residential Neighborhoods: An Evaluation of Data Accuracy." *Social Forces*, 35(2): 142–147.

Greenbaum, Paul E., and Susan D. Greenbaum. 1981. "Territorial Personalization Group Identity and Social Interaction in a Slavic-American Neighborhood." *Environment and Behavior*, 13(5): 574–589.

Hegemann, Werner, and Elbert Peets. 1922. *The American Vitruvius: An Architects' Handbook of Civic Art*. New York: De Facto Publishing.

Herbert, Gilbert. 1963. "The Neighbourhood Unit Principle and Organic Theory." *Sociological Review*, 11(2): 165–213.

Hess, Daniel, Hiroaki Hata, and Ernest Sternberg. 2013. "Pathways and Artifacts: Neighborhood Design for Physical Activity." *Journal of Urbanism: International Research on Placemaking and Urban Sustainability*, 6(1): 52–71.

Holt, John Bradshaw. 1940. *Report of a Reconnaissance Survey of Neighborhoods and Communities of Caswell County, North Carolina with Recommendations*. Washington, DC: U.S. Department of Agriculture.

I Grow Chicago. 2017. *Our Peace House*. Chicago: I Grow Chicago.

Kinney, Nancy T., and William E. Winter. 2006. "Places of Worship and Neighborhood Stability." *Journal of Urban Affairs*, 28(4): 335–352.

Knight Foundation. 2015. *Informed and Engaged Communities Challenge*. Miami: Knight Foundation.

Kolb, John H. 1921. *Rural Primary Groups: A Study of Agricultural Neighborhoods*. Unpublished Ph.D. diss., University of Wisconsin in Madison.

Krier, Léon. 1984. *Léon Krier, Houses, Palaces, Cities*. New York: St. Martin's Press.

Krier, Léon. 2011. *The Architecture of Community*. Washington, DC: Island Press.

Larice, Michael Angelo. 2005. *Great Neighborhoods: The Livability and Morphology of High Density Neighborhoods in Urban North America*. Unpublished Ph.D. diss., University of California, Berkeley.

Marcuse, Peter. 1987. "The Grid as City Plan: New York City and Laissez-Faire Planning in the Nineteenth Century." *Planning Perspectives*, 2: 287–310.

Minnery, John R., Jon Knight, John Byrne, and John Spencer. 2009. "Bounding Neighbourhoods: How Do Residents Do It?" *Planning Practice and Research*, 24(4): 471–493.

Mumford, Lewis. 1949. "Planning for the Phases of Life." *Town Planning Review*, 20(1): 5–16.

Mumford, Lewis. 1937. "What Is a City?" *Architectural Record*, 82(November): n.p.

New York University. 2015. *The Project for the Advancement of our Common Humanity (PACH)*. New York: The Project for the Advancement of Our Common Humanity.

Norris, Nathan. 2015. "The Neighborhood School." In *Charter of the New Urbanism*, edited by Emily Talen, 158–159. New York: McGraw-Hill Professional Publishing.

Osofsky, J. D. 1999. "The Impact of Violence on Children." *The Future of Children, Center for the Future of Children, the David and Lucile Packard Foundation*, 9(3): 33–49.

Patricios, Nicholas. 2002. "Urban Design Principles of the Original Neighborhood Concepts." *Urban Morphology*, 6(1): 21–32.

Peets, Elbert. 1931. "Current Town Planning in Washington." *Town Planning Review*, 14(4): 219–237.

Perry, Clarence Arthur. 1914. *The School as a Factor in Neighborhood Development*. New York: Russell Sage Foundation.

Perry, Clarence Arthur. 1929. *Neighborhood and Community Planning. Regional Survey of New York and Its Environs*. New York: Regional Plan of New York and Its Environs.

Plater-Zyberk, Elizabeth, Gianni Longo, Peter J. Hetzel, Robert Davis, and Andres Duany. 2000. *The Lexicon of the New Urbanism*. Miami: Duany, Plater-Zyberk.

Sanchez-Jankowski, Martin. 2008. *Cracks in the Pavement*. Berkeley, CA: University of California Press.

Sanderson, E. Dwight, and Warren Thompson. 1923. *The Social Areas of Otsego County*. Ithaca, NY: Cornell University Press.

Sinha, Jill Witmer, Amy Hillier, Ram A. Cnaan, and Charlene C. McGrew. 2007. "Proximity Matters: Exploring Relationships among Neighborhoods, Congregations, and the Residential Patterns of Members." *Journal for the Scientific Study of Religion*, 46(2): 245–260.

Sorlien, Sandy. 2015. *Neighborhood Conservation Kit*. Blog.

Talen, Emily, Sunny Menozzi, and Chloe Schaefer. 2015. "What Is a 'Great Neighborhood'? An Analysis of APA's Top-Rated Places." *Journal of the American Planning Association*, 81(2): 121–141.

The Hindu. 2011. "The Hindu (English): RTE Act: Definition of 'Neighbourhood' Unclear." Chennai, India, February 28.

Ward, Colin. 1979. *The Child in the City*. New York: Pantheon.

Ward, Edward J. 1914. "Where Suffragists and Anti's Unite." *American City*, 10(January–June): 519–524.

Warner, Sam Bass. 1962. *Streetcar Suburbs: The Process of Growth in Boston, 1870–1900*. Cambridge, MA: Harvard University Press.

Williams, Albert L. 1912. *A Plan for a Co-Operative Neighborhood*. Madison, WI: State Board of Public Affairs.

SECTION 3

Homeownership

8

DEMOGRAPHIC AND POLICY FACTORS AFFECTING HOMEOWNERSHIP ACROSS OECD COUNTRIES

Mark Calabria and Vanessa Brown Calder

Introduction

There are perhaps few areas of the economy more directed by government supervision and subsidy than the housing market (Bramley 2003; Malpezzi 2005). A number of countries, including the United States, directly provide housing to eligible citizens, while almost all offer some sort of rental or homeownership subsidy, often via the tax code. Most countries provide some limited supply of publicly owned rental housing, such as council housing found in the United Kingdom or *Habitation à Loyer Modéré* of France (Blanc 2010). Obviously, the construction of housing remains highly regulated, from land use to building codes. Housing also played a dominant role in the recent Great Recession, more so than any other asset class. Housing represents the largest single asset holding for most owner-occupied households, and shelter is universally considered a basic need (United Nations 1948, see also the chapter authored by W. Dennis Keating). For these and other reasons, policymakers, economists, demographers, and others devote considerable attention to monitoring housing outcomes, including homeownership and homeownership rates.

We begin this chapter with an in-depth analysis of homeownership trends in the United States and examine important homeownership policy issues, such as racial and ethnic differences in homeownership, and review federal policies to promote homeownership. We then move to a comparison of homeownership rates among members of the Organization for Economic Co-operation and Development (OECD). We present descriptive statistics across countries, primarily as a panel but with some time series, where data permits. We will discuss cross country differences for select demographic, economic, and policy factors, focusing on housing finance and affordability.

We then focus on the relationship between homeownership and demographic factors such as age, marriage, residential mobility, and urbanization, which tend to strongly correlate with homeownership. We then discuss correlations between homeownership and public policy, arguing that the mortgage interest tax deduction has little impact on the homeownership rate, while other policies such as taxes on imputed rent and the mass privatization of public housing do correlate with changes in homeownership rates.[1] We discuss the United States as a case study to highlight the greater importance of the relationship between demographics and homeownership compared to policy factors. We conclude that in most cases homeownership rates are primarily influenced by demographic and economic factors rather than public policies. Outliers are the result of unique policy choices, such as the taxation of imputed rent in Switzerland.

Homeownership in the United States

The United States has experienced a number of distinct phases in regard to homeownership (see Figure 8.1). The first phase was the colonial period to the Civil War (Collins and Margo 2001, 2011). The second phase, 1865 to 1900, witnessed a modest decline in the homeownership rate among White households but a large increase in the homeownership rate among African Americans, from around 8 percent in 1870 to just over 20 percent in 1900 (Collins and Margo 2001, 2011). With the exception of very small programs for shipbuilders and war production workers in 1918, federal policy toward homeownership was largely non-existent from 1870 to 1890, although some private and local government initiatives did exist (Anacker 2015).

The third phase of homeownership witnessed the birth of various New Deal housing programs, although the homeownership rate remained relatively flat during the 1930s and 1940s due to the Great Depression and the following World War II. While New Deal programs made it easier for middle-income families to obtain mortgages, the effects only became apparent after World War II (Anacker 2015).

The fourth phase of homeownership began at the end of World War II and ended in 1960, and led to the development of the interstate highway system, extensive rent controls in a number of cities, and the creation of the Veterans Home Loan program. Fetter (2013b: 29) estimates that "at least 10 percent of the urban increase in homeownership over the first half of the 1940's [sic]" is the result of rent control as this policy motivated landlords to sell at uncontrolled prices rather than renting at the controlled price.

During this phase, the homeownership rate increased from around 45 percent to 62 percent. Fetter (2013a) estimates that 7.4 percent of the overall increase in homeownership from 44 percent in

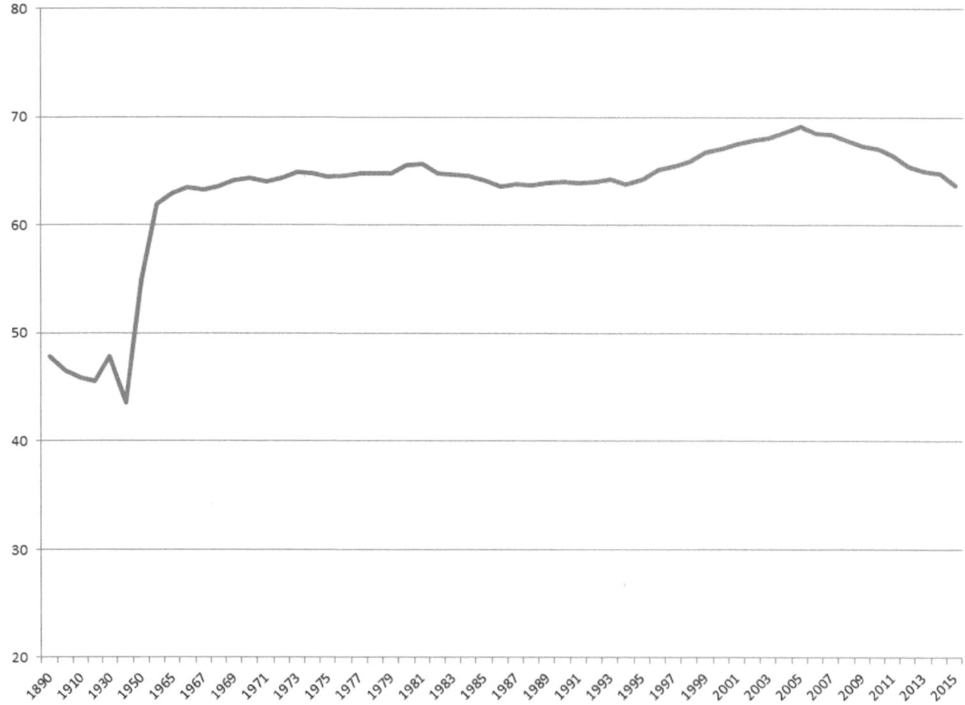

Figure 8.1 Homeownership Rate, United States, 1890–2015.

Source: U.S. Bureau of the Census.

1940 to 62 percent in 1960 resulted from the establishment of the Veterans Home Loan program, granted under World War II and Korean War GI bills (Fetter 2013a). Other scholars attribute the increase in homeownership during this period to innovations in mortgage finance and demographic shifts (Chambers et al. 2016).

Between 1994 and 2005 the homeownership rate increased substantially (see Table 8.1). For people younger than 25, the rate increased by nearly 11 percentage points, compared to around 2 percentage points for households aged 55 to 64. This increase was partly explained by the relaxation of mortgage underwriting standards. For example, down payment requirements declined from more than 20 percent in the 1980s to less than 20 percent in the 1990s (Acharya et al. 2011). However, the homeownership rate has decreased after 2005, especially among young households, who typically have lower credit scores (Fry and Brown 2016).

Since the mid-1990s, however, the African-Americanhomeownership gap has grown despite modest nominal gains in the African-Americanhomeownership rate and despite the Community Reinvestment Act, the Equal Credit Opportunity Act, and the Fair Housing Act, which center on reducing racial and ethnic homeownership gaps. Historically, non-Hispanic White and Asian households have had substantially higher homeownership rates than other racial and ethnic groups, for example, African Americans and Hispanics/Latinos (Russell Sage Foundation 2012; U.S. Bureau of the Census 1993). As of this writing, the homeownership rate for non-Hispanic White households is 72.9 percent, more than 30 points higher than that of African-Americanhouseholds at 41.6 percent. Since the Great Recession, African-Americanhomeownership rates have been at low levels that have not been seen since 1970. Hispanic/Latino homeownership rates have largely mirrored those of African-Americanhouseholds. In 2017 the Hispanic homeownership rate was 46.2 percent (Calderon 2017). Asian homeownership rates have tended to be about 10 percentage points higher than those for Black and Hispanic households. In 2018, the Asian, Native Hawaiian, and Pacific Islander homeownership rate was 57.3 percent (U.S. Bureau of the Census 2018).

Since the 1970s scholars have attempted to empirically estimate the drivers of the racial and ethnic homeownership gap (Coulson 1999; Gabriel and Rosenthal 2005; Jackman and Jackman 1980). While factors such as age, household structure, household income, and housing costs, among others, explain most of the difference, racial and ethnic variables usually show statistical significance. Whether discrimination partially explains different homeownership outcomes among racial and ethnic groups remains a point of continued debate (Bayer et al. 2014; Black et al. 2015).

For example, Hilber and Liu (2007) find that the Black-White homeownership gap disappears after controlling for demographic and socioeconomic characteristics including household income, age of the head of the household, household structure, number of years since household formation, education, wealth, parental wealth, and location preferences. The racial and ethnic homeownership gap will ultimately depend upon future trends in household income and wealth, marriage, education, labor force participation, and immigrant status (Anacker 2015).

Table 8.1 Homeownership Rate in the United States by Age, Race, and Ethnicity, 1994, 2004, and 2014.

	All Households	*Age of Householder*					*Race/Ethnicity of Householder*				
		Under 35	*35–44*	*45–54*	*55–64*	*65 and Over*	*White*	*Hispanic*	*Black*	*Asian Other*	*All Minority*
1994	64.0	37.3	64.5	75.2	79.3	77.4	70.0	41.2	42.5	50.8	43.2
2004	69.0	43.1	69.2	77.2	81.7	81.1	76.0	48.1	49.7	59.7	51.0
2014	64.5	35.8	59.7	70.7	76.3	79.9	72.6	45.4	43.8	56.7	47.2

Source: U.S. Bureau of the Census.

Homeownership in the OECD Countries

Despite homeownership being characterized as part of the American Dream, the homeownership rate in the United States is surprisingly similar to that of other OECD countries. Even at the peak of the U.S. homeownership rate at 69 percent in 2004, half of OECD countries still exceeded this rate, with the (unweighted) OECD average of 70 percent. Only six of the current 34 OECD countries have homeownership rates below that of the United States, which was most recently measured at 64 percent (see Table 8.2), while half a dozen OECD countries have rates clustered around that of the United States.

The distribution of OECD homeownership rates closely resembles a normal distribution with a few interesting outliers. Most prominent is that of Switzerland at 44 percent, followed by Germany at 52.5 percent. At the other extreme is Slovakia at 90.3 percent, followed by Hungary at 89.1 percent, both of which are formerly Communist countries where mass housing privatization programs propelled homeownership rates.

In sum, homeownership rates across the OECD appear to be clustered by region. The highest rates are found in Central and Eastern European, often among former Communist Bloc states. Next are those countries located in Southern Europe. The English speaking countries of the OECD fall in the middle, while the Nordic and Continental European countries have the lowest rates of homeownership. Obviously, there are exceptions that stand out. The biggest exceptions to these regional patterns are Norway (with 84 percent), which far exceeds other Nordic countries, and Belgium (72 percent) and Luxembourg (73 percent), which significantly exceed other countries of Continental Europe. The Pacific members of the OECD (Japan and South Korea) also display some of the lowest homeownership rates (61 and 54 percent, respectively). These differences will be discussed later.

This chapter focuses on the 34 members of the OECD due to the greater availability of longitudinal data for OECD countries.[2] The United Nations Center for Human Settlements (UNCH) has periodically collected data for a larger number of countries with data going back to 1993. The first release of UNCH data provided homeownership rates for 106 countries. Interestingly enough, the distribution of homeownership rates in the UNCH data is similar to that found among OECD countries (Fisher and Jaffe 2003), suggesting that a focus on the OECD will have some lessons to offer for non-OECD countries.[3]

This chapter focuses on OECD countries for several reasons. First, while data on homeownership are broadly available, measures of some important drivers of homeownership are not. Country-specific legal, cultural, political, and economic characteristics diverge more beyond the OECD. Second, housing characteristics are more comparable within the OECD than outside of it. For example, the average size of a home in Australia, Canada, and the United States (all over 2,000 square feet) is more than twice that found in the relatively wealthy countries of Germany, France, or Sweden. However, it is more than four times the size of an average house in India. Housing quality is even more difficult to compare within the OECD and between the OECD and other countries. Thus, we limit our discussion to the 34 OECD countries.

Homeownership Rates Since the 1980s

The change in the homeownership rates among OECD countries since the 1980s is worth exploring. Around the time of the fall of the Berlin Wall countries in Central and Eastern Europe, particularly in Slovakia, Poland, Hungary, Estonia, Slovenia, and the Czech Republic, had average homeownership rates that were around 15 percentage points lower than today. This increase over a quarter century is remarkable, similar to the increase that occurred in the United States between the end

Table 8.2 Homeownership Rates Across OECD Countries.

Country	*Most Recent Year*	*Homeownership Rate Most Recent Year*	*Homeownership Rate 1990*	*Homeownership Rate 1980*
Australia	2013/2014	67%	71%	69%
Austria	2014	57%	50%	48%
Belgium	2014	72%	65%	59%
Canada	2013	68%	61%	61%
Chile	2012	67%	72%[1]	N/A
Czech Republic	2014	79%	40%	38%
Denmark	2014	63%	52%	52%
Estonia	2013	81%	N/A	N/A
Finland	2014	73%	65%	68%
France	2014	65%	54%	47%
Germany	2014	53%	38%	31%
Greece	2014	74%	77%	75%
Hungary	2014	89%	74%	70%
Iceland	2013	78%	80%	88%
Ireland	2013	70%	81%	76%
Israel	2011	68%	73%[2]	N/A
Italy	2014	73%	67%	59%
Japan	2013	62%	62%	61%
Luxembourg	2013	73%	67%	60%
Mexico	2010	76%	78%	66%
Netherlands	2014	67%	45%	42%
New Zealand	2013	65%	73%	71%
Norway	2013	84%	67%	78%
Poland	2014	84%	55%[3]	N/A
Portugal	2014	75%	67%	52%
Slovakia	2014	90%	49%	N/A
Slovenia	2014	77%	61%	N/A
South Korea	2014	54%	53%[4]	N/A
Spain	2014	79%	78%	75%
Sweden	2014	69%	40%	42%
Switzerland	2014	44%	31%	30%
Turkey	2011	67%	N/A	N/A
United Kingdom	2013	65%	67%	55%
United States	2014	64%	64%	64%

Source: authors, based on sources listed in data appendix.

1 2000
2 1995
3 2000
4 1995

of World War II and the early 1960s and in the United Kingdom in the 1980s, although to a lesser degree. All these increases were facilitated by significant policy changes. Post-Soviet countries and the United Kingdom engaged in large-scale housing privatization, the latter based on the Housing Act of 1980 and its Right to Buy program for sitting council tenants (Clapham and Kintrea 1996).

Switzerland's homeownership rate also grew from 30 percent in 1980 to 44 percent in 2014, but for different reasons. The OECD Economic Survey attributes the growth in homeownership to a strong economy, a low rate of rental vacancies, unprecedentedly low mortgage interest rates, and a recent policy change that allows homebuyers to tap into mandated pension savings to make down payments (OECD 2015). Schneider and Wagner (2015) attribute the initial low Swiss rate primarily to the legal framework that first allowed for the purchase of housing units beginning in 1965. Nevertheless, the relatively neutral Swiss taxation scheme, which taxes owner occupiers and landlords equivalently yet also provides a tax subsidy for both groups, combined with conservative lending standards, keeps homeownership rates relatively low (Schneider and Wagner 2015).

While there may be some year-to-year variation, and even an uptick associated with a housing boom in places like the United States, Canada, and Spain, a surprisingly large number of countries have seen a stagnant or even a declining homeownership rate overall since 1980 (Buck 2017; Carlson 2017; see Table 8.2). For instance, despite temporary increases during the late 1990s and early 2000s, the United States currently has a homeownership rate matching that of the late 1960s. France, Finland, and Denmark have seen almost no change in the homeownership rate since at least the early 1990s, if not earlier. The relatively slow growth in disposable household incomes in these countries appears to be the primary driver (Arundel and Doling 2017).

Interestingly, some countries in Continental Europe have seen dramatic increases in their homeownership rates since the 1980s. The largest increase in the rate occurred in Germany, from 36 percent in 1994 to 52 percent in 2014. Although the current homeownership rate is still the second lowest in the OECD, the increase by 16 percentage points over 20 years is the largest. The primary driver of this increase was the privatization of state-owned housing stock in former East Germany after 1989, followed by the strong economy with relatively low unemployment rates in the Southern and Western parts of Germany, and the recent record-low interest rates set by the European Central Bank (Jones and Shotter 2017; Nienaber 2016).

The homeownership rate has also significantly increased in Austria and the Netherlands. Austria's homeownership rate increased from 46 percent in 1987 to just over 57 percent in 2014, and the Netherlands' homeownership rate grew from 47 percent in 1994 to almost 67 percent in 2014. In the latter case this growth may be attributed to state policies promoting homeownership, including subsidies for homeowners and a national mortgage guarantee that was introduced around the beginning of the early 1990s (Arundel and Doling 2017; Schiffer 2002; Vandevyvere and Zenthofer 2012). Some countries in Southern Europe have had a similar growth in the homeownership rate, although starting from a higher base. Italy and Greece currently have rates of approximately 74 percent, around 10 percentage points higher than their rates in the early 1990s. Homeownership rates in Spain and Portugal have long been among the highest in Western Europe, for the past few decades hovering in the high 70s (see Table 8.2).

Some countries such as the U.S. may have higher current homeownership rates for cultural reasons that are difficult to measure across countries. In the U.S., homeownership is often associated with the term "ownership society," i.e., the belief that homeownership leads to taking responsibility for and exercising control over an individual's life and surroundings. As President George W. Bush stated, "Part of being a secure America is to encourage homeownership. So somebody can say, this is my home, welcome to my home" (Bush 2002).

While the 2014 World Values Survey (WVS) did not ask questions related to homeownership, respondents were recently queried for the first time on how much freedom of choice and control over their own lives they believed they had, on a scale of 1 to 10.[4] While only 15 OECD countries were included in the particular wave, results show a weak positive correlation between homeownership rates and the value placed upon freedom of choice and control. Surveying individual homeowners provides a similar result: Homeowners report higher levels of perceived control over their lives than renters (Tighe and Mueller 2013). Future research efforts could focus on the present role of culture on homeownership in OECD countries.

Demographic Factors Affecting Homeownership

The strongest and most consistent predictor of the homeownership rate at the individual level is age. As individuals enter adulthood they often rent due to enrollment in higher education, a more fluid relationship with the labor market, and lack of savings for a down payment. As homeownership generally entails significant fixed transaction costs, individuals and households tend to postpone purchase until their employment has reached some degree of stability (Haurin and Gill 2002). In the U.S., homeownership rates are generally twice as high for individuals at retirement age of 65 and over (79.9 percent in 2014) than for individuals under 35 (35.8 percent in 2014) (see Table 8.1).

The transition from renter to homeowner often accompanies another transition—marriage (Hendershott et al. 2009). There is a positive relationship between homeownership and marriage rates. In the United States, every hypothetical percentage point increase in the proportion of married households raises the homeownership rate by 1.12 percentage points (Andrews and Sánchez 2011a). The delay of marriage and childbearing may place downward pressure on homeownership rates in most OECD countries, particularly in Spain, Italy, Germany, and Australia (Andrews and Sánchez 2011b).

However, there is a negative relationship between homeownership and the average age at first marriage. For example, in the United States the median age at first marriage for men has increased from 22.5 years in 1970 to over 28 years as of this writing. Similar increases in the age at first marriage have occurred in other OECD countries. In Germany marriage rates for men at ages 55 to 59 remain over 4 percentage points lower than in the United States (United Nations 2015).

Interestingly, there is a positive relationship between homeownership and the proportion of the population never getting married. In the United States, the proportion of males between 25 and 35 that were never married literally doubled, from 19 percent in 1950 to 39 percent in 2000. However, the homeownership rate for this group increased due to changes in mortgage underwriting, the growth in female labor force participation, and a larger proportion of unmarried but cohabitating singles who are now able to access mortgage credit. The homeownership rate among unmarried individuals in the United States increased from 36 to 43 percent between 1985 and 2015, as well (Pollock 2016). In conclusion, while marriage has remained an important factor of homeownership and the differences across countries, this relationship is likely more fluid than other variables, such as age (Mulder and Thomas 2016).

Previous research on United Nations Communications Group (UNCH) countries[5] found a strong negative relationship between homeownership and urbanization, i.e., the higher the percentage of a country's population that lives in an urban area, the lower its rate of homeownership (Fisher and Jaffe 2003). For example, a hypothetical 20 percent increase in the urbanization rate, similar to the difference in the urbanization rate between France and Spain, would decrease the homeownership rate by about 5 percent. For example, those OECD countries with the highest rates of homeownership, Slovakia, Poland, and Hungary, also have low urbanization rates. Japan has a relatively low homeownership rate of 62 percent and a relatively high urbanization rate, while Switzerland has a relatively modest level of urbanization and the lowest homeownership rate among OECD countries (Fisher and Jaffe 2003). However, there are exceptions to the negative relationship between homeownership and urbanization. For example, Norway, one of the most highly urbanized OECD countries, has one of the highest homeownership rates among countries in Northern and Central Europe.

Furthermore, there is also a positive relationship between homeownership and ethnic diversity in many countries; sociologists have speculated that homeownership provides a way for populations to segregate themselves from people with different backgrounds (Fisher and Jaffe 2003).

In sum, demographic data alone explain about three-fourths of the variation in homeownership rates, on average (Andrews and Sánchez 2011a). However, this finding does not control for policies that ideally should be controlled for. Some of these policies are outlined in the following text.

Policy Factors Affecting Homeownership

In addition to the strong explanatory power of demographic factors, national policy differences, i.e., differences in national political or legal systems, also explain homeownership rates (La Porta et al. 2008).

Fisher and Jaffe (2003) grouped countries by their legal system and found significant differences across them. Countries with legal systems that are entirely or partially based on British Common Law have higher homeownership rates than countries with legal systems that are based on French civil law, followed by countries with Scandinavian or German civil law traditions. The reason for the relatively high homeownership rates in countries with a British Common Law tradition may be that their legal structures are associated with more efficient dispute resolution and greater judicial independence, resulting in more efficient contract enforcement and stronger private property rights. On the other hand, civil law tradition is associated with the opposite: less efficient contract enforcement and greater public ownership of property.

Aside from different legal systems, governments have proactively or inadvertently used a variety of other policy levers to influence homeownership rates. For example, per capita government spending, taxes and homeownership transfers, the Mortgage Interest Deduction (MID), transfer and registration fees, and changes to loan-to-value ratios on government-backed mortgages are all policy factors that may influence homeownership. Per capita government spending is a policy choice that seems to have implications for the homeownership rate. In a cross-national study a higher per capita spending level was associated with a decrease in the homeownership rate because social spending and homeownership both act as forms of social insurance, and more of one means less of the other (Conley and Gifford 2006). For example, if the United States were to increase the size of per capita government spending on public pensions to match that of the Netherlands' relatively large public expenditure on public pensions, the U.S. homeownership rate would decrease by about 2 percentage points, all else equal. Greece, Italy, and Spain are unique because they have relatively high per capita government spending and relatively high homeownership rates (Castles and Ferrera 1996).

Taxes and homeowner-specific transfers also affect the homeownership rate. Most countries that subsidize homeownership generally do so via the tax code and often do so in rather complex ways. Switzerland—a good example of this complexity—allows mortgage interest payments to be deducted. However, Switzerland imposes a wealth tax on net home equity, or home wealth.[6] This tax is posed in addition to the various, and relatively high, transfer, registration, and capital gains taxes that are levied, depending on the amount of gain and period of ownership of residential real estate in Switzerland. The result is that Switzerland has a very low homeownership rate and that those homes that are owned are highly leveraged, or have a high debt to equity ratio. The effects of Switzerland's tax and transfer policies on homeownership and borrowing can be seen via comparison with neighboring Austria, whose homeownership rate is 12 percentage points higher than Switzerland's, and its mortgage debt to GDP is two-thirds lower. Austrian households hold greater equity in the homes they own.

For many countries the most widely offered subsidy to homeownership is the MID, which causes a difference between mortgage interest rates and the after-tax cost of debt financing. Johansson (2011) has analyzed this difference across OECD countries. For example, while the Netherlands and Denmark offer relatively generous MIDs, Italy and Luxembourg offer very modest deductions. The United States, Sweden, and Belgium fall in the middle of this distribution. Surprisingly, Johansson's results suggest that the MID has very little, if any, impact on homeownership rates as this subsidy is capitalized into the price of homes. Indeed, the MID increases home prices. While the deduction benefits existing homeowners, it may cause housing affordability issues for first-time homebuyers. In addition to downward pressure being put on homeownership rates, Johansson claims that the MID has "adverse efficiency effects on housing and other markets by distorting the allocation of saving and investment" (Johansson 2011, 8). Bourassa et al. (2013) conclude that a removal of the MID would likely increase homeownership rates and estimate that while MID policies lower the user cost of housing, such reductions are canceled out by the increase in prices caused via inflated demand. Indeed, the MID causes the greatest increase in house prices in areas where restrictive zoning or land use regulations discourage homebuilding and homeownership.

Similar to the MID, other government policies may inadvertently act as barriers to homeownership. Transfer and registration fees are a good example as they usually cannot be financed and must be paid out of pocket; they can be a substantial barrier to home purchase. As one would expect, the higher the transfer and registration fees, the lower a country's homeownership rate. Germany, with one of the lowest homeownership rates in the OECD, has uniquely high transfer and registration fees, which range from 3.5 to 5 percent of the house price (Just Landed n.d.). Allison (2013, 54) states that "for young people [who] anticipate moving often, the transaction cost (which is material) of buying and selling houses may overwhelm the economic benefits of owning." In the United States, transfer fees are usually well under 1 percent, although they differ widely across states and counties. Even in states at the high end (Delaware) and in the District of Columbia, transfer and recording fees are approximately 2 percent of the house price, still well below that in Germany (National Conference of State Legislatures n.d.). Policy reforms that reduce transfer and registration fees might have a positive effect on the incentives that drive home purchasing decisions.

Interestingly, differences in allowable loan-to-value (LTV) ratios, i.e., the proportion of the mortgage in relation to the total appraised value of the property, do appear to impact homeownership rates across OECD countries, particularly for younger households. The maximum LTV ratio is generally set by either bank regulators or government mortgage guarantors. In the United States, the Federal Housing Administration has a maximum LTV of 96.5 percent (U.S. Department of Housing and Urban Development 2016). Andrews and Sánchez (2011a) estimate that in a typical OECD country a 10 percentage point increase in LTV, from, for example, 80 to 90 percent, increases the homeownership rate of those aged 25 to 34 by 4.4 percent, although increases are considerably smaller for older individuals. Goodman and Nichols (1997) argue that such increases in the allowable LTV may simply be accelerating homeownership for individuals who would have become homeowners regardless.

Conclusion

There is a large literature on the benefits of homeownership, as well as a smaller literature on the costs (Rohe and Van Zandt 2011). The purpose of this chapter was to examine what factors influence homeownership rates. We compared OECD countries and reviewed relevant explanatory variables. In sum, to a rather large degree, homeownership rates are driven by demographics. The largest factor is age, followed by household structure, particularly marital status and urbanization to a lesser degree. The projected aging of the population in the United States and Europe, all else equal, will lead to lower homeownership rates. Declining familial stability and stagnant, if not declining, real wages will likewise negatively impact homeownership. Continued immigration may accelerate the decline in the homeownership rate as immigrant status has a negative impact on homeownership in the short term (Coulson 1999). However, after the Global Financial Crisis, the decrease and convergence of mortgage interest rates may reverse, and interest rates may increase due to economic factors, for example, increased business investment and a strengthening labor market. Alternatively, as a result of the debt crisis and ongoing political and economic uncertainty in a post-Brexit Europe mortgage costs may increase in some EU countries, such as Greece, Italy, and Spain. The net effect is that homeownership rates are unlikely to grow in most of Europe, with exceptions in select countries.

Demographics may well be destiny when it comes to housing, but policy choices are relevant. The most popular policy, the MID, actually appears to slightly lower homeownership rates, a counterintuitive conclusion, as do other subsidies that broadly increase the amount of mortgage debt in an economy (see Figure 8.2) Glaeser and Shapiro (2002) find that the higher the average value of the

Figure 8.2 Mortgage Debt to GDP Ratio, 2012.

Source: authors, based on European Mortgage Federation Hypostat (2013).

MID subsidy, the lower the country's corresponding homeownership rate. In the United States, the bulk of MID benefits go to borrowers who itemize their deductions and who own homes in locations with high home prices, where the MID further raises home prices and puts homeownership further out of reach (Calabria 2013). These households are likely to own a home even in the absence of the subsidy (Glaeser and Shapiro 2002). Groups across the ideological spectrum have called for its reform or elimination and the Tax Cuts and Jobs Act of 2017 reduced the MID and limited its application, though did not eliminate it[7] (Becker et al. 2017). Subsidies that increase the size of the mortgage sector should also be reconsidered, for example, the non-taxation of imputed rent for homeowners relative to renters, especially when they largely increase housing demand without also resulting in an increase in housing supply.

Relaxing down payment requirements in the United States has had a mixed history. Doing so is strongly associated with higher rates of homeownership among households under the age of 35. Yet such gains appear to be highly cyclical and perhaps transitory. Considerable costs in terms of both public subsidy and financial stability are also associated with lower down payments. Given the already low down payment requirements in the United States, it would appear that this option is only open to other OECD countries and is one that should be considered with great caution.

Although all policy interventions have costs and benefits, large-scale privatization of government-owned housing promises considerable gains to homeownership. Housing privatization in both the United Kingdom and the former Eastern Bloc countries resulted in unprecedented increases in homeownership. The Right to Buy program in the United Kingdom raised the national homeownership rate by around 15 percent (Disney and Luo 2017). Given the high levels of socialized rental housing that remain in many OECD countries, such as Austria, Ireland, and the Netherlands, privatization is an option for about half of the OECD countries. Some countries, such as the Netherlands, are currently engaged in such efforts (Braga and Palvarini 2013; Elsinga et al. 2014, see also the chapter authored by Marja Elsinga and Marietta Haffner). Other countries, such as the United States, that lack a history of large-scale government ownership of housing could only see very modest gains from privatization. Even in the United States, where government ownership of housing has been relatively modest, proposals to privatize that stock have via HOPE VI or the Rental Assistance Demonstration (RAD) programs have been met with controversy (Cohen 2014). Still, if improvements in homeownership rates are desired, privatization is a viable method.

Although policy choices may influence homeownership rates, they are only somewhat effective, very costly, and sometimes only temporary. Given policymakers' limited ability to influence major changes in demographic factors and sidestep costly unintended consequences, the most efficient course is likely to remove artificial incentives for homeownership and reform existing policies that act as barriers to homeownership.

Disclaimer

This chapter represents Mark Calabria's views alone and does not represent any official policy of the federal government.

Data Appendix

For United States homeownership trends, all data are from the U.S. Bureau of the Census www.census.gov/housing/index.html. For decennial years, the rate is taken from the Census of Housing, for inter-Census years, data are derived from the Current Population Survey (which is benchmarked to the decennial). The most recent EU figures are from Eurostat http://ec.europa.eu/eurostat. OECD data are used for non-EU OECD countries. As Eurostat only provides homeownership data back to 2003, the following sources are used for historical rates:

Czech Republic, Estonia, Finland, Hungary, Poland Slovakia, Slovenia, Sweden:

The Hague: Ministry of Interior and Kingdom Relations. (2010) "Housing Statistics in the European Union 2010" pp. 64–66, www.bmwfw.gv.at/Wirtschaftspolitik/Wohnungspolitik/Documents/housing_statistics_in_the_european_union_2010.pdf

Japan, New Zealand, Switzerland:

Atterhog, Mikael. (2005) "Importance of Government Policies for Home Ownership Rates," Royal Institute of Technology Working Paper No. 54, p. 17, www.kth.se/polopoly_fs/1.167143!/Menu/general/column-content/attachment/54.pdf

Israel:

Global Property Guide (2015) "Israeli Housing Prices Rising Faster and Faster," www.globalpropertyguide.com/Middle-East/Israel/Price-History

Mexico and Chile:

Torche, Florencia, and Spilerman, Seymour. (2006) "Household Wealth in Latin America," United Nations University Research Paper No.2006/114, pp. 8–10, www.econstor.eu/bitstream/10419/63394/1/52137930X.pdf

Data for individual countries also collected from the Turkish Statistical Institute, Statistics New Zealand, Finland's National Institute of Statistics and Geography, Statistics Korea, Statistics Bureau of Japan, Statistics Canada, and Australian Bureau of Statistics. Data also gathered from the Housing Finance Information Network, http://hofinet.org/index.aspx.

European Mortgage Federation Hypostat. (2013). www.hypo.org/Content/Default.asp?pageId=578, and European Mortgage Federation Key Figures 2012, www.hypo.org/content/default.asp?PageID=414.

Notes

1 In countries where imputed rent is taxed such as Switzerland, the Netherlands, Iceland, Slovenia, and Luxembourg, it estimates what rent the owner would pay for a comparable rental unit, hence the imputation of rent. The owner is then taxed on some portion of that imputed rent.
2 The OECD includes the following countries: Australia, Austria, Belgium, Canada, Chile, the Czech Republic, Denmark, Estonia, Finland, France, Germany, Greece, Hungary, Iceland, Ireland, Israel, Italy, Japan, Luxembourg, Mexico, the Netherlands, New Zealand, Norway, Poland, Portugal, Slovakia, Slovenia, South Korea, Spain, Sweden, Switzerland, Turkey, the United Kingdom, and the United States.
3 The 1994 UNCH data release displays a median of 69.3 percent. The distribution is heavily concentrated in the high 60s and low 70s. The primary difference is that the global outliers in the right tail of the distribution are outside the OECD. Cambodia heads the list with a homeownership rate just over 95 percent. The left tail of the UNCH distribution is primarily OECD countries, particularly Switzerland, Sweden, and the Czech Republic.
4 The WVS wave 6 question covers the years 2010–2014 and includes 60 countries, 14 of which were OECD countries.
5 United Nations Communications Group countries consist of 106 countries, including non-OECD countries.
6 A wealth tax is levied on current asset owners based on current total value of assets, whereas capital gains taxes are levied at the time of an asset transfer (sale).
7 The deduction previously applied to $1.1 million in mortgage debt and now applies to $750,000 in mortgage debt.

References

Acharya, Viral V., Matthew Richardson, Stijn Van Nieuwerburgh, and Lawrence J. White. 2011. *Guaranteed to Fail: Fannie Mae, Freddie Mac, and the Debacle of Mortgage Finance*. Princeton, NJ: Princeton University Press.

Allison, John A. 2013. *The Financial Crisis and the Free Market Cure: Why Pure Capitalism is the World Economy's Only Hope*. Ashland, OR: Blackstone Audio.

Anacker, Katrin B. 2015. "Saving While Black: From the Freeman's Savings Bank, to Public Policy Missing in Action, to a New Down payment Savings Product and Policy." *Banking and Financial Services Policy Report: A Journal of Trends in Regulation and Supervision*, 34(12): 12–18.

Andrews, Dan, and Aida Caldera Sánchez. 2011a. "The Evolution of Homeownership Rates in Selected OECD Countries: Demographic and Public Policy Influences." *OECD Journal: Economic Studies*, 1: 1–37.

Andrews, Dan, and Aida Caldera Sánchez. 2011b. *Drivers of Homeownership Rates in Selected OECD Countries Economics Department*. Paris: OECD.

Arundel, Rowen, and John Doling. 2017. "The End of Mass Homeownership? Changes in Labour Markets and Housing Tenure Opportunities Across Europe." *Journal of Housing and the Built Environment*, 32(4): 1–24.

Bayer, Patrick, Fernando Ferreira, and Stephen L. Ross. 2014. *Race, Ethnicity and High-Cost Mortgage Lending*. Cambridge, MA: National Bureau of Economic Reseach.

Becker, Bernie, Nancy Cook, and Seung Min Kim. 2017. "Republicans Close in on Historic Tax Deal." *Politico*, December 12.

Black, Harold, Thomas P. Boehm, and Ramon P. DeGennaro. 2015. *Is There Discrimination in Mortgage Pricing? The Case of Overages*. Atlanta: Federal Reserve Bank of Atlanta.

Blanc, Maurice. 2010. "The Impact of Social Mix Policies in France." *Housing Studies*, 25(2): 257–272.

Bourassa, Steven, Donald Haurin, Patric Hendershott, and Martin Hoesli. 2013. "Mortgage Interest Deductions and Homeownership: An International Survey." *Journal of Real Estate Literature*, 21(2): 181–203.

Braga, Michela, and Pietro Palvarini. 2013. *Social Housing in the EU*. Brussels/Belgium: European Parliament.

Bramley, Glen. 2003. "Planning Regulation and Housing Supply in a Market System." In *Housing Economics and Public Policy*, edited by Tony O'Sullivan and Kenneth Gibb, 193–217. Oxford: Blackwell Science.

Buck, Tobias. 2017. "Spain: Boom to Bust and Back Again." *Financial Times*, April 6.

Bush, George. 2002. *President Calls for Expanding Opportunities to Home Ownership*. Washington, DC: White House.

Calabria, Mark A. 2013. *Tax Reform and Residential Real Estate*. Washington, DC: CATO Institute.

Calderon, Marisa. 2017. *State of Hispanic Homeownership Report*. San Diego, CA: National Association of Hispanic Real Estate Professionals.

Carlson, Ben. 2017. "Canada's Housing Bubble Will Burst." Bloomberg.com, June 21.

Castles, Francis, and Maurizio Ferrera. 1996. "Home Ownership and the Welfare State: Is Southern Europe Different?" *Southern European Society and Politics*, 1(2): 163–185.

Chambers, Matthew, Carlos Garriga, and Don Schlagenhauf. 2016. *The Postwar Conquest of the Home Ownership Dream*. St. Louis, MO: Federal Reserve Bank of St. Louis.

Clapham, David, and Keith Kintrea. 1996. "The Patterns of Privatization in Eastern Europe." In *Housing Privatization in Eastern Europe*, edited by David Clapham, József Hegedüz, Keith Kintrea, Ivan Tosics, and Helen Kay, 169–191. Westport, CT: Greenwood Publishing Group.

Cohen, Rachel. 2014. "The RAD-ical Shifts to Public Housing." *The American Prospect*, August 28.

Collins, William J., and Robert A. Margo. 2001. "Race and Home Ownership: A Century-Long View." *Explorations in Economic History*, 38(1): 68–92.

Collins, William J., and Robert A. Margo. 2011. "Race and Home Ownership from the End of the Civil War to the Present." *The American Economic Review*, 101(3): 355–359.

Conley, Dalton, and Brian Gifford. 2006. "Home Ownership, Social Insurance, and the Welfare State." *Sociological Forum*, 21(1): 55–82.

Coulson, N. Edward. 1999. "Why are Hispanic-and Asian-American Homeownership Rates so Low? Immigration and Other Factors." *Journal of Urban Economics*, 45(2): 209–227.

Disney, Richard, and Guannan Luo. 2017. "The Right to Buy Public Housing Program in Britain: A Welfare Analysis." *Journal of Housing Economics*, 35: 51–68.

Elsinga, Marja, Mark Stephens, and Thomas Knorr-Siedow. 2014. "The Privatization of Social Housing: Three Different Pathways, in Social Housing in Europe." In *Social Housing in Europe*, edited by Kathleen Scanlon, Christine Whitehead, and Melissa Fernández Arrigoitia, 389–413. Chichester: John Wiley & Sons.

European Mortgage Federation. 2014. *European Mortgage Markets Quarterly Review Q1–2014*. Brussels/Belgium: European Mortgage Federation.

Fetter, Daniel. 2013a. "How Do Mortgage Subsidies Affect Home Ownership? Evidence from the Mid-Century GI Bills." *American Economic Journal: Economic Policy*, 5(2): 111–147.

Fetter, Daniel. 2013b. *The Home Front: Rent Control and the Rapid Wartime Increase in Home Ownership*. Cambridge, MA: National Bureau of Economic Research.

Fisher, Lynn M., and Austin J. Jaffe. 2003. "Determinants of International Home Ownership Rates." *Housing Finance International*, 18(1): 34–42.

Fry, Richard, and Anna Brown. 2016. *In a Recovering Market, Homeownership Rates Are Down Sharply for Blacks, Young Adults*. Washington, DC: Pew Research Center.

Gabriel, Stuart A., and Stuart S. Rosenthal. 2005. "Homeownership in the 1980s and 1990s: Aggregate Trends and Racial Gaps." *Journal of Urban Economics*, 57(1): 101–127.

Glaeser, Edward L., and Jesse M. Shapiro. 2002. *The Benefits of the Home Mortgage Interest Deduction*. Cambridge, MA: National Bureau of Economic Research.

Goodman, John L., and Joseph B. Nichols. 1997. "Does FHA Increase Home Ownership or Just Accelerate It?" *Journal of Housing Economics*, 6(2): 184–202.

Haurin, Donald R., and H. Leroy Gill. 2002. "The Impact of Transaction Costs and the Expected Length of Stay on Homeownership." *Journal of Urban Economics*, 51(3): 563–584.

Hendershott, Patric H., Rachel Ong, Gavin A. Wood, and Paul Flatau. 2009. "Marital History and Home Ownership: Evidence from Australia." *Journal of Housing Economics*, 18(1): 13–24.

Hilber, Christian A.L., and Yingchin Liu. 2007. *Explaining the Black-White Homeownership Gap: The Role of Own Wealth, Parental Externalities and Locational Preferences*. London: London School of Economics.

Jackman, Mary R., and Robert W. Jackman. 1980. "Racial Inequalities in Home Ownership." *Social Forces*, 58(4): 1221–1234.

Johansson, Åsa. 2011. *Housing Policies in OECD and Candidate for Accession Countries: Survey-Based Data and Implications*. Paris: OECD.

Jones, Claire, and James Shotter. 2017. "German Push for Home Ownership Drives Price Bubble Fears." *Financial Times*, January 4.

Just Landed. n.d. *Costs When Buying Property*. N.l.: Just Landed.

La Porta, Rafael, Florencio Lopez-de-Silanes, and Andrei Shleifer. 2008. "The Economic Consequences of Legal Origins." *Journal of Economic Literature*, 46(2): 285–332.

Malpezzi, Stephen. 2005. "Low Income Housing Markets and Policy: Some International Perspectives, and Possible Lessons for Korea." In *Residential Welfare and Housing Policies: The Experience and Future of Korea*, edited by MoonJoong Tcha, n.p.–n.p. Seoul: Korea Development Institute.

Mulder, Clara H., and Michael J. Thomas. 2016. "Partnership Patterns and Homeownership: A Cross-Country Comparison of Germany, the Netherlands and the United Kingdom." *Housing Studies*, 31(8): 935–963.

National Conference of State Legislatures, n.d. *Real Estate Transfer Taxes*. Denver, CO: National Conference of State Legislatures.

Nienaber, Michael. 2016. "Germans Flock to Property as Interest Rates Fall and Rents Rise." *Reuters*, April 24.
OECD. 2015. *OECD Economic Surveys: Switzerland*. Paris: OECD.
Pollock, Alex J. 2016. "Homeownership Rates: It Depends on Whether You are Married." *Housing Finance International*, Autumn: 13–14.
Rohe, William M., and Shannon Van Zandt. 2011. "The Sustainability of Low-Income Homeownership: The Incidence of Unexpected Costs and Needed Repairs among Low-Income Home Buyers." *Housing Policy Debate*, 21(2): 317–341.
Russell Sage Foundation. 2012. *Chartbook of Social Inequality*. New York: Russell Sage Foundation.
Schiffer, Karel. 2002. "Dutch Guarantee Fund for Homeownership; Model for a Single Guarantee Fund for the European Union?" Paper presented at *Eucocatalyst Conference*, Madrid, Spain, September 22, 2002.
Schneider, Martin, and Karin Wagner. 2015. "Housing Markets in Austria, Germany and Switzerland." *Monetary Policy and the Economy*, n.v.(n.i.): 42–58.
Tighe, Rosie J., and Elizabeth J. Mueller. 2013. *The Affordable Housing Reader*. New York: Routledge.
United Nations. 1948. "Universal Declaration of Human Rights." Articles 17 and 25. New York: United Nations.
United Nations. 2015. *World Marriage Data*. New York: United Nations.U.S. Bureau of the Census. 1993. *Black Americans: A Profile*. Washington, DC: U.S. Bureau of the Census.
U.S. Bureau of the Census. 2018. *Quarterly Residential Vacancies and Homeownership: First Quarter 2018*. Washington, DC: U.S. Bureau of the Census.U.S. Department of Housing and Urban Development. 2016. *Federal Housing Administration: Annual Report to Congress*. Washington, DC: U.S. Department of Housing and Urban.
Vandevyvere, Windy, and Andreas Zenthöfer. 2012. *The Housing Market in the Netherlands*. Brussels/Belgium: European Commission.

9

DECLINING HOMEOWNERSHIP IN LIBERAL, ENGLISH SPEAKING COUNTRIES

Richard Ronald and Christian Lennartz

Introduction

During the latter half of the twentieth century, rates of homeownership advanced significantly among economically liberal, English speaking countries. Indeed, Australia, Britain, Canada, Ireland, New Zealand, and the U.S. were increasingly considered "homeowner societies," with the tenure associated with not only home and housing practices, but also discourses on citizenship, neighborhood improvement, and the wider organization of the economy (Forrest and Hirayama 2015; Kemeny 1981; Kurz and Blossfeld 2004; McCabe 2016; Rohe et al. 2002; Rolnik 2013; Ronald 2008; Saunders 1990, Smith 2015). By the beginning of the twenty-first century, homeownership had also gained ground across Continental Europe, especially in those countries where large-scale public or social rental housing sectors had once flourished, such as Sweden and the Netherlands, not to mention Eastern Transition economies where housing was privatized *en-masse* following the collapse of communism. The advance of homeownership was not confined to the global Northwest, with owner-occupation also becoming the primary mode of housing provision in China and across East Asian economies (Doling and Ronald 2014).

More recently, however, homeownership has been eroded. The Global Financial Crisis (GFC) has been central in this decline, with mortgage financialization and subprime lending practices establishing conditions for first the Credit and then the Financial Crisis, which undermined housing property values and mortgage lending as well as stimulating foreclosures on a mass scale (Immergluck 2015). Nonetheless, an emerging shift in the housing landscape was evident well before the crisis, with declining homeownership rates across a number of homeownership orientated societies—especially, but not only, among the English speaking, economically liberal ones—since the turn of the century. A particular driver of diminishing owner-occupancy rates has been the increasing difficulties experienced by younger people in exiting the rental sector or, in many cases, even leaving the parental home. Post-crisis conditions have, moreover, reinforced this trend.

Homeownership has been traditionally associated with socioeconomic integration and positive externalities. However, since the crisis, it appears to have become a more divisive axis, sustaining the growth of social and economic inequalities, especially intergenerational ones. In this chapter we consider both the rise of and recent declines in homeownership in terms of shifting conditions surrounding the tenure. We pay particular attention to the shifting status of housing markets and policies across a range of countries where homeownership appears to have become a vulnerable pillar of the housing and welfare system. We reflect on the socio-political origins of mass owner-occupation as well the increasing commodification and financialization of the home. We specifically challenge the assumed durability of mass homeownership and its capacity to sustain widespread household wealth accumulation.

The Rise and Fall of Mass Homeownership

While a century ago, most households in industrialized societies rented their home from a private landlord, in the period following World War I, but especially after World War II, the idea that every household could and should buy their own home gained ground. In the Republic of Ireland, for example, the early promotion of mortgages, grants, and subsidies helped rapidly expand homeownership, especially small holdings, which was considered a means to support capital redistribution and spread the ownership of assets, reinforcing family-based welfare subsidiarity (Norris 2016). In the U.S., alternatively, the development of a federal mortgage finance system following the failure of the economy in the early 1930s, and later, state-sponsored mortgage securitization provided by Fannie Mae, provided the basis for a rapid advance in owner-occupation rates after 1945 (Grigsby 1990). While approaches to its promotion varied, in many ways the postwar era reflected a "golden age" for homeownership sustained by economic growth, suburbanization, and a rapidly expanding middle class. Political support for the tenure, nonetheless, was also critical to the emergence of homeowning as the majority tenure (Ronald 2008).

In much of Europe, public rental housing had also been advanced as a solution to postwar housing shortages often alongside the expansion of homeownership, with both contributing to rebuilding and growth efforts. Housing policy often reflected the Keynesian model of public demand management that made the state primarily responsible for the social and economic security of workers. Following the global economic disruption of the early 1970s along with a shift toward a post-industrial service economy, however, the expansion of social policy and state-driven housing supply faded, with the market forwarded as a more suitable mechanism of welfare service provision (Birch and Siemiatycki 2015). Housing was a particularly vulnerable part of social policy as markets remained influential in housing provision and allocation (Harloe 1995). Some countries, like the UK, established more direct means to facilitate the private consumption of owner-occupied housing, with the 1980s' Right to Buy policy stimulating sales of millions of public housing units—mostly at a large discount—to sitting tenants (Forrest and Murie 1988). By the 1990s, many corporatist and social democratic welfare economies had also begun to dismantle social rental housing sectors with a view to enhancing mortgaged owner-occupation across a broader income spectrum of the population (Atterhög 2006; Bengtsson et al. 2017; Kurtz and Blossfeld 2004; Ronald and Dol 2011).

Thus, while the tendency is to associate augmentation in homeownership with increasing affluence and convergence around a global economic regime, its contemporary origins reflect different, but parallel, political rhetoric, policy interventions, and welfare state restructuring. In the UK, for example, Thatcherism established a formidable consensus around housing privatization as a means to achieve a form of "property owning democracy" in which increases in homeowning would stimulate the distribution of wealth and enhance feelings of social inclusion. Labor governments subsequently incorporated owner-occupancy into discourses of an "opportunity society" and sought to expand the tenure to more than 70 percent of households. In the U.S., meanwhile, institutional discourses had long asserted that living in a single-family, owner-occupied dwelling was central to a secure and successful life, with federal policy both reflecting and reinforcing this normative conception (Shlay 2006). Homeownership rates stagnated in the 1980s, but were reignited by HUD initiatives under the Clinton administration (Vale 2007). By the early 2000s homeownership rates were approaching 70 percent, with the Bush administration, under the banner of an "ownership society," seeking to advance the tenure further. In the context of increasing asset values and stagnation in wages, low homeownership rates among low-income and minority communities were identified as drivers of socioeconomic inequity (see also the chapter authored by J. Revel Sims and Carolina S. Sarmiento). Policy thus sought to facilitate more and cheaper borrowing for low-income communities who would, through their homes, derive greater wealth from market growth (Retsinas and Belsky 2005).

The arrival of mass homeownership, then, may be seen as both an integrated feature and outcome of political and economic transformations over the last half of the twentieth century. While policies supporting home buying resonated with conceptions of middle-class inclusion and increasing affluence—along with forms of citizenship in which the owner-occupied home represented a bedrock of ontological security and financial self-reliance—homeownership was also promulgated by political actors on the behalf of the building and finance sectors. In the U.S., significant federal funds were funnelled into private house building projects, while subsidies for homebuyers (most notably mortgage interest tax deductions) became important to the advance of the sector over other, more collective housing strategies. Politicians also sought to promote homeownership abroad, with mass homeownership offering "a critical building block for a larger, more persuasive American model of capitalism that connected open markets to democratic institutions" (Kwak 2015, 2). This was initially achieved through international agencies funded by the U.S., and was taken up in the 1980s by organizations like the World Bank that pressured governments to develop financial institutions that would provide mortgage loans and advance homeownership over other forms of housing provision.

The era in which state interventions were relatively successful in expanding access to homeownership has been considered a characteristically social one (Forrest and Hirayama 2015). The growing consumption of owner-occupied housing was initially seen as a means to compensate for the deeper liberalization of the economy. Housing wealth accumulation both integrated private households into the market and ostensibly provided an economic buffer. At the same time, mortgage lending and housing asset accumulation were understood to be relatively conservative practices. For Schwartz and Seabrooke (2008), however, shifts in the political economy undermined this stability, and while mortgage lending and housing property had been "inert, immobile and illiquid" in the postwar decades, from the 1980s on they became increasingly "live, cashable and liquid" (210). Capital of various origin increasingly flowed into housing and mortgage finance along with deeper global integration (Aalbers 2015). Lending for home purchase became very much an end in itself, with rent seeking and profit making driving the massive gains seen in property buying and house values that led up to the credit and financial crises of the late 2000s.

While mortgage lending had been a relatively stable and regulated sector, from the 1980s on, it was increasingly co-opted by processes of neo-liberalization and financialization. The now notorious subprime market emerged as a relatively minor feature of mortgage finance, but in the climate of exhilaration surrounding owner-occupied housing around the turn of the century, quickly gained a foothold (Langley 2010). Its arrival also reflected a level of exhaustion in the dynamics of expanding homeownership, with the gap created by insecure employment and diminishing wage increases filled by debt and the growing sophistication of lending technologies (García-Lamarca and Kaika 2016). Indeed, failures in the prime sector along with a downturn in the U.S. labor market also contributed to the disaster that became the GFC (Aalbers 2008). While mortgage securitization and its collapse was centered in the U.S., markets worldwide had become subjugated to an international architecture of financial growth built on mortgage credit and real estate. National mortgage and finance institutions beyond the U.S. thus relied on market funds and interbank lending to finance domestic activities (European Central Bank 2009). As such, the Credit Crisis had implications far beyond housing markets.

Although house price bubbles had formed and burst before, and the onset of the GFC marked a watershed in the advancement of homeownership. In the late 2000s and early 2010s, households across most developed societies were exposed to greater housing market risks such as indebtedness and price volatility that were exacerbated by the downturn in economic and labor market conditions (OECD 2011). Meanwhile, the market for housing assets fell away just as households needed to access them in order to weather the economic storm.

In the aftermath of the crisis, however, investment in housing property has not faded. Rather, access to homeownership and the distribution of housing equity and debt have become increasingly distorted. Homeownership has taken on a new political and economic hue, and even though the dynamics have changed, governments remain committed to supporting this form of housing provision and consumption above all others, despite the social and economic disruption that has ensued.

The Post-Crisis Context of Homeownership

The pattern of boom and bust across developed societies leading up to and following 2008 was shaped not only by local exposure to securitized mortgage debt, but also by the structure of housing systems and welfare states, and the extent to which they protected households from the economic downturn. For example, in many Northern European countries, income support for the unemployed and protective measures against foreclosure (such as mortgage guarantees) softened the impact of house price deflation and diminishing employment security. Moreover, many of these countries had large affordable rental housing sectors, and low-income homeownership was relatively limited. On the other hand, in contexts like Spain and Ireland, speculative housing investment had stimulated excessive house building, making the downturn harsher, with subsequent austerity packages and foreclosure practices amplifying the effects of the crisis (García-Lamarca and Kaika 2016; Waldron and Redmond 2015).

Housing market recovery has also been remarkably uneven. In Europe this is exemplified by differences between north and south, and while house prices in Spain are still only 75 percent of 2010 levels, Swedish housing property values are more than 41 percent up (Eurostat 2015). In Northern European contexts, low interest rates and continued demand have helped sustain price increases, despite stricter credit conditions and subdued wage increases. Moreover, residential property appears to have become a more specific target of international financial flows, with cross-border residential property investment increasing by 334 percent between 2009 and 2015 (Savills 2016). In Germany, for example, where house price growth has long been restrained, Kemp and Kofner (2014) identify how the economic crisis enhanced the attractiveness of residential property as a rental prospect, but also as a safe haven for capital among foreign investors. While institutional investment has been significant, private wealth, especially that of high and super-high net worth individuals, has become increasingly mobile and focused on housing investment opportunities abroad (Rogers and Koh 2017). Global cities have been a particular focus of investment, while the regions have often endured continued downturn. In the UK, for example, London prices are up more than 44 percent on pre-crisis levels, although regions like Northern Ireland have experienced little recovery. The overall picture, nonetheless, is one of renewed house price appreciation since 2011 driven by a wider mobilization of wealth around property investment rather than easy credit.

How homeownership and housing wealth are individually understood and managed appears to have adapted to emerging economic realities, post-crisis. Specifically, while homeownership has long been embedded with ontological and economic security (Saunders 1990), its status as a commodity has been normatively enhanced by previous property inflation cycles. Indeed, the treatment of the home and housing finance as a relatively stable and inert domain has ostensibly been supplanted in many contexts by more proactive individual and family strategies to housing property as a store of wealth or means of generating income. The evidence suggests, especially among English speaking homeowner societies, that households have not just become more active in accessing and mobilizing their housing assets (via equity release products and so on), but also more proactive in terms of property accumulation and rent seeking via small-scale letting (Ronald et al. 2017). For example, in the UK, the numbers of private individuals buying one or two extra properties to let began to expand in the early 2000s, but has exploded since the GFC, and while there were less than one million landlords and around 2.5 million rental units in 2002, by 2014 this had expanded to 2.2 million private landlords and around 4.9 million units (Ronald et al. 2017).

In the U.S., there has been a comparable growth in private investment in rental property. This has, furthermore, been associated with an emerging imbalance between generations in ownership and access to housing equity (Bloomberg 2016). Baby boomers in particular have been accused of using their wealth—in large part facilitated by historic access to a more secure, state-sponsored mortgage system and consistent house price increases—to buy more property, often with a view to letting it out. Indeed, in the past few decades, returns on housing assets have been much greater than increases in incomes, which has translated into a disproportionate advantage for the holders of capital who have increasingly adopted urban real estate as a store of wealth (Piketty 2014). For lower-income households meanwhile, especially among younger cohorts, access to homeownership and housing assets has been undermined due a combination of factors including diminishing labor security, renewed house price augmentation, and diminishing access to credit. This dynamic appears to have been reinforced in the aftermath of the GFC (Kennett et al. 2013). It has, thus, been suggested that the process of housing financialization—where homes have been increasingly subjugated to market processes, profit making, and wealth accumulation practices—associated with the growth in homeownership and housing property equity since the early 1990s has deeply enhanced possibilities for rent extraction and greater economic stratification (Aalbers and Christophers 2014), which now seems to be undermining the sector more broadly.

The Polarization of Homeownership

The social project of homeownership that had been advanced in the second half of the past century is argued to have been overridden in the twenty-first century by a "neoliberal project" (Forrest and Hirayama 2015), in which profit making and rent seeking have driven advances and developments in owner-occupied housing sectors. While credit for home buying shrank subsequent to the Credit Crisis, flows of wealth into housing markets, especially urban ones, have continued, with patterns of home buying and owner-occupation becoming remarkably uneven. This trend has had a number of outcomes, not least of which have been diminishing homeownership rates driven by declining access and affordability.

Demand for housing has returned since the crisis, but under conditions that now favor more polarized access. Overall homeownership rates have declined almost 10 percent in the UK and the U.S. (from just over 70 percent of housing in the mid-2000s to around 64 percent by 2015), in large part driven by diminishing entry among younger households. In the UK, for example, the ratio of homeowning households aged 34 and under has fallen from 51 percent in 2005 to 31.2 percent in 2015 (UK Department for Communities and Local Government 2016), while in the U.S. there has been a decline from 43.1 percent in 2005 to 36.7 percent in 2013 (U.S. Bureau of the Census 2016). Similar patterns can be found across homeowning countries like Australia, New Zealand, and Ireland (Norris 2016; Stebbing and Spies-Butcher 2016), but also across many other developed societies, albeit often to a lesser extent (Lennartz et al. 2016). Figure 9.1, using European Union data (Eurostat 2015), illustrates shifting patterns across a selection of European countries. While the UK appears something of an extreme case in terms of declining homeownership among younger people, similar patterns are evident in Denmark and the Netherlands, where home buying has become deeply financialized (and mortgage dependent) since 1990.

While owner-occupation has been diminishing among younger cohorts, rates have remained steady among older people, typically above 70 percent among those above retirement age in the UK and U.S. Moreover, housing equity appears to be increasing overall among established homeowners. Indeed, in 2015, the ratio of owner-occupied housing owned outright in the UK surpassed the proportion of mortgaged property (Savills 2016), in large part as an outcome of the maturation of the boom cohort who entered the sector in the 1970s and 1980s, driving an increase in the tenure share. Many of this generation have not only repaid their home loans, but also accrued significant asset wealth through

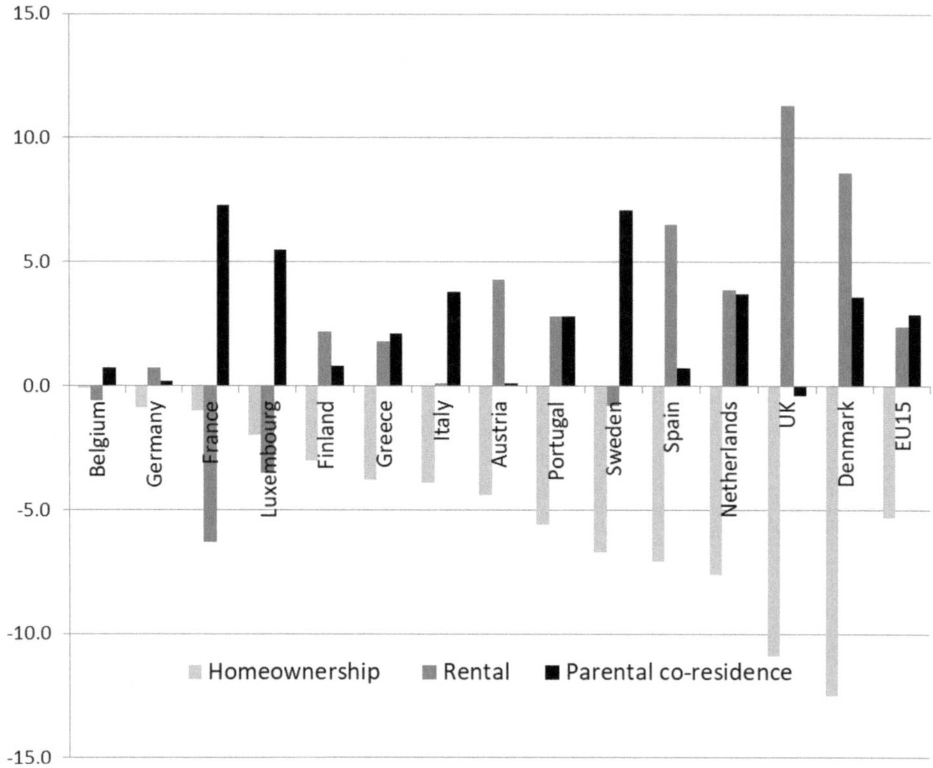

Figure 9.1 Tenure Changes (in Percent) Among 18–34 Year-Olds (2007–2012) in Select Countries in Europe.
Source: authors, based on EU-SILC, Eurostat.

cycles of house price inflation. The proportion of these cohorts buying up extra residential property to rent (often leveraged against existing housing wealth) has also expanded, with around 12 to 15 percent of baby boom households (born 1948 to 1957) now renting out extra properties (Ronald and Kadi 2016). Nonetheless, while the proportion of landlords among subsequent cohorts born in the 1960s and 1970s is lower, the rate of growth in this practice is actually higher (Ronald and Kadi 2016). In the U.S., a similar pattern is emerging (Bloomberg 2016), and while the advance in un-mortgaged homeownership is less pronounced, it has been increasing since the crisis: from 31.7 percent of owner-occupied homes in 2009 to 36.3 percent in 2013 (U.S. Bureau of the Census 2016).

In the context of the emerging divide between younger and older generations, in the longer durée it seems that the historic promotion of homeownership in the latter decades of the twentieth century (Ronald 2008) lies at the heart of a socioeconomic process that is currently undermining homeownership and driving down tenure ratios by undermining access at historically unprecedented rates. Inequalities in homeownership are not driven by cohort alone, and there remain significant intra-generational differences in terms of housing tenure and relative housing equity and debt that largely mirror social class and income distributions. Indeed, there are many low-income households who have been pulled into homeownership in the context of attractive past lending conditions and a deep normalization of the tenure, many of whom have become more economically vulnerable, or even poorer, as mortgaged homeowners (Gurney 1999; Nettleton and Burrows 1998; Shlay 2006).

Growing geographic inequalities have also been increasingly evident in owner-occupied housing wealth between urban regions, but also across neighborhoods. With property markets in global cities attracting greater investment from abroad (Rogers and Koh 2017), and urban middle classes also more effectively capturing their wealth in their homes (Piketty 2014), major cities have seen not only disproportionate augmentation in property values, but also a revival in gentrification processes. The effect has been uneven, but inevitably contributed to diminishing owner-occupation rates among the young, low-paid, and employment insecure, undermining owner-occupation more broadly, but also exacerbating housing and economic security among a growing class of urban renters.

Intra-generational inequalities have also been exacerbated by emerging housing market conditions. While there are long-standing inequalities between renters and owners, as well as between different market segments of owners in terms of differences in equity gains in the ways housing wealth enhances or undermines existing income-based stratification, these are being reinforced by the growing importance of intergenerational transfers. Inter-vivo transfers and inheritance of housing have escalated significantly in the context of the historic rise in owner-occupancy among growing property wealth of baby boomer cohorts (Cigdem and Whelan 2017). For young people, becoming a homeowner, especially since the GFC, has increasingly reflected the housing market position and rent seeking behavior of their parents or even grandparents. The children of secure homeowners (and growing numbers of multiple-property owners) are likely to eventually be helped out when buying a home (Druta and Ronald 2017). The children of renters, over-leveraged mortgage-holders, or aging households who rely on their un-mortgaged property to meet their own needs, however, are likely to remain locked out unless they have a considerable income. Even then, household composition is also playing a more important role, with dual incomes becoming increasingly necessary in order to borrow to buy a home.

Conclusions: Post-Homeownership Societies

While there has been substantial recovery in housing markets, owner-occupied housing sectors have not weathered the economic vicissitudes of the past decade well. Critically, the era in which homeownership grew and its distribution advanced across income classes and ethnicities seems to have passed. In its place, renting has expanded quite rapidly along with intensified demand and, consequently, increasing rents. Another important, but also often understated, development has been cross national increases in younger adults boomeranging back to or staying on in the parental home (Arundel and Lennart 2017).

Various indicators suggest that homeownership rates will continue to diminish, especially among younger cohorts (see, for example, Intermediary Mortgage Lenders Association 2014). Although government policy and discourse continue to focus on owner-occupation as a socioeconomic panacea, it has arguably become more divisive. Along with greater welfare state retrenchment for the very poor, public programs supporting homeownership (such as non-means-tested tax relief, grants, financial assistance, and public guarantees for homebuyers) have been rolled out further, and according to a recent OECD report, now account for as much as 2.3 percent of the GDP (Salvi del Pero et al. 2016). The effects of these measures to prop up homeownership not only are highly inequitable in terms of the income groups who benefit most, they also contribute significantly to the pressure on house prices. Nonetheless, there appears to be no political will to explore the support of other tenures or erode the privileges of homeownership.

Homeownership and getting into it is now widely recognized as not just socially desirable, but also a means of accruing wealth and building an asset critical to security and well-being in later life. Not becoming a homeowner has thus become an important aspect of financial inclusion/exclusion in an era when other sources of economic security, such as the public welfare cover, have faded. For Forrest and Hirayama (2015), the "neo-liberal era" of homeownership is quite distinct from the "social era" in which homeownership was understood to have only positive social, economic, and political externalities. Inclusion in a society of homeowners has thus become more critical to life chances, but also more difficult to achieve or sustain. While social class differences are still important, in housing, intergenerational differences and cohort effects have also come to the fore. Indeed, the differences between homeowning baby boomers and Generation Rent have helped define the economic injustices of the post-crisis era. One suggestion has been that many countries are becoming "post-homeownership" societies (Ronald 2008, 250) where changes in housing conditions are shaping new patterns of wealth accumulation, distributions of welfare risks, and movements through life-courses.

Acknowledgement

This chapter is based on research funded by the European Research Council under the European Union's ERC Grant Agreement No. 283881[HOUWEL].

References

Aalbers, Manuel B. 2008. "The Financialization of Home and the Mortgage Market Crisis." *Competition & Change*, 12(2): 148–166.
Aalbers, Manuel B. 2015. "The Great Moderation, the Great Excess and the Global Housing Crisis." *International Journal of Housing Policy*, 15(1): 43–60.
Aalbers, Manuel B., and Brett Christophers. 2014. "Centering Housing in Political Economy." *Housing, Theory and Society*, 31: 373–394.
Arundel, Rowan, and Christian Lennartz. 2017. "Returning to the Parental Home: Boomerang Moves of Younger Adults and the Welfare Regime Context." *Journal of European Social Policy*, 27(3): 276–294.
Atterhög, Mikael. 2006. "The Effect of Government Policies on Home Ownership Rates: An International Survey." In *Home Ownership: Getting in, Getting from, Getting out: Part 2*, edited by John Doling and Marja G. Elsinga, 7–34. Amsterdam: IOS Press.
Bengtsson, Bo, Hannu Ruonavarra, and Jardar Sørvoll. 2017. "Home Ownership, Housing Policy, and Pateh Dependence in Finland, Norway, and Sweden." In *Housing Wealth and Welfare*, edited by Caroline Dewilde and Richard Ronald, 60–84. Cheltenham: Edward Elgar Press.
Birch, Kean, and Matti Siemiatycki. 2015. "Neoliberalism and the Geographies of Marketiziation: The Entangling of State and Markets." *Progress in Human Geography*, 40(2): 177–198.
Bloomberg. 2016. "Landlord Nation: Boomer's New Retirement Plan is Millennials Paying Rent."
Cigdem, Melek, and Stephen Whelan. 2017. "Intergenerational Transfers and Housing Tenure: Australian Evidence." *International Journal of Housing Policy*, 17(2): 227–248.
Doling, John, and Richard Ronald, eds. 2014. *Housing East Asia: Socioeconomic and Demographic Challenges*. Basingstoke: Palgrave Macmillan.
Druta, Oana, and Richard Ronald. 2017. "Young Adults' Pathways into Homeownership and the Negotiation of Intra-Family Support: A Home, the Ideal Gift." *Sociology*, 51(4): 783–799.
European Central Bank. 2009. *Housing Finance in the EURO Area*. Frankfurt: European Central Bank.
Eurostat. 2015. *Housing Statistics Brussels*. Belgium: Eurostat.
Forrest, Ray, and Murie Alan. 1988. *Selling the Welfare State. The Privatisation of Public Housing*. New York: Routledge.
Forrest, Ray, and Yosuke Hirayama. 2015. "The Financialisation of the Social Project: Embedded Liberalism, Neoliberalism and Home Ownership." *Urban Studies*, 52(2): 233–244.
García-Lamarca, Melissa, and Maria Kaika. 2016. "'Mortgaged lives': The Biopolitics of Debt and Housing Financialization." *Transactions of the Institute of British Geographers*, 41(3): 313–327.
Grigsby, William G. 1990. "Housing Finance and Subsidy in the United States." *Urban Studies*, 27(6): 831–845.
Gurney, Craig M. 1999. "Pride and Prejudice: Discourses of Normalisation in Private and Public Accounts of Home Ownership." *Housing Studies*, 14(2): 163–183.
Harloe, Michael. 1995. *The People's Home? Social Rented Housing in Europe and America*. Oxford: Blackwell.
Immergluck, Dan. 2015. *Preventing the Next Mortgage Crisis: The Meltdown, the Federal Response, and the Future of Housing in America*. Lanham, MD: Rowman and Littlefield.
Intermediary Mortgage Lenders Association. 2014. *Reshaping Housing Tenure in the UK: The Role of Buy-to-Let*. London: Intermediary Mortgage Lenders Association.
Kemeny, Jim. 1981. *The Myth of Home Ownership: Public Versus Private Choices in Housing Tenure*. New York: Routledge.
Kemp, Peter A., and Stefan Kofner. 2014. "Germany." In *Private Rental Housing: Comparative Perspectives*, edited by A. Tony, T. Crook, and Peter A. Kemp, 27–47. Cheltenham: Edward Elgar.

Kennett, Patricia, Ray Forrest, and Alex Marsh. 2013. "The Global Economic Crisis and the Reshaping of Housing Opportunities." *Housing, Theory and Society*, 30(1): 10–28.

Kurz, Karin, and Hans-Peter Blossfeld, eds. 2004. *Home Ownership and Social Inequality in Comparative Perspective*. Stanford, CA: Stanford University Press.

Kwak, Nancy H. 2015. *A World of Homeowners: American Power and the Politics of Housing Aid*. Chicago, IL: University of Chicago Press.

Langley, Paul. 2010. "The Performance of Liquidity in the Subprime Mortgage Crisis." *New Political Economy*, 15(1): 71–89.

Lennartz, Christian, Rowan Arundel, and Richard Ronald. 2016. "Younger Adults and Home Ownership in Europe through the Global Financial Crisis." *Population, Space, and Place*, 22(8): 823–835.

McCabe, Brian J. 2016. *No Place Like Home: Wealth, Community and the Politics of Home Ownership*. New York: Oxford University Press.

Nettleton Sarah, and Roger Burrows. 1998. "Mortgage Debt, Insecure Home Ownership and Health: An Exploratory Analysis." *Sociology of Health and Illness*, 20(5): 731–753.

Norris, Michelle. 2016. *Property, Family and the Irish Welfare State*. Cham, Switzerland: Palgrave Macmillan.

OECD. 2011. *Housing and the Economy: Policies for Renovation: Economic Policy Reforms 2011: Going for Growth*. Paris: OECD.

Piketty, Thomas. 2014. *Capital in the Twenty-First Century*. Cambridge, MA: Harvard University Press.

Retsinas, Nicolas P., and Eric S. Belsky, eds. 2005. *Building Assets: Building Credit: Creating Wealth in Low-Income Communities*. Washington, DC: Brookings Institution Press.

Rogers, Dallas, and Sin Yee Koh. 2017. "The Globalisation of Real Estate: The Politics and Practice of Foreign Real Estate Investment." *International Journal of Housing Policy*, 17(1): 1–14.

Rohe, William, Shannon van Zandt, and George McCarthy. 2002. "Home Ownership and Access to Opportunity." *Housing Studies*, 17(1): 51–61.

Rolnik Raquel. 2013. "Late Neoliberalism: The Financialization of Homeownership and Housing Rights." *International Journal of Urban and Regional Research*, 37(3): 1058–1066.

Ronald, Richard. 2008. *The Ideology of Home Ownership: Homeowner Societies and the Role of Housing*. Lanham, MD: Palgrave Macmillan.

Ronald, Richard, and Kees Dol. 2011. "Housing in the Netherlands Before and After the Global Financial Crisis." In *Housing Markets and the Global Financial Crisis*, edited by Ray Forrest and Ngai-Ming Yip, 93–112. Cheltenham: Edward Elgar.

Ronald, Richard, and Justin Kadi. 2016. *The Revival of Private Landlords in Britain's Post Homeownership Society*. Amsterdam: University of Amsterdam.

Ronald, Ronald, Justin Kadi, and Christian Lennartz. 2017. "Whatever Happened to Asset-Based Welfare? Shifting Approaches to Housing Wealth and Welfare Security." *Policy & Politics*, 45(2): 173–193.

Salvi del Pero, Angelica, Willem Adema, Valeria Ferraro, and Valérie Frey. 2016. *Policies to Promote Access to Good-Quality Affordable Housing in OECD Countries*. Paris: OECD.

Saunders, Peter. 1990. *A Nation of Homeowners*. London: Unwin Hyman.

Savills. 2016. *Around the World in Dollars & Cents: Trends in International Real Estate Trading*. London: Savills.

Schwartz Herman, and Leonard Seabrooke. 2008. "Varieties of Residential Capitalism in the International Political Economy: Old Welfare States and the New Politics of Housing." *Comparative European Politics*, 6: 237–261.

Shlay, Anne B. 2006. "Low-Income Homeownership: American Dream or Delusion?" *Urban Studies*, 43(3): 511–531.

Smith Susan J. 2015. "Owner Occupation: At Home in a Spatial, Financial Paradox." *International Journal of Housing Policy*, 15(1): 61–64.

Stebbing, Adam, and Ben Spies-Butcher. 2016. "The Decline of a Home Owning Society? Asset-Based Welfare, Retirement and Intergenerational Equity in Australia." *Housing Studies*, 31(2): 190–207.

U.K. Department for Communities and Local Government. 2016. *English Housing Survey 2016*. London: UK Department for Communities and Local Government.

U.S. Bureau of the Census. 2016. *Housing*. Washington, DC: U.S. Bureau of the Census.

Vale, Lawrence J. 2007. "The Ideological Origins of Affordable Homeownership Efforts." In *Chasing the American Dream: New Perspectives on Affordable Homeownership*, edited by William M. Rohe and Harry L. Watson, 15–40. Ithaca, NY: Cornell University Press.

Waldron, Richard, and Declan Redmond. 2015. "Stress in Suburbia: Counting the Costs of Ireland's Property Crash and Mortgage Arrears Crisis." *Tijdschrift voor Economische en Sociale Geografie*, 107(4): 484–501.

SECTION 4

Rental Housing

10

SUBSIDIZED RENTAL HOUSING PROGRAMS IN THE U.S.

A Case of Rising Expectations

Kirk McClure

Introduction

The United States is a country with about 134 million housing units for 117 million households (U.S. Bureau of the Census 2016). Even with such a large surplus of housing, there are too few units offered at prices low enough to permit extremely low-income people to consume good quality housing in high-opportunity neighborhoods without paying a very high percentage of their income for that housing. The poorest households in the United States tend to be renters rather than homeowners. For this reason, the federal government funds a set of programs designed to help eligible low-income renter households live in good quality housing priced within their means. Sadly, the amount of assistance falls well short of the needs. The federal government provides assistance to about six million households (McClure and Johnson 2015). Even with this level of effort, there remain about 20 million renter households who pay in excess of 30 percent of income on rent. Nearly ten million of these renter households have an annual income below $20,000 (U.S. Bureau of the Census 2016).

Over time, these federal affordable housing programs have changed. The goals served by the programs have expanded and evolved, and the expectations have changed as well. In this chapter, the origins of the different U.S. housing strategies will be described. This discussion is followed by an identification of how these strategies are changing and the path that they may follow in the future. The chapter concludes by calling for reforms to the two active strategies in operation today. The production strategy, the Low-Income Housing Tax Credit (LIHTC) program, needs to be revised to ensure that new units are added to markets only where there is a need for additional housing at the price point served by the program and, to the extent possible, where the location offers access to good schools, gainful employment, and a safe living environment. The voucher strategy, the Housing Choice Voucher (HCV) program, should be guided by the Gautreaux and Moving to Opportunity experiments. These experiments suggest strongly that the HCV program needs to be revised so that it better helps extremely low-income households, especially families with children, to enter into high-opportunity neighborhoods offering racially, ethnically, and economically diverse populations (see also the chapter authored by Howard Husock).

Legislative History of Programs

The U.S. federal government funds a set of affordable housing programs for low- or moderate-income renter households. The goals of the programs have changed over time, evolving from a focus on increasing the supply of low-cost units to a focus on helping eligible low-income households afford existing rental units. The programs initially focused on producing rental housing for low-income occupancy. For many years, the nation followed a single strategy of attaching subsidy to units to be occupied by low-income renters. In the mid-1970s, the approach shifted to a twin strategy of sponsoring both project-based supply-side programs and tenant-based demand-side programs simultaneously. Over time the expectations of these twin strategies have expanded beyond addressing just the problem of housing affordability to serving the goals of poverty deconcentration and breaking the cycle of poverty.

The Origins of Supply-Side Programs

Supply-side programs subsidize the development of dwelling units for occupancy by eligible low-income households. The early form of this approach only assisted publicly owned units. Over time, the approach changed, attaching subsidy to privately owned units. In all cases, the subsidy flows to the property owner with the condition that the occupants must meet certain income restrictions.

Public Housing

The U.S. public housing program began in 1937 with little private participation. Public housing is owned and managed by public housing authorities (PHAs), and these agencies are branches of municipal government. The initial development of public housing and most subsequent capital improvements are financed by the federal government. The only private role in the process is found in the construction of the units, which is performed by private firms contracted by the PHA.

There are approximately 1.1 million public housing units in the U.S., owned and managed by some 3,200 PHAs (Council of Large Public Housing Authorities 2016). These days, public housing is generally housing for the poorest of the poor. All tenants must have low income, defined as income less than 80 percent of the metropolitan area's median family income (AMI), but most households have extremely low income (less than 30 percent of the AMI) (Council of Large Public Housing Authorities 2016). A tenant in a public housing unit pays 30 percent of income toward rent or a flat rent if the flat rent is higher. The typical household in public housing has an income of about $12,000 per year. Thus, the tenant contribution toward rent is 30 percent of the monthly income, or about $330 (Council of Large Public Housing Authorities 2016). However, many households have no income but must pay the flat rent, which is often a figure such as $50 per month (Center for Budget and Policy Priorities 2016).

The national portfolio of public housing has been contracting slowly. Over time units are being demolished as the projects age and cease being serviceable. The City of Atlanta recently completed the planned demolition of all of its public housing (Oakley et al. 2013). Small numbers of units are being added through various initiatives such as the HOPE VI program and its successor, the Choice Neighborhood Program. These programs seek the removal of severely distressed public housing coupled with replacement of some of the units. Usually, the replacement units are in a mixed-income configuration and often with mixed-ownership whereby the PHA partners with other private for-profit and nonprofit organizations (Popkin et al. 2004). The goal is to transform struggling neighborhoods with distressed public or other HUD-assisted housing into healthy and desirable neighborhoods through replacing the distressed housing with high-quality mixed-income housing (U.S. Department of Housing and Urban Development 2016a). These efforts have met with decidedly mixed success as some developments created a type of integration that produced marginalization and alienation among the relocated public housing residents as well as among those who were able to move back (Chaskin and Joseph 2015).

Below-Market Interest Rate Financing

Over time, discomfort with public housing has increased as many projects have become troubled. Thus, the federal government sought private sector partners in the hope that their expertise could remedy some of the problems. In the 1960s, the federal government created the Below Market Interest Rate (BMIR) programs to increase housing supply by combining the Federal Housing Administration (FHA) mortgage insurance with a direct mortgage interest subsidy. These programs offered insured loans usually with interest rates of only 1 percent, in exchange for the developer promising to charge low rents and limit the income of the eligible households who could occupy the units (U.S. Government Accountability Office 1978). These programs included the Section 236 program for family housing, the Section 202 program for elderly housing owned by a nonprofit sponsor, and the Section 811 program for the disabled. There are about 500,000 units left from this era of programs, with most of these units subsidized under the Section 236 program, which ran from 1968 to 1973 (U.S. Department of Housing and Urban Development 2016c).

Section 8 New Construction/Substantial Rehabilitation

The Housing and Community Development Act of 1974 was a watershed act that ended the urban renewal program and began the Section 8 program. Section 8 had two parts, one a project-based component and the other a tenant-based component. The project-based component leased newly developed or substantially rehabilitated units, guaranteeing the rent for a period of 30 or 40 years. With that guarantee, the developer could leverage financing for the development from either private or public lenders. The lease stipulated that the government would pay the difference between the tenant's contribution (30 percent of income) and the rent on the apartment. Tenants had to be low-income, and the landlord was required to maintain low-income occupancy on the building for a period of at least 20 years.

Developers competed for these very lucrative leases, causing them to choose locations and offer designs that would win the lease. The result was housing that, while not luxurious, was generally seen as an improvement compared to public housing (McClure and Johnson 2015). Formally known as the Section 8 New Construction/Substantial Rehabilitation program, it added units to the affordable housing stock. The program was most active during the Carter administration, when it was viewed as a counter-cyclical tool to boost employment through construction as well as a procedure to add to the stock of affordable housing. The Reagan administration brought the program to a halt by zero-funding it. As the portfolio of these units grows older, the units are being removed from service, generally through demolition, although some projects in strong markets are being converted to market-rate housing (National Low-Income Housing Coalition 2015). As of this writing, there are about 750,000 units remaining, down from a peak of about 900,000 units.

Low-Income Housing Tax Credit

The LIHTC program was created by the Tax Reform Act of 1986, an unlikely place for the birth of a housing program, filling the void created by the Reagan administration when it stopped funding for the Section 8 New Construction/Substantial Rehabilitation program. Because the tax act ended a great many of the preferential tax benefits given to real estate in general and housing in particular, supporters of affordable housing pushed for inclusion of some tax benefits that could restore a housing production program within the new income tax rules created by the act. The LIHTC is, at is core, a tax benefit. If investors put money in a housing project developed for low- or moderate-income occupancy, they can receive credits against their federal income tax liability. The process of allocating the tax credits is complicated, and the process of converting the tax credits into equity is convoluted (Khadduri et al. 2012). Despite all its complexities, the LIHTC program is currently the federal government's primary program for adding rental units to serve the needs of eligible low-income households. There are about 2.8 million LIHTC units in service across the nation (U.S. Department of Housing and Urban Development 2016b).

The Origins of Demand-Side Programs: Rent Supplement Program and Rental Assistance Program

The Rent Supplement and Rental Assistance programs were the earliest forms of tenant-based rental assistance, supplementing some of the BMIR programs. These programs helped to cover the costs of operating rental properties for eligible low-income occupants when the financing subsidy was insufficient to hold the rents down to a level affordable to low-income people. These programs were precursors to the later and much larger certificate/voucher program and are being phased out, with only about 14,000 households still assisted under these programs (U.S. Department of Housing and Urban Development 2016c).

Section 8 Certificates/Housing Choice Vouchers

The tenant-based component of the Section 8 program attached the housing subsidy to the household, rather than the housing unit. An eligible low-income household admitted to the program received a certificate permitting it to select a unit for rent in the marketplace. If the rent was below a regulatory ceiling and the unit passed physical inspection, the household was allowed to lease the unit, paying 30 percent of income toward rent, and the government paid the difference between the tenant's contribution and the rent on the unit.

This tenant-based approach of rental housing assistance is premised on the notion that the existing stock of privately owned rental units offers a sufficient number of units to absorb the eligible low-income households with certificates. It is also premised on the notion that the participating households would use the rental assistance to locate in good quality neighborhoods (Winnick 1995).

This certificate program transitioned into the currently active Housing Choice Voucher program. With the voucher, rather than a ceiling on rent, the tenant works with a ceiling on the subsidy amount that can be given to an eligible household in each market. Thus, the household could spend more than 30 percent of income on housing if it chose to do so. However, the household could spend no less than 30 percent but no more than 40 percent of income on housing, giving the household freedom to choose to consume more housing than the 30 percent limit would permit, if that meets the household's needs. The vast majority of households with vouchers do not exercise this option to spend more. Rather, most households tend to consume a unit with rents at the normal rent ceiling and contribute 30 percent of income toward payment of that rent (McClure 2005). Currently, there are over 2.4 million households in the HCV program (U.S. Department of Housing and Urban Development 2016c).

Performance of the Programs

Housing Market Problems Changed

Over time, the goals served by public housing have changed. When Congress created public housing in 1937 it sought to achieve two different goals, one for slum removal, and the other to house the working poor. First, the public housing program obligated PHAs to meet a one-for-one replacement requirement; i.e., for each unit of public housing that was created, one unit of substandard housing had to be demolished. Obviously, the construction of public housing was seen as a mechanism to remove housing in poor condition and to replace it with housing in good physical condition. However, the replacement housing was never meant to be of high quality as public housing was limited to being housing of modest design (Mitchell 1985). Given these requirements, public housing developments generally have been unattractive developments located in the least desirable neighborhoods. The one-for-one replacement requirement was dropped with the passage of the Housing Act of 1949. The urban renewal program took over as the program to demolish substandard housing, leaving public housing to augment the supply of affordable housing, but the housing tended to remain in low-income neighborhoods and to be rather bland in appearance.

Second, the public housing program was not originally intended to be housing for the poorest of the poor. In its original design, the federal government provided funding for the development and construction of the public housing, but the local housing authority had to charge rents sufficient to cover all operating expenses including maintenance, management, insurance, and property taxes. Therefore, the rents were below-market-rate rents given that there was no need to cover debt service on the development, but the rents had to be high enough to pay for the other expenses, which could be considerable. Eligible tenants needed sufficient income to pay those rents, making public housing beyond the reach of the poorest of the poor. However, the ability to attract tenants with incomes at the upper end of the eligible low-income households became increasingly difficult. Given its lack of desirable design and poor location, public housing came to be occupied by the lowest-income households, making it difficult, if not impossible, for local housing authorities to charge the rents needed to properly maintain the developments (Vale 2013). With the passage of the Brooke Amendment in 1969, the purpose of public housing changed dramatically. From then on, housing authorities received operating subsidies from the federal government, permitting these authorities to charge rents to tenants that were no more than 25 percent of the tenant's income. This figure was changed to 30 percent in 1981 (Feins et al. 1994). With this shift in rent calculations and the flow of operating subsidies, public housing became housing for the poorest of the poor.

Problems Shifting From Adequate Numbers of Units to the Affordability of Units

Public housing began in an era when rental units, though often adequate in number, were inadequate in physical condition. The public housing program was designed to remedy this problem by replacing poor quality with good quality units. Over time, housing policy experts recognized that high rent levels rather than the number of good quality rental units was at the heart of the rental housing affordability problem (U.S. Department of Housing and Urban Development 1974).

Beginning in the 1970s, affordability became the dominant problem among low-income renter households, not the availability of units. Within the U.S. Department of Housing and Urban Development (HUD) there was a recognition that nationally the housing supply had expanded and that the quality of the housing stock had increased. If the stock of housing is adequate in size and quality, then the focus of affordable housing policy and programs no longer needs to be on adding units to the supply or even improving their physical condition. Rather, the focus needs to be on helping low-income households afford units that already exist (Winnick 1995).

As part of the recognition that affordability had become the dominant problem, HUD sponsored the Experimental Housing Allowance (EHAP) program (Struyk and Bendick 1981). This program was an experiment launched in multiple cities to test the capacity of the poor to use income supplements earmarked for housing consumption. The experiment was designed to track the responses of the households, the landlords, and the program administrators to answer several wide-ranging questions. For example, would households move to better neighborhoods, deconcentrating poverty? Would racial and ethnic minority households seek out locations that are racially and ethnically integrated, desegregating urban neighborhoods? Would landlords improve properties to qualify for entry into the program, improving the quality of the stock? The answers were not always encouraging, but the answers did demonstrate that such a program could help the poor to consume housing in the private rental market without imposing an undue burden on their incomes. Among housing policy experts, there appeared to be a major shift away from reliance on project-based affordable housing production programs and toward tenant-based assistance programs. Winnick (1995) argues that the shift was not due to the successes or failures of the EHAP but due to the many failures of the assisted housing production programs. He argues that the shift was going to take place, independent of the outcomes of the EHAP, due to changing partisan politics, with liberal Democrats on the production side and conservative Republicans on the voucher side joined by leading economists favoring the voucher approach.

Affordability may be the dominant housing problem confronted by low-income renters, but this does not mean that the federal government has shifted entirely to tackling just that problem. Since the inception of the tenant-based approach in the Section 8 program, there has only briefly been a period when the nation endorsed only the tenant-based approach, rejecting the project-based approach. When the Reagan administration ended the project-based component in 1982, there was no large-scale production program in place until 1986, when the LIHTC program was brought into being. From 1986 to the present, the United States has continued to follow a dual path of funding both the tenant-based HCV program and the project-based LIHTC program.

Providing Access to High-Opportunity Neighborhoods

The expectations of the federal housing programs are now higher. Originally the programs were intended to provide affordable housing for the participating households as efficiently as possible. With the rise of the tenant-based programs, it has become clear that in most cases vouchers are the most cost-effective means to resolve housing affordability problems. Vouchers generally operate at a lower cost per household than project-based rental assistance programs (Deng 2005; U.S. Government Accountability Office 2002). In the past few decades, the expectations have moved to a new and higher level. Not only is assisted housing to provide cost-effective affordable shelter but the housing is also to serve as a platform to lift households out of poverty and into safe neighborhoods providing access to good schools and gainful employment (U. S. Department of Housing and Urban Development 2011).

As the supply of rental housing improved in physical condition, public housing as well as privately developed BMIR financed and Section 8 housing units were increasingly seen as problems rather than solutions. The developments were too often poorly maintained and managed and were sources of criminal activity, creating problems for the surrounding neighborhoods. While these distressed developments may be a small proportion of the developments in the nation's portfolio of public and assisted housing, the reputation of these supply-side programs has suffered severe damage. Nevertheless, these problems have led to calls for the adoption of policies that reduce the concentration of the very poor in public housing in particular and project-based housing in general (Schill 1993).

Reducing poverty concentration is now a driving force in U.S. affordable housing policy. Many low-income housing developments can be socially viable if they are well funded and well managed. However, developments entirely occupied by the poor too often become problem projects. Unemployment and limited opportunities are the norm, with little meaningful opportunity to rise out of poverty (Massey and Kanaiaupuni 1993). Residents of these projects often live in fear of crime, and many of the assisted housing projects have deteriorated to the point that the housing itself is dangerous to the health of the residents (National Commission on Severely Distressed Public Housing 1992). Project-based housing tends to concentrate the poor in neighborhoods that offer low levels of opportunity. Thus, tenant-based housing programs are now consciously designed to deconcentrate the poor through guidance of households to low-poverty, high-opportunity neighborhoods.

Tenant-based programs offer greater potential for mobility and, therefore, moving out of concentrated poverty than project-based programs. National studies of HCV households find that assisted households tend to reside in tracts with lower levels of poverty, as well as other indicators of distress, than do residents of public housing, but they also find that HCV households are underrepresented in neighborhoods with low levels of poverty compared to all renter households (McClure and Johnson 2015; Newman and Schnare 1997; Turner 2003). Even with this limited success in deconcentrating poverty, the inability to promote racial and ethnic integration continues to be a problem even with the freedom of choice offered by the HCV program. Vouchers have been relatively ineffective in countering racial and ethnic segregation (Pendall 2000). Minority HCV households tend to live in neighborhoods with a high proportion of minorities (Devine et al. 2003; McClure et al. 2015; Pendall 2000).

The Gautreaux and Moving to Opportunity experiments provide instruction for the future of assisted housing policy. In the 1970s, the Gautreaux program was established based on a court consent decree when the Chicago Housing Authority (CHA) was sued for racially discriminatory practices. The consent decree called for HUD to provide a special allocation of rental housing vouchers for use by households from the CHA's public housing projects or its waiting list. The vouchers were granted to households in two groups. One group of vouchers required that the participating household move to low-poverty, racially integrated neighborhoods, often out of the city in nearby suburbs. The other group of vouchers had no special locational constraints; the households were free to locate anywhere including staying in the racially impacted South Side of Chicago. Rosenbaum (2012) found that participants who moved to suburban locations experienced more substantial improvements in educational attainment, employment, and income than did those households who remained in the inner city.

Many housing scholars criticized the findings from the Gautreaux program because the screening of households for receipt of the vouchers that moved households to the integrated suburbs may have created a selection bias. Those who passed the screening may have been a more motived population, skewing the results. To correct for the selection bias, the Moving to Opportunity (MTO) experiment was carried out in five markets, with three very carefully constructed samples of households in each market. One sample received vouchers with the condition that these households must locate in low-poverty neighborhoods. One sample received vouchers with no locational constraints. The last sample remained in public housing.

The final report on the program found that the MTO experiment had no significant effect on the educational attainment among the participating children or on the employment or the income of the parents (Sanbonmatsu et al. 2011). However, the program did improve the participants' sense of safety as well as certain aspects of mental health, physical health, and overall sense of well-being (Briggs et al. 2010). Interestingly, later studies have found more positive outcomes from the MTO program for families with children, the target group for poverty deconcentration efforts. Turner et al. (2012) controlled for the length of time that a household remained in a low-poverty neighborhood and found better outcomes for health, work, and schooling for those who stayed longer. Most recently, Chetty et al. (2015) found a very positive effect on educational outcomes and income levels for those MTO households who moved to low-poverty neighborhoods while their children were very young. Where this was the case, the earnings gains were sufficient to cover the costs of the voucher program. These research results from the MTO experiment suggest that breaking the cycle of poverty will require a very long time frame. The benefits of poverty deconcentration may not be realized by the adults in the families who move, but rather, in successful life outcomes for the children in the families who move.

The dispersal strategy is widely accepted, but the strategy has its critics. Imbroscio (2008) contends that the proponents of dispersal are attempting to override the freedom of choice granted in the HCV program by steering people to neighborhoods selected by program administrators rather than the participating households. He suggests that the proponents of poverty deconcentration strategies are overselling the evidence of the gains to be achieved from dispersal and are slighting community development alternatives that are available to improve the neighborhoods where the poor reside. Goetz (2003) examined the performance of many efforts to tear down public housing, relocating the residents and rebuilding mixed-income housing on the site. He finds that it is easy to move forward with demolition, but it is very difficult to successfully relocate the hard to house tenants who were moved out of the public housing and even harder to build replacement units in high-opportunity neighborhoods. While poverty deconcentration may be a worthy goal, Goetz suggests that we have yet to learn how to implement these strategies effectively.

The differences between the Gautreaux and the MTO experiments demonstrate the complexity of designing strategies to move low-income households to high-opportunity neighborhoods. The experimental group in the Gautreaux initiative was required to move to a neighborhood with a low level of poverty that was racially integrated as well. These households experienced significantly lower unemployment than those households who located in poor, inner city tracts (Rosenbaum 2012). The experimental group in the MTO initiative was required to move to a neighborhood with a low level of poverty but with no requirement that the neighborhood be racially integrated. The employment gains found with Gautreaux households were not found with the MTO households (Sanbonmatsu et al. 2011). Perhaps the program results are highly sensitive to the criteria included in the definition of a high-opportunity neighborhood or perhaps the emphasis on racial integration is critical. More research is needed to address these concerns.

Vouchers have been both welcomed and feared in many neighborhoods. Research provides conflicting evidence on the impact of vouchers on the values of nearby properties. Galster et al. (2010) examined the use of vouchers in Baltimore County, finding that vouchers could have a positive impact in higher-valued, predominantly White neighborhoods experiencing appreciation in property values. However, the authors also found that in low- or moderately valued neighborhoods experiencing declining values, vouchers could have a substantial, negative impact, accelerating the declining values. In order for the vouchers to have a positive impact, the authors suggested that they should be well dispersed, with no more than a few households at any one location. However, the HCV program is not dispersing households appropriately. Indeed, Wang et al. (2008) found that vouchers often cluster spatially, creating new pockets of concentrated poverty. The extent of clustering varies across metropolitan areas, but where it occurs, it can exacerbate problems in neighborhoods experiencing racial and economic transition. Where the HCV program worsens neighborhood conditions, support for the program can be very fragile. Goering (2005) found that vouchers can generate beneficial effects if used to help eligible extremely low-income households locate in low-poverty neighborhoods, but community opposition from residents of the neighborhoods where these voucher households locate can be swift and powerful.

Going Forward

Public housing has been in transformation. It is slowly being demolished. Some of the demolished public housing is being replaced with developments where units occupied by low-income people will be only a small portion of all the units in the new development. Many of the households in the public housing that is being demolished will be relocated to the private market using vouchers, and the neighborhoods where the public housing was located will be the beneficiaries of a variety of programs to remedy low-quality schools, high crime rates, and the inability to find gainful employment (U.S. Department of Housing and Urban Development 2017). Mixed-income housing and dispersal strategies are now the center piece of federal rental assistance policy (Popkin et al. 2000).

The strategies currently driving federal affordable housing policy are mobility and poverty deconcentration. Mobility is at the heart of the voucher program, with the participating household able to move to the neighborhood of its choice. Poverty deconcentration is also well served by the voucher approach as it seeks to disperse the poor. To the extent that public housing is rebuilt, it is likely to be in a mixed-income format.

Even with their many problems, the strategies to disperse low-income people and to produce mixed-income housing seem to be the dominant ideas driving the implementation of the HCV and the transformation of public housing program. They do not seem to be driving the LIHTC program. The program contains no incentives to produce mixed-income housing. Rather, developers have an incentive to maximize the number of tax credits awarded to their projects, resulting in the vast majority of these projects being structured with 100 percent of the units designated for low-income occupancy (McClure 2000). The LIHTC program serves a less poor population than does either the public housing or the Section 8 project-based housing. The households assisted by the LIHTC program have incomes in the range of 30 to 60 percent of AMI, rather than those below 30 percent who live in public housing or receive vouchers. The LIHTC program is not only serving a more well-off population, but adding units to a market segment that is not suffering from a shortage of units. The LIHTC program is not locating units where there is a shortage of units serving the target population of renter households with incomes ranging from 30 to 60 percent of AMI. Rather, the units are being developed where there is already a surplus of units serving the 30 to 60 percent of AMI households (McClure 2010). However, there is evidence that the LIHTC units attract a sizable share of HCV households. It is estimated that 19 percent of vouchers locate in LIHTC developments (Williamson et al. 2011). The presence of voucher households helps to serve the purposes of deconcentrating the poor because often the LIHTC development is located in low-poverty neighborhoods. The LIHTC program is performing on par with the HCV program in terms of locating units in low-poverty, especially suburban, high-opportunity neighborhoods (McClure 2006). The LIHTC program is locating units well despite containing incentives for developers to locate in Qualified Census Tracts (QCTs) that are areas with high concentrations of poverty. This incentive exacerbates the problems of poverty concentration. The high incidence of LIHTC developments in these QCTs attracts disproportionately more HCV households, further working against the notion of poverty dispersal and mixed-income housing.

The LIHTC program has remained surprisingly unchanged since its inception in 1986. After 30 years, enough is known about its performance that it is ready for major revisions. Revisions to the LIHTC program should include the adoption of incentives to the provision of mixed-income housing so as to reduce the incidence of developments that concentrate poverty. Further revisions should also include a more rigorous market analysis of development proposals so that the developments are only built where there is a demonstrable shortage of rental units in the price point served by the program to reduce the practice of adding units to already glutted markets. The program can and should be revised to incentivize the renovation of existing housing or the replacing of deteriorated housing rather than adding units to markets where ample units exist. Finally, the program should be revised to require LIHTC property managers to actively market units located in high-opportunity neighborhoods to households living in low-opportunity neighborhoods so as to affirmatively further fair housing.

The Housing Choice Voucher program could benefit from revisions as well. HUD is already working toward adjusting the rent limits within each metropolitan area to reflect higher rent levels in more desirable, high-opportunity neighborhoods. Further revisions could enhance the process. Greater outreach to landlords needs to take place so as to expand the participation in the program. The program should also be coordinated across housing authorities within a metropolitan area. The coordination could pool resources to identify high-opportunity neighborhoods and to reach out to landlords in these neighborhoods. Metropolitan-wide coordination could offer households with vouchers more options, improving their search for good units in high-opportunity neighborhoods, again helping to deconcentrate poverty and promote racial integration (see also the chapter authored by Andrew J. Greenlee).

Affordability continues to be the biggest problem in housing markets across the nation. The housing programs are up to the task of resolving this problem, but the lack of funding ensures that unmet need will be the norm (Turner 2003). Both the tenant-based and project-based approaches remain viable and part of the array of federal housing programs. The nation shows little evidence of abandoning its dual strategy. The U.S. contains many housing markets with a great range of problems, calling for many different types of programs to intervene so as to resolve these problems.

As these programs are refined for the future, there appears to be growing evidence that spatial mobility and poverty deconcentration will be the driving goals of the subsidized rental housing programs in the U.S. for some time to come.

References

Briggs, Xavier de Sousa, Susan J. Popkin, and John Goering. 2010. *Moving to Opportunity: The Story of an American Experiment to Fight Ghetto Poverty*. Oxford: Oxford University Press.

Center for Budget and Policy Priorities. 2016. *Policy Basics: Public Housing*. Washington, DC: Center for Budget and Policy Priorities.

Chaskin, Robert J., and Mark L. Joseph. 2015. *Integrating the Inner City: The Promise and the Periods of Mixed-Income Public Housing Transformation*. Chicago: University of Chicago Press.

Chetty, Raj, Nathaniel Hendren, and Lawrence F. Katz. 2015. "The Effects of Exposure to Better Neighborhoods on Children: New Evidence from the Moving to Opportunity Project." *American Economic Review*, 106(4): 855–902.

Council of Large Public Housing Authorities. 2016. *Facts about Public Housing*. Washington, DC: Council of Large Public Housing Authorities.

Deng, Lan. 2005. "The Cost-Effectiveness of the Low Income Housing Tax Credit relative to Vouchers: Evidence from Six Metropolitan Areas." *Housing Policy Debate*, 16(3/4): 469–511.

Devine, Deborah. J., Robert J. Gray, Lester Rubin, and Lydia B. Taghavi. 2003. *Housing Choice Voucher Location Patterns: Implications for Participant and Neighborhood Welfare*. Washington, DC: U.S. Department of Housing and Urban Development.

Feins, Judith D., Sally R. Merrill, Nandinee Kutty, Kathleen Heintz, and Gretchen P. Locke. 1994. *Revised Methods of Providing Federal Funds for Public Housing Agencies: Final Report*. Washington, DC: U.S. Department of Housing and Urban Development.

Galster, George C., Peter Tatian, and Robin Smith. 2010. "The Impact of Neighborhoods Who Use Section 8 Certificates on Property Values." *Housing Policy Debate*, 10(4): 879–917.

Goering, John. 2005. "Expanding Housing Choice and Integrating Neighborhoods: The MTO Experiment." In *The Geography of Opportunity: Race and Housing Choice in Metropolitan America*, edited by Xavier de Sousa Briggs. 127–149. Washington, DC: Brookings Institution Press.

Goetz, Edward G. 2003. *Clearing the Way: Deconcentrating the Poor in Urban America*. Washington, DC: The Urban Institute Press.

Imbroscio, David. 2008. "[U]nited and Actuated by Some Common Impulse of Passion: Challenging the Dispersal Consensus in American Housing Policy Research." *Journal of Urban Affairs*, 30(2): 111–130.

Khadduri, Jill, Kimberly Burnett, Carissa G. Climaco, Laurie Gould, and Louise Elving. 2012. *What Happens to Low-Income Housing Tax Credit Properties at Year 15 and Beyond?* Washington, DC: U.S. Department of Housing and Urban Development.

Massey, Douglas S., and Shawn M. Kanaiaupuni. 1993. "Public Housing and the Concentration of Poverty." *Social Science Quarterly*, 74(1): 109–144.

McClure, Kirk. 2000. "The Low-Income Housing Tax Credit as an Aid to Housing Finance: How Well Has It Worked?" *Housing Policy Debate*, 11(1): 91–114.

McClure, Kirk. 2005. "Rent Burden in the Housing Choice Voucher Program." *Cityscape*, 8(2): 5–20.

McClure, Kirk. 2006. "The Low-Income Housing Tax Credit Program Goes Mainstream and Moves to the Suburbs." *Housing Policy Debate*, 17(3): 19–446.

McClure, Kirk. 2010. "Are Low-Income Housing Tax Credit Developments Locating Where There is a Shortage of Affordable Units?" *Housing Policy Debate*, 20(2): 153–171.

McClure, Kirk, and Bonnie Johnson. 2015. "Housing Programs Fail to Deliver on Neighborhood Quality Re-Examined." *Housing Policy Debate*, 25(3): 463–496.

McClure, Kirk, Alex Schwartz, and Lydia B. Taghavi. 2015. "Housing Choice Voucher Location Patterns a Decade Later." *Housing Policy Debate*, 25(2): 215–233.

Mitchell, J. Paul. 1985. *Federal Housing Policy and Programs: Past and Present.* New Brunswick, NJ: Center for Urban Policy Research, Rutgers University.

National Commission on Severely Distressed Public Housing. 1992. *The Final Report of the National Commission on Severely Distressed Public Housing.* Washington, DC: National Commission on Severely Distressed Public Housing.

National Commission on Severely Distressed Public Housing. National Low Income Housing Coalition. 2015. *New Tenant Protection Voucher Provision Distinguishes Replacement and Relocation Vouchers.* Washington, DC: National Low Income Housing Coalition.

National Low-Income Housing Coalition. 2015. *New Tenant Protection Voucher Provision Distinguishes Replacement and Relocation Vouchers.* Washington, DC: National Low-Income Housing Coalition.

Newman, Sandra J., and Ann B. Schnare. 1997. "'. . . And a suitable living environment': The Failure of Housing Programs to Deliver on Neighborhood Quality." *Housing Policy Debate,* 8(4): 703–741.

Oakley, Deirdre, Erin Ruel, and Lesley Reid. 2013. "Atlanta's Last Demolitions and Relocations: The Relationship between Neighborhood Characteristics and Resident Satisfaction." *Housing Studies,* 28(2): 205–234.

Pendall, Rolf. 2000. "Why Voucher and Certificate Users Live in Distressed Neighborhoods." *Housing Policy Debate,* 11(4): 881–910.

Popkin, Susan J., Larry F. Buron, Diane K. Levy, and Mary K. Cunningham. 2000. "The Gautreaux Legacy: What Might Mixed-Income and Dispersal Strategies Mean for the Poorest Public Housing Tenants?" *Housing Policy Debate,* 11(4): 911–942.

Popkin, Susan J., Bruce Katz, Mary K. Cunningham, Karen D. Brown, Jeremy Gustafson, and Margery Austin Turner. 2004. *A Decade of HOPE VI: Research Findings and Policy Challenges.* Washington, DC: The Urban Institute.

Rosenbaum, James. 2012. "Gautreaux Program." In *The Encyclopedia of Housing,* edited by Andrew T. Carswell. 239–240. Los Angeles: Sage Publications.

Sanbonmatsu, Lisa, Jens Ludwig, Lawrence F. Katz, Lisa A. Gennetian, Greg J. Duncan, Ronald C. Kessler, Emma Adam, Thomas W. McDade, and Stacy Tessler Lindau. 2011. *Moving to Opportunity for Fair Housing Demonstration Program. Final Impacts Evaluation.* Washington, DC: U.S. Department of Housing and Urban Development.

Schill, Michael H. 1993. "Distressed Public Housing: Where Do We Go from Here?" *The University of Chicago Law Review,* 60(2): 497–554.

Struyk, Raymond J., and Marc Bendick. 1981. *Housing Vouchers for the Poor: Lessons from a National Experiment.* Washington, DC: Urban Institute Press.

Turner, Margery A. 2003. "Strengths and Weaknesses of the Housing Voucher Program." Testimony Before the Subcommittee on Housing and Community Opportunity, Committee on Financial Services, United States House of Representatives, June.

Turner, Margery A., Austin Nichols, and Jennifer Comey. 2012. *Benefits of Living in High-Opportunity Neighborhoods: Insights from the Moving to Opportunity Demonstration.* Washington, DC: The Urban Institute.

U. S. Bureau of the Census. 2016. *American Community Survey. 2010–2014 5-Year Estimates.* Washington, DC: U.S. Bureau of the Census.

U.S. Department of Housing and Urban Development. 1974. *Housing in the Seventies: A Report of the National Housing Policy Review.* Washington, DC: U.S. Department of Housing and Urban Development.

U.S. Department of Housing and Urban Development. 2011. "Informing the Next Generation of Rental Housing Policy." *Evidence Matters,* (Spring): n.p.

U.S. Department of Housing and Urban Development. 2016a. *Choice Neighborhoods.* Washington, DC: U.S. Department of Housing and Urban Development.

U.S. Department of Housing and Urban Development. 2016b. *Low-Income Housing Tax Credits.* Washington, DC: U.S. Department of Housing and Urban Development.

U.S. Department of Housing and Urban Development. 2016c. *Picture of Subsidized Households.* Washington, DC: U.S. Department of Housing and Urban Development.

U.S. Department of Housing and Urban Development. 2017. *About HOPE VI.* Washington, DC: U.S. Department of Housing and Urban Development.

U.S. Government Accountability Office. 1978. *Section 236 Rental Housing: An Evaluation with Lessons for the Future. PAD-78-13.* Washington, DC: U.S. Government Accountability Office.

U.S. Government Accountability Office. 2002. *Federal Housing Assistance: Comparing the Characteristics and Costs of Housing Programs. GAO-02-76.* Washington, DC: U.S. Government Accountability Office.

Vale, Lawrence J. 2013. "Public Housing in the United States: Neighborhood Renewal and the Poor." In *Policy, Planning, and People: Promoting Justice in Urban Development,* edited by Naomi Carmon and Susan S. Fainstein, 285–306. Philadelphia: University of Pennsylvania Press.

Wang, Xinhao, David Varady, and Yimei Wang. 2008. "Measuring the Deconcentration of Housing Choice Voucher Program Recipients in Eight U.S. Metropolitan Areas Using Hot Spot Analysis." *Cityscape*, 10(1): 65–90.

Williamson, Anne R., Marc T. Smith, and Marta Strambi-Kramer. 2011. "Housing Choice Vouchers, the Low-Income Housing Tax Credit, and the Federal Poverty Deconcentration Goal." *Urban Affairs Review*, 45(1): 119–132.

Winnick, Louis. 1995. "The Triumph of Housing Allowance Programs: How a Fundamental Policy Conflict Was Resolved." *Cityscape*, 1(3): 95–121.

11

REDEFINING RENTAL HOUSING CHOICE IN THE HOUSING CHOICE VOUCHER PROGRAM

Andrew J. Greenlee

Introduction

The federal Housing Choice Voucher program (HCVP) serves more than two million low-income households in the U.S. (Center on Budget and Policy Priorities 2017). Unlike place-based subsidized strategies such as public housing, tax credit developments, and project-based Section 8, where subsidies are attached to housing units, the Housing Choice Voucher moves with the tenant, who can use their voucher to shop for housing on the private market. Theories of housing choice rooted in neoclassical economics suggest that given perfect information, tenants will choose a new residential location that best fits their household needs (Rossi 1955; Tiebout 1956). Housing Choice Vouchers are designed to tap into this principle by increasing the share of income that households will devote to housing, resulting in housing opportunities that low-income tenants could otherwise not afford. In practice, outcomes from the program show that while housing unit quality is often improved by the program subsidy, they are more mixed regarding whether voucher-subsidized households actually move to better quality neighborhoods (McClure et al. 2015).

The primary goal of the HCVP is to meet the goal of providing safe, decent, and affordable housing, set by the U.S. Department of Housing and Urban Development (HUD). The second major policy focus is the deconcentration of poverty and the desegregation of historic patterns of racially segregated neighborhoods (Basolo and Nguyen 2005; Chaskin 2013; Goetz 2010; Vale 2013). This goal responds to concerns regarding racial and ethnic segregation and poverty concentration in both public housing and federally subsidized private project-based housing (Goetz 2013; Goetz and Chapple 2010).

Working toward this goal in the HCVP entails the following logic model. The voucher subsidy should open up access to a more diverse range of housing opportunities and neighborhoods, expanding household location choice, which should result in households choosing to live in neighborhoods that reflect a net reduction in household exposure to racially, ethnically, and economically segregated conditions. The exposure to neighborhoods featuring better resources should help households secure a stronger footing to pursue better jobs, better education, and greater household stability. To this end, vouchers have been used both as a legal remedy and as an experimental tool to better understand how housing and neighborhood choice matter for economic and social outcomes for low-income households. Empirical research indicates, however, that "the 'free market' choice assumptions behind the HCV program do not hold in reality" (DeLuca et al. 2013, 271).

This chapter questions some of the shortcomings of the program design of the HCVP in terms of achieving program goals by reviewing demographic and socioeconomic outcomes of housing choice for program participants as well as evidence from research about the program. Drawing upon these observations, I call for an expanded *ecology of choice* that accounts for the role of market actors, i.e., landlords and the housing authority administration, in shaping housing choice outcomes and program outcomes for voucher-assisted tenants.

Housing Vouchers and Housing Choice

Housing vouchers were officially authorized as part of the 1974 Housing and Community Development Act, which modified Section 8 of the United States Housing Act of 1937 to allow for tenant-based rental certificates. Over the next 40 years, the program has grown substantially due to widespread legislative popularity given the subsidy's efficiency, the ability for households to choose their housing unit and neighborhood, and the administrative efficiency of the program (U.S. Department of Housing and Urban Development 2001). A major consolidation in 1998 as part of the Quality Housing and Work Responsibility Act (QHWRA) merged multiple voucher-based programs into the current HCVP.

The HCVP is delivered through collaboration between public housing authorities and private market landlords, meaning that voucher-assisted households are only one of several entities making choices that can impact household outcomes. HUD's *Housing Choice Voucher Program Guidebook* (2001) defines four important relationships that dictate program structure (see Figure 11.1).

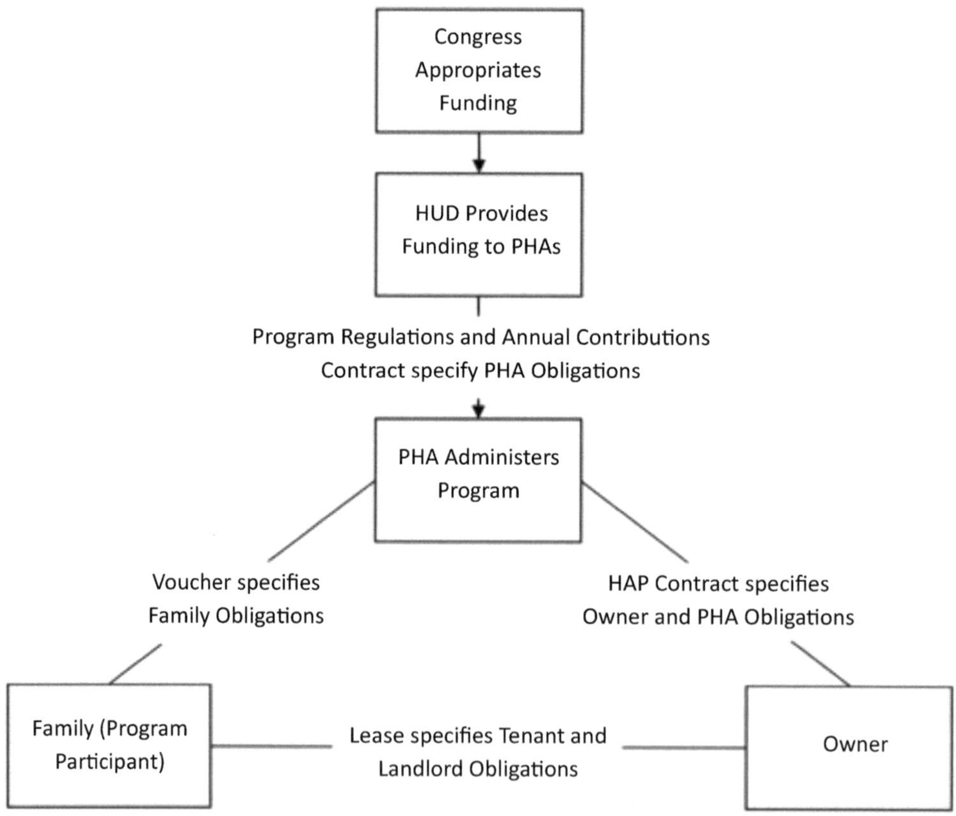

Figure 11.1 Housing Choice Voucher Program Structure.

Source: Greenlee 2012, modified from the *Housing Choice Voucher Program Guidebook* (2001).

After Congress appropriates funding, HUD allocates this funding to local public housing authorities (PHAs) via budget appropriations. In FY2014, the federal government appropriated approximately $19.18 billion for the HCVP, of which $16.34 billion (85.19 percent) was spent in the form of housing assistance payments to private market landlords (Center on Budget and Policy Priorities 2015).[1] Federal funds flow to 2,230 PHAs that administer local HCVPs. PHAs work with program-eligible tenants to identify housing opportunities with private market landlords. PHAs maintain lists of landlords who are willing to participate in the program; however, eligible households can search for housing at their discretion, as long as a housing unit meets basic program requirements. PHAs are also responsible for inspecting units to meet HUD housing quality standards, for negotiating the final rental price and housing assistance payment, and for conducting an annual reexamination of units and households for continued program eligibility. A landlord's relationship with the HCVP is determined by two contractual documents—the lease, which is a rental agreement with the tenant; and the housing assistance payment contract, which is an agreement with the housing authority. Households participating in the program are responsible for identifying an appropriate housing opportunity, negotiating an initial lease with the landlord (who then will undergo further negotiations with the housing authority), and paying an affordable share (30 percent) of their income toward rent.

The HCVP is the largest low-income housing program in the nation. In 2015, the HCVP served 2.2 million households (Center on Budget and Policy Priorities 2017). Evolving alongside the expansion of the program as a strategy for delivering affordable housing, vouchers have played a central role in two types of choice experiments that link the housing mobility of low-income households to neighborhoods and their impacts on low-income households (Sampson 2008). The next section describes these choice experiments and their implications for housing choice within the HCVP.

Experimenting With Choice Using Housing Vouchers

Housing vouchers have been used as a legal remedy to institutional bias and discrimination for several decades. This type of solution was first proposed as a result of the 1976 Supreme Court decision on *Hills v. Gautreaux* (425 US 284), which upheld that HUD violated both the U.S. Constitution's Fifth Amendment and the 1964 Civil Rights Act. The case began in 1974 with *Gautreaux v. Chicago Housing Authority*, in which the plaintiff, Dorothy Gautreaux (a resident of Chicago's Altgeld Gardens public housing development), alleged that the Chicago Housing Authority violated HUD and Civil Rights Act mandates by concentrating public housing units in low-income minority neighborhoods—an act that the plaintiff's attorney Alexander Polikoff later called "residential apartheid" (*Chicago Reporter* 1978, 2).

The 1976 Supreme Court decision prescribed a regional solution to the systematic segregation of African-Americanpublic housing residents. The Gautreaux Assisted Housing program used vouchers to move more than 7,500 low-income households to racially integrated neighborhoods in Chicago and its suburbs. Evaluation research showed that Gautreaux movers showed improvement in terms of safety, employment status, and public assistance dependence rates (Rosenbaum and DeLuca 2000). However, the positive impacts of the program on households were overshadowed by racial discrimination, social isolation, and, on a few occasions, racially motivated violence faced by participants who moved into certain suburban neighborhoods. Rubinowitz and Rosenbaum (2002) note that "no family that experienced such attacks felt comfortable in the suburbs, and some remained very anxious for a long time" (99). The prospect of social isolation and the threat of physical and psychological violence prompted many Gautreaux movers to move back to low-income central city neighborhoods following their experiences in the suburbs (Rubinowitz and Rosenbaum 2002).

Chicago's Gautreaux Consent Decree became a template for many other consent decrees that used housing vouchers as a remedy for many forms of discrimination, including *Walker v. HUD* in Dallas and *NAACP v. City of Yonkers* in New York, which became the context of the 2015 docudrama series *Show Me a Hero.*[2] Ongoing programs, including Baltimore's Special Mobility Housing Choice Voucher program (a result of *Thompson v. HUD* (220 F.3d 241 2000)), focus on regional solutions to housing segregation, involving cooperation between regional housing agencies to help households move to and remain in more advantageous locations.

Directly influenced by the Gautreaux Consent Decree, HUD's ten-year Moving to Opportunity for Fair Housing Demonstration (MTO) began in 1992 amid a particularly vibrant conversation around public and subsidized housing. During that same year, the National Commission on Severely Distressed Public Housing identified 86,000 severely distressed public housing units based upon a combination of maintenance, management, and neighborhood issues (Schill 1993). Journalist Alex Kotlowitz released *There Are No Children Here* (1992), a gripping portrait of Pharaoh and Lafayette Walton's life[3] growing up in Chicago's Henry Horner housing projects. In *The Truly Disadvantaged* (1990), sociologist William Julius Wilson described patterns of middle-class African-Americanresidential mobility as a leading cause of the isolation in urban poor communities that contributed to a "tangle of pathology" (U.S. Department of Labor 1965, 29) in terms of socially maladaptive behavior and outcomes. The 1992 acquittal of the California Highway Patrol officers involved in the Rodney King beating sparked riots in Los Angeles that brought into stark relief the broader state of race relations in the U.S. (Matheson and Baade 2004; Serrano 1992).

Amid this national conversation and drawing inspiration from the Gautreaux Consent Decree, the federal MTO demonstration program sought to find out whether moving low-income households away from concentrated poverty could significantly impact other aspects of their lives.[4] A group of low-income experimental subjects who resided in public housing were given housing vouchers but were constrained in their location choices to neighborhoods with less than 10 percent poverty (treatment group 1). A similar group was given housing vouchers but was not constrained in their location choice (treatment group 2). A third group remained in public housing and was given no additional assistance (control group), but was tracked to provide a baseline comparison to the other two groups. Evaluation of program location outcomes over 10 to 15 years showed that movers with no assistance lived in areas with an average annualized poverty rate of 40 percent. At the beginning of the experiment the poverty rate of treatment groups 1 and 2 was 9 and 7 percentage points lower, respectively. By the end of the experiment, the poverty rates of treatment groups 1 and 2 were 18 and 11 percentage points lower, respectively (Sanbonmatsu et al. 2011).

Within the MTO demonstration, vouchers helped both treatment groups move to neighborhoods with lower poverty rates (although the poverty rates in these neighborhoods are still above average when compared to the nationwide neighborhood poverty rate). Movers in the treatment group also lived in better quality housing, felt safer in their neighborhoods, and had more social ties with more affluent people. Movers in both treatment groups saw a slight reduction in racial segregation, although, on average, these households remained in neighborhoods that were majority minority. The MTO experiment also tested for effects in a range of other areas, including adult and youth physical and mental health, risky behavior, educational outcomes, and economic outcomes (Briggs et al. 2010). While significant improvements were observed in terms of housing quality, health, safety, and poverty for experimental group participants who moved to low-poverty neighborhoods, the experiment did not result in a significant improvement for educational outcomes, earnings, or employment. As the researchers stated in their final evaluation, "MTO was more successful in improving mental and physical health in poor families that signed up to participate in the program than in bringing about the improvements in education and labor market outcomes needed to boost family incomes" (Sanbonmatsu et al. 2011, xxxi). Research on intergenerational outcomes for MTO participants finds more positive impacts, particularly for women with children (Chetty et al. 2016; see the chapter authored by Howard Husock). Taken together, these findings point to the pernicious ways in which structural disadvantage impacts individuals and families, resulting in the need for sustained intervention across generations to achieve desired results.

Location Choices in the Housing Choice Voucher Program

The design and outcomes of these experimental mobility programs provide an important point of reference for understanding the choices of voucher-assisted households participating in the HCVP. Data from HUD's *Picture of Subsidized Households*, published in 2015, provides estimates of the number of voucher-subsidized households by Census tract, which are frequently used as proxies for neighborhoods (Coulton et al. 2001; Sperling 2012).[5] Examining these data provides a portrait of where HCVP households not participating in experimental mobility programs reside. Combining those data with additional demographic and socioeconomic information from the 2010–2014 American Community Survey provides important information for analyzing the neighborhood context for voucher households, and provides the opportunity to compare the location context for voucher-subsidized households to that for all households in the United States.

Table 11.1 compares the neighborhood demographics and socioeconomics of the Census tracts in which voucher-subsidized households live to neighborhood demographics and socioeconomics of all households in the United States. Estimates are weighted based upon the number of voucher-subsidized households and the total number of households (respectively) in each tract. Voucher-subsidized households constitute 2.05 percent of all households nationally. When one looks at a series of indicators that are associated with neighborhood deprivation (Messer et al. 2006), voucher-subsidized households on average live in neighborhoods that are worse off when compared to all households in the United States. More specifically, voucher-subsidized households live in neighborhoods with higher poverty rates (24.81 percent of households in poverty compared to 15.43 percent of all households), a greater share of non-White households (44.19 percent of non-White households compared to 25.08 percent of all households), and higher unemployment rates (19.52 percent compared to 9.25 percent for all households).

The question of how to measure neighborhood level opportunity has sparked much work (Briggs 2006; Galster and Killen 1995). Looking at the neighborhood demographic thresholds applied in the Gautreaux Consent Decree (less than 30 percent neighborhood racial minority population) and MTO experiment (less than 10 percent neighborhood poverty rate) provides a means of assessing neighborhood choice for the nation's voucher-assisted population. These thresholds are frequently applied within both policy and research contexts (McClure 2008).

Applying the Gautreaux neighborhood racial diversity criteria, nearly 66 percent of HCVP households reside in a neighborhood with less than 30 percent minority population when compared to almost 88 percent of all households (see Table 11.2). Looking at the demographic and

Table 11.1 Household-Weighted Census Tract Demographics and Socioeconomics.

Tract-Level Indicator	*HCVP Households*	*All Households*
Households	2,387,501	116,211,092
Poverty Rate	24.81%	15.43%
Non-White Households	44.19%	25.08%
Female Headed Households	20.15%	13.03%
Households Making > $30K	40.11%	28.33%
Households With Public Assistance Income	5.04%	2.82%
Overcrowded Households	5.59%	3.32%
Unemployment Rate	12.78%	9.25%
Adults With No High School Degree	19.52%	13.27%
Cost Burdened Households	43.71%	34.86%

Source: author, derived from the *HUD Picture of Subsidized Households* (2015) and 2010–2014 American Community Survey Estimates.

Table 11.2 Household-Weighted Census Tract Demographics and Socioeconomics for Racially Mixed Tracts (Less Than 30 Percent African-American Population).

Tract-Level Indicator	*HCVP Households*	*All Households*
Households	1,569,297 (65.72%)	101,743,470 (87.55%)
Poverty Rate	21.23%	13.86%
Non-White Households	28.00%	19.00%
Female Headed Households	15.73%	11.40%
Households Making < $30K	35.77%	26.37%
Households With Public Assistance Income	4.43%	2.56%
Overcrowded Households	6.02%	3.22%
Unemployment Rate	10.54%	8.46%
Adults With No High School Degree	18.54%	12.45%
Cost Burdened Households	41.32%	33.67%

Source: author, derived from the *HUD Picture of Subsidized Households* (2015) and 2010–2014 American Community Survey Estimates.

Table 11.3 Household-Weighted Census Tract Demographics and Socioeconomics for Low-Poverty Tracts (Less Than 10 Percent Poverty).

Tract-Level Indicator	*HCVP Households*	*All Households*
Households	316,906 (13.27%)	46,299,469 (39.48%)
Poverty Rate	6.72%	5.75%
Non-White Households	25.50%	17.34%
Female Headed Households	11.35%	8.92%
Households Making < $30K	19.04%	16.04%
Households With Public Assistance Income	2.08%	1.50%
Overcrowded Households	2.34%	1.49%
Unemployment Rate	7.26%	6.56%
Adults With No High School Degree	8.85%	6.77%
Cost Burdened Households	33.96%	30.28%

Source: author, derived from the *HUD Picture of Subsidized Households* (2015) and 2010–2014 American Community Survey Estimates.

socioeconomic attributes for the subset of voucher-assisted households and all households that live in neighborhoods meeting the Gautreaux neighborhood diversity criteria, on average, the sociodemographic indicators for voucher-assisted households in these neighborhoods are still less favorable when compared to all households. On average, HCVP households that live in neighborhoods that meet the Gautreaux neighborhood diversity criteria live in neighborhoods with higher poverty rates (21.23 versus 13.86 percent for all households). Despite being considered diverse, voucher-subsidized households live in neighborhoods with a higher share of non-White households (28 versus 19 percent for all households) on average.

Applying the MTO neighborhood poverty criteria, i.e., a poverty rate of less than 10 percent in the Census tract, 13 percent of HCVP households live in low-poverty Census tracts, compared to almost 40 percent of all households nationwide (see Table 11.3). Again, comparing the average demographic and socioeconomic conditions for those voucher-assisted households living in low-poverty neighborhoods to all households living in those neighborhoods, demographic and socioeconomic conditions are less favorable for voucher-assisted households, although the differential is smaller for indicators, especially poverty, income characteristics, and cost burden, when compared to the differential based on the racial diversity criterion.

While these measures focus on neighborhood conditions and not on individual household characteristics, voucher-assisted households are on average more likely to be exposed to less favorable demographic and socioeconomic conditions in the neighborhoods they live in, even under the circumstances where they are living in neighborhoods that meet definitions of being "high opportunity." Although the gap decreases in the case of low-poverty neighborhoods, a much smaller proportion of voucher-subsidized households are able to access these neighborhoods when compared to accessing racially diverse neighborhoods. These choices could simply reflect preferences or housing market constraints, but other factors may also influence these outcomes.

Program Design, Choice, and Outcomes

Housing policy experiments and consent decrees help to illustrate the potential impact of certain neighborhood choices for low-income households that can be facilitated by housing vouchers. Whether looking within the context of experimental mobility programs or the HCVP in general, outcomes for voucher-assisted households tend to fall short of the goals set by HUD (Galvez 2010; Goetz and Chapple 2010; Kleit et al. 2016). Empirical evidence shows that housing quality and search barriers both explain a portion of this gap (Wang 2018). Voucher-assisted households may prefer to live in neighborhoods other than the ones they actually choose, but these preferences are ultimately subordinated to other needs including access to existing job networks, access to educational institutions, and access to friend and family networks. Similarly, voucher-assisted households may lack knowledge or familiarity with certain neighborhoods, thereby limiting the set of choices that they will look at. A portion of this gap is also explained by market competition and specific trade-offs between affordability and accessibility (McClure 2015; Tremoulet et al. 2016).

An alternate explanation could be that researchers and evaluators are missing or improperly measuring housing market constraints faced by voucher-assisted households. For instance, the MTO program evaluators noted that "the success of voucher-based assisted housing mobility programs, like that of the housing voucher program generally, hinges on a chain of cooperative action by landlords, tenants, housing agencies, and sometimes organized interest groups" (Briggs et al. 2010, 65).

Either scenario implies a level of unevenness in the rental housing market that puts voucher-assisted households at a disadvantage, and a need to look more holistically at the influence of program design and the choices of program actors on spatial location outcomes for voucher-subsidized households.

Looking a bit deeper at these potential explanations, it is important to think carefully about what types of assumptions housing policy researchers and evaluators tend to make about housing choice. A first assumption is that choosing where to live happens within a rational choice framework. Theories of residential mobility often presuppose that landlords and tenants are operating in a manner that involves utility maximization, and the weighing of costs and benefits associated with changing location (Kleit et al. 2016; Turner and Briggs 2008). Within this type of framework, households move when they are sufficiently dissatisfied with their current location to accept the opportunity cost of moving, and landlords are sufficiently compelled to offer a housing opportunity to a searching household when the household can pay an acceptable price. Furthermore, as a form of housing subsidy, the housing voucher would induce an economically rational landlord to rent to a household whom they otherwise would not be compelled to rent to based on income, or the risk that the tenant would not be able to reliably meet rent demands. Housing vouchers mitigate a portion of a landlord's financial risk through the delivery of the subsidy as a guaranteed form of rent. While the rational choice framework presents a conceptually tidy way for thinking about how choices are made within the program, some of the assumptions involved may not be appropriate (Kleit et al. 2016).

Other approaches emphasize the importance of housing fit across the lifespan or housing career (Payne and Payne 1977; Rex et al. 1967; Skobba and Goetz 2013). Again, the choice process is driven by tenants who seek changes to their housing conditions at such times in their lives when their housing unit characteristics no longer adequately meet the needs of the household. An example of a housing career-based change might include upsizing housing to accommodate the birth of a new child, or downsizing after retirement. Within a housing career framework, efficiency of the housing unit itself might factor into a mobility decision as much as the unit's price or the opportunity cost of moving (Lansing et al. 1969).

Yet other approaches examine the oftentimes unmeasured determinants of choice. Neighborhood demographic and socioeconomic features may be an important proxy for the opportunities that households will have in a new residential location, but other, hard-to-measure factors such as social networks may also prove instrumental to understanding where low-income households move (Clampet-Lundquist 2003; Dominguez and Watkins 2003; Skobba and Goetz 2013). Hard-to-measure factors including past housing histories and pressures to leave previous housing environments and situations can also prove important for understanding what choices low-income households are confronted with (Skobba 2008).

Examining these different approaches together suggests a more complicated model of choice for voucher-assisted households that should account not only for household choices, but also for the choices of other important actors, including housing authorities and landlords. As Turner and Briggs (2008) note, the outcome of a decision to move with a voucher is dependent upon an interconnected series of factors, many of which only get measured as outcomes:

> [The target low-income] families will *want* to move to better locations, given their motivations and the information about choices that are offered; they will be *able* to move, given counseling, search assistance, and other supports; [and] they will be able to *stay* in new locations long enough to benefit from relocating.
>
> *(Turner and Briggs 2008, 27, italics in original)*

Within the previously described choice frameworks, tenants maintain a primacy of choice regarding their mobility decisions. Even in cases where there may be substantial push factors that motivate the decision to move, low-income households still maintain a substantial degree of control as to whether to move. However, as DeLuca et al. (2013) argue, "poor families rarely choose the social contexts that affect their lives, and the dynamic of life in poverty and structural forces they face severely limit their ability to choose and what the choice set looks like" (5). Looking beyond tenant choice reveals the important roles of other actors directly involved in providing the voucher subsidy. In particular, the strategic choices of housing authority officials and landlords can play an important role in explaining where voucher-subsidized households live.

Housing authorities establish the rules that program participants must follow during their tenure in the program, including standards for housing search times. Amid a series of housing rationing processes (Moore 2016), housing authorities may knowingly or unknowingly put pressure on tenants to lease up in an available unit, regardless of neighborhood context, often due to the scarcity of available housing opportunities in some housing markets and the pragmatic perspectives of housing authority case managers (Greenlee 2011). Experimental mobility programs like Gautreaux and MTO integrated some level of in-depth housing mobility counseling into the housing search process to help families make initial moves to communities that met program demographic and socioeconomic requirements (Feins et al. 1997). For most participants in the HCVP, however, mobility counseling is lacking (Briggs and Turner 2006; Cunningham and Sawyer 2005). While each voucher-assisted household does have a relationship with a housing authority case manager, case managers frequently report that voucher-assisted tenants did not seek case manager assistance with housing searches, and instead viewed these case managers as providing a "rubber stamp" on their leases (Greenlee 2011, 2012).

Landlords play an important yet largely unexamined role in shaping household location choice, characteristic of an overall lack of empirical knowledge about housing supply (DiPasquale 1999). Among other non-tenant actors, landlords represent the housing market, and serve as gatekeepers to specific housing opportunities. Within a rational choice framework, the voucher subsidy is designed to induce landlords who may be risk-averse to rent to low-income households by providing a guarantee of payment of the voucher subsidy as well as vetting tenants who must abide by rules and regulations specified in their lease and by the PHA. However, landlords tend to be less monolithic and less rational in their motivations (Varady et al. 2017). Many landlords participate in the HCVP to obtain a profit, to hedge potential losses on late rent from market-rate tenants, to keep units filled, because they inherited voucher-subsidized tenants when they purchased a property, or due to a sense of personal social commitment to helping households in need (Greenlee 2014).

While some landlords find relative comfort in their collaborations with housing authorities and voucher-assisted tenants, other landlords perceive program requirements, housing inspections, and navigating multiple contractual relationships as being too onerous to motivate continued or expanded program participation (Greenlee 2014). With regard to overall access to housing opportunities, many landlords maintain the right to explicitly discriminate against voucher-assisted households who are looking to rent from them.[6] Around 25 percent of voucher-assisted households reported experiencing some form of landlord discrimination during their housing search process, a finding that likely underestimates the level of actual discrimination during the search process (Varady and Walker 2007).

Contributing to a negative bias against voucher-assisted households is a general stigma associated with the voucher subsidy, as well as subsidized housing in general (Greenlee 2012). Not only can general bias against subsidized households create hostile environments for low-income households, such as that observed by Rubinowitz and Rosenbaum (2002) for Gautreaux movers, but the perception of bias can also influence voucher-subsidized households' housing choice considerations in ways that make certain landlords and housing choices paths of least resistance to successfully finding housing. This is well exemplified by the following quote from a housing authority official who explains the concentration of voucher-subsidized households within one private rental development: "I think a lot of them [voucher households] go to one place in town here, and they kind of get based in one area. . . . And they usually end up there because other people in our community don't really choose that area to be in. So in a way, they have their own little community there" (Greenlee 2011, 510).

This statement reflects the ambiguity of the role of landlords in shaping residential location patterns—households may be choosing a certain development because they feel safe or perceive the unit as a good opportunity, or they may be choosing a unit knowing that they will face discrimination from other landlords.

Three Principles for Program Reform

As a way forward, I suggest three strategies to address the need for an *ecology of choice* approach to program evaluation and reform. First, I suggest that housing authorities and local governments should collaborate to support the choices that voucher-assisted tenants make. Second, I suggest that housing program designers should reduce administrative barriers to choice. Third, I suggest that voucher-subsidized households should be more involved in defining ways to measure the efficacy of their choices. I describe each strategy in turn.

Housing Authorities and Local Governments Should Collaborate to Support the Choices That Voucher-Assisted Tenants Make

A main focus of experimental mobility programs and consent decrees is to help move low-income households facing structural disadvantage into more advantageous neighborhood environments. Since neighborhood attractiveness contributes to price points for housing, the housing voucher subsidy should increase access to housing options that the market views as bundling more attractive neighborhood resources. While this is an implicit goal of the HCVP, there are no set definitions for what a more advantageous neighborhood environment is. Experimental mobility programs do, however, define such standards for the purpose of evaluating program impacts. Applying the standards used in experimental mobility programs reveals that for a large share of voucher-assisted households, actual housing choices fall somewhere in between the neighborhood demographic and socioeconomic conditions prior to participating in the program. In thinking about conditions in these neighborhoods, federal and local public officials face the challenge of determining how to allocate scarce resources for neighborhood improvement. A complementary strategy promoting mobility to high-opportunity neighborhoods would target scarce public resources in those communities in which voucher-assisted households choose to live. Two main challenges remain to realizing this approach. First, there is a lack of consensus regarding what the best combination of place-based and tenant-based investment strategies might be. Second, even if questions regarding the strategic mix of programs was resolved, the sheer scarcity of resources available at the local level is overwhelmed by increased need—a challenge that has only grown worse under local government fiscal austerity in the aftermath of the 2008 housing and financial crisis.

Housing Program Designers Should Reduce Administrative Barriers to Choice

After years of devolution, wavering financial support due to federal sequestration, and increased operating costs due to an aging public housing portfolio, many housing authorities find themselves struggling to effectively serve as many low-income households as their mission allows (Kleit and Page 2008). While doing more with less may currently be a politically popular mantra across all levels of government, implementing this approach without diminishing service provision to existing recipients would be extremely challenging. At the same time, many housing authorities are looking for ways to innovate to most effectively use their scarce resources without duplication of services, and to think about the causes and consequences of poverty more holistically (Bratt 2008). Reducing the types of red tape and administrative overhead that make it difficult for housing authorities and other agencies to collaborate around service provision could reduce barriers to choice for voucher-assisted households. Programs such as the Baltimore Special Mobility Housing Choice Voucher program (Engdahl 2009) and Chicago's Regional Housing Initiative (Semuels 2015) institutionalize regional cooperation between local housing authorities by pooling resources and standardizing administration of vouchers across regional housing authority jurisdictions. Reducing the administrative burden on housing authorities (such as those associated with the billing of housing voucher portability) and other service providers not only frees up resources for more proactive housing counseling and other services, but also frees up households to search more freely for housing within their region.

Voucher-Subsidized Households Should Be More Involved in Defining Ways to Measure the Efficacy of Their Choices

A more difficult area to deal with particularly within the context of program evaluation is whether the housing choice goals defined by HUD are good proxies for the types of opportunities that low-income households view as valuable or a priority within their lives. Measures of housing choice including poverty deconcentration and racial integration should be complemented by a more nuanced understanding of individual household needs and goal attainment. As corroborated within the literature on housing careers, housing choices are not monolithic, and are contextualized by evolving life goals and needs. Existing HUD programs such as the Family Self-Sufficiency (FSS) program work closely with voucher-subsidized households to develop a personalized framework for measuring life goals while receiving the voucher subsidy. While the FSS program model has faced substantial challenges in its implementation (Rohe and Kleit 1999), the program's counseling model provides a framework for defining household needs and choice frameworks in a more holistic manner. Although each household approaches program participation differently, feedback from this type of counseling approach would help confirm when and how locations matter for the improvement of low-income households' lives (Anthony 2005; Rohe and Kleit 1999).

Conclusion: Toward an Ecology of Choice

This chapter argues that housing policy analysts and program designers need to think differently about the nature of housing choice within the HCVP. The program is designed to provide economic leverage to low-income households to increase the degree of housing choice available to them. These housing and neighborhood choices when combined with the ongoing housing subsidy should constitute the basis for increased household self-sufficiency. The evidence presented in this chapter looks at program design and points to some of the limits that voucher-assisted households face in actualizing expanded choice from the subsidy. Describing and analyzing a broader *ecology of choice* by evaluating the constraints and motivations of housing authorities and landlords alongside those of tenants offers new potential to both understand spatial outcomes for voucher-assisted households and think about ways to address the challenges associated with collaboration between tenants, housing authorities, and landlords to deliver the voucher subsidy. Engagement with the ecology of choice in the program can also serve as a basis for thinking more holistically about barriers to assisted households' self-sufficiency and long-term goals. As a starting point, finding additional ways to support the choices that people make, reducing administrative barriers to choice, and empowering participants to define relevant measures for their choices reflect first steps toward this goal. Addressing these challenges will potentially improve the program's effectiveness at assisting low-income households in moving to neighborhoods of true choice.

Notes

1 Federal budget sequestration efforts in 2014 reduced the national capacity to serve households using vouchers by 85,697 households. Of 2,230 agencies, 1,603 (71.88 percent) administering HCVs saw a reduction in the number of households that they served as a result of budget sequestration (Center on Budget and Policy Priorities 2015).

2 See Popkin et al. (2003) for an evaluation of consent decree programs using housing vouchers.

3 Kotlowitz chose to change their last name to Rivers within his book.

4 Briggs et al. (2010) note that Gautreaux ultimately "turned out to be a quite limited guide" (65) for MTO program design and implementation given the acute level of disadvantage faced by MTO participants and the complicated nature of partnerships with nonprofit social service organizations that were a core component of the MTO demonstration.

5 For the sake of simplicity and the comparability of data, I limit my analysis to voucher households in the 50 states and the District of Columbia.

6 Some state, county, and local governments have adopted source of income protection statutes that prevent landlords from explicitly discriminating against households who use housing vouchers as an income support. See Tighe et al. (2017) for a review of state source of income protection laws and Poverty and Race Research Action Council (2017) for a list of jurisdictions with source of income protection laws.

References

Anthony, Jerry. 2005. "Family Self-Sufficiency Programs: An Evaluation of Program Benefits and Factors Affecting Participants' Success." *Urban Affairs Review*, 4(1): 65–92.

Basolo, Victoria, and Mai Thi Nguyen. 2005. "Does Mobility Matter? The Neighborhood Conditions of Housing Voucher Holders by Race and Ethnicity." *Housing Policy Debate*, 16(3–4): 297–324.

Bratt, Rachel G. 2008. "Viewing Housing Holistically: The Resident-Focused Component of the Housing-Plus Agenda." *Journal of the American Planning Association*, 74(1): 100–110.

Briggs, Xavier de Souza. 2006. *The Geography of Opportunity: Race and Housing Choice in Metropolitan America*. Washington, DC: Brookings Institution Press.

Briggs, Xavier de Souza, Susan Popkin, and John Goering. 2010. *Moving to Opportunity*. New York: Oxford University Press.

Briggs, Xavier de Souza, and Margery Austin Turner. 2006. "Assisted Housing Mobility and the Success of Low-Income Minority Families: Lessons for Policy, Practice, and Future Research." *Northwestern Journal of Law and Social Policy*, 1(1): 25–61.

Center on Budget and Policy Priorities. 2015. *United States Fact Sheet: The Housing Choice Voucher Program*. Washington, DC: Center on Budget and Policy Priorities.

Center on Budget and Policy Priorities. 2017. *United States Fact Sheet: Federal Rental Assistance*. Washington, DC: Center on Budget and Policy Priorities.

Chaskin, Robert. J. 2013. "Integration and Exclusion: Urban Poverty, Public Housing Reform, and the Dynamics of Neighborhood Restructuring." *Annals of the American Academy of Political and Social Science*, 647(1): 237–267.

Chetty, Raj, Nathaniel Hendren, and Lawrence Katz. 2016. "The Effects of Exposure to Better Neighborhoods on Children: New Evidence from the Moving to Opportunity Project." *American Economic Review*, 106(4): 855–902.

Chicago Reporter. 1978, March. Polikoff Combats "Residential Apartheid" with Gautreaux, Sacrifices Inner City Shelter Needs for Open Housing Goal.

Clampet-Lundquist, Susan. 2003. "Finding and Keeping Affordable Housing: Analyzing the Experiences of Single-Mother Families in North Philadelphia." *Journal of Sociology and Social Welfare*, 30: 123–140.

Coulton, Claudia J., Jill Korbin, Tsui Chan, and Marilyn Su. 2001. "Mapping Residents' Perceptions of Neighborhood Boundaries: A Methodological Note." *American Journal of Community Psychology*, 29(2): 371–383.

Cunningham, Mary K., and Noah Sawyer. 2005. *Moving to Better Neighborhoods with Mobility Counseling (A Roof Over Their Heads: Changes and Challenges for Public Housing Residents)*. Washington, DC: The Urban Institute.

DeLuca, Stefanie, Philip M. E. Garboden, and Peter Rosenblatt. 2013. "Segregating Shelter: How Housing Policies Shape the Residential Locations of Low-Income Minority Families." *Annals of the American Academy of Political and Social Science*, 647(1): 268–299.

DiPasquale, Denise. 1999. "Why Don't We Know More About Housing Supply?" *Journal of Real Estate Finance and Economics*, 18(1): 9–23.

Domínguez, Silvia, and Celeste Watkins, Celeste. 2003. "Creating Networks for Survival and Mobility: Social Capital among African-American and Latin-American Low-Income Mothers." *Social Problems*, 50(1): 111–135.

Engdahl, Lora. 2009. *New Homes, New Neighborhoods, New Schools: A Progress Report on the Baltimore Housing Mobility Program*. Washington, DC: Poverty and Race Research Action Council.

Feins, Judith D., Deborah McInnis, and Susan Popkin. 1997. *Counseling in the Moving to Opportunity Demonstration Program*. Washington, DC: U.S. Department of Housing and Urban Development.

Galster, George C., and Sean P. Killen. 1995. "The Geography of Metropolitan Opportunity: A Reconnaissance and Conceptual Framework." *Housing Policy Debate*, 6(1): 7–43.

Galvez, Martha M. 2010. *What Do We Know about Housing Choice Voucher Program Location Outcomes?* Washington, DC: The Urban Institute.

Goetz, Edward G. 2010. "Desegregation in 3D: Displacement, Dispersal and Development in American Public Housing." *Housing Studies*, 25(2): 137–158.

Goetz, Edward. G. 2013. *New Deal Ruins: Race, Economic Justice, and Public Housing Policy*. Ithaca, NY: Cornell University Press.

Goetz, Edward G., and Karen Chapple. 2010. "You Gotta Move: Advancing the Debate on the Record of Dispersal." *Housing Policy Debate*, 20(2): 209–236.

Greenlee, Andrew J. 2011. "A Different Lens: Administrative Perspectives on Portability in Illinois' Housing Choice Voucher Program." *Housing Policy Debate*, 21(3): 377–403.

Greenlee, Andrew J. 2012. *A Relational Analysis of Mobility Within Illinois' Housing Choice Voucher Program*. Doctoral Dissertation. University of Illinois at Chicago.

Greenlee, Andrew J. 2014. "More Than Meets the Market? Landlord Agency in the Illinois Housing Choice Voucher Program." *Housing Policy Debate*, 24(3): 500–524.

Kleit, Rachel G., Seungbeom Kang, and Corianne P. Scally. 2016. "Why Do Housing Mobility Programs Fail in Moving Households to Better Neighborhoods?" *Housing Policy Debate*, 26(1): 188–209.

Kleit, Rachel G., and Stephen B. Page. 2008. "Public Housing Authorities under Devolution." *Journal of the American Planning Association*, 74(1): 34–44.

Kotlowitz, Alex. 1992. *There Are No Children Here: The Story of Two Boys Growing Up in The Other America*. New York: Doubleday.

Lansing, John B., Charles W. Clifton, and James N. Morgan. 1969. *New Homes and Poor People: A Study of Chains of Moves*. Ann Arbor, MI: Institute for Social Research, University of Michigan.

Matheson, Victor A., and Robert A. Baade. 2004. "Race and Riots: A Note on the Economic Impact of the Rodney King Riots." *Urban Studies*, 41(13): 2691–2696.

McClure, Kirk. 2008. "Deconcentrating Poverty with Housing Programs." *Journal of the American Planning Association*, 74(1): 90–99.

McClure, Kirk. 2015. "The Future of Research on Assisted Housing for the Poor." *Housing Policy Debate*, 25(4): 802–805.

McClure, Kirk, Alex F. Schwartz, and Lydia B. Taghavi. 2015. "Housing Choice Voucher Location Patterns a Decade Later." *Housing Policy Debate*, 25(2): 215–233.

Messer, Lynne C., Barbara A. Laraia, Jay S. Kaufman, Janet Eyster, Claudia Holzman, Jennifer Culhane, Irma Elo, Jessica G. Burke, and Patricia O'Campo. 2006. "The Development of a Standardized Neighborhood Deprivation Index." *Journal of Urban Health*, 83(6): 1041–1062.

Moore, Kathleen M. 2016. "Lists and Lotteries: Rationing in the Housing Choice Voucher Program." *Housing Policy Debate*, 26(3): 474–487.

Payne, Judy, and Geoff Payne. 1977. "Housing Pathways and Stratification: A Study of Life Chances in the Housing Market." *Journal of Social Policy*, 6(2): 129–156.

Popkin, Susan J., George C. Galster, Kenneth Temkin, Carla Herbig, Diane K. Levy, and Elise K. Richer. 2003. "Obstacles to Desegregating Public Housing: Lessons Learned from Implementing Eight Consent Decrees." *Journal of Policy Analysis and Management*, 22(2): 179—199.

Poverty and Race Research Action Council. 2017. *Expanding Choice: Practical Strategies for Building a Successful Housing Mobility Program: Appendix B: State, Local, and Federal Laws Barring Source-of-Income Discrimination*. Washington, DC: Poverty and Race Research Action Council.

Rex, John, Robert S. Moore, Alan Shuttlework, and Jennifer Williams. 1967. *Race, Community and Conflict: A Study of Sparkbrook*. London: Institute of Race Relations.

Rohe, William M., and Rachel G. Kleit. 1999. "Housing, Welfare Reform, and Self-Sufficiency: An Assessment of the Family Self-Sufficiency Program." *Housing Policy Debate*, 10(2): 333—369.

Rosenbaum, James E., and Stefanie DeLuca, Stefanie. 2000. *Is Housing Mobility the Key to Welfare Reform? Lessons from Chicago's Gautreaux Program*. Washington, DC: The Brookings Institution.

Rossi, Peter Henry. 1955. *Why Families Move; A Study in the Social Psychology of Urban Residential Mobility*. Washington, DC: Free Press.

Rubinowitz, Leonard. S., and James E. Rosenbaum. 2002. *Crossing the Class and Color Lines: From Public Housing to White Suburbia*. Chicago: University of Chicago Press.

Sampson, Robert J. 2008. "Moving to Inequality: Neighborhood Effects and Experiments Meet Structure." *American Journal of Sociology*, 114(11): 189–231.

Sanbonmatsu, Lisa, Jens Ludwig, Lawrence F. Katz, Lisa A. Gennetian, Greg J. Duncan, Ronald C. Kessler, Emma Adam, Thomas W. McDade, and Stacy Tessler Lindau. 2011. *Moving to Opportunity for Fair Housing Demonstration Program: Final Impacts Evaluation*. Washington, DC: U.S. Department of Housing and Urban Development.

Schill, Michael H. 1993. "Distressed Public Housing: Where Do We Go from Here?" *University of Chicago Law Review*, 60(2): 497–554.

Semuels, Alana. 2015. "How Chicago Is Trying to Integrate Its Suburbs." *The Atlantic*, July 17.

Serrano, Richard A. 1992. "All 4 Acquitted in King Beating: Verdict Stirs Outrage: Bradley Calls It Senseless Trial: Ventura County Jury Rejects Charges of Excessive Force in Episode Captured on Videotape. A Mistrial Is Declared on One Count against Officer Powell." *Los Angeles Times*, April 30.

Skobba, Kimberly. 2008. *The Influence of Section 8 Vouchers on the Housing Careers of Working Poor Families*. Bethesda, MD: ProQuest.

Skobba, Kim, and Edward G. Goetz. 2013. "Mobility Decisions of Very Low-Income Households." *Cityscape*, 15(2): 155–171.

Sperling, Jonathan. 2012. "The Tyranny of Census Geography: Small-Area Data and Neighborhood Statistics." *Cityscape*, 14(2): 219–223.

Tiebout, Charles M. 1956. "A Pure Theory of Local Expenditures." *Journal of Political Economy*, 64(5): 416–424.

Tighe, J. Rosie, Megan E. Hatch, and Joseph Mead. 2017. "Source of Income Discrimination and Fair Housing Policy." *Journal of Planning Literature*, 32(1): 3–15.

Tremoulet, Andrée, Ryan Dann, and Arlie Adkins. 2016. "Moving to Location Affordability? Housing Choice Vouchers and Residential Relocation in the Portland, Oregon, Region." *Housing Policy Debate*, 26(4–5): 692–713.

Turner, Margery Austin, and Xavier de Souza Briggs. 2008. *Assisted Housing Mobility and the Success of Low-Income Minority Families: Lessons for Policy, Practice, and Future Research*. Washington, DC: The Urban Institute.

U.S. Department of Housing and Urban Development. 2001. *Housing Choice Voucher Program Guidebook*. Washington, DC: U.S. Department of Housing and Urban Development.

U.S. Department of Labor. 1965. *The Negro Family: The Case for National Action*. Washington, DC: Office of Policy Planning and Research, United States Department of Labor.

Vale, Lawrence. J. 2013. *Purging the Poorest: Public Housing and the Design Politics of Twice-Cleared Communities*. Chicago: University of Chicago Press.

Varady, David P., Joseph Jaroscak, and Reinout Kleinhans. 2017. "How to Attract More Landlords to the Housing Choice Voucher Program: A Case Study of Landlord Outreach Efforts." *Urban Research and Practice*, 10(2): 1–13.

Varady, David P., and Carole Walker. 2007. *Neighborhood Choices: Section 8 Housing Vouchers and Residential Mobility*. New York: Routledge.

Wang, Ruoniu. 2018. "Tracking 'Choice' in the Housing Choice Voucher Program: The Relationship between Neighborhood Preference and Locational Outcome." *Urban Affairs Review*, 54(2): 302–331.

Wilson, William J. 1990. *The Truly Disadvantaged: The Inner City, the Underclass, and Public Policy*. Chicago: University of Chicago Press.

12

ACCOMMODATING AND ACCUMULATING

How the Property Interests of Homeowners and Renters Impact Housing Satisfaction

Allison Freeman and Kimberly Manturuk

Introduction

A large body of research has found that homeowners are more satisfied with their housing than renters. Housing satisfaction is one of the key benefits associated with homeownership, and promoting housing satisfaction is one of the reasons given to support policies aimed at increasing access to homeownership. While there have been many studies of the drivers of housing satisfaction overall, and a number of studies on the satisfaction of owners versus renters in particular, to date there has been little work devoted to understanding the extent to which the different functional interests of owners and renters drive differences in housing satisfaction. This chapter aims to fill this research gap.

As we examine what drives the housing satisfaction of homeowners and renters, our hypothesis is that insofar as owners and renters share *interests* in their property, they will also share drivers of housing satisfaction. For example, both owners and renters share a use-interest in their property; i.e., both groups live in and make use of their homes; and we suspect that property characteristics (for example, the number of bathrooms) will play a similar role in facilitating both groups' housing satisfaction. Where use-interest differs between the two groups is in owners' and renters' differing levels of control over their property: Renters are less able than owners to change things that make them dissatisfied with where they live. Because of this differing level of control, we anticipate that renters will be more affected by housing-related stressors—those factors that might cause the home to be a stressful place to live in—than owners are. Finally, where owners and renters differ most obviously in their interests is that owners have a financial stake in their homes while renters do not. We therefore believe that owners will be more heavily affected by neighborhood characteristics, since these will ultimately have some effect on the value of their homes.

To test our hypothesis, we use data from the Community Advantage Panel Study (CAPS). CAPS, which began in 2003, follows a group of low- and moderate-income (LMI) homeowners and renters throughout the United States in order to understand the impacts that homeownership has on LMI people's lives. To assess drivers of housing satisfaction, we include a diverse range of variables in our models, which can be roughly categorized into three groups: (1) housing stressors, (2) housing physical characteristics, and (3) neighborhood characteristics. We then construct OLS regression models to predict housing satisfaction. Because we are looking to identify the drivers of housing satisfaction that are unique to both homeowners and renters, we run models separately for each group. Our results indicate that both owners and renters are affected by their dwelling's physical characteristics, and there is some evidence that homeowners may be more affected by neighborhood conditions. However, the evidence from our analysis is mixed.

Literature Review

Our analysis of the determinants of housing satisfaction is situated in the body of research examining the drivers of housing satisfaction. We examine this literature in the following text. Galster (1987) assesses both dwelling and neighborhood satisfaction using a unique dataset focused on homeowners in Wooster, Ohio, and Minneapolis, Minnesota. His independent variables include homeowner characteristics (education, income, marital status, etc.), dwelling characteristics (rooms per occupant, space per occupant, age of structure, etc.), and neighborhood characteristics (racial composition, socioeconomic status, etc.). He finds that owners further along the age-spectrum are progressively more satisfied with their homes; that more rooms per person and more bathrooms per person are related to increased satisfaction, though there are decreasing returns to these things and the effect varies across household type; and that the age of one's home is negatively correlated with one's satisfaction.

Galster's finding that the drivers of housing satisfaction differ systematically between groups (older residents versus younger residents, those with young children versus those with teens) is not surprising; much of the other research into the topic of housing satisfaction reveals the same. Christensen et al. (1992) looked specifically at the drivers of housing satisfaction for older people and discovered that structural adequacy and quality of maintenance were significant predictors of participants' evaluation of their housing. For the housewives interviewed in Amérigo and Aragones' (1990) study of satisfaction with public housing, attachment to neighborhood and relationships with neighbors explained the greatest variance in residential satisfaction. In Dupuis and Thorns' (1998) study of the relationship between housing, homeownership, and well-being (subtly different from housing satisfaction, of course) of older New Zealand homeowners, the authors concluded that homeownership and well-being were related through the economic benefits residents believe their homes convey, through the routines and habits associated with living in a familiar spatial context, and through residents' understanding of an interweaving of home and family relationships.

All of these studies suggest that whom one is—one's gender or age or place in the life cycle—will affect one's experience of housing satisfaction. It is reasonable to assume that one's relationship to one's dwelling—that is, one's tenure status—might do the same. And indeed, this is the case: Research has shown that the drivers of housing satisfaction differ between homeowners and renters.

While the physical characteristics of one's home are found to affect both owners' and renters' housing satisfaction, research has revealed that the effect differs between the two tenure groups. Lane and Kinsey (1980) examined how dwelling characteristics, tenure, and household demographics affect residents' satisfaction with their housing, and discovered that while housing characteristics (dwelling type, physical quality of home, number of rooms per person, distance to city center) affect both renters' and owners' housing satisfaction, they affect the satisfaction of owners to a greater degree than they do that of renters. Interestingly, among the variables having to do with housing characteristics, dwelling *quality* was less important in explaining owners' satisfaction than it was in explaining renters' satisfaction, which might possibly lend credence to our hypothesis that owners will be less affected by aspects of housing that they can control.

O'Bryant and Wolf (1983) examined housing satisfaction among a group of older homeowners and renters, controlling for demographic variables, housing characteristics, and subjective factors measuring attachment to the home. They found that while housing quality was important to owners, it was the primary driver of renters' housing satisfaction: "for owners, as opposed to renters, housing satisfaction increased less gradually as housing quality improved" (227). In the end, these authors conclude that psychological attachment to the home plays the largest role in owners' housing satisfaction, while dwelling quality is the primary driver of renters' housing satisfaction.

Diaz-Serrano (2009) uses data from the European Community Household Panel (ECHP) to look specifically at the transition to ownership and its effects on the housing satisfaction of those who used to rent. His analysis reveals that renters who became homeowners experienced a significant increase in housing satisfaction; importantly, these results hold even for those who did not change their actual dwellings, but transitioned from renting to owning the same unit. Diaz-Serrano further determines that "another more general finding that holds true across alternative samples and models is that homeowners tend to perceive a lower . . . disutility from the same housing context than renters do. That is, homeownership smooths the impact of the housing characteristics on satisfaction" (754–755). In other words, owners seem to be less bothered by issues associated with their dwellings than renters are.

The studies cited here suggest that the drivers of housing satisfaction differ between homeowners and renters, but they do not consistently explain why. In the next section, we will explore a somewhat different theory about the housing satisfaction of homeowners and renters. We anticipate that the housing satisfaction of these tenure groups is indeed driven by different factors, which are determined in part by owners' and renters' different interests in their property and by their differing abilities to advance these interests and promote their housing satisfaction. Our hypotheses are:

H1: Renters' housing satisfaction will be more heavily affected by housing-related stressors than homeowners' satisfaction, because renters have less control over these housing stressors.

H2: The housing satisfaction of homeowners and renters will be equally affected by physical characteristics of the dwelling, because the groups share a use-interest in their dwellings.

H3: Homeowners' housing satisfaction will be more heavily affected by neighborhood characteristics than renters' satisfaction, because homeowners have an investment interest in their housing and this investment interest will be affected by neighborhood characteristics.

Data and Methods

We test these hypotheses using data from CAPS, a study that began in 2003 and that follows a nationwide panel of lower-income homeowners in order to better understand the impacts that homeownership has on their lives (UNC Center for Community Capital 2014). CAPS also follows a panel of renters, added to the study in 2004, who were selected to match the homeowners in terms of geography and income levels. When CAPS began, the study included 3,743 homeowners and 1,530 renters. The data we use in the current analysis were collected in 2010, a year in which panelists completed a survey asking about a variety of housing and financial topics. In 2010, the sample of those who completed the CAPS survey included 2,088 homeowners and 875 renters.[1]

Before we begin our analysis, we note that CAPS is not a random sample, but we believe it is generally representative of the experiences of lower-income families.[2] The CAPS homeowners were selected from a larger pool of people whose mortgages were included in the Community Advantage Program (CAP), a secondary mortgage market demonstration for affordable home loans; the demonstration was enabled through a joint effort between Self Help (a community development financial institution in Durham, NC), the Ford Foundation, and Fannie Mae. The CAP portfolio contains prime, fixed-rate loans made to LMI borrowers whose income levels, credit scores, small downpayments, and high debt-to-income ratios would otherwise have pushed them into subprime mortgages. The CAP loans were purchased by Self Help and, using a grant from the Ford Foundation to underwrite risk, were sold to Fannie Mae. The portfolio now contains over 45,000 mortgages; the subset of survey respondents was selected from the portfolio just before the study began in 2003.

Measures

Our dependent variable is housing satisfaction, which we measure with the question, "On a scale of 1 to 10, where 1 is the worst and 10 is the best, how would you rate your dwelling as a place to live?" The variable is skewed right with a mean of 8 and a standard deviation of 1.5. Owners are more satisfied with their homes than renters: The mean for renters is 7.6, while the mean for owners is 8.2.

Because we are interested in understanding as many of the causes of housing satisfaction as possible, we include a diverse range of independent variables in our models. These can roughly be categorized into three groups: (1) housing stressors, (2) housing physical characteristics, and (3) neighborhood characteristics. Descriptive statistics for all the variables are shown in Tables 12.1 (homeowners) and 12.2 (renters).

Housing stressors are factors that might cause the home to be a stressful place to live and thereby affect housing satisfaction. We control for a number of these, for example, for housing type, because it might affect stress in any number of ways: People in single-family homes are likely to have less contact with neighbors, so might feel greater stress from a sense of isolation, yet they also have more privacy and so may feel reduced stress as a result of that; people in mobile homes have the privacy of their own dwelling, yet they tend to live close enough together that they are more in other people's space than those in single-family homes, etc. In our models, single-family, detached dwellings are the reference category. To further probe the issue of housing stressors, we ask whether, in the prior year, respondents felt that paying for housing is stressful, home maintenance is stressful, or their neighbors are stressful. People who reported stress were coded 1, while those who reported no stress

Table 12.1 Descriptive Statistics for Sample of Homeowners (n = 1,623).

Variable	*Percent*	*Number*	*Mean*	*Std. Dev.*	*Min*	*Max*
Detached dwelling	87.50	1,421			0	1
Attached dwelling	10.16	163			0	1
Other dwelling type	2.34	38			0	1
Children in the home	65.83	1,069			0	1
Percent of income spent on housing			30.26	16.13	0	60.86
Paying for housing is stressful	36.08	586			0	1
Home maintenance is stressful	50.43	819			0	1
Neighbors are stressful	21.31	346			0	1
Housing payment is too high	38.24	621			0	1
Years in the home			8.65	5.76	0.5	61.92
People per bedroom			1.06	0.51	0.2	4
Number of bathrooms			1.84	0.68	1	5.5
Tract percent single parents			10.33	5.50	0.53	55.95
Tract percent unemployed			5.64	4.23	0	45.07
Tract percent receiving public assistance			3.47	3.50	0	41.14
Tract percent below poverty line			12.31	9.03	0.69	64.24
Rating of the neighborhood (1–10)			7.91	1.50	1	10
Would recommend the neighborhood	91.13	1,480			0	1
Neighborhood is good place for kids	79.43	1,290			0	1
Neighborhood has good schools	75.18	1,221			0	1
Neighborhood has gotten better	14.96	243			0	1
Neighborhood has gotten worse	9.24	150			0	1
Neighborhood has stayed the same	73.89	1,200			0	1
Moved to a new neighborhood	1.91	31			0	1

Source: authors.

were coded 0.[3] To look more fully into the issue of housing cost and stress, we incorporated two additional variables: a measure of the percent of household income that is spent on housing, and an indicator (coded 1/0) of whether respondents believe their housing payment is too high. Because research documents that people with minor children experience higher levels of stress (Deater-Deckard and Scarr 1996; Deater-Deckard 1998), we include a 1/0 measure indicating whether there are children in the home. The final housing stressor variable is a continuous variable measuring the number of years people have lived in the same dwelling. We anticipate that people will feel more comfortable in their homes the longer they have lived in them. However, this could vary between the homeowners and renters: Even long-term renters usually have to renew their lease periodically, so they may not gain satisfaction and a sense of stability in the home over time.

Looking at descriptive statistics for the housing stressor variables, we see that a greater share of homeowners than renters reported experiencing some form of stress related to their housing. As shown in Tables 12.1 and 12.2, homeowners were most likely to say that maintenance was stressful, while renters were more likely to say that paying for housing was stressful. The second group of variables in our analysis is related to housing physical characteristics. Here we include two continuous variables: the ratio of dwelling residents to bedrooms, and the number of bathrooms.[4] Both capture the extent to which the home is physically comfortable, and the former assesses the adequacy of the space given the size of the household.

Finally, we included a group of variables that measure different aspects of the neighborhood and surrounding community. We include four tract-level measures from the 2010 Census: percent single

Table 12.2 Descriptive Statistics for Sample of Renters (n = 549).

Variable	*Percent*	*Number*	*Mean*	*Std. Dev.*	*Min*	*Max*
Detached dwelling	34.79%	191			0	1
Attached dwelling	57.92%	318			0	1
Other dwelling type	7.29%	40			0	1
Children in the home	48.45%	266			0	1
Percent of income spent on housing			45.32	44.73	0	100
Paying for housing is stressful	36.25%	199			0	1
Home maintenance is stressful	24.23%	133			0	1
Neighbors are stressful	24.23%	133			0	1
Housing payment is too high	37.70%	207			0	1
Years in the home			6.61	7.47	0.5	44.25
People per bedroom			1.12	0.54	0.25	5
Number of bathrooms			1.42	0.56	1	4
Tract percent single parents			11.57	6.10	0	45
Tract percent unemployed			6.36	4.41	0	25.35
Tract percent receiving public assistance			4.31	4.07	0	25.83
Tract percent below poverty line			15.07	10.32	0.89	67.08
Rating of the neighborhood (1–10)			7.54	1.94	1	10
Would recommend the neighborhood	82.51%	453			0	1
Neighborhood is good place for kids	66.85%	367			0	1
Neighborhood has good schools	73.22%	402			0	1
Neighborhood has gotten better	15.48%	85			0	1
Neighborhood has gotten worse	14.57%	80			0	1
Neighborhood has stayed the same	57.74%	317			0	1
Moved to a new neighborhood	12.20%	67			0	1

Source: authors.

parents; percent unemployed; percent receiving public assistance; and percent of residents below the poverty line. These four measures are often used when examining the extent to which neighborhood conditions impact individual outcomes (Sampson 1991; Sampson and Groves 1989). From the CAPS data, we also incorporate four measures of how respondents perceive their neighborhood conditions: how people rate satisfaction with their neighborhood (on a scale from 1 (lowest) to 10);[5] whether they would recommend the neighborhood to a friend (coded 1/0); whether the neighborhood is a good place to raise children (coded 1/0); and whether the neighborhood has good schools (coded 1/0).[6] The homeowners generally felt more favorably about their neighborhoods than the renters did: 91 percent of the homeowners would recommend their neighborhoods to a friend, while only 82 percent of the renters would. Similarly, 79 percent of homeowners felt that their neighborhood was a good place to raise children, while only 67 percent of the renters did. We also include in the models a series of categorical variables that indicate the direction in which people feel their neighborhood is moving, i.e., whether they think it is improving or declining (coded 1/0).[7] We include these perceptual measures in addition to the Census measures because it is possible that people's housing satisfaction could be more influenced by their perceptions of the neighborhood than by objective neighborhood conditions. One noticeable group of measures not included in our models is sociodemographic control variables. We included variables for employment status, marital status, race/ethnicity, and education in our preliminary regression analyses. Since none of these variables were statistically significant, we excluded them from the final analysis.

Method

This analysis uses OLS regression models to predict housing satisfaction. Because we are looking to identify the drivers of housing satisfaction that are unique to both homeowners and renters, we run models separately for each group. We run each group's models three times, first with just the housing stress variables, then adding the housing physical characteristics, and finally the full model, which includes neighborhood characteristics. We use list-wise deletion of observations with missing data on any measure to ensure a consistent sample size across all three models.

Results

To recap, our underlying hypothesis is that the housing satisfaction of homeowners and renters is driven by their different interests in property and by their differing abilities to advance these interests. We assume that owners and renters share a use-interest in their properties, and because of this commonality, we anticipate that both groups will be affected by physical characteristics of their dwellings. However, because owners can control and make changes to their properties—they can promote their use-interest by painting rooms or insulating walls that are too thin, for example—we suspect that their housing satisfaction will be less affected than renters' housing satisfaction by stressful property characteristics.

We further anticipate that owners' housing satisfaction will be driven by factors affecting their financial interest in their properties, an interest that is not shared by their renter counterparts. Most owners expect that their investment in the home will eventually provide them with wealth gains via an increase in equity. It follows that owners' housing satisfaction might be driven by neighborhood-level factors that either increase or deflate home value. While renters *use* neighborhood amenities, they do not view those amenities in terms of how they affect housing value, and we suspect that while neighborhood amenities might affect renters' *neighborhood* satisfaction, they will have little effect on renters' *housing* satisfaction. However, we expect that neighborhood factors will disproportionately drive homeowners' housing satisfaction. In the following text we present what our analysis revealed on these matters.

Homeowners and Housing Satisfaction

Table 12.3 presents the results for the sample of homeowners. Model 1 (housing stressors) finds that living in a multifamily dwelling is associated with a 0.418 lower housing satisfaction score. Similarly, those who report living in an "other dwelling type"—typically a mobile home—were less satisfied than those in single-family houses (beta −0.565). We also find that people who say that home maintenance is stressful are less satisfied (−0.439), as are those who find their neighbors to be stressful (−0.392). Interestingly, the financial stress measures did not significantly predict housing satisfaction for homeowners.

Model 2 adds the measures for the physical attributes of the home and shows that each additional bathroom is associated with a 0.396 point higher housing satisfaction score. Interestingly, the ratio of people to bedrooms has no effect on homeowners' housing satisfaction.

When we add the neighborhood characteristics in Model 3 we find several significant predictors. The most unexpected finding is that Census tracts with more single parents are associated with an increase in housing satisfaction. However, we note that this effect is very small, just over two one-hundredths of a point.

Model 3 shows that homeowners' perceptions of their neighborhoods significantly affect their housing satisfaction: For each point increase in owners' rating of their neighborhoods, the measure of housing satisfaction increases 0.546. This is consistent with expectations, since there is likely some overlap between how satisfied people are with their dwelling and how they view their neighborhood. Similarly, homeowners who feel that their neighborhood is a good place to raise children are more satisfied with their dwelling, but the effect is smaller at 0.169. Somewhat surprisingly, we find that owners who say their neighborhood has gotten worse over the past year are more satisfied with their dwelling compared to those who say their neighborhood has stayed the same. However, upon exploring the literature, we find a possible explanation: In declining neighborhoods, residents face what is termed an "exit or voice" decision (Farrell 1983; Manturuk et al. 2009; Orbell and Uno 1972) in which they must decide between leaving the community or becoming active in trying to make improvements. Homeowners are more likely than renters to stay and seek change (Manturuk et al. 2017), which could contribute to feeling more satisfied with where they live.[8]

It is informative to look at the r-square statistics for each of the homeowner models. The model including only the housing stressors explains just 9 percent of the variation in housing satisfaction. When the housing characteristics variables are added, the model accounts for 10 percent of the variation in housing satisfaction. When we add the neighborhood-level measures, the r-square increases dramatically, indicating that the model is now explaining 49 percent of the variation in housing satisfaction. Thus we feel that this third model is fairly effective in capturing the drivers of housing satisfaction for homeowners.

Renters and Housing Satisfaction

Table 12.4 shows the regression models predicting housing satisfaction for the renters. In Model 1 we looked at the effect of stressors on renters' housing satisfaction. The renters have a few things in common with their homeowner counterparts. As was the case for the homeowners, renters living in attached, multifamily dwellings were less satisfied with their housing (when compared with renters living in single-family, detached dwellings) by just under half a point on the 10-point scale (−0.451). Also consistent with the homeowners, housing satisfaction is correlated with finding dwelling maintenance stressful and finding neighbors stressful. However, unlike the homeowners, renters who felt that their housing payment was too high were less likely to be satisfied with their dwelling (−0.484).

Model 2 adds variables representing physical attributes of the residence. The physical housing characteristic variables operate the same for renters as for homeowners. The people-to-bedroom

Table 12.3 OLS Regression Model Predicting Housing Satisfaction (1–10 Scale) for Homeowners.

	Model 1	*Model 2*	*Model 3*
Attached dwelling	−0.418***	−0.439***	−0.299***
	[−3.73]	[−3.98]	[−3.34]
Other dwelling type	−0.565*	−0.616**	−0.851***
	[−2.56]	[−2.84]	[−4.84]
Children in the home	−0.071	−0.096	−0.101
	[−0.99]	[−1.08]	[−1.40]
Percent of income spent on housing	0.036	0.034	0.017
	[1.75]	[1.71]	[1.05]
Paying for housing is stressful	−0.010	−0.021	−0.071
	[−0.12]	[−0.27]	[−1.13]
Home maintenance is stressful	−0.439***	−0.366***	−0.233***
	[−6.06]	[−5.11]	[−4.03]
Neighbors are stressful	−0.392***	−0.364***	0.073
	[−4.77]	[−4.52]	[1.09]
Housing payment is too high	0.016	−0.037	−0.057
	[0.22]	[−0.50]	[−0.97]
Years in the home	−0.005	0.004	0.009
	[−0.87]	[0.68]	[1.84]
People per bedroom		−0.071	−0.107
		[−0.88]	[−1.63]
Number of bathrooms		0.396***	0.260***
		[7.85]	[6.25]
Tract percent single parents			0.023**
			[3.09]
Tract percent unemployed			0.015
			[1.51]
Tract percent receiving public assistance			0.006
			[0.46]
Tract percent below poverty line			0.000
			[0.06]
Rating of the neighborhood (1–10)			0.546***
			[25.03]
Would recommend the neighborhood			−0.103
			[−0.94]
Neighborhood is good place for kids			0.169*
			[2.09]
Neighborhood has good schools			0.126
			[1.91]
Neighborhood has become better			-0.037
			[-0.50]
Neighborhood has become worse			0.313**
			[3.03]
Moved to a new neighborhood			0.050
			[0.26]
Constant	8.665***	7.937***	3.225***
	[95.64]	[54.07]	[14.55]
R-square	0.09	0.10	0.49
N	1623	1623	1623
t statistics in brackets			

Source: authors.

* $p < 0.05$, ** $p < 0.01$, *** $p < 0.001$

Table 12.4 OLS Regression Model Predicting Housing Satisfaction (1–10 Scale) for Renters.

	Model 1	*Model 2*	*Model 3*
Attached dwelling	−0.451**	−0.398*	−0.261
	[−2.63]	[−2.29]	[−1.93]
Other dwelling type	−0.358	−0.365	−0.075
	[−1.16]	[−1.18]	[−0.32]
Children in the home	0.160	0.027	−0.030
	[1.00]	[0.14]	[−0.20]
Percent of income spent on housing	0.053	0.055	−0.045
	[0.66]	[0.68]	[−0.73]
Paying for housing is stressful	0.024	0.059	0.084
	[0.13]	[0.32]	[0.59]
Home maintenance is stressful	−0.563**	−0.576**	−0.377*
	[−2.91]	[−2.98]	[−2.54]
Neighbors are stressful	−0.602**	−0.579**	0.032
	[−3.25]	[−3.12]	[0.22]
Housing payment is too high	−0.484**	−0.526**	−0.347**
	[−2.78]	[−3.00]	[−2.60]
Years in the home	−0.005	−0.001	0.010
	[−0.49]	[−0.11]	[1.18]
People per bedroom		0.034	−0.004
		[0.21]	[−0.03]
Number of bathrooms		0.331*	0.267*
		[2.21]	[2.25]
Tract percent single parents			0.034*
			[2.35]
Tract percent unemployed			0.040*
			[2.03]
Tract percent receiving public assistance			−0.016
			[−0.66]
Tract percent below poverty line			−0.002
			[−0.16]
Rating of the neighborhood (1–10)			0.609***
			[14.77]
Would recommend the neighborhood			0.411
			[1.95]
Neighborhood is good place for children			−0.053
			[−0.31]
Neighborhood has become better			0.110
			[0.65]
Neighborhood has become worse			0.264
			[1.39]
Moved to a new neighborhood			0.023
			[0.12]
Constant	8.311***	7.809***	2.051***
	[42.67]	[24.35]	[4.89]
R-square	0.06	0.10	0.42
N	549	549	549
t statistics in brackets			

Source: authors.

* $p < 0.05$, ** $p < 0.01$, *** $p < 0.001$

ratio is not significant, while the variable for the number of bathrooms is significant. Each additional bathroom is associated with a 0.331 increase in the housing satisfaction score. Finally, Model 3, which includes the neighborhood-level measures, produces some similar results. As was the case for the homeowners, there is an association between the percent of single parents in the neighborhood and housing satisfaction, but again the effect size is small enough (.034) to essentially be immaterial. This model also finds a similar association between the tract percent of unemployed residents, but again it is an extremely small effect (0.04). The rating of the neighborhood is correlated with housing satisfaction: Each point increase in the neighborhood rating scale is associated with a 0.609 increase in housing satisfaction.

One place where the renters' Model 3 diverges from the homeowners' Model 3 is when we look at the variable for neighborhood change. Recall that homeowners who felt that their neighborhoods had gotten worse were more satisfied with their dwellings. We find no significant association between this variable and renters' housing satisfaction, lending support to the theory that in neighborhoods that are declining, renters might be more likely than homeowners to choose to move away or do nothing. It is the homeowners, due to their investment in the community, who are more apt to take action to change the neighborhood (Manturuk et al. 2017), and this involvement may make them feel more connected to and satisfied with where they live.

Finally, we see a similar pattern in the r-square statistics for the renters as we saw with the homeowners. Models 1 and 2 in combination explain just 10 percent of the variation in housing satisfaction. Once the neighborhood measures are included, however, the model explains 42 percent of the variation.

Discussion

Our first hypothesis was that renters' housing satisfaction would be more heavily affected by housing-related stressors than homeowners' satisfaction, because renters have less control over mitigating housing stressors. Indeed, we found that the two groups had more in common than we had anticipated: Both homeowners and renters were affected by experiencing stress related to maintaining the home and dealing with neighbors, and both groups experienced less satisfaction when living in a multifamily unit. Where our hypothesis might have merit is around financial stress and housing satisfaction: Only the renters showed a decline in housing satisfaction related to the feeling that their housing payment is too high. As we noted previously, it is plausible that renters have less control over changes in their housing payments, and therefore how satisfied they are with where they live is impacted by whether the cost of their unit has increased beyond what can be comfortably afforded.

Our second hypothesis was that the housing satisfaction of homeowners and renters would be equally affected by physical characteristics of the dwelling, because the groups share a use-interest in their dwellings. Since both owners and renters use their properties for accommodation, it stands to reason that both groups would be affected by how well their dwellings do, indeed, accommodate them. We find some support for this hypothesis: The housing satisfaction of both owners and renters increases as the number of bathrooms in their homes increases. Interestingly, our other measure of physical characteristics, i.e., number of persons per bedroom, was not a significant contributor to housing satisfaction for either group.

Our third hypothesis was that homeowners' housing satisfaction would be more heavily affected by neighborhood characteristics than renters' satisfaction, because homeowners have an investment interest in the neighborhood, and neighborhood factors can affect that investment. The evidence is mixed on this. First, we find that very few individual neighborhood-level measures are good predictors of housing satisfaction for either group, yet collectively the neighborhood measures explain roughly half the variation in satisfaction for both homeowners and renters. But we also find that owners and renters have more in common than we had anticipated as regards the effect of neighborhood factors on housing satisfaction. The one measure that stands out for both groups is neighborhood rating: The higher owners and renters rated their neighborhoods, the greater their reported housing satisfaction. Where the groups differ is in the relationship between perceived neighborhood change and housing satisfaction: We find that homeowners, but not renters, feel more favorably about their dwellings when they think the neighborhood has become worse in the prior year. We will discuss this counterintuitive finding, and all of our findings, in the conclusion.

Conclusion

In this chapter, we set out to examine the drivers of owners' and renters' housing satisfaction. We hypothesized that, insofar as owners and renters share *interests* in their property, they would also share drivers of housing satisfaction. Because both owners and renters share a use-interest in their property—i.e., both groups live in and make use of their homes—we anticipated that property characteristics would play a similar role in driving both groups' housing satisfaction. However, because renters have less ability to advance this use-interest—i.e., they cannot necessarily address aspects of the property they find stressful—we expected that housing stressors would have greater impact on renters' housing satisfaction than on homeowners' housing satisfaction. Finally, where owners and renters differ most obviously in their interests is that owners have a financial stake in their homes while renters do not. We therefore anticipated that neighborhood characteristics would drive owners' housing satisfaction to a greater degree than they drive renters' housing satisfaction, since neighborhood factors have some effect on home values.

Our analysis resulted in mixed findings, which we review here, touching on their implications for both affordable housing policy and program design. First, we found some support for our hypothesis that renters' housing satisfaction would be more heavily affected by housing-related stressors than homeowners' satisfaction. While we found that the housing satisfaction of both homeowners and renters was affected by stress related to home maintenance and dealing with neighbors—similarities that refute our hypotheses—we also found that the housing satisfaction of renters, but not owners, was affected by feeling that their housing payment was too high, and this difference lends some support to our hypothesis. As we noted previously, it is very plausible that renters have less control over changes in their housing payments than owners do and that the unpredictability of rental increases might affect their sense of housing satisfaction. Additionally, owners may have a sense that their housing payment is a form of "paying themselves," whereas renters watch their payment flow to another party each month. The forced savings associated with the housing payment might lead to satisfaction on the part of owners, while renters might be dissatisfied to pay this money to other people. The implications of this finding are as follows.

Insofar as renters experience less housing satisfaction than owners owing to their inability to "pay themselves" with their rental payments, program design can have an effect. At least one innovative program is attempting to link rental housing with both the wealth gains and control over space that we typically associate with homeownership. The Cornerstone Corporation for Shared Equity (Cincinnati, Ohio) takes as its mission the provision of "affordable housing and economic opportunities that enable low-income people to live in safe and affordable housing, build financial assets, and develop management skills needed to be self-sufficient and engaged in the community."[9] To achieve this goal, Cornerstone developed its Renter Equity program, which includes several benefits traditionally associated with homeownership: Residents can remain in their units as long as they wish (providing they fulfill the terms of their occupancy agreements); residents contribute to maintenance and upkeep of their community's common areas; residents may decorate, paint, and upgrade their units as they wish (with approval from management); residents help manage their communities and create and vote upon community rules; and, importantly, residents earn credits in a financial fund, and after a five-year vesting period, they can withdraw this money or use it as collateral for a loan.

It is also reasonable to assume that the cause of renters' feeling that their housing payment is too high might be that their housing payment is, indeed, too high. Since the mortgage finance crisis that began 2006 and the Great Recession that followed, there has been unprecedented pressure on rental housing markets as owners have transitioned back to renting and potential buyers have delayed home purchase. The Joint Center for Housing Studies of Harvard University (2011) notes that, "from 2006 to 2010, the number of renter households jumped by 692,000 annually on average, to 37 million, while the number of owner households fell on net by 201,000 annually. This is a complete reversal from the preceding decade and a half, when homeowners drove the vast majority of household growth and the number of renters stagnated" (22). The tightening of U.S. rental markets and the increased costs of renting have not abated since this time: According to the Joint Center for Housing Studies of Harvard University (2015), the national rental vacancy rate dropped in 2014 to 7.6 percent, the lowest it had been for two decades, and rents rose at a rate of 3.2 percent in the same year, which is twice the rate of inflation overall. Given these trends, it is no surprise that renters feel their housing payments are too high and that this feeling might affect their housing satisfaction. The many ways to develop new or preserve existing affordable rental stock are too numerous to detail in this chapter, but it is an area of tremendous opportunity for both national and local policy and program development (see also the chapter authored by Ingrid Gould Ellen).

To return to our results, the analysis provided clear support for our second hypothesis, that the housing satisfaction of homeowners and renters would be equally affected by physical characteristics of the dwelling. Interestingly, while our measure of use of physical space (number of persons per bedroom) was not a significant contributor to housing satisfaction for either owners or renters, both groups experienced significant increases in housing satisfaction as the number of bathrooms in their homes increased. This finding provides support for the importance of housing design, suggesting that it matters to renters and owners alike. Developers and builders of affordable housing who are concerned with promoting occupants' housing satisfaction would do well to remember Wright's (2014) conclusion that the "skillful organization of interiors, views, public areas, outdoor spaces, and even facades is especially important when budgets and square footage are at a premium" (70).

The analysis provided limited, and somewhat unexpected, support for our third hypothesis, i.e., that homeowners' housing satisfaction would be more heavily affected by neighborhood characteristics because homeowners have an investment interest in their neighborhoods. Surprisingly, we found that homeowners, but not renters, feel more favorably about their dwellings when they think the neighborhood has *declined* in the prior year. This seemingly contradictory finding makes more sense than might be expected, since previous studies have documented that homeowners in declining neighborhoods often become more invested in their neighborhoods and dwellings: For example, Manturuk et al. (2009) found that homeowners in disadvantaged neighborhoods were more likely to vote in local elections than homeowners in more affluent areas. While renters may choose to move from declining neighborhoods, the transaction costs associated with homeowners selling their properties and moving are higher; homeowners may be more likely than renters to stay and seek change, which could translate into a greater feeling of belonging and increased housing satisfaction.

While this difference between the two groups lends some support to our hypothesis that homeowners' housing satisfaction would be more heavily affected by neighborhood characteristics, we found again that owners and renters had more in common than we had anticipated. The one measure that stands out for *both* groups is neighborhood rating. The more highly owners and renters rated their neighborhoods, the greater their housing satisfaction. The power of this effect is undeniable: For each group, an increase in the neighborhood rating scale was correlated with more than a half-point increase in the housing satisfaction measure. The similar effect of neighborhood on housing satisfaction between groups suggests that owners and renters have more in common than we had hypothesized.

Perhaps the most interesting aspect of our analysis was the *extent* to which neighborhood factors affected the housing satisfaction of both homeowners and renters. When we controlled for just housing stressors and dwelling physical characteristics, our models predicted only 10 percent of the variation in both owners' and renters' housing satisfaction. When we added in neighborhood characteristics, however, the predictive power of our models improved substantially; by adding in neighborhood factors, we were able to predict 42 percent of the variation in renters' and 49 percent of the variation in owners' housing satisfaction. This means that neighborhood characteristics affect not just the *neighborhood* satisfaction of owners and renters, but their actual *housing* satisfaction as well. Although our analysis was not structured to reveal exactly what factors led people to rate their neighborhoods more highly, we still take a moment to consider what the link between neighborhood satisfaction and housing satisfaction might imply for policies and programs concerned with increasing the housing satisfaction of lower-income owners and renters.

First, if neighborhood satisfaction influences housing satisfaction, then policies and programs concerned with promoting housing satisfaction should take into account those factors that have been shown to increase residents' neighborhood satisfaction. Studies have shown that any number of factors influence neighborhood satisfaction. Many of these pertain to neighborhood design, including landscape structure (Lee et al. 2008), openness (Hur et al. 2010), attractive appearance (Lovejoy et al. 2010), green space (Kaplan and Austin 2004), and the opportunity to enjoy shared outdoor space (Kearney 2006). These issues of neighborhood design have clear implications for both the siting and design of affordable housing, and should be incorporated into the process of planning new and improving existing housing. Other factors that have been shown to influence neighborhood satisfaction—though the effect of these varies by location and by the particular population under study—include perceived safety of the neighborhood (Lovejoy et al. 2010), the absence of conditions deemed undesirable (such as street noise or trash)[10] (Spain 1988), interacting socially with other residents (Hur and Morrow-Jones 2008), and having strong social bonds within the neighborhood (Oh 2003). While some of these things—fighting crime by creating neighborhood watch programs, for example—can be addressed by programs at the local level, others—having strong social bonds in the neighborhood, for example—speak more to increasing lower-income households' options around where they can live. Programs that aim at increasing and promoting residential choice, such as housing voucher programs, might be most efficacious in addressing these elements of neighborhood, and thereby residential, satisfaction.

The second implication of our finding that neighborhoods matter to lower-income owners' and renters' housing satisfaction picks up from this final point. If policies and programs are concerned with increasing households' housing satisfaction, it seems reasonable to assume that promoting owners' and renters' neighborhood choice would increase their likelihood of having access to neighborhoods that they like and that will promote their housing satisfaction. This connection has implications both for the siting of affordable housing—both owned and rented—and for rental housing programs aimed at allowing participants to choose where they live. An examination of the literature on the pitfalls and successes of federal programs designed to promote housing access and neighborhood choice is beyond the scope of this chapter.[11] Instead, we propose that our study is one more piece of evidence that place really does matter (see the chapter authored by Andrew Greenlee). Neighborhoods affect housing satisfaction, and insofar as greater housing satisfaction leads to greater residential stability,[12] this link may have important implications for the development of policies and programs aimed at promoting regional well-being.

We conclude by suggesting a number of areas for future research. First, our models leave unexplained more than 50 percent of the variation in housing satisfaction. Future research should continue to probe into this issue in order to elucidate what combination of factors drives homeowners' and renters' housing satisfaction. Second, our analysis revealed that neighborhood characteristics play a much larger role in *both* homeowners' and renters' housing satisfaction than we had expected.

Future research should be structured to tease out the connection between neighborhood characteristics and housing satisfaction. Finally, our analysis has concentrated on LMI families, and there is a need for more far-reaching study of the intersection between housing and place to determine how the intricacies of housing satisfaction work and might be promoted more generally.

Notes

1 As with all longitudinal studies, there has been attrition over time. Between year one and year seven, the overall attrition rate was 44 percent for the homeowners and 43 percent for the renters. Most of the attrition happened in the first year; of those who completed the 2009 survey, 94 percent completed the 2010 survey. Attrition analysis indicates that men and Hispanic participants had higher attrition rates, while people with a college degree and those with higher credit scores were less likely to drop out of the panel.

2 To assess the generalizability of the CAPS findings, Riley et al. (2009) compared the CAPS participants to the 2003 Current Population Survey (CPS) respondents in the same income bracket. The analysis showed that CAPS homeowners are very similar to a random national sample, but there are a few differences. CAPS homeowners skew younger than a random sample, likely because people tend to enter homeownership in their 20s and 30s and the baseline survey occurred shortly after people bought their first homes. Along similar lines, more of the CPS participants were retired or out of the labor force than is found among the CAPS participants. Among the renters, CAPS includes a higher proportion of women than a random sample would, and the CAPS renters are more likely to be married than the CPS sample. (For details, see Riley et al. 2009.) In spite of these differences, we believe that the CAPS sample is largely similar to a random sample of lower-income households.

3 The response categories are very stressful, somewhat stressful, or not at all stressful. We collapsed the categories so people who felt that a given factor was somewhat or very stressful were coded 1 and those reporting no stress were coded 0.

4 We initially captured this by including a ratio of the number of residents to the square footage of the home. However, we found that over 30 percent of the renters could not give an answer when asked to estimate their home's square footage. We therefore relied on the bedroom-to-resident ratio and number of bathrooms instead.

5 Like the housing satisfaction variable, this measure also skews right with a mean of 7.9 for homeowners and 7.5 for renters.

6 For the questions about whether the neighborhood is a good place to raise children and whether the neighborhood has good schools, participants were asked to rate the neighborhood and the schools as excellent, good, fair, and poor. We collapsed excellent and good and coded those respondents as 1, while those who felt their schools and neighborhoods were fair or poor were coded 0.

7 We asked respondents, "Since we talked to you last, how has this neighborhood changed as a place to live? Has it gotten a lot better, gotten somewhat better, stayed about the same, gotten somewhat worse, or gotten a lot worse?" The response options were: gotten a lot better, gotten somewhat better, stayed the same, gotten somewhat worse, and gotten a lot worse. We created a 1/0 variable combining the two responses indicating improvement, and we created another indicator variable combining the two responses indicating decline. To this, we added one indicator variable to identify people who felt that their neighborhood had stayed the same. We also created an indicator variable for people who had moved in the prior year and therefore could not evaluate whether the neighborhood had changed since the last survey; we did so in order not to have these people drop out of the larger analysis.

8 We thank one of the editors for emphasizing the counterintuitive nature of the positive relationship between owners' perceptions about their neighborhood getting worse and their increased housing satisfaction. Further research is needed to assess the extent to which this finding is a replicable result or an artifact of the specific predictors used in this analysis. We respectfully disagree with this editor that our supposition concerning this relationship is unconvincing. Interested readers may want to see Manturuk et al. (2017) for a large body of research on the social impacts of affordable homeownership. One of the chapters in that book provides insight into the long-running debate over whether homeownership promotes civic participation or locks lower-income families in disadvantaged areas where they are likely to withdraw from community life: The analysis shows that entry into homeownership can provide LMI people with a catalyst for increasing goal-seeking types of civic engagement.

9 See http://cornerstone-equity.org/about/. For additional examples of rental programs that include a savings component, see Abromowitz and Edelman (2014).

10 In this article, "neighborhood conditions" was defined as the absence of "street noise; streets in need of repair; crime; trash; abandoned buildings; industrial, nonresidential activity; odors, smoke, gas."
11 See, for a start, recent articles by Horn et al. (2014), Walter et al. (2015), McClure et al. (2015), and Hendey et al. (2015).
12 There are mixed findings on the association between neighborhood characteristics and mobility; see, for example, Rabe and Taylor (2010).

References

Abromowitz, David M., and Sarah Edelman. 2014. *As More Households Rent, How Can We Encourage Them to Save?* Washington, DC: Center for American Progress.

Amérigo, Maria, and Juan Ignacio Aragones. 1990. "Residential Satisfaction in Council Housing." *Journal of Environmental Psychology*, 10(4): 313–325.

Christensen, David L., Frances M. Carp, Galen l. Cranz, and James A. Wiley. 1992. "Objective Housing Indicators as Predictors of the Subjective Evaluations of Elderly Residents." *Journal of Environmental Psychology*, 12(3)3: 225–236.

Deater-Deckard, Kirby. 1998. "Parenting Stress and Child Adjustment: Some Old Hypotheses and New Questions." *Clinical Psychology: Science and Practice*, 5(3): 314–332.

Deater-Deckard, Kirby, and Scarr, Sandra. 1996. "Parenting Stress among Dual-Earner Mothers and Fathers: Are there Gender Differences?" *Journal of Family Psychology*, 10(1): 45–59.

Diaz-Serrano, Luis. 2009. "Disentangling the Housing Satisfaction Puzzle: Does Homeownership Really Matter?" *Journal of Economic Psychology*, 30(5): 745–755.

Dupuis, Ann, and David C. Thorns. 1998. "Home, Home Ownership and the Search for Ontological Security." *The Sociological Review*, 46(1): 24–47.

Farrell, Dan. 1983. "Exit, Voice, Loyalty, and Neglect as Responses to Job Dissatisfaction: A Multidimensional Scaling Study." *Academy of Management Journal*, 26(4): 596–607.

Galster, George C. 1987. *Homeowners and Neighborhood Reinvestment*. Durham, NC: Duke University Press.

Hendey, Leah, George C. Galster, Susan J. Popkin, and Chris Hayes. 2015. "Housing Choice Voucher Holders and Neighborhood Crime A Dynamic Panel Analysis from Chicago." *Urban Affairs Review*, 52(4): 471–500.

Horn, Karen Mertens, Ingrid Gould Ellen, and Amy Ellen Schwartz. 2014. "Do Housing Choice Voucher Holders Live Near Good Schools?" *Journal of Housing Economics*, 23(March): 28–40.

Hur, Misun, and Hazel A. Morrow-Jones. 2008. "Factors that Influence Residents' Satisfaction with Neighborhoods." *Environment and Behavior*, 40(5): 619–635.

Hur, Misun, Jack L. Nasar, and Bumseok Chun. 2010. "Neighborhood Satisfaction, Physical and Perceived Naturalness and Openness." *Journal of Environmental Psychology*, 30(1): 52–59.

Joint Center for Housing Studies of Harvard University. 2011. *The State of the Nation's Housing 2011*. Cambridge, MA: Joint Center for Housing Studies of Harvard University.

Joint Center for Housing Studies of Harvard University. 2015. *The State of the Nation's Housing 2015*. Cambridge, MA: Joint Center for Housing Studies of Harvard University.

Kaplan, Rachel, and Maureen E. Austin. 2004. "Out in the Country: Sprawl and the Quest for Nature Nearby." *Landscape and Urban Planning*, 69(2): 235–243.

Kearney, Anne R. 2006. "Residential Development Patterns and Neighborhood Satisfaction Impacts of Density and Nearby Nature." *Environment and Behavior*, 38(1): 112–139.

Lane, Sylvia, and Jean Kinsey. 1980. "Housing Tenure Status and Housing Satisfaction." *Journal of Consumer Affairs*, 14(2): 341–365.

Lee, Sang-Woo, Christopher D. Ellis, Byoung-Suk Kweon, and Sung-Kwon Hong. 2008. "Relationship between Landscape Structure and Neighborhood Satisfaction in Urbanized Areas." *Landscape and Urban Planning*, 85(1): 60–70.

Lovejoy, Kristin, Susan Handy, and Patricia Mokhtarian. 2010. "Neighborhood Satisfaction in Suburban versus Traditional Environments: An Evaluation of Contributing Characteristics in Eight California Neighborhoods." *Landscape and Urban Planning*, 97(1): 37–48.

Manturuk, Kim, Mark Lindblad, and Roberto G. Quercia. 2009. "Homeownership and Local Voting in Disadvantaged Urban Neighborhoods." *Cityscape*, 11(3): 213–230.

Manturuk, Kimberly R., Mark R. Lindblad, and Roberto G. Quercia. (2017) *A Place Called Home: The Social Dimensions of Homeownership*. New York City: Oxford University Press.

McClure, Kirk, Alex F. Schwartz, A. F., and Lydia B. Taghavi. 2015. "Housing Choice Voucher Location Patterns a Decade Later." *Housing Policy Debate*, 25(2): 215–233.

O'Bryant, Shirley L., and Susan M. Wolf. 1983. "Explanations of Housing Satisfaction of Older Homeowners and Renters." *Research on Aging*, 5(2): 217–233.

Oh, Joong-Hwan. 2003. "Social Bonds and the Migration Intentions of Elderly Urban Residents: The Mediating Effect of Residential Satisfaction." *Population Research and Policy Review* 22(2): 127–146.

Orbell, John M., and Toru Uno. 1972. "A Theory of Neighborhood Problem Solving: Political Action vs. Residential Mobility." *The American Political Science Review* 66(2):471–489.

Rabe, Birgitta, and Mark Taylor. 2010. "Residential Mobility, Quality of Neighbourhood and Life Course Events." *Journal of the Royal Statistical Society: Series A (Statistics in Society)* 173(3): 531–555.

Riley, Sarah F., HongYu Ru, and Roberto G. Quercia. 2009. "The Community Advantage Program Database: Overview and Comparison With the Current Population Survey." *Cityscape* 11(3): 247–256.

Sampson, Robert J. 1991. "Linking the Micro- and Macrolevel Dimensions of Community Social Organization." *Social Forces* 70(1): 43–64.

Sampson, Robert J., and W. Byron Groves. 1989. "Community Structure and Crime: Testing Social-Disorganization Theory." *The American Journal of Sociology* 94(4): 774–802.

Spain, Daphne. 1988. "The Effect of Changing Household Composition on Neighborhood Satisfaction." *Urban Affairs Quarterly*, 23(4): 581–600.

UNC Center for Community Capital. 2014. *Research Brief: Community Advantage Panel Study*. Chapel Hill, NC: Center for Community Captial. Available at http://ccc.unc.edu/contentitems/community-advantage-panel-study-sustainable-approaches-to-affordable-homeownership/

Walter, Rebecca J., Yanmei Li, and Serge Atherwood. 2015. "Moving to Opportunity? An Examination of Housing Choice Vouchers on Urban Poverty Deconcentration in South Florida." *Housing Studies*, 30(7): 1064–1091.

Wright, Gwendolyn. 2014. "Design and Affordable American Housing." *Cityscape: A Journal of Policy Development and Research*, 16(2): 69–86.

SECTION 5

Social Housing

13
THE U.S. APPROACH TO "SOCIAL HOUSING"

Rachel G. Bratt

The provision of social housing is at the heart of a right to housing agenda: The ability to provide long-term housing that is high quality and permanently affordable is the central mission of those who embrace this idea. Important, too, is the availability of options to deal with the need for emergency shelter, when individuals or households are, for whatever reason, suddenly displaced from their homes. The subject of this chapter, however, focuses only on the provision of long-term housing that is affordable to households at or below 60–80 percent of area median income (AMI). It should be underscored that a right to housing would be applicable to everyone, but the key challenge is providing housing for poor and lower-income people.

The term "social housing" is not frequently used in the U.S. In contrast, it is a well-known and familiar concept elsewhere across the world, particularly Europe (Houard 2011; Reeves 2014; Scanlon et al. 2014). The relative rarity of the term's usage in the U.S. notwithstanding, the U.S. social housing sector includes about five million units in two major sectors: public housing and nonprofit-owned housing.

"Public housing" refers primarily to housing that has been built under the U.S. Public Housing Act of 1937 or, in a few cases, through a similar program operated at the state level.[1] Under this program, local housing authorities (LHAs) produce units with a federal subsidy for the development and management of the public housing; long-term ownership is retained by the LHA, with regulatory oversight provided by the U.S. Department of Housing and Urban Development (HUD). A relatively small number of housing units that are owned and operated for military personnel by the U.S. Defense Department are also considered "public housing." While they are not typically considered part of the public housing stock, they are included in Figure 13.2.

Housing that is owned by nonprofit organizations constitutes the second part of the social housing sector and is the primary focus of this chapter. This housing can have a variety of unique characteristics, depending on the particular type of nonprofit that is sponsoring the development. The intent is for social housing units to be affordable over the long term, typically meaning more than the 30 years that is currently required in the federal Low-Income Housing Tax Credit (LIHTC) program. Since nonprofit and public owners strive for affordability in perpetuity, with one exception noted in the section "Public Housing," only units owned by these sponsors are viewed as part of the social housing sector.

Using this definition, housing units that are developed and owned by for-profit firms and that are either subsidized by the federal government (e.g., through LIHTC or an earlier public-private partnership program) or produced as a result of local regulatory provisions, such as inclusionary zoning, are not viewed as "social housing." This definition may underestimate the total number of social housing units, since it excludes: (1) housing built by for-profit developers through local regulatory programs that do, in fact, require long-term affordability; and (2) housing built by a handful of for-profit developers, such as McCormack Baron Salazar (Bratt 2016), that have a commitment to the long-term affordability of their units.[2] However, in view of the scope of this chapter, it is not possible to ascertain how much housing in these two categories should be counted as long-term social housing, and the size of the overall social housing stock may be underestimated.

Also excluded from the "social housing" category is a major type of assisted housing—private market housing that is rented to income-eligible households using a Section 8 Housing Choice Voucher. As with private for-profit project-based housing, this housing also is not protected through long-term affordability restrictions.

Although some public housing units have, over time, been lost from the public housing stock, the intent has been for such units to be affordable in perpetuity. Figure 13.1 shows that as of 2016, there were about 8.6 million active assisted housing units. Many of the project-based units are also included in the tally of social housing sector units shown in Figure 13.2. The latter presents an overview of the approximately five million units that are part of the social housing sector.

A further observation is that, in keeping with the definition of social housing, only some of the nonpublic housing project-based units noted in Figure 13.1 are included in Figure 13.2; the project-based units that are included were built by nonprofit developers (rather than for-profits) whose

Project-Based Programs	*Number. of Units*	*% of Total Assisted Housing Stock*	*% of Project-Based Programs*
Public housing (all types, as noted on Figure 13.2)	1,516,260	17.5%	24.5%
Below-market interest rate programs (public subsidy/private ownership)	1,404,589	16.2%	22.7%
Section 8 production programs (public subsidy/private ownership)	590,774	6.8%	9.6%
LIHTC (public subsidy/private ownership)	2,100,000[1]	24.3%	34.0%
USDA below-market interest rate programs	571,000	6.6%	9.2%
Sub-total project-based programs	**6,182,623**	**71.4%**	**100%**
Rental Subsidies			*% of Rental Subsidies*
Housing Choice Vouchers (also known as Section 8 vouchers)	2,200,000[11]	25.4%	89.0%
USDA Rental Assistance	273,000	**3.2%**	11.0%
Sub-total rental subsidies	**2,473,000**	**28.6%**	**100%**[111]
TOTAL ASSISTED HOUSING STOCK	**8,655,623**	**100%** [111]	

Figure 13.1 Major Components and Approximate Size of the U.S. Assisted Housing Sector, 2016

Notes:

★ Includes only still active units; many public housing and other project-based units are no longer in the assisted housing inventory.

★★ This number comes from the National Low Income Housing Coalition (2016, 4–34). Schwartz (2015) had estimated slightly more units in this category; N = 2,235,180.

★★★ Errors due to rounding.

Source: author, based on notes.

	Number of Units	*% of Total Social Housing Sector*	*% of Public Housing*
Federal (family and elderly) (LHAs, N = 3,095)[ii]	1,156,839[ii]	22.9%	76.3%
Department of Defense (for military households)	300,000[iii]	5.9%	19.8%
Other programs (state and local)	59,421[iv]	1.2%	3.9%
Sub-total public housing	**1,516,260**	**30.0%**	**100%**
			% of Nonprofit Housing (Adjusted Total)
Community development corporations (N = 4,600)[v]	1,252,000[v]	24.8%	35.5%
Housing Partnership Network members (N = 97)[vi]	219,300[vi]	4.3%	6.2%
Community land trusts (N = 220)[vii]	59,813[vii]	1.2%	1.7%
Limited equity cooperatives	425,000[viii]	8.4%	12.1%
Federally subsidized housing for the elderly	400,000[ix]	7.9%	11.3%
Habitat for Humanity (N = 1,500)[x]	130,523[xi]	2.6%	3.7%
Mutual housing and other nonprofit organizations	240,000[viii]	4.8%	6.8%
Sub-total nonprofit housing	**2,726,636**	**54.0%**	
Estimated additional production data collected on which above figures are based	800,000[xii]	16.0%	22.7%
Adjusted sub-total nonprofit housing	**3,526,636**		**100%**
TOTAL ALL SOCIAL HOUSING UNITS	**~ 5,042,896**	**100%**	

Figure 13.2 Components and Approximate Size of the U.S. Social Housing Sector, 2016.[i]

Notes:

i. Table updated from Bratt (2012, 442 and from Bratt 2015). The various notes that follow describe some variations from those prior presentations. Attempting to compile this type of data is very difficult and subject to errors. For many of the types of social housing, approximations are provided.

ii. Cited in Schwartz (2015, 164).

iii. www.acq.osd.mil/housing/ct97_congress.htm (accessed October 14, 2016).

iv. Stone (2006) had estimated that there were some 700,000 other state and local public housing units, and this was the number presented previously (Bratt 2012, 442). However, my subsequent efforts to confirm this number have proven unsuccessful. Based on the information I have been able to find, I believe that a more accurate number of "other state and local public housing" is 59,421 units. According to Linda Couch, Senior Vice President for Policy, National Low Income Housing Coalition, there are only four states with state-based public housing programs, Connecticut, Hawaii, Massachusetts, and New York (email communication, February 26, 2015). Connecticut = 11,500 units (Michael C. Santoro, CD Specialist, Office of Policy, Research and Housing Support, Connecticut Department of Housing, Private email communication, March 3, 2016). Hawaii = 864 units. A report states that there is "no subsidy" associated with these units (Hawaii Public Housing Authority 2015, 10). However, in a follow-up phone call to the agency a staff member, Mari Car, clarified that this means that there is no federal subsidy (March 4, 2016). Massachusetts = 45,484 units (number provided by Margot LeClair, Massachusetts Department of Housing and Community Development, March 1, 2016 (telephone call)). As of June 30, 1997, there were 50,285 state-aided public housing units (Citizens' Housing and Planning Association 1999), meaning that 90 percent of this portfolio has remained under state control. New York = 1,573. In all, New York State financed the construction of some 66,123 public housing units (New York State, Homes and Community Renewal 2016). Thus, nearly all of New York State's state-funded public housing units have been "federalized"—transferred to the federal public housing stock, www.nyshcr.org/Programs/PublicHousing/ (accessed October 14, 2016).

v. This number comes from the last national census of CDCs (National Congress for Community Economic Development 2005). However, according to the National Alliance of Community Economic Development Associations (2010), the production by CDCs, large nonprofit housing producers, and other types of nonprofits totaled 1,614,000 units. In this Figure, an effort is made to break out the different types of nonprofits. However, there may be some double counting in the number of units listed for CDCs, Housing Partnership Network members, and other types of nonprofits. This is one reason why the numbers in the Figure should be viewed as approximations.

vi. Housing Partnership Network, Membership Profile (2014 data).

Figure 13.2 (Continued)

vii. The total number of CLTs is an estimate offered by Emily Thaden (private email communication, February 16, 2016). However, the production data are based only on the number of members in the CLT Network, N = 136. However, since 24 of these organizations did not have any rental units in their portfolio as of December 31, 2014, the production number in the Figure is based only on 112 CLTs. "Of members with residential units (N = 112), one member stewarded 32,000 cooperative units, which accounted for 53.5 percent of all member units" (Thaden and Webb 2015, 9). The production figure of 59,813 was calculated from this information.

viii. The numbers for "limited equity cooperatives" and "mutual housing and other nonprofit organizations" are from Stone (2006).

ix. See the National Low Income Housing Coalition (2016). Schwartz (2015) cites a slightly lower number, 360,000, and also notes that some sources indicate that the number could be as low as 270,000.

x. www.habitat.org/how/affiliates (accessed October 14, 2016).

xi. Private email communication from Angie Kuhlmann, Habitat for Humanity, March 10, 2015.

xii. Several of the numbers in the Figure are outdated, particularly production by CDCs. CDCs produced some 602,000 units in the seven years between 1998 and 2005, for an annual average of 86,000/year. In view of the continuing sophistication of CDCs as housing producers, one estimate could be that about this number of units has been produced by CDCs in each of the 11 years from 2006 through 2016, since the last CDC census (National Congress for Community Economic Development 2005). Additional nonprofit production, by groups other than CDCs, is also not captured in the figure. In particular, as noted previously, the data for "limited equity cooperatives" and "mutual housing and other nonprofit organizations" are more than ten years old. On the other hand, nonprofit production was likely reduced significantly during the worst years of the recession, so it is reasonable to assume that annual nonprofit production was reduced for several years. On balance, I am estimating that the nonprofit sector has added at least 800,000 units, since the data cited for this Figure were collected. Assuming that there have been at least 4,000 CDCs producing housing, across the country, over the past ten plus years, this would come to an average annual production of some 20 units/year/CDC. The total size of the social housing sector is roughly the same as presented previously (Bratt 2012), but some of the production, by program or organizational type, has shifted. And, as noted in several places throughout the paper, the final tally is, by necessity, an approximation.

Source: author, based on notes.

explicit mission typically involves a commitment to long-term affordability. The specific program through which nonprofits have built housing is not, however, specified in Figure 13.2 since the available information on nonprofit production is not broken down in this way. Therefore, in making the calculations of social housing built by CDCs, for example, we know that a large amount of their social housing production includes project-based developments, but we do not know how many CDC units can be attributed to any specific program.

While most of the social housing specified in Figure 13.2 was built using some form of federal housing assistance, most of the nonpublic housing project-based units are not, as noted earlier, considered social housing since they were built by private for-profit developers and the affordability of the units is not guaranteed over the long-term.

Consistent with many other countries, the U.S. federal government has been retreating from directly subsidizing housing for lower-income households since the 1980s, when President Reagan started to severely cut back the amount of new budget authority for housing assistance—new subsidy dollars to provide support for additional households. Between 1977 and 2012 new budget authority declined 52 percent, in constant dollars (Schwartz 2015, 56).[3] The decline over more than four decades has been significant and damaging[4] and perhaps more importantly, appropriations have not kept pace with the level of need.[5]

In addition to reduced direct federal funding levels, involvement has been waning in a number of other ways as well. Many of the functions historically carried out by government—production, management, the provision of subsidies, and long-term ownership—are, in many programs, now provided by non-governmental bodies (see also the chapter authored by Lawrence Vale and Yonah Freemark). First, many public housing developments have been redeveloped privately under the HOPE VI program or are being transformed to a Section 8 platform through the Rental Assistance

Demonstration (RAD) program, discussed later. Second, new public housing production has been replaced by affordable housing development by private for-profit and nonprofit sponsors. Third, inclusionary housing programs completely shift the burden of creating affordable units onto the private sector. And, fourth, unlike many of the other federal housing programs, the LIHTC program, while providing the major federal incentive for affordable housing production, only provides a portion of the subsidy needed to make units affordable to eligible low-income households, thereby necessitating the layering of other public and private resources.

As a tax incentive program that provides tax savings to private investors in affordable housing development, it is hard to estimate exactly how much money is lost each year due to LIHTC. This is in contrast to the availability of precise information concerning HUD funding through direct Congressional appropriations. However, based on estimates published by the Joint Committee on Taxation, in 2018 the LIHTC subsidy cost the federal government about $9 billion.

In an effort to fill the direct federal funding shortfalls, there has been increasing state and local government involvement in promoting affordable housing (Bratt 2009a; Davis 2006; Goetz 1993). Nonprofit organizations are often viewed as key partners in a locale's affordable housing agenda.

At the current time, the social housing sector, particularly nonprofits, is facing formidable challenges in trying to meet the housing needs of very low-income households, particularly renter households (Joint Center for Housing Studies 2018). Thus, it is important to better understand the record of the social housing sector—its history, major attributes, and challenges. The following sections discuss the key components: public housing and nonprofit-owned housing. The chapter concludes with some ideas for how the nonprofit sector could be strengthened through a set of policies that are more "nonprofit-centric"—geared to supporting and helping to develop this part of the social housing sector. While there is much room for the nonprofit housing sector to grow, there has been little support for an expansion or new funding of the public housing program for many decades. The major current challenge for the public housing program is to continue to stabilize the existing stock by placing it on a sound financial footing and striving to manage and modernize the buildings so that they can continue to provide shelter to their very low- income occupants (see also the chapter authored by Lawrence Vale and Yonah Freemark).

Public Housing

In 1937, the U.S. embarked on a social housing agenda in the form of public housing, albeit without much enthusiasm. The primary goal of the public housing program enacted that year was to jump-start a depressed economy and to stimulate employment (Marcuse 1986); it was only secondarily aimed at meeting housing needs. Facing a great deal of opposition from the private real estate and homebuilders lobbies, because of their concern about competition with market-rate housing, it has been noted that President Franklin D. Roosevelt, himself, had to be coaxed to support the bill (Friedman 1968). Over the years, the public housing program has faced obstacles and challenges, and it has continued to be shaped by a number of issues not directly related to housing needs (Marcuse 1998). Nevertheless, it has been limping along,[6] with inadequate funding, ongoing management problems, and a generally negative public image except, perhaps, on the part of households who live there (for a summary of several surveys of public housing tenants, see Bratt 1986).

Public housing was set up as an option for local governments to pursue; it was not a requirement. State enabling legislation created a framework for local governments that chose to create local housing authorities, which were given the responsibility of developing and managing public housing. The first locales to participate in the program were the largest cities of the Northeast, Midwest, and West Coast. Smaller cities and suburban locales lagged behind, with many municipalities never choosing to form a local housing authority. Locating suitable sites for public housing proved difficult; many of the most desirable locations were already built out. Often, whatever land that was "left over" and of little interest to the private for-profit development community became the site of public housing. For example, Boston's Columbia Point public housing development was built on the site of a former city dump, on an isolated peninsula, apart from an abutting neighborhood or public services.

In many locales, efforts to build public housing in an established residential area were greeted by vehement opposition, often led by private real estate owners and developers, who were concerned that the new low-cost housing would make their market-rate housing comparatively unappealing, due to the higher rents of the market-rate units without, perhaps, much difference in quality. A desire to contain costs, as well as the efforts on the part of the homebuilders and real estate lobbies to ensure that public housing would look different and have fewer amenities than private market housing, resulted in the easily recognizable "project" look associated with early public housing developments. In addition, the architectural starkness can also be attributed, in part, to prevailing views about minimalism and the clean lines associated with the "modern" design movement.[7]

Public housing has a complex racial history, with the early developments explicitly segregated. During the 1960s and 1970s, racial changes in the large cities resulted in many developments having only a few non-White residents. Concurrently, there were complaints about the inadequacy of the public services provided, such as police and fire protection and trash collection (Epp 1996; Jones et al. 1979; as cited in Huth 1981). The result was that many developments in need of significant repairs and populated by a very low-income, non-White tenantry were often viewed as "housing of last resort."

Production under the public housing program was strong in the pre-World War II years, with over 61,000 units built in 1941. Public housing development continued after the war, and production was robust in the 1950s and 1960s, peaking in 1971, with some 91,539 new units added that year (data cited in Bratt 1986). However, virtually no new public housing has been built for several decades, except under special programs such as HOPE VI. This program, as well as the several reasons behind the demise of the public housing approach to developing social housing, is discussed in the following text.

Public housing was built through a federal commitment to cover the annual principal and interest payments on housing authority bonds. The original funding formula did not cover any maintenance or repair costs; these were to be borne by the tenants. However, already by the 1960s, this strategy was proving unworkable; a new group of poorer residents was moving into the housing, as the economic status of the original residents improved. In 1950, the income of public housing residents was over 60 percent of the national median income; in 1989, the proportion had fallen to 28 percent (President's Commission on Housing 1982). Not only were the newer households in a weaker financial position and increasingly unable to cover the costs of day-to-day maintenance and repairs, but the housing was now aging and in need of major capital repairs. The Brooke Amendments, enacted in 1969–1971, capped the amount of rent that a household was required to pay at 25 percent of income,[8] and also provided operating subsidies to cover the shortfall. However, the history of federal appropriations for both public housing maintenance and capital improvements reveals a pattern of consistent underfunding.

In 1992, the National Commission on Severely Distressed Public Housing released its final report (1992). Although it found that "only" 6 percent of the public housing stock met the definition of being severely distressed, it underscored that even this percentage, which translated into 86,000 units, was far too high. Moreover, it warned that, without public intervention, many more units would almost certainly fall into the "severely distressed" category. Among the key recommendations of its proposed National Action Plan was significant new funding for both resident services and the physical rehabilitation of the developments.

The major direct outcome of the Commission's report was the appropriation of $300 million in FY 1993 for a new Urban Revitalization Demonstration, specifically focusing on severely distressed public housing.[9] Funding for what became known as the HOPE VI program was put toward major redevelopment of the housing, into high-quality mixed-income communities. While many of the long-standing problems of the old developments disappeared, at the same time, thousands of very low-income public housing residents were displaced. One of the key problems is that often the old tenants were not able to meet the requirements set by the new tenant committees and management entities. According to the National Low Income Housing Coalition, some 98,000 public housing units were demolished, but only 48,348 public housing units were built (2015, 4–14; for an additional critique of the HOPE VI program see Chaskin and Joseph 2015). Congress funded the HOPE VI program between FY 1994 and FY 2011; in FY 2010 President Obama launched a variation on HOPE VI, known as the Choice Neighborhoods initiative program.

While HOPE VI redeveloped only deteriorated public housing developments, Choice Neighborhood grants have gone to both public housing and project-based Section 8 projects. Unlike the HOPE VI program, HUD's Choice Neighborhood guidelines call for a one-for-one replacement of the affordable units. If public housing units are replaced by units owned by private for-profit developers, unless they are permanently safeguarded as affordable, there would be a net reduction in the social housing stock, given the definition of social housing used in this chapter. Even though the number of affordable units may remain the same, the issue is whether there is a strong likelihood that these units will remain affordable long term. Only when the housing is owned by a nonprofit organization or a public housing authority, or otherwise under a permanent affordability restriction, is the long-term affordability virtually assured, and the units can be counted as social housing.

Another initiative aimed at trying to solve the many problems facing public housing, particularly the estimated more than $55 billion backlog of needed repairs,[10] is the RAD program, authorized by Congress in 2012. This program is an option for maintaining and improving the public housing stock and other project-based assisted housing[11] in order to have a better chance at providing needed upgrades and, at the same time, ensuring long-term affordability.

Through the RAD program, local housing authorities and other private owners of HUD-assisted housing have the opportunity to raise public and private debt and equity to invest in the public housing stock. With the public and assisted housing developments converted to Section 8 housing, the new owners can then access various types of funds, including LIHTCs. The new RAD units are then covered by long-term contracts by using either Project-Based Vouchers or Project-Based Rental Assistance that, by law, must be renewed, thereby ensuring permanent affordability to low-income households.[12] Since the units under the RAD program would provide long-term affordable housing whether they are owned by public housing authorities, nonprofits, or for-profits, they would continue to be counted as social housing units, and are thus the one exception to the definition of social housing (referenced earlier), which generally excludes subsidized developments owned by for-profit firms.

Although it is noteworthy that the federal government developed a strategy for dealing with the public housing repair backlog, it is important to underscore that the RAD solution does not rely on direct appropriations for operating subsidies and repairs. Instead, the chosen strategy is to provide assistance by leveraging private investments through the tax incentives offered through the LIHTC program, as well as through long-term Section 8 subsidies.

Certainly, many public housing developments have been and are being upgraded through the various programs discussed in the preceding text, as well as through other initiatives, but the more than one million federal public housing units across the country continue to face serious challenges. In particular, as noted earlier, the aging public housing stock is in need of significant repairs while annual appropriations to address these demands fall short (National Low Income Housing Coalition 2018; see also Abt Associates, Inc. 2010).

This reality—the enormous costs involved in maintaining an aging housing stock—is one reason behind the lack of new public housing production. Whatever funds for public housing that Congress has been willing to appropriate have been needed to maintain the existing stock, not to build new housing. Indeed, as noted earlier, the original funding formula for the program did not include operating subsidies and requires annual infusions of new money to keep the projects viable.

In addition, there are at least four further reasons why the original public housing program is no longer producing new units. First, there has been a growing consensus on the part of academics and policymakers that concentrating large numbers of poor people is bad public policy and that a mixed-income approach is far preferable.[13]

Second, as private for-profit affordable housing developers have become increasingly important (originally in the public-private partnership programs of the 1960s and, later, under the LIHTC program starting in 1986), their lobbying power for the public-private approach has grown, while support for traditional public housing has dropped precipitously.

Third, and noted previously, with the considerable backlog of repair needs in many public housing developments, and a persistent image of the "projects" being a place where crime and other social pathologies fester, the program's reputation has continued to decline.

And, fourth, since the Reagan era there has been a diminished concern about the needs of the poor. Instead, whatever support there has been for housing lower-income households, the emphasis, starting in 2007, has been on problems facing the middle class, the need for workforce housing, and the many problems associated with the mortgage crisis.

There are a number of persistent questions about the future of the public housing program. In particular, will there be sufficient federal funds to maintain and upgrade the existing federally subsidized units (more than 1.1 million units, as shown in Figure 13.2) so that they can continue to be a long-term resource for current and future generations of households who need this assistance? What will be the outcomes of the RAD program, and will it, indeed, succeed in rehabilitating units while still maintaining them as long-term affordable housing, as the program mandates? And, finally, what will be the role of local housing authorities, and will they contract in size if more traditional public housing units are shifted to private management under RAD?

Nonprofit Housing

The other major component of the U.S. social housing stock includes housing that has been built and is owned by nonprofit organizations. Federal funding for these organizations has always been piecemeal and sporadic, with no long-standing commitment to supporting this sector. As shown in Figure 13.2, total cumulative production by various types of nonprofit housing organizations comprises about 70 percent of the overall U.S. social housing sector; public housing comprises the remaining total of the approximately five million social housing units. Among nonprofits, community development corporations (CDCs) are the single largest type of nonprofit housing producer.

The origin of CDCs dates back to the 1960s, as a community-based response to the serious housing problems facing low-income people and a concern about inadequate services and opportunities within poor, mostly urban communities. Grassroots protest and advocacy movements, which often formed around arson, redlining, displacement due to urban renewal, and dilapidated housing, were transformed into a more positive set of initiatives aimed at producing new or rehabilitated affordable units for local residents. Also, an important stimulus behind the creation of CDCs was the War on Poverty, a President Johnson administration initiative, which provided a range of programs and services to low-income populations in mostly urban areas. One of the most defining features of that era was the call for the "maximum feasible participation of the poor" in publicly funded community development efforts, as a way to empower this population. Thus, among the CDCs that were created during this period (as well as for those that followed), there was an emphasis on engaging local residents as board members and ensuring that community planning efforts would be bottom up, rather than top down, which had characterized the earlier urban renewal program.

During the 1970s and 1980s, the nonprofit housing movement began to take shape, with large numbers of organizations forming during this period. As of late 1990, there were an estimated 2,000 CDCs; seven years later, the approximate number of these organizations skyrocketed to 3,600 (National Congress for Community Economic Development 1991, 1999). The last census of CDCs, which is now more than a decade old, placed the number of CDCs at 4,600 (National Congress for Community Economic Development 2005). While the net number of organizations has grown considerably during the past four decades, many organizations have gone out of business or merged with other nonprofits. The numbers presented do not capture the kind of more nuanced fluctuations that have occurred.

Momentum for CDC growth was also fueled by the creation, between 1978 and 1982, of three national intermediary organizations, which provided technical and financial resources to nonprofits (e.g., NeighborWorks America, formerly Neighborhood Reinvestment Corporation; Local Initiatives Support Corporation; and Enterprise Community Partners, formerly The Enterprise Foundation). The current period, dating from the 1990s, has seen increasing professionalism of the sector, a greater focus on training programs, and efforts to work closely with for-profit firms, social services providers, and public officials in collaborative development efforts. The various entities and supports for CDCs have been characterized as being part of a community development system (Bratt 1989; Frisch and Servon 2006; Mayer and Keyes 2012).

Although CDCs are the single most prevalent type of nonprofit housing organization, the nonprofit component of the social housing sector encompasses several different forms, as noted in Figure 13.2, each of which has various attributes, strengths, and weaknesses. (For more details and comparisons of the approaches that different types of social housing initiatives embrace, see Davis (1994) and Stone (2006)). Even among CDCs, there are many different types of organizations, from small "mom and pop" storefront groups with only a few employees to large groups with hundreds of staff members (Vidal 1992). There is also a wide variation in the level of productivity, competence, and professionalism among both CDCs and housing nonprofits more generally, with some having several thousand units in their portfolios, while many others have less than 100. Many, if not most, of the largest housing nonprofits, in terms of both size of the organization and units produced or under management, are members of the Housing Partnership Network (HPN).[14]

Figure 13.2 lists estimates of production by a number of other types of nonprofit housing organizations that contribute to the overall stock of social housing. Community land trusts (CLT) own land in perpetuity, with the housing built on the land subject to long-term affordability restrictions. In limited equity cooperatives and mutual housing association developments, residents own a share of the development, but not their actual unit, and the value of the home is based on inflation and other factors, such as improvements to the home. Owners of these types of units, as with CLT units, cannot sell the house at prevailing private market rates.

Federally subsidized housing for the elderly, built under the Section 202 program, is the nonprofit program closest to public housing; the federal government makes a long-term commitment to maintain the units as affordable. Finally, Habitat for Humanity, a homeownership model that started in the U.S., but that has spread throughout the world, depends on local public and private contributions of material and labor, as well as "sweat equity" on the part of the new homeowners, to produce low-cost homes that are required to be affordable for lower-income households in perpetuity.

Despite the variations in nonprofit-based social housing models, most of the research on nonprofit housing organizations has focused on CDCs. A number of researchers have explored the various challenges facing these groups and the kinds of capacities needed for them to be effective and efficient producers and managers of affordable housing (Bratt 2006; Erickson 2009; Glickman and Servon 1998; Stoutland 1999). At the same time, CDCs are attempting to fulfill a broad mission pertaining to neighborhood revitalization,[15] citizen engagement, and providing social supports and other services aimed at promoting economic security and well-being for their residents.

Some of the most compelling challenges facing CDCs include:

- The long hours of work required and the frequency of staff "burn-out," accompanied by generally lower salaries than would be available in other settings
- The need for strong leadership by both the board and the executive director
- The need to have on staff, or have access to, a team of highly skilled development professionals, who are able to carry out a varied set of tasks that include legal work, financial analysis, community organizing, and delivery of resident services
- The frustration of having to work in an environment of scarce resources, both to support organizational operations and to fund development deals
- The scarce resources of most nonprofits typically mean that development deals are likely to take a long time to put together and that a number of funding resources will need to be assembled, all having different requirements
- An awareness that long timeframes are needed to observe any appreciable change in the local environment
- The need to collaborate with a range of public, private, and nonprofit organizations in order to bring deals to fruition and to provide the desired level and types of services

An additional challenge is that the very structure of CDCs may be problematic. The potential conflict has been summarized in the following question, posed by the author as a "straw man" type of argument: "To what extent can CDCs, organizations whose names includes the word 'corporation,' and who are forced to work closely with financial institutions and city governments, really be advocates for low income interests?" (Bratt 2006, 348).

Although the various challenges facing CDCs have been well understood for a number of years, they continue to be concerns for housing nonprofits. Of course, the extent to which any one group will encounter specific challenges and be able to capably manage them, is likely dependent on its size, asset base, organizational capacity, and the local context in which it works. For example, since most nonprofit deals are negotiated through the private real estate market, various obstacles can make it more difficult for them to undertake deals or to manage existing properties. Weak markets are characterized by high vacancy rates, abandoned or boarded up buildings, vandalism/crime in the area, and long periods that homes and apartments stay on the market, unrented or unsold. In such areas, construction and management costs also may be higher due to the need for additional security. Further, in weak markets, low-cost market-rate housing is likely to compete with any existing or planned social housing in terms of attracting tenants or homeowners. In contrast, in hot markets, desirable land and building are costly and rarely stay on the market for long, with sales prices often exceeding asking prices, thereby making it prohibitive or very difficult for a nonprofit to compete with for-profit developers (Bratt 2009b).

Given the complexity of the challenges facing CDCs and other housing nonprofits, it is reasonable to question whether the current strategy is, indeed, a viable one. What are the compelling reasons for supporting nonprofit organizations as producers and managers of affordable housing?

First, the basic concepts on which the nonprofit movement is based are still valid. Nonprofits have a clear mission to produce and maintain affordable housing over the long term. While it is essential that they be able to cover their costs, the desire to maximize returns and to generate profits is absent. This means that their work is focused on the outcome for the residents of the housing, and for the larger community, rather than on returns to the organization.

Second, and in direct response to the "straw man" argument noted above, despite the potential for a built-in conflict, CDCs appear to be doing a good job of mediating between the demands of "capital and community." They are typically able to maintain their commitment to serving low-income people and communities while working closely with the financial and public sectors (Bratt, 2006).

Third, nonprofits that are community-based also provide opportunities for residents to become deeply involved with the housing production or rehabilitation effort (Bratt 2008a). The expectation is that this will not only increase the likelihood that the housing will fit in with the general neighborhood, but also promote resident involvement in day-to-day decision-making and, in some cases, may even provide ultimate control of the housing.[16]

Fourth, although the provision of resident services is not restricted to nonprofit housing, this is typically a central concern of these housing providers who work hard to coordinate or, in some cases, directly provide a range of services and programs that will help promote household economic security and well-being.

And, fifth, since there is no new funding for public housing development, the only way to increase the number of social housing units is through the nonprofit sector. As competition for housing continues to be intense in strong market areas, social housing is a key mechanism (along with aggressive regulatory controls) for providing low-income residents with security against unwanted displacement (see also the chapter authored by Dennis Gale). Unless housing is separated from the private, speculative market, low-income tenants are faced with the virtual certainty of increasing rents and a real likelihood that they will need to move. How, then, could the nonprofit housing sector be better supported?

Moving the U.S. Nonprofit Housing Agenda Forward

Based on the most recent available production numbers, as well as the estimated amount of production that has taken place since these numbers were compiled (as shown in Figure 13.2), nonprofit housing organizations have managed to produce over 3.5 million units of housing affordable to lower-income households, constituting about 70 percent of the total social housing sector. To accomplish this, however, the groups have had to rely on a patchwork of financing and technical support programs. Contemporary nonprofits have not been operating within a framework of what I have called a "nonprofit-centric system" for affordable housing development. Instead, past programs to support the work of nonprofits "have formed a discontinuous series of initiatives, which have not generally built on the experiences of their predecessors . . . we have never had a coherent set of federal policies aimed at supporting nonprofit programs" (Bratt 1998, 140 and 153). To improve upon the disconnected and partial array of programs, an affordable housing system is needed that reflects an understanding of the unique attributes and strengths of nonprofit housing organizations and creates the appropriate interventions to help overcome constraints and challenges, thereby promoting more optimal productivity. At the very least, a model support system (Bratt 1989) or, in other terminology, a "nonprofit-centric system" would include:

- Technical assistance and operating support to help address the organizational needs of nonprofits
- A clearinghouse of experts in land use, architecture, engineering, law, real estate development, and housing management, who, for example, would be available to help nonprofits package deals, at low (or no) cost
- Land-banking programs that would encourage local and state governments to set aside land and buildings suitable for affordable housing development
- Construction and long-term debt financing at attractive interest rates
- Syndicators that can package the deals at low costs
- Adequate development and operating subsidies, both to bring the projects to fruition and to enable developments to be managed efficiently over the long term
- Project-based subsidies that will offer long-term affordable opportunities to very low-income households

- On-site resident services programs that would promote opportunities for households to move toward greater economic stability
- Opportunities for nonprofits to share information and lessons learned through trade, education, and support networks

To be sure, many of these components exist in many parts of the country;[17] some exist at the national level (e.g., support and technical assistance provided by the nonprofit intermediaries such as LISC, NeighborWorks, and Enterprise Community Partners), while others are offered through various state or local governments. But, in no part of the country are all the pieces in place, particularly the deep subsidies needed to make affordable housing deals work, without layering a number of financial resources.

This approach would go beyond the "business as usual" way in which the federal government has dealt with nonprofit housing organizations. Specifically, while it has not provided much in the way of direct organizational assistance and supports it has, on occasion, given preferential treatment or priority to nonprofit housing producers (for a summary of some of these initiatives see Bratt 2008b).[18]

Although the general trend in the affordable housing arena is to provide ever increasing roles for the private for-profit sector, while spending as few federal dollars as possible, this chapter argues for increased support for both types of social housing strategies—public housing and nonprofit-owned housing. The social housing sector has the skills and capacity to play a much larger role in meeting affordable housing needs, particularly in view of the significant support provided by the national nonprofit intermediaries. With the types of supports outlined earlier for nonprofits, and with consistent and robust levels of funding for public housing, the social housing stock can be preserved and expanded. This would constitute important and significant steps in advancing a right to housing agenda.[19]

On a less optimistic note, however, the increasing reliance on the private for-profit sector is having the subtle but clear result of giving the federal government a "pass" on fulfilling the national housing goal, first articulated by the U.S. Congress in 1949: "a decent home and a suitable living environment for every American family." Although the public-private partnership programs are important, without a strong and committed federal partner, they are not sufficient to fully address the nation's housing needs. Advocates need to keep articulating the message that decent housing is a federal responsibility and that the government must be held accountable for assuming the leadership role in defining a responsive, comprehensive affordable housing agenda. Ideally, this would place nonprofits in a central position, but it would also embrace and support the work of high performing, mission-oriented for-profit developers who are similarly committed to assuring top quality, long-term affordable housing. If, indeed, we could guarantee that affordable housing built by for-profit developers would stay affordable beyond the program-mandated minimums, we could see an increase in the social housing stock and come closer to fulfilling the goal of a right to housing.

Notes

1 Note #iv, accompanying Figure 13.2, identifies the four states that have their own public housing programs.

2 In this context, it should be noted that for-profit developers have always produced the majority of units under the various federal public-private partnership programs, such as the below-market interest rate programs of the 1960s, the Section 8 New Construction and Substantial Rehabilitation program of 1974, and the LIHTC program of 1986 (for a summary of for-profit activity under these various programs, see Bratt 2016).

3 Schwartz (2015) also points out that while there has been a marked decrease in new budget authority, actual annual expenditures for housing, or outlays, have increased over the same period. This is because outlays include prior budgetary commitments. The level of new budget authority is, however, a more accurate gauge of the direction that federal support for housing is going. Only by providing new funding

for housing—funding that is not already committed to supporting existing households and programs—can there be any progress in reducing the number of households needing housing assistance.

4 While there have been overall declines in new funding for low-income housing assistance over the past several decades, subsidies for individuals who are eligible for various kinds of tax breaks through the income tax system have been on an upward trajectory. Until passage of the 2017 Tax Cuts and Jobs Act, the loss to the U.S. Treasury for homeowners' deductions for property taxes and mortgage interest climbed to about $90 billion per year (Joint Committee on Taxation 2017). However, the new tax law increased the standard personal deduction and reduced the number of homeowners itemizing their deductions. As a result, the homeowners' deduction is estimated to decline to about $30 billion per year (Joint Committee on Taxation 2018). Homeowners' deductions disproportionately benefit higher-income taxpayers (Bernstein 2013; Fischer and Sard 2013).

5 The Joint Center for Housing Studies of Harvard University (2018) describes one serious aspect of housing needs as follows: "increases in federal rental assistance have lagged far behind growth in the number of renters with very low incomes, the group typically eligible for subsidies. Between 1987 and 2015, the number of very low-income renters grew by 6 million while the number assisted rose only 950,000, reducing the share with assistance from 29 percent to 25 percent" (p. 5).

6 Indeed, even when public housing was relatively new, one of its original supporters wrote about "The Dreary Deadlock of Public Housing" and noted: "After more than two decades [public housing] still drags along in a kind of limbo, continuously controversial, not dead but never more than half alive" (Wurster 1957, 140).

7 For a good discussion of the opposition to public housing on the part of the private for-profit real estate and homebuilding sectors, see von Hoffman (2008). In particular, he notes that: "By following USHA [United States Housing Authority] guidelines for project design and construction costs, for example, the local authorities produced standardized versions of modernist architecture monotonously executed with the favorite building material of the bricklayers unions—easily recognizable and therefore stigmatized as poor people's housing (10)." For additional details on the design of public housing and how it was connected to problems with that program, see Wurster (1957).

8 The Omnibus Budget and Reconciliation Act of 1981 raised the percent of income to be paid in rent in federally assisted housing to 30 percent. This is the standard today.

9 A key example of this early effort was the transformation of Boston's Columbia Point public housing project to Harbor Point, a federally subsidized mixed-income community.

10 In 2010, it was estimated that there was a need for more than $25 billion in capital repair needs. It was further estimated that additional capital repair needs would accure at the rate of $3.4 billion per year (Abt Associates, Inc. 2010).

11 The other older programs that can also participate in RAD, in addition to the traditional public housing program, and that can be converted to what is known as a "Section 8 platform" are: the Moderate Rehabilitation program, the Rent Supplement program, and the Rental Assistance program.

12 If a developer uses Project-Based Vouchers, the housing authority will retain major involvement in administering the property and the vouchers. Using Project-Based Rental Assistance results in a contractual relationship between the private owner and HUD (Roller and Cassella, 2018).

13 Although the mixed-income approach has received a great deal of academic attention and is often the preferred approach for developers, questions have been raised about the actual benefits for low-income households (see, for example, Brophy and Smith 1997; DeFilippis and Fraser 2010; Joseph et al. 2007; Levy et al. 2013; Moore and Glassman 2007; Schwartz and Tajbakhsh 1997).

14 Some members of the HPN are not housing developers and are, instead, community development financial institutions (CDFIs), or perform other non-development roles connected to affordable housing. CDFIs are alternative types of financial entities that focus on assisting lower-income households and communities. For more information see: www.cdfifund.gov/Pages/default.aspx (accessed April 19, 2019). The HPN was founded in 1992, is based in Boston, and provides a number of services including peer support, assistance in raising needed capital, and management of innovative programs. It partners with its nearly 100 member nonprofit organizations that work locally, regionally, and nationally on housing and community development issues. A few of the largest CDCs in the country are members of the HPN. For more information see: https://housingpartnership.net/about (accessed April 19, 2019).

15 Varady et al. (2015) observe that there is only very limited information about the impact of CDCs on community revitalization and conclude that, based on the available empirical research, the various efforts have not been able to reverse decline.

16 For example, a well-established community-based CDC in Boston, Urban Edge, shares management control with residents. The organization has about 1,200 units; about 40 percent are owned in collaboration with an organized resident group, serving as the general partner.

17 A number of cities and/or states have been singled out as being particularly supportive of nonprofit housing. See, for example, Goetz (1993), von Hoffman (2003), Davis (2006), Erickson (2009), and Mallach (2009).

18 For example, the HOME Investment Partnership Program, a block grant for affordable housing initiatves, directs each participating jurisdiction to allocate at least 15 percent of its funding to nonprofit community housing development organizations (CHDOs). In addition to nonprofits being given preferential status and creating a comprehensive system of supports for nonprofit housing organizations, a number of changes in the overall approach to providing housing subsidies are needed: developing financing mechanisms, particularly changes in the LIHTC program, to be able to meet the needs of small-size development projects (e.g., fewer than about 25–30 units); clarifying why investments in affordable housing are essential; and continuing to work on how housing policy can promote fair housing and equal opportunities across racial groups.

19 Certainly, a number of questions would arise and be debated in relation to adopting a right to housing. For example, how much and what quality housing should be available to the lowest-income households, given that housing quality, size, and locations are so varied? In addition, how much does the direct provision of housing tend to discourage households from seeking work and striving toward their own economic improvement and stability? Indeed, the substantial literature on the extent to which the provision of subsidized housing impacts participation in the workforce would likely shape this conversation (see, for example, Olsen et al. 2005). Despite the importance of exploring questions such as these, they are beyond the scope of this chapter.

References

Abt Associates, Inc. 2010. *Capital Needs in the Public Housing Programs*. Washington, DC: U.S. Department of Housing and Urban Development.

Bernstein, Jared. 2013. *Tax Expenditures: How Cutting Spending Through the Tax Code Can Lower the Deficit, Improve Efficiency, and Boost Fairness in the US Tax Code. Testimony Before U.S. Senate Budget Committee*. Washington, DC: Center for Budget and Policy Priorities.

Bratt, Rachel G. 1986. "Public Housing: The Controversy and Contribution." In *Critical Perspectives on Housing*, edited by Rachel G. Bratt, Chester Hartman, and Ann Meyerson, 335–361. Philadelphia, PA: Temple University Press.

Bratt, Rachel G. 1989. *Rebuilding a Low-Income Housing Policy*. Philadelphia, PA: Temple University Press.

Bratt, Rachel G. 1998. "Nonprofit Developers and Managers: Their Evolution of their Role in U.S. Housing Policy." In *Shelter and Society: Theory, Research, and Policy for Nonprofit Housing*, edited by C. Theodore Koebel, 139–156. Albany, NY: State University of New York Press.

Bratt, Rachel G. 2006. "Community Development Corporations: Challenges in Supporting a Right to Housing." In *A Right to Housing: Foundation for a New Social Agenda*, edited by Rachel G. Bratt, Michael E. Stone, and Chester Hartman, 340–359. Philadelphia, PA: Temple University Press.

Bratt, Rachel G. 2008a. "Viewing Housing Holistically: The Resident-Focused Component of the Housing-Plus Agenda." *Journal of the American Planning Association*, 74(1): 100–110.

Bratt, Rachel G. 2008b. "Nonprofit and For-Profit Developers of Subsidized Rental Housing: Comparative Attributes and Collaborative Opportunities." *Housing Policy Debate*, 19(2): 323–365.

Bratt, Rachel G. 2009a. "Promoting Housing Affordability." In *Local Planning: Contemporary Principles and Practice*, edited by Gary Hack, Eugenie L. Birch, Paul H. Sedway, and Mitchell J. Silver, 377–385. Washington, DC: ICMA Press.

Bratt, Rachel G. 2009b. "Challenges for Nonprofit Housing Organizations Created by the Private Housing Market." *Journal of Urban Affairs*, 31(1): 67–94.

Bratt, Rachel G. 2012. "The Quadruple Bottom Line and Nonprofit Housing Organizations in the United States." *Housing Studies*, 27(4): 438–456.

Bratt, Rachel G. 2015. "The Role of Nonprofits in Meeting the Housing Challenge in the U.S." Paper presented at the conference *The Future of Public Housing*, Vancouver, British Columbia, May 20–23, 2015.

Bratt, Rachel G. 2016. *Affordable Rental Housing Development in the For-Profit Sector: A Case Study of McCormack Baron Salazar*. Cambridge, MA: Joint Center for Housing Studies, Harvard University.

Brophy, Paul C., and Rhonda N. Smith. 1997. "Mixed-Income Housing: Factors for Success." *Cityscape: A Journal of Policy Development and Research*, 3(2): 3–31.

Chaskin, Robert J., and Mark L. Joseph. 2015. *Integrating the Inner City: The Promise and the Perils of Mixed-Income Public Housing Transformation*. Chicago, IL: University of Chicago Press.

Citizens' Housing and Planning Association. 1999. *Housing Guidebook for Massachusetts: A Comprehensive Guide to State and Federal Housing Programs and Resources*. Boston, MA: Citizens' Housing and Planning Association.

Davis, John E. 1994. "Beyond the Market and the State: The Diverse Domain of Social Housing." In *The Affordable City: Toward a Third Sector Housing Policy*, edited by John E. Davis, 75–106. Philadelphia, PA: Temple University Press.

Davis, John E. 2006. "Between Devolution and the Deep Blue Sea: What's a City or State to Do?" In *A Right to Housing: Foundation for a New Social Agenda*, edited by Rachel G. Bratt, Michael E. Stone, and Chester Hartman, 364–398. Philadelphia, PA: Temple University Press.

DeFilippis, James, and James Fraser. 2010. "Why Do We Want Mixed-Income Housing and Neighborhoods?" In *Critical Urban Studies: New Directions*, edited by Jonathan S. Davies and David L. Imbroscio, 135–148. Albany, NY: SUNY Press.

Epp, Gayle. 1996. "Emerging Strategies for Revitalizing Public Housing Communities." *Housing Policy Debate*, 7(3): 563–588.

Erickson, David J. 2009. *The Housing Policy Revolution: Networks and Neighborhoods*. Washington, DC: The Urban Institute Press.

Fischer, Will, and Barbara Sard. 2013. *Chart Book: Federal Housing Spending Is Poorly Matched to Need. Tilt Toward Well-Off Homeowners Leaves Struggling Low-Income Renters without Help*. Washington, DC: Center on Budget and Policy Priorities.

Friedman, Lawrence M. 1968. *Government and Slum Housing*. Chicago, IL: Rand McNally.

Frisch, Michael, and Lisa J. Servon. 2006. "CDCs and the Changing Context for Urban Community Development: A Review of the Field and Environment." *Community Development Journal*, 37(4): 88–108.

Glickman, Norman J., and Lisa J. Servon. 1998. "More than Bricks and Sticks: Five Components of Community Development Corporation Capacity." *Housing Policy Debate*, 9(3): 497–539.

Goetz, Edward G. 1993. *Shelter Burden: Local and Progressive Housing Policy*. Philadelphia, PA: Temple University Press.

Hawaii Public Housing Authority. 2015. *Annual Report Fiscal Year 2014–2015*. Honolulu, HI: Hawaii Public Housing Authority.

Houard, Noémie. 2011. *Social Housing across Europe*. Paris: La Documentation Française.

Huth, Mary Jo. 1981. "An Examination of Public Housing in the United States after Forty Years." *Journal of Sociology and Social Welfare*, 8(3): 471–488.

Joint Center for Housing Studies. 2018. *The State of the Nation's Housing 2014*. Cambridge, MA: Harvard University Press.

Joint Committee on Taxation. 2017. *Estimates of Federal Tax Expenditures for Fiscal Years 2016–2020*. Washington, DC: Congress of the United States.

Joint Committee on Taxation. 2018. *Estimates of Federal Tax Expenditures for Fiscal Years 2018–2022*. Washington, DC: Congress of the United States.

Jones, Ronald, David Kaminsky, and Michael Roanhouse. 1979. *Problems Affecting Low-Rent Public Housing Projects: A Field Study*. Washington, DC: U.S. Department of Housing and Urban Development.

Joseph, Mark L., Robert J. Chaskin, and Henry S. Webber. 2007. "The Theoretical Basis for Addressing Poverty through Mixed-Income Development." *Urban Affairs Review*, 42(3): 369–409.

Levy, Diane K., Zach McDade, and Kassie Bertumen. 2013. "Mixed-Income Living: Anticipated and Realized Benefits for Low-Income Households." *Cityscape: A Journal of Policy Development and Research*, 15(2): 16–28.

Mallach. Alan. 2009. *A Decent Home: Planning, Building, and Preserving Affordable Housing*. Chicago, IL: American Planning Association.

Marcuse, Peter. 1986. "Housing Policy and the Myth of the Meddling State." In *Critical Perspectives on Housing*, edited by Rachel G. Bratt, Chester Hartman, and Ann Meyerson, 248–258. Philadelphia, PA: Temple University Press.

Marcuse, Peter. 1998. "Mainstreaming Public Housing: For a Comprehensive Approach to Housing Policy." In *New Directions in Urban Public Housing*, edited by David P. Varady, Wolfgang F. E. Preiser, and Francis P. Russell, 23–44. New Brunswick, NJ: Center for Urban Policy Research.

Mayer, Neil, and Langley Keyes. 2012. "City Government's Role in the Community Development System." In *The Community Development Reader*, edited by James DeFilippis and Susan Saegert, 158–166. New York: Routledge.

Moore, Sandra M., and Susan K. Glassman. 2007. *The Neighborhood and Its School in Community Revitalization: Tools for Developers of Mixed-Income Housing Communities*. St. Louis: Urban Strategies, Inc.

National Alliance of Community Economic Development Associations. 2010. *Rising Above: Community Economic Development in a Changing Landscape*. Washington, DC: National Alliance of Community Economic Development Associations.

National Commission on Severely Distressed Public Housing. 1992. *The Final Report*. Washington, DC: National Commission on Severely Distressed Public Housing.

National Congress for Community Economic Development. 1991. *Changing the Odds: The Achievements of Community-Based Development Corporations*. Washington, DC: National Congress for Community Economic Development.

National Congress for Community Economic Development. 1999. *Coming of Age: Trends and Achievements of Community-Based Development Organizations*. Washington, DC: National Congress for Community Economic Development.

National Congress for Community Economic Development. 2005. *Reaching New Heights: Trends and Achievements of Community-Based Development Organizations*. Washington, DC: National Congress for Community Economic Development.

National Low Income Housing Coalition. 2018. *Advocates' Guide*. Washington, DC: National Low Income Housing Coalition.

New York State, Homes and Community Renewal. 2016. *About the Public Housing Program*. Albany, NY: New York State.

Olsen, Edgar O., Catherine A. Tyler, Jonathan W. King, and Paul E. Carrillo. 2005. "The Effects of Different Types of Housing Assistance on Earnings and Employment." *Cityscape: A Journal of Policy Development and Research*, 8(2): 163–187.

President's Commission on Housing. 1982. *The Report of the President's Commission on Housing*. Washington, DC: President's Commission on Housing.

Reeves, Paul. 2014. *Affordable and Social Housing: Policy and Practice*. New York: Routledge.

Roller, Shamus and Jessica Cassella. 2018, July 30. "The Promise and Perils of HUD's RAD Program." *Shelterforce*, n.p.

Scanlon, Kathleen, Christine Whitehead, and Melissa Fernandez Arrigoitia, eds. 2014. *Social Housing in Europe*. Hoboken, NJ: Wiley-Blackwell.

Schwartz, Alex F. 2015. *Housing Policy in the United States*. New York: Routledge.

Schwartz, Alex, and Kian Tajbakhsh. 1997. "Mixed-Income Housing: Unanswered Questions." *Cityscape: A Journal of Policy Development and Research*, 3(2): 71–92.

Stone, Michael E. 2006. "Social Ownership." In *A Right to Housing: Foundation for a New Social Agenda*, edited by Rachel G. Bratt, Michael E. Stone, and Chester Hartman, 240–260. Philadelphia, PA: Temple University Press.

Stoutland, Sara E. 1999. "Community Development Corporations: Mission, Strategy, and Accomplishments." In *Urban Problems and Community Development*, edited by Ronald F. Ferguson and William T. Dickens, 193–240. Washington, DC: The Brookings Institution Press.

Thaden, Emily, and Jason Webb. 2015. *National Community Land Trust Network: 2015 Member Report*. Portland, OR: National Community Land Trust Network.

Varady, David, Reinhout Kleinhans, and Marten van Ham. 2015. "The Potential of Community Entrepreneurship for Neighbourhood Revitalization in the United Kingdom and the United States." *Journal of Enterprising Communities*, 9(3): 253–276.

Vidal, Avis C. 1992. *Rebuilding Communities*. New York: Community Development Research Center, New School for Social Research.

von Hoffman. Alexander. 2003. *House by House, Block by Block: The Rebirth of America's Urban Neighborhoods*. New York: Oxford University Press.

von Hoffman. Alexander. 2008. *Enter the Housing Industry, Stage Right: A Working Paper on the History of Housing Policy*. Cambridge, MA: Joint Center for Housing Studies, Harvard University.

Wurster, Catherine B. 1957. "The Dreary Deadlock of Public Housing." *Architectural Forum*, 106(5): 140–142, 219, 221.

14

THE PRIVATIZATION OF AMERICAN PUBLIC HOUSING

Leaving the Poorest of the Poor Behind

Lawrence J. Vale and Yonah Freemark

Introduction: Pathways to Privatization

In recent years, the public sector has played a shrinking role in direct housing provision for eligible lowest-income Americans, leading to what critics frequently decry as "the privatization of public housing" (Arena 2012, xxix, also see Hackworth 2007). The Rental Assistance Demonstration (RAD) program, authorized by the U.S. Congress in 2011, has brought this language to the fore—even though supporters of the program insist that its sole purpose is to "save" public housing by providing a mechanism that raises substantial funds to repair or redevelop it so that it can remain an asset for the lowest-income households. Detractors, however, see RAD as an unwelcome blurring of the public-private boundary since the program permits conventional public housing apartments to be converted to long-term project-based contracts that no longer require public ownership or management. Privatization, in this sense, is a term of derision proffered by those who see RAD as no more than another component in the neoliberal takeover of American cities—another way to push out the poorest to make way for private investment. This critique, however, misses the larger picture: American public housing has been on the pathway to privatization since its very inception. Rather than assess whether privatization has been a good thing or a bad thing for the actual management and operation of deeply subsidized housing—certainly a vitally important question—this chapter instead focuses on the extent to which privatization has failed to forestall the shortfall in public housing supply.

We argue that public housing in the United States is more than the stereotyped federal public housing program of the 1950s and 1960s, with its series of frequently vilified superblock neighborhoods and discredited policies. Rather, the ever-expanding notion of "public housing" since the 1930s provides windows into evolving attitudes about which of the American poor ought to be housed with assistance, either directly or indirectly, from the government. Over time, the initial idea behind "conventional" public housing—housing financed, owned, and operated by the public sector—has undergone so many changes that it makes increasingly little sense to describe all of it with the same name (Marcuse 1995).

Public Action in Service of the Private Sector

The basic "public"-ness of public housing in the United States has been contested right from the start. Truly "public" housing would have a governmental entity design, fund, construct, own, operate, and maintain buildings—take on, in other words, all of the tasks necessary to make and manage a home. But little if any of the public housing ever built in the U.S. fits this definition precisely. Projects funded by the Public Works Administration in the mid-1930s were designed by private architectural firms; they were bid out to private contractors; and they benefited from private bond financing (extolled as "high grade investments of the future") (*The New York Times* 1937, 35). The basic justification for public housing was, from the beginning, rooted in the goal of stimulating the private building industry (Marcuse 1995). Has public housing *ever* been truly public?

Throughout its history, promoters of "public" housing needed to find ways to accommodate the needs and preferences of the private sector. Private property interests have molded the evolving shape of the public housing program in many important ways (Vale 2000). Most fundamentally, the Housing Act of 1937 targeted public housing access to low-income families through a formula that tied low rents to low incomes. By limiting eligibility to families earning no more than five times the fixed rent (six times the rent in the case of large families), the legislation thereby ensured that any significant rise in income would spur their departure into private market-rate housing. The bill also appeased private homebuilders and Congressional conservatives with provisions stipulating dollar limits on per-room and per-dwelling construction costs, intended to minimize the chance that public housing would compete with privately supplied housing. Further reducing competitive threats were stipulations that federal money would go only to projects that were without "elaborate or expensive design or materials" and that cost less than the "average construction cost of dwelling units . . . currently produced by private enterprise, in the locality" (U.S. Housing Act 1937, Section 2 (1), Section 15 (5)). Despite such stipulations, real estate interests frequently blamed early public housing for running unfair competition with the private sector. Writing in 1944, former U.S. Housing Authority Administrator Straus did his best to debunk such claims:

> Public housing throughout every phase of its production is actually private business. . . . The sites are appraised by qualified private appraisers. The buildings are designed by private architectural firms. The projects are constructed by private building contractors chosen on a competitive bid basis. The labor employed works under normal conditions of private business.
>
> *(Straus 1944, 167–168)*

Despite its accommodation of private sector concerns, public housing faced persistent opposition from private realtors and developers. Writing in 1935, Stout, president of the Atlanta Real Estate Board, cast the public housing question in familiar ideological terms of individualism, self-reliance, and frontier manhood:

> The working classes of this country will rue the day when they are housed in Government-owned, Government-built, and Government-regulated houses. Masters house their slaves, but free men house themselves. Those who are descendants of pioneer American stock will not regard as "Home" a unit in a fine building, built at taxpayers' expense, in a slum clearance project.
>
> *(Stout 1935, 72–94, capitalization in the original)*

Such private sector complaints notwithstanding, American-style public housing was never entirely public, and soon became even less so.

Starting in the late 1950s, all manner of private and not-for-profit entities entered the fray even more directly. New social partnership experiments kept public housing in "project-based" form but declined to saddle those particular projects—which were publicly subsidized but developed and managed by the private sector—with the "public housing" name, choosing instead to use names that were indistinguishable from market-rate housing developments. Many public housing authorities (PHAs) engaged in large-scale experiments to re-conceptualize public housing as tenant-based vouchers dispersed across a landscape of private landlords. There have also been partnerships that offered somewhat less deep housing subsidies utilizing the tax code through Low-Income Housing Tax Credits (LIHTC), administered by the federal Internal Revenue Service (IRS), and thereby engaged both investors and a variety of nonprofit organizations and financial intermediaries. Imperceptibly, but incontrovertibly, what once could be confidently described as "public housing" became transmuted into a continuum of public-private ventures.

The United States has never gone as far as the large-scale explicit privatization of council housing launched in the United Kingdom by Thatcher in 1979—and the future of the program, once supported by both Conservative and Labour parties, is now in question (Hodkinson and Robbins 2013).[1] American policymakers have consistently found new and creative ways to incorporate the private sector, yielding what may be more aptly described as a form of *public-private housing*.[2] With public-private partnerships (PPPs) proliferating across nearly every domain of urban development—from wastewater and sewage treatment works to schools, airports, subways, government offices, fire and police stations, hospitals, and prisons, to telecommunications infrastructure, toll roads, bridges, and tunnels—it is hardly surprising that the phenomenon has come to encompass public housing as well (Sagalyn 2012).

Others have noted this phenomenon using the term "hybridization" (Billis 2010), with some focusing specifically on housing subsidy provision (Bratt 2012; Mullins et al. 2012; Nguyen et al. 2012) to describe the transitions in housing policy and practice in the U.S. and European contexts.[3] Marcuse has argued that American-style public housing will remain marginalized as long as it is seen as failing to serve "a broad spectrum of residents"; by contrast, "mainstreaming" public housing would entail treating all manner of housing subsidies, from vouchers to tax deductions, as part of a more inclusive concept of public provision (Marcuse 1998, 31–34). That approach might expand the reach of public housing but remains firmly at odds with the current public understanding of the conventional public housing program. Thus the importance of a new term, "public-private housing," that responds to the reality of a federal deeply subsidized housing program that is in many ways an evolving public-private partnership.

In this chapter, our goal is to integrate the saga of American public housing into a larger cultural narrative of housing initiatives. Housing scholars and planning practitioners typically draw sharp and tight boundaries around what counts as "public housing," treating it as an entirely separate genus from programs that utilize vouchers or that deploy federal subsidies to assist rental housing that is privately owned. This taxonomic separatism can be weakly justified on the grounds that these other programs are "public" in some ways but not in others, yet all of them have rapidly evolved and descended from a common ancestor—the Housing Act of 1937. We argue that they ought to be included within the broader ambit of *public-private housing*, defined to encompass all forms of publicly assisted deep subsidies intended to help the poorest Americans afford their housing. Seeing public housing this way illuminates the long trajectory of its privatization.

Shifting From Conventional Public Housing to New Modes

Just as public housing now incorporates a wide variety of built environments, fiscal programs, management regimes, and social constructs and is itself "public" in some ways but not in others, those alternatively named governmental policies that were designed to spread housing affordability through tenant- or project-based vouchers or loans to private developers were developed as evolutions of the initial public housing program, rarely as dramatic departures from it. "Conventional" public housing is no longer so conventional, and the U.S. Department of Housing and Urban Development (HUD) has made its public housing units increasingly fungible with deeply subsidized housing units funded through other programs. It is time, we argue, to have a single term that describes the larger American phenomenon of deeply subsidized housing.[4]

Public-private housing for low-income households comes with a subsidy from the federal government that may be defined as either "deep" or "shallow." Public housing, as the original "deep subsidy" program for low-income Americans, shares several important characteristics with the newer hybrid forms of deep housing subsidy that have come to supplement it.[5] At base, a deep subsidy links the size of rent payment to a percentage of household income (typically 30 percent at present). The subsidy covers the difference between this percentage of tenant income and the "fair market rent" for that unit and can apply to public housing, project-based vouchers, or tenant-based vouchers.[6] By contrast, "shallow subsidy" programs, such as LIHTC, also charge below-market rents, but those rents do not vary according to the income of the tenant. Our principal focus here is on the deeply subsidized portion of the public-private housing stock.

Looking across 80 years of experimentation in public housing and its public-private offspring, the federal government has launched many different kinds of policies that subsidize a small fraction of the nation's poorest households. Cumulatively, despite the terrible reputation that "public housing" has held for decades, the nation's overall commitment to deeply subsidized housing rose steadily for five decades—even though it assuredly remained well short of the actual need. Since the early 2000s, however, the stock of deeply subsidized housing, including vouchers, has stopped growing; given that the national population has risen by about 15 percent during the same period, on a per capita basis the stock of such housing has lost considerable ground. Figure 14.1 reveals the place of the

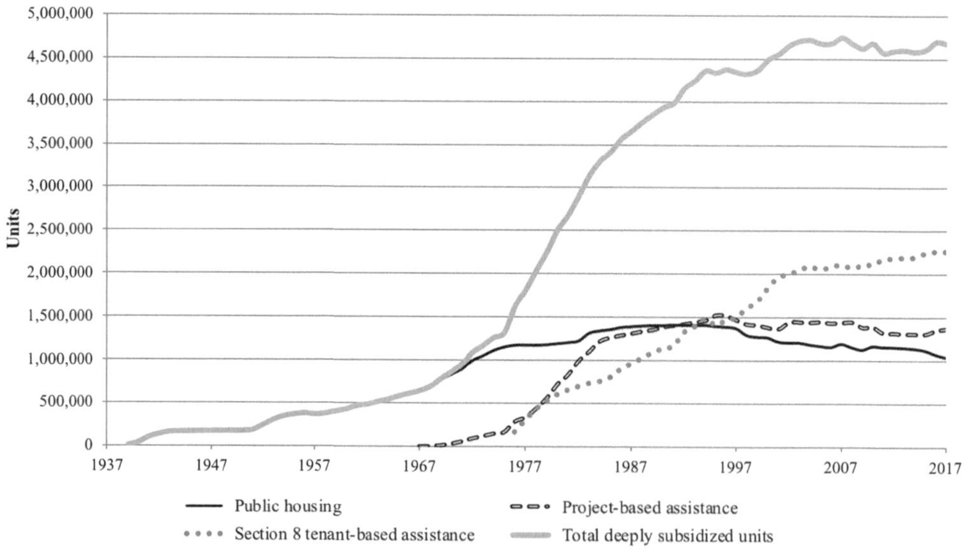

Figure 14.1 Trends in Deep Subsidies for Federally Assisted Rental Housing. Overall, these subsidies expanded significantly over several decades, but have slowed dramatically during most of the last 25 years, in tandem with a decreasing reliance on the conventional public housing program.

Source: authors, based on Bratt (1989); Olsen (2003); U.S. Department of Housing and Urban Development (2002, 2005, 2008, 2010, 2011a, 2012, 2016b, 2018a).

public housing program in the larger realm of federally sponsored efforts to generate deep subsidies for low-income families through public-private housing. The clear message is that conventional public housing, while certainly the pioneer program, has long since been eclipsed by other efforts that have grown larger even as the number of public housing units has declined by 370,000 since the program's peak in the early 1990s. Conventional public housing now accounts for less than a quarter of all deeply subsidized apartments and stands out far more starkly in peoples' minds than it does in the world of actual housing supply.

Its lingering hold on the public imagination—undergirded by an enormous scholarly literature coupled with a much larger presence in the popular media—is both due to its distinctive visibility on the urban landscape and because it is the nation's most noticeable reminder of persistent urban poverty in a wealthy nation. Public housing makes evident the tensions of race, ethnicity, and gender relations, and both strains and clarifies disputes over class by revealing how self-definitions of "poor" and "middle class" are constructed in the American city. In doing so, moreover, it demonstrates the power of private markets to shape even that which is explicitly termed "public."

Even at the times when Congress passed the major legislation in 1937 and 1949 that placed public housing at the center of the nation's low-income housing policy, it was advanced only thanks to the support of the building trades, which saw new affordable housing construction as a "counter-cyclical" public work that, ironically, supported private contractors (Winnick 1995, 102). During the mid-1950s, once the constituency seeking family public housing started to shift away from the White working class, policymakers sought new ways to serve other portions of the poor judged to have particular worth. Instead of a program designed to support young two-parent upwardly mobile working-class households, public housing quietly shifted into a social experiment in housing the elderly (typically at least age 62) and/or the disabled. The Housing Act of 1956 redefined public housing eligibility to include the single elderly, and both this legislation and the Housing Act of 1961 offered local housing authorities an annual premium for each elderly household housed in its developments. The elderly poor clearly benefited from a broadened definition of public housing. Section 202 of the Housing Act of 1959 provided federal capital grants to nonprofits to complete and operate apartments for very low-income seniors, for the first time offering federal funds to non-governmental entities for the provision of deeply subsidized housing (Schwartz 2015a).[7] By the 1960s and 1970s, Meehan (1979, 36) comments that the elderly poor were "the favored darlings of the public housing program." Today, those over age 62 head about one-third of public housing households, mostly located in buildings specifically constructed to house seniors (Schwartz 2015a).

Seniors, however, were hardly the only new population seeking public housing. From the late 1950s to the 1980s, largely as a reaction to the pressures of broader American societal trends, i.e., the Civil Rights Movement and the Second Great Migration, the leadership of most large American cities gradually re-tenanted "the projects" with the least advantaged and most economically desperate urban dwellers as their initial White working-class constituency chose to leave them behind for other housing alternatives in the private sector. In 1960, the average public housing household earned 41 percent of the national median income, but by 1970 this had dropped to 29 percent of the median, and it continued to plummet further, sinking well below 20 percent of the median by the early 1990s (Nenno 1996). By 1987 Congress had enacted explicit preferences for a broad array of disadvantaged applicants, encompassing the homeless, the disabled (including the mentally ill), and those paying more than 50 percent of their income in rent. The Reagan administration defended these policies as a necessary recognition of the greater impoverishment among those seeking housing assistance (Schill 1993; Vale 1992).

As public housing was systematically turned into "housing of last resort," federal policymakers sought reform that expanded the role of private sector actors beyond senior-focused developments and toward family units. They also moved private actors past the construction and financing phase and into management and ownership of units. This evolution inexorably moved the American commitment to deeply subsidized housing toward public-private housing, taking it well outside the bounds of conventional projects built, managed, and maintained by local housing authorities. The period from the early 1960s to the mid-1980s—from the Kennedy administration to the heart of the Reagan revolution—marks a crucial ideological battleground over the nature and direction of the national obligation to deeply subsidized low-income housing. Although it is certainly correct to regard the Nixon administration moratorium on new federal housing finance, announced in January 1973, as a watershed moment in U.S. public housing history, and correct to view the Housing and Community Development Act of 1974 as a significant move away from production-oriented approaches (such as the conventional public housing program) and toward consumer-based housing vouchers, much of this shift had occurred years earlier.

Though both Presidents Kennedy and Johnson called for the expansion of the conventional public housing program, both administrations pursued reforms that encouraged private investment in public housing and in housing that served those with somewhat higher incomes. The 221(d)(3) program of the 1961 Housing Act provided insurance and subsidized interest rates for mortgages on multifamily units for moderate-income households constructed by nonprofit and limited-profit developers, thereby permitting subsidized rents. During the Great Society, the effort to engage private groups in the public housing production process also sped up; "We are striving . . . to get private business to take on some of this development," President Johnson (1966) averred. Developers would be able to design and build "public housing" projects and then sell them to housing authorities under the Turnkey system, or do so in later years through acquisition. The Rent Supplement program introduced in 1965 allowed local authorities to distribute low-income households into privately held units (particularly those subsidized by Section 221(d)(3)) and restrict their rental payments to a low percentage of their incomes, while making up the rest via federal grants. Conversely, the 1965 Housing Act also allowed some tenants to buy apartments in conventional public housing buildings through the Turnkey III program, though this pilot measure was rarely used. A similar attempt at private ownership of individual units passed in 1988, HOPE I, which also failed to take hold (Schill 1990). Privatization proceeded apace, just more subtly.

Meanwhile, authorities could lease some existing private apartments under the new Section 23 program and then provide them to the poor, who would pay rents as if they were living in conventional public housing. Local housing authorities referred to these units as yet another form of "public housing," an appropriate appellation considering their place in the federal program, but clear evidence that the program had evolved past the conventional financing and ownership paradigms of the past (Orlebeke 2000). Section 23 made possible direct private investment in what would become public-private housing.

In 1968, Johnson's ambitions further expanded as he signed into law a ten-fold expansion in funding for low- and moderate-income housing through a bevy of new subsidies for public housing and other financing programs. The new Section 236 program, which replaced Section 221(d)(3), provided a major boost to moderate-income housing construction; when combined with the Rent Supplement program, it allowed for the provision of a limited number of deeply subsidized units for low-income households since those tenants needed to pay only 25 percent of their income in rent (later increased to 30 percent). The president's aim was not only to build more housing, but to do so "to enable private industry not only to build low-income housing developments, but also to manage them" (Johnson 1968; Schwartz 2015a, 159). The members of the Kaiser Committee (1968, 72), appointed by Johnson to study the nation's housing, suggested that conventional public housing had "been transformed by so many new variations" that it could imagine—but did not endorse—a situation in which "the federal government could develop a single housing subsidy technique" incorporating vouchers for both public and private housing.

President Nixon followed through on the funding promises of the 1968 bill, at least at first. In his first term, the government produced the largest investment in new federally subsidized low-income housing ever, building more units in four years than had been produced by the 1949 Housing Act. In 1964, conventional units accounted for 88 percent of all new public housing produced; by 1972, their share had declined to just 15 percent. HUD transformed what "public" meant when it came to public housing, financing three-and-a-half times as many units of turnkey, acquisition, and leased housing as conventional public housing (U.S. Department of Housing and Urban Development 1973). In the process, it showed that public housing could come in a very private form. The mold was set.

The success of these alternative approaches in providing more units serving low-income households at a decreased cost—though several of the programs struggled with poor quality and foreclosure by the mid-1970s (Winnick 1995)—emboldened conservatives, even as conventional public housing complexes began to struggle under the weight of their welfare mission. If private construction of new "public" housing worked, why not simply rely entirely on the private market? Housing authorities could negotiate with developers to allow leasing of an adequate number of units for low-income people. Some even suggested providing direct vouchers to individuals. Disenchanted by the progress of the expanded housing programs and concerned by violence and incivility at places like St. Louis' Pruitt-Igoe complex, HUD Secretary Romney had by 1972 been convinced. "We can no longer afford $100-billion mistakes," he stated. The media and the executive branch increasingly publicized problems with federal housing programs, but Congressional paralysis and disagreements about the value of programs like public housing made legislative reform of the 1968 Act difficult to foresee (Herbers 1972, 22). In January 1973, with some ambivalence, Romney announced a complete moratorium on new funding for federal housing subsidies and subsequently resigned (Orlebeke 2000). Nixon (1973, n.p.), called government housing investment "boondoggling," and suggested that housing programs "have been turning the Federal Government into a nationwide slumlord."

In late 1973, President Nixon proposed ending traditional federal housing programs and replacing them with either housing or cash allowances based on income, the latter a goal he had been developing under his welfare reform program since 1969. Congress did not go that far but eventually passed the Housing and Community Development Act in summer 1974, dramatically reducing funding for the construction of new conventional public housing (decreasing the six-year total subsidy to just 30,000 new units). The Section 8 construction program, with 400,000 new or rehabilitated private units by 1981, would provide most of the new low-income housing. Meanwhile, thousands of rental certificates would go directly to households, letting them choose apartments in existing buildings—a demonstration program that Nixon wished to expand (Varady and Walker 2007).[8] This method of subsidizing the poor, focusing on choice and private initiative rather than public service provision, fell ideologically in line with conservatives in the Nixon and Ford administrations (Ford 1974). It also addressed the sense that the most significant housing problem facing American low-income families was no longer building quality—overcrowding and dilapidation had declined substantially since the 1940s—but affordability (President's Commission on Housing 1982). With public housing in most large cities now seen as ghettoized projects that degraded the poor rather than as aspirational living environments for the upwardly mobile lower middle class,[9] both major parties favored a continued evolution of a program that had once relied almost entirely on public initiatives for funding and management into one that incorporated increasing private involvement.

Seeking Private Landlords

Proponents of the new rental certificate program did not dare adopt a name that bore any obvious relationship to "public housing," even though the program had emerged from the original public housing legislation and even though it resembled the Section 23 leasing and Rent Supplement programs introduced a decade before. Instead, legislators (and the landlords and tenants who implemented the policy) kept its policy wonk identity intact as "Section 8," referring to the piece of the Housing Act of 1937 that left the legislation open to future alteration. Section 8 simply provided a convenient way to append a large number of new initiatives. Tellingly, the piece of Section 8 that exploded into the nation's largest public housing program was not even the one most noticed in 1974. By the early 1990s, the housing subsidized by this program quietly came to outnumber conventional public housing units in most large cities (other than New York), sometimes quite dramatically (Vale and Freemark 2012). The certificate program (initially called the Section 8 Existing Housing program) that provided portable rental subsidies for use by tenants in the private market eventually merged with another similar voucher program in 1999, to constitute what are now formally called Housing Choice Vouchers. That terminology, with its explicit emphasis on empowering tenants to decide their own residential location, could be directly counterposed to the presumed entrapment—choicelessness—of the earlier mode of public housing project (Finkel and Buron 2001). Both ways of naming the new policy—Section 8 and Housing Choice Vouchers—neatly masked the fact that it provided a level of deep public subsidy to tenants that was nearly identical to that of public housing projects. This enabled public housing to expand even during its most stigmatized welfare stage, without actually having to name the new parts of it "public housing."[10]

This strategic political choice did not stop landlords from refusing to participate (Varady et al. 2017) or prevent nearby residents from resenting their deeply subsidized neighbors, but the stealth name and the furtive decentralization of the program still permitted accelerating growth—much faster than the contemporaneous program of scattered-site public housing. In 1994 the voucher program surpassed the unit volume of the conventional public housing program itself, and is now more than twice its size (see Figure 14.1).

Seeking Private Developers

The other, and initially better used, main component of the Section 8 programs, introduced in 1974, retained the "projects" piece of the public housing concept from 1937, but shifted the ownership of such projects to the private sector, thereby making it more ideologically palatable. By 1977, Congress favored this new housing production aspect of Section 8, and developers began to master its "lucrative profit potential" (Orlebeke 2000, 504). The program enjoyed a brief heyday before the Reagan administration, arguing that the general housing supply had improved sufficiently to make it unnecessary, applied the brakes and stopped it, preferring instead to expand the tenant-based voucher portion of Section 8. Despite the relatively short period of implementation, the Section 8 New Construction and Substantial Rehabilitation program left a lasting impact on the metropolitan landscape, resulting in nearly a million deeply subsidized apartments. As Figure 14.2 suggests, the deep subsidies that support apartments in privately developed housing projects are more numerous than conventional public housing in many large cities. Like tenant-based Section 8 units, these project-based units serve a population that is almost entirely extremely low income.[11]

Seeking Community Partners and Private Investors

Other important federal initiatives launched since the mid-1970s gained strength during the 1990s, undergirding the important contributions of more than 4,000 community development corporations (CDCs), which financed housing units through federally provided Community Development Block Grants in 1974, HOME funds from the National Affordable Housing Act of 1990, and

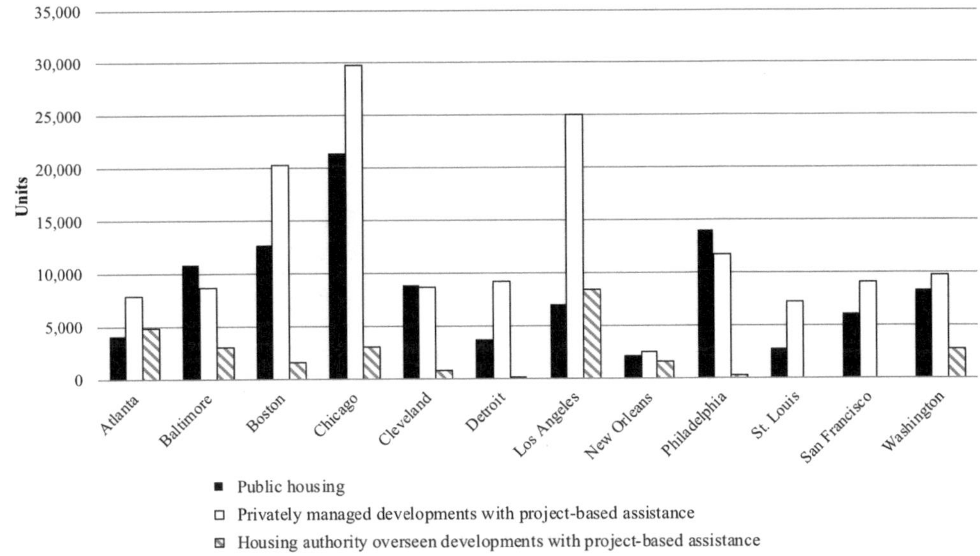

Figure 14.2 Other "Projects" on the Rise in Large U.S. Cities: Privately Developed Public Housing (2015). In many large cities, the number of Section 8-subsidized project-based housing units is greater than the public housing stock. The bulk of project-based assistance has centered on the privately developed projects constructed in the 1970s and 1980s, but more recently some housing authorities have also gained the flexibility to convert some of their own public housing to project-based assistance, including through RAD, though many of these units are not owned by the authorities themselves.

Source: authors, based on housing authorities (for public housing data and project-based units run by agencies, 2015–2016, including moderate rehabilitation units); U.S. Department of Housing and Urban Development (2016b).

LIHTC (introduced in 1986, as part of the tax code). As shown in Figure 14.3, more than twice as many apartments receive LIHTC contracts as are in the conventional public housing program, and, unlike the deeply subsidized programs, LIHTC has seen a constant rate of increased production for the past three decades; it quietly became the nation's largest housing subsidy program. Erickson (2009, xiv, 35) calls this "stealth housing program" a "revolution from below," and notes that these projects have frequently been of higher quality and greater political popularity than either public housing or project-based Section 8 efforts. Unfortunately, it is also true that the more than two million units constructed by public, nonprofit, and private entities with LIHTC subsidies have not reached many renters with extremely low incomes, except for those units that house residents with additional tenant-based Section 8 vouchers (Varady 2006). Of LIHTC households *without* additional rental assistance, only 31.3 percent are extremely low income (i.e., with incomes less than 30 percent of area median income), far less than comparable welfare-era public housing or Section 8 project-based programs, though the remaining households are largely very low income (i.e., with incomes below 50 percent of AMI) (O'Regan and Horn 2013). In short, most LIHTC units reach the poor, but not the poorest, and they do not provide the deep subsidy that comes with limiting tenant rent to 30 percent of income. Still, they are surely an important part of the emergent public-private housing landscape.

These various efforts enacted under Section 8 and the tax code to bring the private sector more fully into the business of deeply subsidized low-income housing did not, however, obviate the need

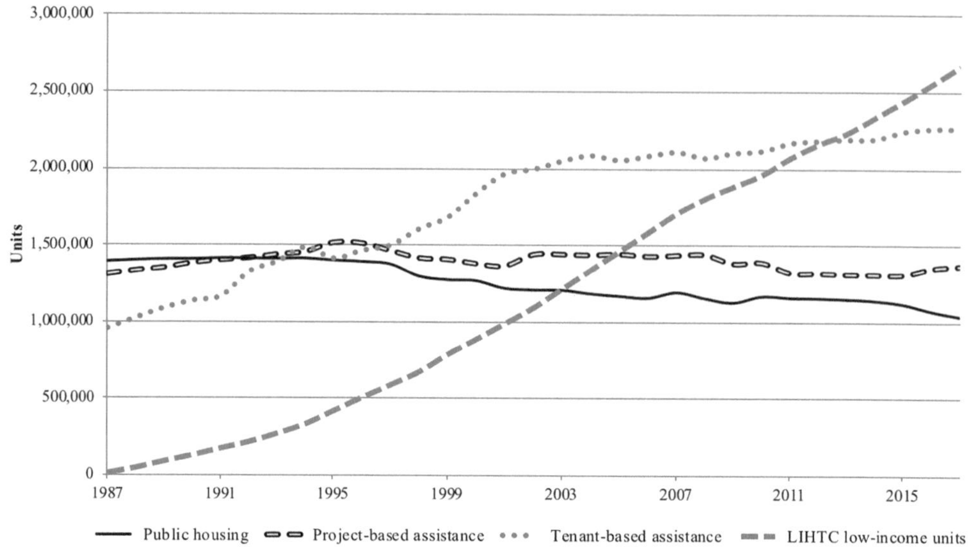

Figure 14.3 The Low-Income Housing Landscape: Three Decades of Dramatic Change. The LIHTC has become the nation's most popular funding mechanism for new subsidized housing, with more than two million units created since 1986. Rents may be set at a rate affordable to families making 60 percent of the metropolitan area's median income, but that means that those with less than that income will end up paying more than 30 percent of their income for rent. This means that LIHTC units do not provide the same sort of "deep subsidy," thereby limiting the program's ability to serve the lowest-income households unless they also hold Section 8 vouchers. The increase in Section 8 tenant-based vouchers beginning in 1995 reflects Congress' decision to authorize more than 630,000 of these between 1995 and 2009. But only 25 percent of these actually served new households; the other 465,000 were issued to assist households being transferred from public housing due to redevelopment, or from other project-based subsidy programs when private owners had opted out of continuing.

Notes: LIHTC figures pre-2001 are contracted units, not completed apartments, so the graph overstates their early rate of growth to some degree; later figures, however, represent apartments in use, though roughly 77,000 units are not included because of a lack of HUD data on the year those projects were completed. Project-based units include Section 8 project-based, Sections 202 and 811, and rent supplement units. Units that receive Section 8 tenant-based assistance and are funded under the LIHTC program are necessarily counted twice; O'Regan and Horn (2013) show that 68 percent of LIHTC units occupied by households with extremely low incomes are also provided some sort of deep subsidy. LIHTC figures for 2014 through 2017 are projected based on recent trends. In addition, the LIHTC total does not include roughly 300,000 units built under the program for non-low-income households.

Source: authors, based on U. S. Department of Housing and Urban Development (2018b) for LIHTC. For public housing, project-based, and tenant-based units, same as Figure 14.1.

to deal with the existing conventional public housing stock, still caught in the impoverished throes of the welfare housing social experiment.

Seeking the Return of the Worthy Poor

As the 1990s began, city leaders and their private sector counterparts eventually embarked on a new public housing social experiment. In an effort to remove "severely distressed" public housing, viewed as both a social danger and a political liability, they embraced HUD's HOPE VI program and championed other housing finance initiatives. They aimed to replace traditional public housing with new communities premised upon a broader socioeconomic constituency, often mixing conventional public housing units, tax credit-financed units, and market-rate units, while simultaneously expanding the earlier policy initiative to supply more of the poor with portable tenant-based vouchers. HOPE VI came in two guises: a redevelopment program that began in 1992, and a separately administered demolition-only program that lasted from 1996 to 2003.

The overall supply of conventional public housing peaked at 1.4 million units in 1991, and has declined steadily ever since. Taken together, HUD has authorized the demolition of about 20 percent of the national public housing stock, with about half of this loss of 370,000 units coming from the HOPE VI program and its associated demolition grants (Cisneros and Engdahl 2009; Goetz 2013). HOPE VI typically involved hiring a private developer, appeasing the needs of private corporate investors who did not trust local PHAs, and making sure that the small percentages of public housing residents who were granted the chance to return to the newly rebuilt communities would face the enhanced scrutiny and strict rule enforcement of a private sector manager (Chaskin and Joseph 2015; Vale 2013).

HOPE VI ushered in new forms of public-private redevelopment partnerships linked to broader policy trends that have coalesced since the early 1990s. Once the politically conservative Republicans captured the Congress in the 1994 mid-term election, public housing reform and welfare reform—a key component of the Clinton administration's federal policy in the mid-1990s—proceeded in tandem. In 1995, the legislature suspended the federal one-for-one replacement requirement, thereby permitting much larger numbers of demolished units to be replaced with vouchers.

More recently, the Obama administration's Choice Neighborhoods Initiative, first funded in 2010, has replaced and expanded on HOPE VI by not only funding the demolition and replacement of public housing, but also providing federal dollars for the redevelopment of privately owned and built projects, such as those covered by the project-based Section 8 program, and in some cases combining them with new middle-income and market-rate units. All of the initially funded projects took advantage of LIHTC for a portion of capital expenditures (Pendall and Hendey 2013).

With the Quality Housing and Work Responsibility Act of 1998 (QHWRA), public housing reform came into full bloom. Among many other provisions, QHWRA formally embraced the new reliance on mixed-finance development, permitting housing authorities to issue bonds or otherwise borrow funds for the renovation or redevelopment of public housing (with repayment premised on future receipt of federal capital funds). At the same time, QHWRA facilitated (or at least proposed) a significant demographic shift over time, rooted in a belief that the least advantaged working-age families would be better off with vouchers and, correspondingly, that future housing developments should have fewer extremely low-income households. By 2001, public housing residents had a mean income equivalent to 23.8 percent of the area median income, still well within the bounds of the "extremely low-income" category, but a significant increase from the 17 percent of area median income recorded in the early 1990s (Council of Large Public Housing Authorities 2006, Quality Housing and Work Responsibility Act of 1998 1998, Riccio 2008; Schwartz 2015a; Vale 1993).

Meanwhile, as housing authorities sort through which of the poor they want to assist in which kinds of places, the number of unserved households facing dire housing needs has continued to grow. Despite the growth of shallower housing subsidies, such as those supplied by LIHTC units, Figure 14.4 shows that the provision of deeper housing subsidies has entirely stalled for the past two decades, falling further behind population growth. The number of deeply subsidized units, both project-based and voucher-based, has declined by almost 20 percent on a per capita basis since the early 1990s.

As already pointed out earlier, the provision of public housing—and, more broadly, the supply of public-private housing—has remained far short of demand. Unlike other social programs that have been treated as entitlements for the very poor, such as Medicaid, which provides health insurance, and the Supplemental Nutrition Assistance Program (SNAP), housing subsidies have been parceled out parsimoniously (Priemus et al. 2005). In 2015, 7.7 million American renter households faced "worst case housing needs," meaning that they received no housing assistance despite having incomes

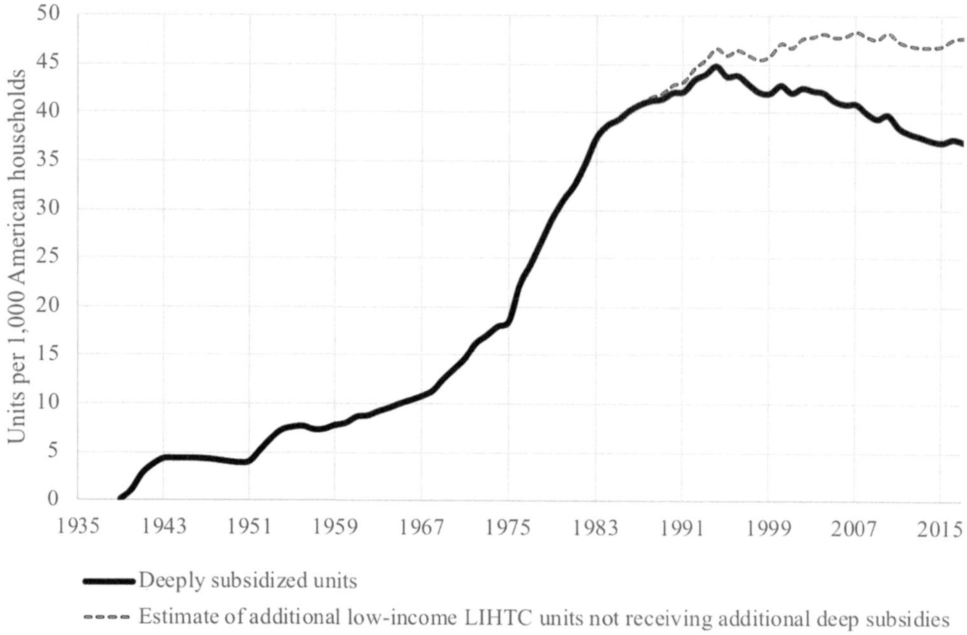

Figure 14.4 Deeply Subsidized Rental Housing in the United States: A Declining Commitment? The provision of deeply subsidized housing units in the United States exceeded the growth of overall housing units for much of the period since the Great Depression, especially in the spurts following the 1937, 1949, and 1968 Housing Acts. Since the early 1990s, however, the number of deeply subsidized units has declined from 45 to 37 per 1,000 households. Even when accounting for the rapid growth in the number of units produced through the LIHTC program that are targeted to low-income households but that do not receive deeper subsidies, the number of federally subsidized low-income units per household has still declined slightly since the early 2000s.

Note: This estimate is generated based on O'Regan and Horn (2013), who study LIHTC data in ten states. They show that, of all households living in LIHTC apartments, 49.6 percent are also provided rental assistance, usually in the form of Section 8 tenant-based vouchers. This chart assumes that this figure can be extrapolated for the entire country.

Source: authors, based on U.S. household tabulation of housing unit annual change by U.S. Bureau of the Census (2018). For units, same as Figure 14.1.

less than 50 percent of the median in their metropolitan area and paying more than half of that income for housing, or else lived in homes with severe problems such as lack of plumbing or heat (U.S. Department of Housing and Urban Development 2015a). Only about one in four qualified households receive some form of deep housing subsidy (Sard and Fischer 2013). Despite extensive public housing experimentation in the United States, most of the eligible public has not been invited to take part.

Moving to Work, Moving to RAD

In 1999, HUD began implementing the Moving to Work (MTW) demonstration program, first legislated in 1996 and designed to encourage flexibility, eliminate burdensome regulations, and reduce expenditures among local PHAs. Families with a working head of household are provided financial incentives and encouraged to increase their incomes while retaining their federal rent subsidies. In addition, participating agencies (there were 38 as of 2015, representing 11 percent of American public housing units and 13 percent of vouchers) are allocated funds for public housing capital and operating expenditures, as well as Housing Choice Vouchers (HCV), but unlike non-MTW agencies, they can move federal allocations freely between those accounts. As such, many MTW agencies have significantly shifted their funding away from public housing (which constituted 53 percent of their units in 1998) and toward the HCV program (which accounted for 71 percent of their units in 2009), though this change was accompanied by a 47 percent increase in deeply subsidized apartments supported by federal funds (Cadik and Nogic 2010; Schwartz 2015a; U.S. Department of Housing and Urban Development 2016a).

In 2013, under the Obama administration, HUD introduced the RAD program, which has further blurred the boundaries between public housing and project-based contracts. In the context of declining federal support for capital improvements and steadily aging buildings, RAD opens up new sources of funding by allowing local housing authorities to convert their conventional public housing apartments to long-term project-based Section 8 rental assistance, given a commitment to maintain each subsidized housing unit on a one-to-one basis. In 2012, Choice Neighborhood Initiative grantees were allowed to apply for RAD-based conversion of their public housing units to Section 8 contracts (U.S. Department of Housing and Urban Development 2015b). RAD also allows PHAs to use LIHTC financing for renovation, adding to public agencies' funding portfolio a revenue source typically used by nonprofit or private developers, and effectively shifting public housing maintenance costs from HUD to the IRS. In a unique melding of programs, RAD allows tenants to exchange their public housing units for rental vouchers for use in private units elsewhere (Schwartz 2015b).

RAD allows the transfer of public housing properties to nonprofit or even for-profit entities, although local housing agencies must retain a controlling stake. Though initially conceived as a demonstration limited to 65,000 units, RAD has quickly expanded nationwide. In July 2018, HUD implemented a new cap of 225,000 units, and some policymakers recommended there be no limit at all. Virtually all small housing authorities engaged with the program thus far have converted all of their public housing into RAD (Schwartz 2015b; National Low Income Housing Coalition 2016). Even New York City has plans to transition 15,000 of its almost 180,000 conventional public housing units through RAD (New York City Housing Authority 2016). RAD could portend a future in which conventional public housing in the United States disappears entirely.

Public Housing: W(h)ither the Public?

After 80 years of policies, ever-increasing numbers of American public housing operations have become completely privatized. In many cities, housing authorities are regularly turning over their conventional housing stock to private managers and often own nothing more than the land beneath their redevelopment endeavors. In this context, even the basic definitional reason for calling some housing "public housing" now comes into question. Indeed, it is not altogether obvious that any PHAs will continue to manage a large number of conventional units. Future research should explore the question of the remaining role for PHAs in providing subsidized housing.

As housing authorities increasingly reinvent themselves as "asset managers," they move more fully into the mainstream of private property practice. Instead of an "owner and manager of public housing," for example, the reinvented Chicago Housing Authority now self-identifies as "a facilitator of housing opportunities" (Chicago Housing Authority 2000). If "public housing" is not distinguished by public ownership and management, what basis is still there for distinguishing it from the other kinds of federally assisted deep-subsidy efforts to house low-income renters? Purists and concerned advocates will wish to retain a firm distinction between the "hard units" of conventional public housing supported by permanently available public subsidies and everything else. They will note the perennial vulnerability of vouchers (which depend on Congressional renewal of funding), the tenuousness of project-based units (which have expiring contracts that may or may not get renewed), and the time limitations of LIHTC units (which may revert to market rates after 15 years). These concerns are all valid, and certainly worrisome, but no more worrisome than the fact that support for conventional public housing also depends on Congressional appropriations, and it is actually these "hard units" that have been disappearing the fastest. The good news is that a half century of aversion to talking about the full panoply of public-private housing as "public housing" quietly facilitated significant growth in the nation's stock of deeply subsidized units between the mid-1960s and mid-1980s. Unfortunately, that growth in the deeply subsidized portion of what is called public-private housing has slowed dramatically since the late 1980s.

Ultimately, then, the pathways to privatization have thus far collectively failed to yield greater provision of housing for the least economically advantaged Americans. This failure has both supply and demand dimensions. First, despite many decades of new programs, the total number of deeply subsidized housing units—including not only conventional public housing, but also vouchers and programs—now lags behind population growth. Instead, fueled by the preferences of private investors, the major growth in subsidized housing through the LIHTC program has targeted the less poor among the poor, rather than the poorest. Moreover, much of the remaining deeply subsidized housing stock remains vulnerable to conversion away from low-income occupancy. This leaves the lowest-income households with a double problem in many American urban areas. Their incomes lag behind housing affordability, even as the affordable housing supply lags well behind demand. For renters, the dearth of housing subsidies has been particularly devastating in the aftermath of the Great Recession that lasted from December 2007 until June 2009; the number of rent-burdened households has sextupled since then, and the results have been increases in evictions, growing concentrations of poverty, and rising homelessness in major cities (Joint Center for Housing Studies 2016). This remains the most disheartening dimension of all.

At base, the privatization of public housing has a much longer history than many of those who treat it as a symptom of neoliberalism care to acknowledge. But that just means that the inadequacy of public housing supply is rooted even more deeply.

Notes

1 Changes to the legal status of council housing have made it easier to sell it off (Wiles 2015).

2 We do not mean to suggest that the term "public-private housing" will gain widespread acceptance, but it can aid our understanding of the overall evolution of housing policy.

3 Hybridity arguments typically stress the role of nonprofit entities in funding housing, largely facilitated in the U.S. through LIHTC. Our chapter de-emphasizes the role of nonprofits since (at least in the U.S. context) they are less associated with projects that reach those with the very lowest incomes and receive deep subsidies.

4 It is certainly possible to disagree with this chapter's argument—that vouchers are evolutions of previous subsidy techniques—but the evolution of the subsidized housing policy environment, particularly between 1964 and 1972, pre-figured much of the move away from conventional public housing by embracing a gradual inclusion of more forms of privatization. This argues against the notion of a truly "dramatic departure" in 1974 (Vale and Freemark 2012).

5 Although, using this definition, public housing can be considered the first "deep subsidy" program, it is worth noting that most public housing did not actually employ this precise kind of income-calibrated rents until passage of the first Brooke Amendment, in 1969, which set rent for a particular apartment at 25 percent of its household's income (raised to 30 percent of income in 1981). The Housing Act of 1937 had required that household incomes be no more than five or six times the rent charged for an apartment (depending on family size), but rents were linked to apartments and did not vary with tenant income. By 1940 the United States Housing Authority slightly altered this policy so that income limits gave way to a maximum rent standard in which rent could not exceed 20 percent of tenant income. Starting in the early 1940s, however, some public housing providers, such as the Chicago Housing Authority, introduced apartment unit rents that did vary by income.

6 Fair market rents vary according to the jurisdictional location and the number of bedrooms in the unit. Rent setting varies according to several factors such as family size and depends on whether utility costs are included. For shallow subsidies, rents may be set at 30–40 percent of the maximum tenant income associated with that unit. For tenants earning less than that maximum income, this means that they may be asked to pay more than 30–40 percent of their income in rent, thereby making it less affordable.

7 Section 202 of the Housing Act of 1959 is not part of the conventional public housing program itself. That program, though still smaller than the number of public housing apartments devoted to seniors, gradually grew to encompass about 270,000 units, reaching about 11 percent of the eligible population of low-income seniors (Schwartz 2015a). Beginning in the early 1990s, some of these units (as well as senior-dedicated Section 8 project-based units) were rented out to income-eligible younger people with mental health or substance abuse histories. Though the number of Section 202 units has not expanded since the 1980s, senior units in cities like Atlanta and Chicago have survived even as many of their family public housing counterparts have been demolished (Heumann 1996; Vale and Freemark 2012).

8 Housing vouchers were then being studied by the Experimental Housing Allowance Program being run by HUD, but "the fact is, Congress enacted the housing allowance's basic legislation before EHAP was launched and was still supplementing that legislation even as EHAP's first raw data trickled in" (Winnick 1995, 96). Formal experiments continued to be outflanked by informal policy experimentation.

9 This view of the most problematic cases overlooks many important success stories (Stockard 1998).

10 Voucher programs continue to reach those with the lowest incomes. Of the nation's Section 8 tenant-based voucher recipients in 2012, 76 percent had an extremely low income (below 30 percent of area median income), more than is true for public housing residents (61 percent). An additional 21 percent of Section 8 voucher holders, compared to 26 percent of public housing residents, reported very low incomes (between 30 and 50 percent of area median income). Nationwide, 28 percent of Section 8 voucher households include at least one non-elderly disabled family member, and 19 percent are elderly, compared to 21 percent and 31 percent, respectively, of public housing families (U.S. Department of Housing and Urban Development 2012).

11 As of 2012, 81 percent of current inhabitants in project-based Section 8 housing are extremely low income. Average monthly rental payments of households were just $265 in 2012, less than the $294 for Section 8 voucher recipients or $318 for public housing inhabitants. In addition, project-based Section 8 units are oriented toward the elderly (36 percent of units) and the non-elderly disabled (28 percent of units) (U.S. Department of Housing and Urban Development 2012). It is worth noting, however, that these "projects" are less likely than conventional public housing to serve families. In 1981, 80 percent of the program's units were designated for the elderly; in 2012, nearly two-thirds of residents were elderly or non-elderly disabled (U.S. Department of Housing and Urban Development 2012; President's Commission on Housing 1982).

References

Arena, John. 2012. *Driven From New Orleans: How Nonprofits Betray Public Housing and Promote Privatization.* Minneapolis, MN: University of Minnesota Press.

Billis, David, ed. 2010. *Hybrid Organizations and the Third Sector: Challenges for Practice, Theory and Policy.* New York: Palgrave Macmillan.

Bratt, Rachel. 1989. *Rebuilding a Low-Income Housing Policy.* Philadelphia, PA: Temple University Press.

Bratt, Rachel G. 2012. "The Quadruple Bottom Line and Nonprofit Housing Organizations in the United States." *Housing Studies,* 27(4): 438–456.

Cadik, Emily, and Amanda Nogic. 2010. *Moving to Work: Interim Policy Applications and the Future of the Demonstration.* Washington, DC: U.S. Department of Housing and Urban Development.

Chaskin, Rorbert J., and Mark L. Joseph. 2015. *Integrating the Inner City: The Promise and Perils of Mixed-Income Public Housing Transformation.* Chicago, IL: University of Chicago Press.

Chicago Housing Authority. 2000. *2000 Moving to Work Report.* Chicago, IL: Chicago Housing Authority.

Cisneros, Henry, and Lora Engdahl, eds. 2009. *From Despair to Hope: Hope VI and the New Promise of Public Housing in America's Cities.* Washington, DC: Brookings Institution Press.

Council of Large Public Housing Authorities. 2006. *Public Housing Residents.* Washington, DC: CLPHA. http://www.clpha.org/publichousingresidents.

Erickson, David J. 2009. *The Housing Policy Revolution: Networks and Neighborhoods.* Washington, DC: Urban Institute Press.

Finkel, Meryl, and Larry Buron. 2001. *Study on Section 8 Voucher Success Rates. Vol. I: Quantitative Study of Success Rates in Metropolitan Areas.* Cambridge, MA: Abt Associates.

Ford, Gerald. 1974. *Statement on the Housing and Community Development Act of 1974.* Santa Barbara, CA: The American Presidency Project.

Goetz, Edward G. 2013. *New Deal Ruins: Race, Economic Justice, and Public Housing Policy*. Ithaca, NY: Cornell University Press.

Hackworth, Jason. 2007. *The Neoliberal City: Governance, Ideology, and Development in American Urbanism*. Ithaca, NY: Cornell University Press.

Herbers, John. 1972. "Romney Recommends Abolition of Federally Subsidized Housing." *The New York Times*, October 24, 22.

Heumann, Leonard F. 1996. "Assisted Living in Public Housing: A Case Study of Mixing Frail Elderly and Younger Persons with Chronic Mental Illness and Substance Abuse Histories." *Housing Policy Debate*, 7(3): 447–471.

Hodkinson, Stuart, and Glyn Robbins. 2013. "The Return of Class War Conservatism? Housing under the UK Coalition Government." *Critical Social Policy*, 33(1): 57–77.

Johnson, Lyndon. B. 1966. "The President's News Conference." May 21.

Johnson, Lyndon B. 1968. "Special Message to the Congress on Urban Problems: The Crisis of the Cities." February 22.

Joint Center for Housing Studies. 2016. *The State of the Nation's Housing*. Cambridge, MA: Joint Center for Housing Studies.

Kaiser Committee. 1968. *A Decent Home: The Report of the President's Committee on Urban Housing*. Washington, DC: U.S. Department of Housing and Urban Development.

Marcuse, Peter. 1995. "Interpreting 'Public Housing' History." *Journal of Architectural and Planning Research*, 12(3): 240–258.

Marcuse, Peter. 1998. "Mainstreaming Public Housing: A Proposal for a Comprehensive Approach to Housing Policy." In *New Directions in Urban Public Housing*, edited by David P. Varady, Wolfgang F. E. Preiser, and Francis P. Russell, 23–44. New Brunswick, NJ: CUPR Press.

Meehan, Eugene. 1979. *The Quality of Federal Policymaking: Programmed Failure in Public Housing*. Columbia, MO: University of Missouri Press.

Mullins, David, Darinka Czischke, and Gerard van Bortel. 2012. "Exploring the Meaning of Hybridity and Social Enterprise in Housing Organisations." *Housing Studies*, 27(4): 405–417.

National Low Income Housing Coalition. 2016. "Memo to Members: RAD Update." January 11.

Nenno, Mary K. 1996. *Ending the Stalemate: Moving Housing and Urban Development into the Mainstream of America's Future*. Lanham, MD: University Press of America.

New York City Housing Authority. 2016. *PACT-RAD Initiative*. New York: New York City Housing Authority.

The New York Times. 1937. "Public Housing Bonds Visioned by Post as High Grade Investments of the Future." *The New York Times*, October 8, 35.

Nguyen, Mai T., William M. Rohe, and Spencer M. Cowan. 2012. "Entrenched Hybridity in Public Housing Agencies in the USA." *Housing Studies*, 27(4): 457–475.

Nixon, Richard M. 1973. Radio Address about the State of the Union Message on Community Development, March 4.

Olsen, Edgar O. 2003. *Housing Programs for Low-Income Households*. Cambridge, MA: National Bureau of Economic Research.

O'Regan, Katherine M., and Keren M. Horn. 2013. "What Can We Learn about the Low Income Housing Tax Credit Program by Looking at the Tenants?" *Housing Policy Debate*, 23(3): 597–613.

Orlebeke, Charles J. 2000. "The Evolution of Low-Income Housing Policy, 1949 to 1999." *Housing Policy Debate*, 11(2): 489–520.

Pendall, Rolf, and Leah Hendey. 2013. *A Brief Look at the Early Implementation of Choice Neighborhoods*. Washington, DC: The Urban Institute.

President's Commission on Housing. 1982. *A Report to the President of the United States*. Washington, DC: President's Commission on Housing.

Priemus, Hugo, Peter A. Kemp, and David P. Varady. 2005. "Housing Vouchers in the United States, Great Britain, and the Netherlands: Current Issues and Future Perspectives." *Housing Policy Debate*, 16(3–4): 575–609.

Quality Housing and Work Responsibility Act of 1998. 1998. *Sections 512, 539*. Washington, DC: U.S. Department of Housing and Urban Development.

Riccio, James A. 2008. "Subsidized Housing and Employment: Building Evidence of What Works." In *Revisiting Rental Housing*, edited Nicolas P. Retsinas and Eric S. Belsky, 191–224. Washington, DC: Brookings Institution Press.

Sagalyn, Lynne B. 2012. "Public-Private Engagement: Promise and Practice." In *Planning Ideas That Matter*, edited by Bishwapriya Sanyal, Lawrence J. Vale, and Christina D. Rosan, 233–255. Cambridge, MA: MIT Press.

Sard, Barbara, and Will Fischer. 2013. *Chart Book: Federal Housing Spending is Poorly Matched to Need*. Washington, DC: Center on Budget and Policy Priorities.

Schill, Michael H. 1990. "Privatizing Federal Low Income Housing Assistance: The Case of Public Housing." *Cornell Law Review*, 75(4): 878–948.

Schill, Michael H. 1993. "Distressed Public Housing: Where Do We Go From Here?" *University of Chicago Law Review*, 60: 497–554.

Schwartz, Alex F. 2015a. *Housing Policy in the United States*. New York: Routledge.

Schwartz, Alex F. 2015b. "The Rental Assistance Demonstration Program (RAD): Public Housing Salvation or Doom?" Presented at *RC43 Conference*, Chicago, IL, September 25.

Stockard, James G. 1998. "Public Housing: The Next 60 Years." In *New Directions in Urban Public Housing*, edited by David P. Varady, Wolfgang F. E. Presider, and Francis P. Russell, 237–264. New Brunswick, NJ: CUPR Press.

Stout, G. M. 1935. "Against Public Housing." *Bulletin of the Georgia Institute of Technology*, 32(4): 72–94.

Straus, Nathan. 1944. *The Seven Myths of Housing*. New York: Knopf.

U.S. Bureau of the Census. 2018. *Population and Housing Unit Estimates*. Washington, DC: U.S. Bureau of the Census. https://www.census.gov/programs-surveys/popest.html

U.S. Department of Housing and Urban Development. 1973. *Housing in the Seventies: A Report*. Washington, DC: U.S. Department of Housing and Urban Development.

U.S. Department of Housing and Urban Development. 2002. *Performance and Accountability Report*. Washington, DC: U.S. Department of Housing and Urban Development.

U.S. Department of Housing and Urban Development. 2005. *Performance and Accountability Report*. Washington, DC: U.S. Department of Housing and Urban Development.

U.S. Department of Housing and Urban Development. 2008. *Performance and Accountability Report*. Washington, DC: U.S. Department of Housing and Urban Development.

U.S. Department of Housing and Urban Development. 2010. *Agency Financial Report*. Washington, DC: U.S. Department of Housing and Urban Development.

U.S. Department of Housing and Urban Development. 2011. *Agency Financial Report*. Washington, DC: U.S. Department of Housing and Urban Development.

U.S. Department of Housing and Urban Development. 2012. *Resident Characteristics Report*. Washington, DC: U.S. Department of Housing and Urban Development.

U.S. Department of Housing and Urban Development. 2015a. *Worst Case Housing Needs: 2015 Report to Congress*. Washington, DC: U.S. Department of Housing and Urban Development.

U.S. Department of Housing and Urban Development. 2015b. *Choice Neighborhoods: Baseline Conditions and Early Progress*. Washington, DC: U.S. Department of Housing and Urban Development.

U.S. Department of Housing and Urban Development. 2016a. *Moving to Work: Participating Sites*. Washington, DC: U.S. Department of Housing and Urban Development.

U.S. Department of Housing and Urban Development. 2016b. *Multifamily Assistance and Section 8 Contracts Database*. Washington, DC: U.S. Department of Housing and Urban Development.

U.S. Department of Housing and Urban Development. 2018a. *Assisted Housing: National and Local*. Washington, DC: U.S. Department of Housing and Urban Development.

U.S. Department of Housing and Urban Development. 2018b. *LIHTC Database*. Washington, DC: U.S. Department of Housing and Urban Development.

U.S. Housing Act of 1937. 1937. Section 2 (1); Section 15 (5). Washington, DC: Government Printing Office.

Vale, Lawrence J. 1992. *Occupancy Issues in Distressed Public Housing: An Outline of Impacts on Design, Management, and Service Delivery*. Washington, DC: National Commission on Severely Distressed Public Housing.

Vale, Lawrence J. 1993. "Beyond the Problem Projects Paradigm: Defining and Revitalizing 'Severely Distressed' Public Housing." *Housing Policy Debate*, 4(2): 147–174.

Vale, Lawrence J. 2000. *From the Puritans to the Projects: Public Housing and Public Neighbors*. Cambridge, MA: Harvard University Press.

Vale, Lawrence J. 2013. *Purging the Poorest: Public Housing and the Design Politics of Twice-Cleared Communities*. Chicago, IL: University of Chicago Press.

Vale, Lawrence J., and Yonah Freemark. 2012. "From Public Housing to Public-Private Housing: 75 Years of Social Experimentation." *Journal of the American Planning Association*, 78(4): 379–402.

Varady, David P. 2006. "Comment on Kirk McClure's 'The Low-Income Housing Tax Credit Program goes Mainstream and Moves to the Suburbs'." *Housing Policy Debate*, 17(3): 461–472.

Varady, David P., and Carole A. Walker. 2007. *Neighborhood Choices: Section 8 Housing Vouchers and Residential Mobility*. New Brunswick, NJ: CUPR Press.

Varady, David P., Joseph Jaroscak, and Reinout Kleinhans. 2017. "How to Attract More Landlords to the Housing Choice Voucher Program: A Case Study of Landlord Outreach Efforts." *Urban Research and Practice*, 10(2): 143–155.

Wiles, Colin. 2015. "This Move into the Public Sector Could Signal the End for Housing Associations." *The Guardian*, October 30.

Winnick, Louis. 1995. "The Triumph of Housing Allowance Programs: How a Fundamental Conflict Was Resolved." *Cityscape*, 1(3): 95–121.

15

PUBLIC HOUSING AUTHORITIES AS SOCIAL ENTERPRISES?

Rachel Garshick Kleit

Introduction

Devolution in federal housing policy since the 1980s has altered federal regulation and financial support for public housing authorities (PHAs), providing the opportunity for them to become surprisingly flexible organizations that, in the right local circumstances, could become key community actors in affordable housing delivery and preservation. Additionally, the growing market-based policy emphasis since the 1980s is hastening PHAs' widespread evolution into hybrid organizations; that is, more of them are becoming mission-driven organizations with characteristics of public, nonprofit, and private sector organizations, as authors who write about hybrid organizations describe (Billis 2010; Hoffman and Badiane 2012; Nguyen et al. 2012; Smith and Skelcher 2014; Skelcher and Smith 2015). While most authors discuss nonprofit or third-sector organizations becoming hybrids (Billis 2010; Smith and Skelcher 2014), others suggest that quasi-public[1] PHAs are also becoming hybrids as they employ private sector methods and financing to meet their public missions (Nguyen et al. 2012).

Similar dynamics have occurred among social housing organizations in Europe as they, too, have begun privatizing ownership and depending more on market dynamics for the provision of housing (Czischke et al. 2012; Mullins et al. 2012). Furthermore, Czischke et al. (2012) argue that in Europe, social housing organizations are beginning to act like social enterprises, which they conceptualize as hybrid entities of varying legal forms that act in accordance with the competing interests of public, market, and community forces and values. This stance at the nexus of these rival interests causes them to take actions that may be entrepreneurial rather than "bureaucratic," (Czischke et al. 2012, 428) and "conflicts between principles are played out in organizational strategies and day-to-day decisions" (Czischke et al. 2012, 428).

In this chapter, I argue that in the U.S. policy changes and devolution have moved PHAs toward becoming hybrid organizations that resemble social enterprises, working at the nexus of public, market, and social concerns. Unlike social housing organizations in Europe, PHAs in the U.S. are quasi-public agencies. In comparison, social housing organizations in, for example, the Netherlands are private entities that no longer receive any financial support from the government. Instead, they depend on private market demand and funding to renovate and manage their housing stock (see also the chapter authored by Marja Elsinga and Marietta Haffner). The present chapter suggests that the ability of PHAs to work as true social enterprises varies, in part, according to their local charter under state enabling legislation. The chapter uses a national survey of PHAs and a content analysis of enabling legislation in the 48 contiguous U.S. states to provide a description of these variations and speculation about the future of PHAs as social enterprises. Furthermore, the chapter goes on to suggest the potential role of the Residential Assistance Demonstration (RAD) program in enhancing the ability of PHAs to act like social enterprises.

Policy Devolution and the Quality of Affordable Housing

From the 1930s through the present, U.S. funding for affordable housing has moved from providing government-funded, publicly owned housing to increasingly depending on the private sector for the provision of affordable housing, thus intermingling market resources with a social mission to provide affordable housing. The most pointed illustration of this transformation is the change in the funding streams for affordable housing in a quest to provide affordable housing and sustain its quality over the long run.

At the start of the history of public housing authorities, the federal government and sometimes state and local governments (Bloom 2008, 2012) provided funding for PHAs to build housing that they then managed for public purpose (see also the chapter authored by Lawrence Vale and Yonah Freemark). By the 1960s, the well-publicized problems of public housing quality in the largest cities in the U.S. encouraged policymakers to look toward the private sector for the provision of affordable housing (Hays 2012). Whether through the incentive programs for the construction of privately owned rental housing in the 1960s, the growth of Section 8 project-based programs since the 1970s, the federal Low-Income Housing Tax Credit beginning in 1987 (see also the chapter authored by Kirk McClure), or the eventual growth of the Housing Choice Voucher (HCV) program to rent units in the private rental market (see also the chapter authored by Andrew J. Greenlee), using the resources and market capacity of the private sector, by the 1990s, had become a key element of U.S. federal housing policy.

It is the quest to provide quality housing that eventually turned policy to blend private sector resources with public housing program resources. The public housing program continued in parallel with these market-based programs until the 1990s. The well-publicized quality deficiencies of large city public housing (albeit in a small minority of units; see also the chapter authored by Lawrence Vale and Yonah Freemark) continued, exacerbated by the rent formula and difficulties funding ongoing capital needs (Schwartz 2015). The funding formula from 1949 through 1969 called for the federal government to pay for the construction, while tenant rents covered the operation of the buildings. This led to PHAs either deferring maintenance to avoid raising rents or charging rents so that tenants paid more than 40 percent of their incomes. In the 1950s, Congress mandated that the capital reserve funds for public housing be used to pay the debt service for construction bonds, leaving PHAs with no source of funds for capital improvements or repairs. By the late 1960s, Congress capped rents at 25 percent of income (raised later to 30 percent) and instituted a system of operating subsidies to make up the difference (Bratt 1989; Hays 2012; Schwartz 2015). Over time, a lack of full funding for operating subsidies, leading to inadequate management, along with sometimes poor local site selection, preparation, and construction, resulted in a pool of some 83,000 units of highly visible, troubled, publically owned housing by the 1990s—roughly 6 percent of the over 1.5 million units at the time. In 1990, the Cranston-Gonzales National Affordable Housing Act made concrete the U.S.'s dependence on private, socially driven nonprofit organizations—called community housing development organizations (CHDOs)—for the construction of affordable housing (see also the chapter authored by Lawrence Vale and Yonah Freemark).

Private Sector Solutions for Public Sector Problems

Meanwhile, troubled public housing motivated local experimentation, including a focus on mixed-income approaches as a solution. The argument was that mixed-income housing developments would result in better housing quality over the long run, role models for people living in poverty, residential stability, resident services, the investment of institutions, public agencies, and businesses, and the potential leverage of the private market (Corcoran 1980; Epp 1996; Goody 1993; Rosenbaum et al. 1996; Ryan et al. 1974). The *Report of the Commission on Severely Distressed Public Housing*, published in 1992, led to the creation of the Urban Redevelopment Demonstration (URD), which subsequently became the HOPE VI grant program (1993–2010) and, later, Choice Neighborhoods (2010–present) (see also the chapter authored by Lawrence Vale and Yonah Freemark). Over their duration, these programs evolved to focus on the redevelopment of troubled public and assisted housing into mixed-income housing, leveraging private sector financial resources and the strength of the private housing market.

Public Housing Authorities as Local Actors in the Private Market

With the onset of HOPE VI, the federal government for the first time specifically asked public housing authorities to leverage private investment using federal grants, public housing capital dollars, and private investment—called mixed finance (Turbov 2006). For some PHAs, this was the first time they had been involved in developing affordable housing, since funds for the public housing program had been exhausted in the 1970s. Other PHAs, however, had been engaged at various times in their histories in the development and ownership of affordable housing outside of the public housing program since their inception (e.g., NYCHA) using state and local funds. In the Pacific Northwest some housing authorities have owned existing rental housing outright since the 1980s (Bloom 2008, 2012; Kleit and Page 2015).

At the federal level and building on the experience of HOPE VI's ability to leverage private capital for investment in redeveloping public housing, the Quality Housing and Work Responsibility Act (QHWRA) of 1998 made what became known as mixed finance generally more available for all PHAs (Kleit, Airgood-Obrycki, Yerena 2019; Kleit and Page 2008). Mixed finance allows PHAs to use federal capital funds to leverage private investment for the redevelopment of public housing, often into mixed-income housing. That is, when a PHA seeks to revitalize a development, it is able to use capital funds as its contribution to the project or as security for loans. Often the ownership of the property after redevelopment includes a partnership of investors, with the PHA as the lead partner with private investors, either due to their involvement in the Low-Income Housing Tax Credit Program (LIHTC; see also the chapter authored by Kirk McClure) or due to other private capital investment. Private firms have an interest in such investment because they are seeking the sort of stable return such housing can offer (Kleit and Page 2015).

QHWRA also mandated that PHAs alter how they manage their stock. Rather than the previous centralized management and waiting lists for housing, PHAs had to implement project-based asset management and site-based waitlists that were more like the portfolio approach used by other private subsidized housing and market-rate rental housing (Schwartz 2015). Under such a portfolio approach, housing owners make choices concerning which properties are worth reinvestment and which are better off being sold off to create cost-efficiencies. Based on the recommendation of what became known as the Harvard Cost Study (Harvard University Graduate School of Design 2003), federal regulations that set the new operating subsidy amounts arose from a multivariate model based upon the operating costs of privately managed affordable housing with FHA mortgage insurance, on a project-by-project basis. Congress, however, has not been consistent in funding these operating subsidies (Kleit and Page 2008; Schwartz 2015). Nonetheless, these changes created a new operating environment for PHAs that has endorsed their engagement in the private market beyond Housing Choice Voucher program management (i.e., subsidies to rental housing owned by private landlords). An example of changed behavior under asset management is one PHA in the Pacific Northwest region of the U.S., where a PHA sold off scattered-site public housing because it proved to be too expensive to manage; this PHA planned to use the proceeds to construct additional units of subsidized housing (Kleit and Page 2015). In another example, the Columbus (Ohio) Metropolitan Housing Authority (CMHA) sold a 100-unit public housing building in a gentrifying neighborhood of Columbus to a private owner for $14 million; the goal was to use the proceeds to construct three times the number of units elsewhere. As CMHA's president commented about the sale, "with shrinking federal funds it is not efficient for a housing authority to own properties that require long-term capital investments" (Ferenchik 2016). The reasons offered for these decisions—the expense of housing maintenance and the possibility of building three times as many units—indicate that whether or not to maintain affordable housing units in a particular neighborhood is a strategic decision for these two PHAs given their missions and housing portfolio. In short, the introduction of asset management and mixed finance—private market methods—is enhancing PHAs' ability to use private market techniques to meet their missions and thus their hybridity. Whether such flexibility will lead to PHAs being social enterprises likely depends on other factors as well.

How Local Control and State Legislation Influence the Hybridity and Social Entrepreneurship of Public Housing Authorities

Although federally regulated due to the receipt of federal dollars both for public housing and for the management of the HCV program, PHAs are actually quasi-public, locally chartered entities with local commissions that govern them. The concurrent devolution of federal housing policy decision-making to local level decision-makers has enhanced PHAs' actions as hybrid housing organizations, defined as employing combinations of the public and the private market and nonprofit strategies to provide services (Czischke et al. 2012; Mullins et al. 2012; Nguyen et al. 2012; Skelcher and Smith 2015; Smith and Skelcher 2014). Furthermore, PHAs' engagement with private or nonprofit resources and strategies varies greatly, depending on local governance and decision-making. One hypothesis is that state enabling legislation is a key lever in the ability of local PHAs to develop and purchase affordable housing outside of the HUD-assisted stock, and thus they use market means to meet their missions, and act like social enterprises.

Diversity of Engagement With Affordable Housing

While funds received from HUD place housing authorities in the federal regulatory environment, PHAs vary widely in their engagement with state and local programs and governments and with respect to private sector investment for the provision of affordable housing. Some PHAs fit the stereotype of the large government bureaucracy under the strict control of the U.S. Department of Housing and Urban Development (HUD). For others, the strategy of using financial resources outside of those provided by HUD goes as far back as the 1930s. A relatively well-known example is the New York City Housing Authority; since the 1930s its projects have grown so that currently 59 percent of its housing units come from either state (37 percent) or city (22 percent) funds (Bloom 2012). Others began purchasing or developing housing outside of the HUD-assisted stock during a wave of funding instability for public housing in the 1980s (Kleit and Page 2015). As Kleit and Page (2015) discuss, PHAs in Washington State and Oregon began to purchase or build units of low-rent, private market housing with the goal of maintaining long-term affordability. Four of these PHAs—the Department of Housing Services of Washington County (OR), the Housing Authority of Portland, now called Home Forward (OR), the Housing Authority of Snohomish County (WA), and the King County Housing Authority (WA)—each own over 2,000 units of unsubsidized rental housing where the rents are at the lower end of the local rental housing market. Seven of the 13 PHAs in their sample purchase homes and renovate them, either to operate as unsubsidized low-cost housing or to sell to first-time homebuyers.

The variety of different ways PHAs engage with the private housing market and private financing are emblematic of what Saul Ramirez, the former Chief Executive Officer of the National Association of Housing and Redevelopment Officials (NAHRO), has described as a widespread change in the world of public housing authorities since the 1990s (Ramirez 2015). Distinct since 1998 is the inclusion of explicit private market resources to finance affordable, quality rental housing. Given a goal of maintaining the social good of affordable rental housing, with or without government subsidies, these decisions move PHAs toward not only hybridity but a social enterprise model. However, state enabling legislation governs the powers of PHAs, and thus their ability to take on such market activities.

The author's survey (2015) and review of varied data sources about public housing authorities nationally illustrate the variations among PHAs with regard to their engagement in both HUD-assisted affordable housing and private market activities. For most PHAs, public housing and vouchers remain their mainstay, although the numbers of units or vouchers managed may be relatively few (see Figure 15.1). The majority (58 percent) of housing authorities in the U.S. manage fewer than 250 units of some combination of public housing and vouchers, while another third of PHAs manage between 250 and 1,250 public housing units and vouchers. A relatively few housing authorities own or manage most public housing and vouchers. Only about 12 percent of housing authorities have 1,250 or more federally assisted units. Large PHAs, which comprise only 11 percent of all PHAs, own another 39 percent of all public housing and vouchers combined. Extra-large PHAs, with over 10,000 federally assisted units, each comprise only 1 percent of PHAs. However, these extra-large PHAs own or manage 34 percent of all public housing and vouchers. Thus, the majority of PHAs have fewer than 1,249 units of public housing and vouchers combined, but own only 18 percent of

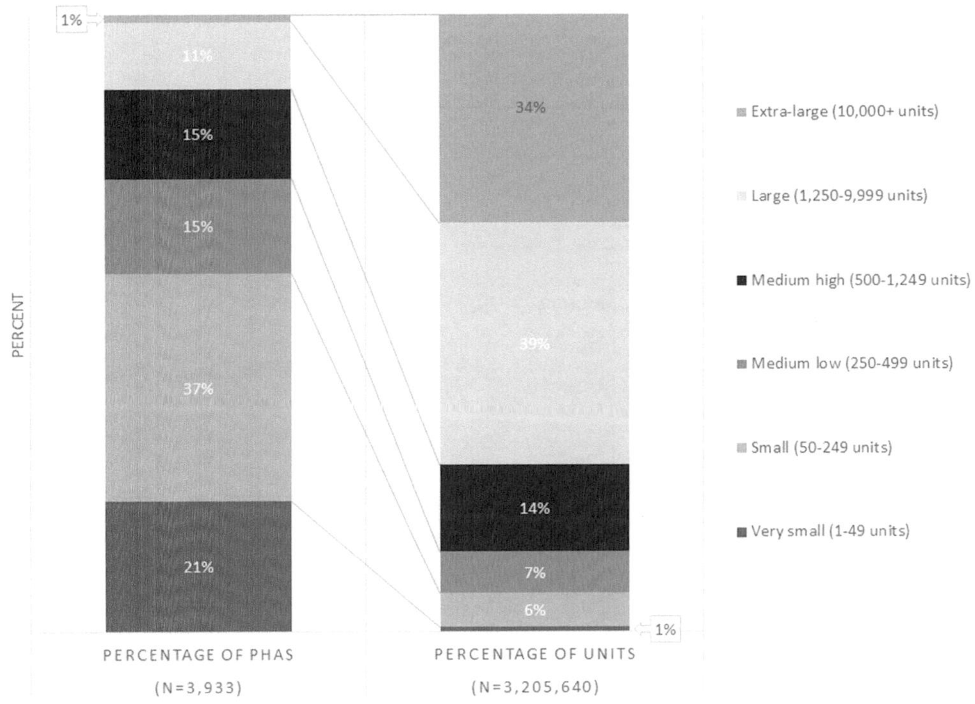

Figure 15.1 PHAs by Size and Percent of Units for that Size, Vouchers and Public Housing Units Combined.

Source: author, based on U.S. Department of Housing and Urban Development 2012, National Housing Preservation Database 2014, Public Housing Authorities Directors Association member records.

all units. The majority of PHAs do manage public housing: 41 percent only manage public housing, while another 37 percent manage both public housing and vouchers (see Table 15.1). Currently, just 18 percent of PHAs only manage vouchers, while 4 percent neither manage vouchers nor own public housing. These latter agencies tend be very small and usually own and manage senior housing.

PHAs vary in their tendency toward hybridity and activities that might indicate their becoming a social enterprise. Just 23 percent of PHAs nationally own housing that is not part of the HUD-assisted stock. These units may represent a step into both hybridity and social enterprise, as they not only arise from HOPE VI redevelopment or from state and local subsidy programs, but also may include LIHTC deals and lower-cost, market rental housing (Bloom 2008; Kleit and Page 2015). Others may engage in activities that are symptomatic of social enterprises by sponsoring an affiliate organization to conduct development. Of the 1,059 public housing authorities that responded to the author's survey of public housing directors in 2015, only 30 percent had such an affiliate organization. Sixteen percent of PHAs are integrated with local government—either as part of community development departments, or as part of offices of housing or redevelopment authorities. This integration provides some latitude in the ownership and construction of affordable housing, and can actually provide more flexibility in the ownership of non-federally subsidized housing. In their Pacific Northwest regional sample, Kleit and Page (2015) found that such integration diversified the financial resources available to a PHA and their ability to finance the ownership of affordable units outside the HUD stock. In total, the 1,059 PHAs in the survey report owning approximately 150,000 units of rental housing that are not part of public housing or project-based vouchers. Many of these are owned outright as non-subsidized, affordable rental housing by PHAs.

Table 15.1 Variations in Public Housing Authorities, 2015.

Activity		*Percent*
Among All PHAS (n = 3,933)		
Own public housing only		41
Manage vouchers only		18
Own public housing and manage vouchers		37
Neither manage vouchers nor own public housing		4
Own housing outside of the HUD-assisted stock		23
Among Sample of PHAs (n = 1,059)		
Stand-alone organization[1]		79
Integrated with local government[2]		16
Have affiliated organization that does development		30
Geographic scope of jurisdiction	Single jurisdiction	77
	County	17
	City-county	2
	Multi-county	2
	Multi-city	2
Estimated number of units outside of HUD-assisted stock		~150,880

Note: Figures are estimates.

Source: author, based on U.S. Department of Housing and Urban Development 2012, National Housing Preservation Database 2014, author survey of public housing authorities.

1 n = 1,049
2 n = 1,012

Policy Ambiguity May Enhance the Impact of State Enabling Legislation

Housing policy in the United States is inherently ambiguous and conflicted as a result of piecemeal legislation and fluctuating goals, as well as the devolution of policy control to localities (Kleit and Page 2008; Quercia and Galster 1997). Initially, federal housing policy emphasized the building of public housing units, and over time it shifted toward demand-side vouchers and scattered-site approaches to affordable housing provision (see also the chapter authored by Lawrence Vale and Yonah Freemark). With this transition came a greater focus on providing additional services and greater access to educational and employment opportunities to assisted tenants. As a result, PHAs have been increasingly expected to meet both housing and service needs for the lowest-income populations with insufficient federal resources (Quercia and Galster 1997), while the need for affordable housing has been concurrently increasing (Joint Center for Housing Studies of Harvard University 2016).

State Enabling Legislation's Potential Influence on PHAs Acting as Social Enterprises

Variations in response to the ambiguity and conflicts of federal policy depend, in part, on the state enabling legislation that governs the charter of each PHA. States used a basic template for their original legislation, and modified it for state ratification. Since the 1930s and 1940s, however, some states have amended their enabling legislation several times to alter the capacities of PHAs. Some parts of enabling legislation govern the commissions' operation, and, therefore, the PHAs' relationship with the locality (see Table 15.2). A content analysis of the enabling legislation for the 50 states demonstrates these variations. For example, in all the states surveyed, the commissions that govern PHAs are locally appointed and controlled. In some jurisdictions, the commission's composition overlaps completely with the city council; 29 percent of states allow commissioners to be employees or officers of the body that governs the jurisdiction where the PHA operates. Sixty-three percent mandate at least one tenant commissioner. In 33 percent of states, commissioners may be compensated. In 98 percent of the states, commissioners are appointed by the executive of the governing body or the mayor.

State legislation also governs aspects of PHA activity that influence innovation, collaboration, and the quality of the developments they control, all theoretically important to their ability to act as social enterprises. With regard to the geographic scope and cooperation across jurisdictions, 38 percent of states allow consolidated PHAs—public housing authorities that cooperate or combine across jurisdictions. The great majority (88 percent) allow county PHAs, while 42 percent allow regional PHAs. Enabling legislation also governs elements that contribute to vital public housing communities. For example, while nearly all states (92 percent) allow PHAs to include community facilities in the developments they build, 52 percent allow for the inclusion of commercial properties along with their housing.

For organizations to act as social enterprises, the ability to construct non-subsidized affordable housing alongside the HUD-assisted stock is essential. Fifty-eight percent of states allow PHAs to create a subsidiary or collaborate with other organizations to develop housing; all of these permit a nonprofit subsidiary, while 44 percent of states allow a for-profit subsidiary. These subsidiaries are private organizations that are not bound by public regulations and thus have a great deal of latitude in how they go about raising capital—through either mortgage debt or investment—for the construction of affordable housing. Additionally, a for-profit subsidiary could make a profit on development investments and pass that on to investors, making such an investment attractive to investors; some profits could also be invested in additional housing.

Without capital or subsidies, it is nearly impossible to build housing affordable to low-income households (Mallach 2009). The cost of construction exceeds the amount of income that would be received from affordable rental prices. For states and local governments, housing bonds may serve as

Table 15.2 Variations in Public Housing Authority Enabling Legislation.

Enabling Legislation Topic		*Number of States*	*Percent*
Annual report required to governing body about financial situation or housing portfolio		42	88
Geographic scope allowed	County	42	88
	Regional	20	42
	Consolidated (combine or cooperate with other HAs)	19	40
Allowable included uses	Community facilities	44	92
	Commercial properties	25	52
Affiliated organization	Create or work with organization	29	60
	Nonprofit	29	60
	For-profit	22	46
Commissioners	Commissioners appointed by executive of governing body or mayor*	47	98
	Commissioners can be compensated	16	33
	Commissioner can be employees or officers of governing body	14	29
	Tenant commissioner required	30	63
Raising funds	Governing body issues bonds, not HA	5	10
	Bonds sold at public or private sale	37	77
	Bond requires permission of governing body or holding a prior public hearing	11	23
	Ability to levy taxes	2	4
Projects receive permission from governing body or hold public hearing prior to undertaking project or acquiring land		20	42
Population targets	Allow higher or moderate income residents	27	56
	Rural preference	19	40
	Elderly preference	18	38
	Persons with disabilities preference	13	27
	Veteran/military preference	23	48
Different regulations for jurisdictions of different sizes, or including different regulations for specific jurisdictions		16	33

Note: Table shows variations in HA enabling legislation for the 48 contiguous states based on content coding.

Source: author's analysis.

* Massachusetts is the only state that has appointed city commissioners and elected town commissioners.

a flexible source of capital. Nearly all PHAs have the capacity to issue housing bonds; just 10 percent must use the governing body (that is, the local government in its jurisdiction) to do so. Whether the sale of the bonds is to private or public investors varies among states—77 percent allow either. Some PHAs have very limited bonding capacity: 23 percent of states require the permission of the governing body or a public hearing prior to issue. To the extent that a local government is not supportive of the PHA, this lack may hamper substantially the raising of capital.

In order to act as a social enterprise, PHAs need some flexibility in terms of the population served, and throughout the history of the public housing program, the focus on the poorest households has varied. While the focus began with housing for those households who needed just a little help to become firmly middle class, the activism of the 1960s saw emphasis turn to the poorest households (Marcuse

1998); today federal regulation allows considerably more variation in population (Schwartz 2015). With the trend toward mixed-income housing, enabling legislation becomes especially important: 55 percent of states allow PHAs to include high- or moderate-income residents along with their focus on low-income residents in the housing they own. State legislation also provides other population foci for each PHA's housing efforts: preferences for rural areas (40 percent), preferences for the elderly (38 percent), preferences for persons living with disabilities (27 percent), and preferences for veterans or people affiliated with the military (33 percent). Thus, while federal regulations shape the populations that a PHA may serve, additional mandates arise from state enabling legislation. Such mandates influence the mix of clientele that each PHA may serve and shape how each PHA is able to raise private capital creatively to reach social goals and thus act as social enterprises.

A New World for Public Housing Authorities

Over the past 40 years, as public housing has experienced reductions in funding, and varied federal initiatives have encouraged PHAs to divest themselves of federal assisted housing units, the future of public housing has become uncertain. As Goetz (2012) argues, HUD's programs have strongly promoted the privatization of public housing via inadequate subsidies for project-based public housing and a preference for mixed tenure and income redevelopment. Beginning with QHWRA, passed in 1998, federal law has encouraged PHAs to broadly to step out of the HUD-regulated public housing system, undertaking substantial demolition and redevelopment (mixed tenure and income) with HOPE VI federal grants or mixed finance, or shifting units into housing vouchers. These actions have enabled PHAs to garner new private investment, but have eroded the stock of permanently affordable public housing by more than 250,000 dwellings since 1993 (Goetz 2012; Schwartz 2015, see also the chapter authored by Lawrence Vale and Yonah Freemark).

The emphasis on leveraging the private market for the redevelopment and preservation of units continues with the Rental Assistance Demonstration (RAD) program, which Congress enacted in 2012 (National Housing Law Project 2015). RAD allows PHAs to leverage private investment to renovate public housing structures and to convert the subsidies on those units to long-term project-based subsidies under the Section 8 program, thus exiting the public housing program. As such, it is a new tool enhancing PHAs' ability to attract private investment in affordable housing.

Such investment has traditionally been difficult for PHAs to find. With its 1993 birth, the HOPE VI program provided much-needed capital that enabled receiving PHAs to leverage investment for the redevelopment of public housing sites. Outside of HOPE VI, however, PHAs had severe problems finding capital for renovation. Schwartz (2015) comments that the RAD program represents a new approach to mixed finance, allowing PHAs and not-for-profit housing agencies to use their capital and operating subsidies to provide 80 percent backing to take out private loans for housing renovation. RAD also allows the use of Low-Income Housing Tax Credits (LIHTC) (see also the chapter authored by Kirk McClure) as an additional source of capital for renovation. Participating in the RAD program allows for much higher investment and thus higher debt levels than previously available. The program has been very successful in obtaining new private investment and facilitating immediate renovation. Indeed, it is oversubscribed; by the end of December 2014, 60,000 units had been approved, another 120,000 were on a waitlist for conversion, and 185,000 units were approved for conversion (National Housing Law Project 2015). While RAD does represent a privatization of affordable rental housing, it also solves the problem of the provision of quality housing over the long run by tying private investment into the life of the development. Expectations for returns by investors create expectations on their part for quality. The resulting development is then owned by a limited-liability corporation (LLC) or limited-liability limited partnership (LLLP) that includes the PHA and private investors.

RAD takes advantage of the flexibility of the PHA model—that is, as PHAs are local, quasi-public entities RAD offers the possibility that PHAs can remain mission-driven organizations that own affordable housing but not outright and not as only public housing in its original form. As RAD transfers public housing units, which have had little political support, to long-term Section 8 project-based contracts that have been less vulnerable to funding reductions, in developments owned by LLCs or LLPs, the hope is that the subsidized units will be more sustainable in a more popular program (Cohen 2015). However, if the process ends up removing the best public housing from the public housing stock, the public housing units that remain could end up in the worst condition, eroding support for the program altogether (Cohen 2015). Nonetheless, this policy is treating PHAs as continued purveyors of affordable housing, and as such allowing them to grow into hybrid social enterprises that own housing renovated with private dollars, owned by a private corporation, and subsidized with public funds.

As this new world expands, PHAs may take on other roles depending on the local context (Kleit, Airgood-Obrycki, Yerena 2019; Kleit and Page 2008). One role is as a manager of HCVs—just 55 percent of all PHAs (see Table 15.1) currently manage the HCV program, which allows households to rent private market housing units and have a portion of their rent paid by the federal government. At this time, PHAs have the flexibility to project base up to 20 percent of their vouchers, based upon current needs in the local housing market. When public housing authorities choose to project base vouchers they are collaborating with the private sector in making choices to balance their engagement in the housing market based upon the local climate, which is the action of a social enterprise. The LIHTC also is an opportunity to leverage private capital for a social goal, although only a small number of housing authorities own LIHTC units. An examination of the National Housing Preservation Database shows that roughly 172 public housing authorities own LIHTC units in about 350 developments containing about 29,000 housing units under affordability restrictions. A portion of these, although not all, are certainly HOPE VI developments. In about half of these developments PHAs own the LIHTC units as would a nonprofit housing developer; private LLC corporations own the other half of these LIHTC units (often with the PHA as the managing partner). These private LLCs may be nonprofit or for-profit corporations. For many of these rental units, the manager is not the PHA, but rather a private, non- or for-profit manager. PHAs own units with other subsidies as well (see Table 15.3): roughly 24,000 under long-term Section 8 project-based contracts,[2] about 4,000 using FHA insurance, and about 9,500 through the Department of Agriculture's Rural Housing Service 515 program.

Innovation for PHAs is not simply occurring through development. Some PHAs are creating comprehensive service models and have become agencies that not only focus on affordable housing, but also support services for employment through either demonstration programs (Metropolitan

Table 15.3 Subsidies Other than Public Housing for Units Under PHA Ownership.

Subsidy Type	*Number of Developments*	*Number of Units*
Section 8 Project-Based Contract	526	23,745
Section 202 Elderly	4	164
FHA Insured	33	4,299
Rural Housing Service 515	89	9,507
HOME Investment Partnerships	42	724

Note: Totals may be undercounted due to inconsistencies in identifying public housing authority owners for nonpublic housing units and inaccuracies in the database.

Source: author, based on National Housing Preservation Database.

Housing and Communities Policy Center 2016) or the Move to Work (MTW) program's relief from federal regulations, freeing individual PHAs from HUD regulations to allow such flexibility (Kleit, Airgood-Obrycki,Yerena, 2019; Kleit and Page 2008, 2015).

Kleit and Page (2015) suggest that PHAs are potentially changing their missions as this new operating environment progresses. Case studies of PHAs in the Pacific Northwest region suggest that PHAs may follow five possible strategic directions, some of which would put them on the path toward hybridization and social entrepreneurship. For example, one strategic direction places primacy on the survival of the agency rather than a primary mission of serving the poorest households, which may cause PHA leadership to consider varied combinations of services and populations jointly with a second strategic direction of diversifying their housing stock, and, thus, may cause a shift away from serving the poorest of the poor.

Alternatively, other approaches (Kleit and Page 2015) discussed may provide diverse paths toward becoming social enterprises. For example, a third strategic direction might have PHAs focusing on a slightly less poor population by producing workforce housing and leveraging the equity of those units to provide housing for the poorest households sometime in the future; a fourth direction would be a focus on poverty alleviation, rather than housing construction and management; or a fifth strategic direction would involve integrating with other government agencies to provide more flexibility for the construction and maintenance of affordable housing. Kleit and Page (2015) note that these paths may occur within their case studies concurrently or solely, and represent the diversity of response to this new policy environment. These variations in path are emblematic of the lack of certainty about whether PHAs can become social enterprises.

Thus, housing authorities vary in their capacity to act as social enterprises, and such activities take place in a local context of housing markets and a local affordable housing industry. Uniform performance is certainly not a hallmark of PHAs—that is why HUD's Public Housing Assessment System (PHAS) and Section 8 Management Assessment Program (SEMAP) both track performance in the hopes of improving it in these two areas. While these innovations may be possible for high-performing agencies, limitations due to charter or capacity may mean that PHAs do not necessarily reach their potential. Regardless, to retain PHAs as providers of housing of the last resort for the very poorest households requires subsidies, either for hard units or to support household-based housing vouchers. The U.S. has experienced extraordinary growth in housing need over the past decade with both a lack of investment in affordable housing for the very poor and a political demise of public housing. Therefore, deliberately refocusing and supporting PHA's efforts on housing preservation and construction using combinations of federal, state, and local resources may be one under-utilized path for the provision of affordable housing in the U.S. Fostering their ability to be locally controlled social enterprises, with the appropriate enabling legislation, leadership, and funding, will allow PHAs to be more effective actors in the affordable housing delivery system.

Acknowledgement

The author would like to thank Whitney Airgood-Orbrycki for her able assistance in this research.

Notes

1 Quasi-public organizations are private corporations that have the backing of government.

2 Project-based Section 8 is distinct from project-based housing vouchers. In the case of project-based Section 8, usually a private for- or nonprofit owner has a long-term contract for federal subsidies that are attached to the units. In the cases discussed here, the PHA is the owner, rather than a private owner (see also the chapter authored by Lawrence Vale and Yonah Freemark)

References

Billis, David. 2010. *Hybrid Organizations and the Third Sector.* New York: Palgrave MacMillan.

Bloom, Nicholas Dagen. 2008. *Public Housing that Worked: New York in the Twentieth Century*. Philadelphia: University of Pennsylvania Press.

Bloom, Nicholas Dagen. 2012. "Learning From New York." *Journal of the American Planning Association*, 78(4): 418–431.

Bratt, Rachel G. 1989. *Rebuilding a Low-Income Housing Policy*. Philadelphia: Temple University Press.

Cohen, Rachel. 2015. Goodbye Public Housing? *The American Prospect*, November 12, n.p.

Corcoran, Joseph E. 1980. "King's Lynne: Public Housing Becomes Private Housing." *Urban Land*, 39(2): 6–12.

Czischke, Darinka, Vincent Gruis, and David Mullins. 2012. "Conceptualizing Social Enterprise in Housing Organizations." *Housing Studies*, 27(4): 418–437.

Epp, Gayle. 1996. "Emerging Strategies for Revitalizing Public Housing Communities." *Housing Policy Debate*, 7(3): 563–588.

Ferenchik, Mark. 2016. "CMHA Sells Senior Building." *Columbus Dispatch*, February 16, n.p.

Goetz, Edward G. 2012. "The Transformation of Public Housing Policy, 1985–2011." *Journal of the American Planning Association*, 78(4): 452–463.

Goody, Joan E. 1993. "From Project to Community: The Redesign of Columbia Point." *Places*, 8(4): 20–33.

Harvard University Graduate School of Design. 2003. *Public Housing Operating Cost Study. Final Report*. Cambridge: Harvard University Graduate School of Design.

Hays, R. Allen. 2012. *The Federal Government and Urban Housing*. Albany: State University of New York Press.

Hoffman, Andrew J., Krista K. Badiane, and Nardia Haigh. 2012. "Hybrid Organizations As Agents of Positive Social Change: Bridging the For-Profit and Non-Profit Divide." In *Using a Positive Lens to Explore Social Change and Organizations: Building a Theoretical and Research Foundation*, edited by Karen Golden-Biddle and Jane E. Dutton. New York: Routledge.

Joint Center for Housing Studies of Harvard University. 2016. *The State of the Nation's Housing 2016*. Cambridge, MA: Joint Center for Housing Studies of Harvard University.

Kleit, Rachel Garshick, Whitney Airgood-Obrycki., Anaid Yerena. 2019. Public Housing Authorities in the Private Market. *Housing Policy Debate*. 1-23.

Kleit, Rachel Garshick, and Stephen B. Page. 2008. "Public Housing Authorities under Devolution." *Journal of the American Planning Association*, 74(1): 34–44.

Kleit, Rachel Garshick, and Stephen B. Page. 2015. "The Changing Role of Public Housing Authorities in the Affordable Housing Delivery System." *Housing Studies*, 30(4): 621–644.

Mallach, Alan. 2009. *A Decent Home*. Chicago: APA Planners Press.

Marcuse, Peter. 1998. "Mainstreaming Public Housing: A Proposal for a Comprehensive Approach to Housing Policy." In *New Directions in Urban Public Housing*, edited by David P. Varady, Wolfgang F. D. Preiser, and Francis P. Russell, 23–44. New Brunswick, NJ: Center for Urban Policy Research. Rutgers University.

Metropolitan Housing and Communities Policy Center. 2016. *HOST: Housing Opportunity and Services Together*. Washington, DC: Urban Institute.

Mullins, David, Darinka Czischke, and Gerard van Bortel. 2012. "Exploring the Meaning of Hybridity and Social Enterprise in Housing Organizations." *Housing Studies*, 27(4): 405–417.

National Housing Law Project. 2015. *Rental Assistance Demonstration 2015*. San Francisco, CA: National Housing Law Project.

National Housing Preservation Database. 2014.

Nguyen, Mai Thi, William M. Rohe, and Spencer Morris Cowan. 2012. "Entrenched Hybridity in Public Housing Agencies in the USA." *Housing Studies*, 27(4): 457–475.

Quercia, Roberto G., and George C. Galster. 1997. "The Challenges Facing Public Housing Authorities in a Brave New World." *Housing Policy Debate*, 8(3): 535–569.

Ramirez, Saul N. 2015. "Managing Risk in Contemporary Community Housing Practice: National and International Perspectives." *Presented at the Australian National Housing Conference*, October 28–30, 2015. Perth, Australia.

Rosenbaum, James E., Linda K. Stroh, and Cathy Flynn. 1996. *Lake Parc Place: A Study of the First Four Years of a Mixed-Income Housing Program*. Evanston, IL: Working Paper. Center for Urban Affairs Research at Northwestern University.

Ryan, William, Allan Sloan, Maria Seferi, and Elaine Werby. 1974. *All in Together: An Evaluation of Mixed-Income Multi-Family Housing*. Boston: Massachusetts Housing Finance Agency.

Schwartz, Alex F. 2015. *Housing Policy in the United States*. New York: Routledge.

Skelcher, Chris, and Steven R. Smith. 2015. "Theorising Hybridity: Institutional Logics, Complex Organizations, and Actor Identities—The Case of Nonprofits." *Public Administration Review*, 93(2): 443–448.
Smith, Steven R., and Chris Skelcher. 2014. "Hybridity and Nonprofit Organizations: Next Questions." *The Nonprofit Quarterly*, Spring: 80–87.
Turbov, Mindy. 2006. "Public Housing Redevelopment As a Tool for Revitalizing Neighborhoods: How and Why Did It Happen and What Have We Learned?" *Northwestern Journal of Law and Social Policy*, 1(1): 167–201.
U.S. Department of Housing and Urban Development. 2012. A Picture of Subsidized Households 2009–2012. US Department of Housing and Urban Development Office of Policy Development and Research. 2012 based on 2010 Census.

16

HOW THE EUROPEAN COMMISSION AFFECTED SOCIAL RENTAL HOUSING IN THE NETHERLANDS AND GERMANY

Marja G. Elsinga and Marietta E.A. Haffner

Introduction

The European Commission (EC) aims to encourage an open market in Europe and therefore has a Directorate General focusing on fair competition in Europe. Social policies are the responsibility of its 28 member states, not the EC (Elsinga et al. 2008; König 2015). Therefore, policies toward social housing and housing allowances schemes are the responsibility of the member states. In the field of housing, however, there appears to be a tension between the EC's aim for open markets and fair competition and social policies concerning housing. After all, housing is a field that is part of the global real estate markets and subject to open competition within the European Union (EU), and, therefore, a responsibility of the European Commission. At the same time, social housing is the responsibility of member states.

The EC aims for fair competition between social and private landlords and developers in providing rental housing to residents, and unintendedly has had quite an impact on social housing systems. In this chapter, we discuss how a competition policy meant to encourage open markets in the EU has influenced social housing models in Europe. We focus on Germany and the Netherlands and how EC policies have impacted social housing models in these countries. We discuss the roots of these two social housing models and explain how discussions regarding EC policies have played out in the two countries. Finally, we discuss how changes initiated by the EC have impacted affordable housing in the two countries. Although the right to adequate and affordable housing is a human right recognized by the European Commission, EC directives have had some adverse unintended consequences.

How Did the Legislation by the European Commission on Fair Competition Impact Social Housing?

One of the main EU objectives has traditionally been to stimulate competitive markets and trade between and within member states. To achieve this objective, the EC has developed market regulations, which in principle prevent national governments from distorting competition, for example, by giving social landlords unfair advantages, called "State aid" in EC terminology. State aid must be targeted primarily toward public and social services that would not be delivered adequately under regular market operations and that concern public objectives.

The EC (European Commission 2011) distinguishes three categories of public and social services: (1) services of general interest (SGI), for example, health care; (2) services of general economic interest (SGEIs) such as social housing; and (3) other services (non-SGEIs) such as rental dwellings provided by institutional investors. At first sight, this distinction may look clear, but categories 1 and 2 are social and category 3 is commercial and, thus, not the subject of State aid.

Some of the Dutch and German landlords who provide social or subsidized housing also provide commercial housing for middle-income households to promote socially mixed neighborhoods that encourage the social inclusion of vulnerable households. And profits of social housing organizations are reinvested in building new social housing units for a broad target group (Committee on Employment and Social Affairs 2013; European Commission 2011). Serving the middle-income market, along with the opposition to doing so by some landlords and politicians in the Netherlands, is one of the reasons why the EC asked the Dutch government to clarify whether the landlords that receive State aid are only involved in traditional social housing production or whether they are involved in commercial housing as well.

In recent years, policymakers have been engaged in an extensive debate concerning the need for a mixed population in social housing. For example, Karima Delli (European Economic and Social Committee 2012), a member of the European Parliament, believes that social housing should include not only vulnerable households, but also middle-income ones as well (Committee on Employment and Social Affairs 2013). Furthermore, the Committee on Employment and Social Affairs (2013) advocated for a broad definition of social housing, including low-income and middle-income renters. As indicated earlier, however, many landlords and their advocacy groups oppose such mixing, thus, illustrating the clash between competition policy and social policy. The remainder of this chapter focuses on how this clash, initiated at the EC level, has affected social housing models in the Netherlands and Germany.

The Dutch Case

History of Subsidized Housing

EU legislation has influenced the vibrant history of the Dutch housing system. The liberal-led government intervened in the housing market with the Housing Act of 1901, establishing a system enabling registered private nonprofit housing providers or organizations, called housing associations or social landlords, to obtain financial assistance from the government. After World War II, large subsidies and strict rent regulations were introduced to increase housing production in order to alleviate the enormous housing shortage. These policies and regulations led to an increase in the social rental stock from around 10 percent in 1945 to 42 percent in 1986, ultimately engendering low housing costs that contributed to making the country competitive. The result was a broad social housing model for lower- as well as middle- and even higher-income groups. This model had been growing since World War II, but has declined in recent decades, leading to a decrease in the social rental stock from 42 percent in 1986 to 32 percent in 2016.

This decrease had been caused by a reduction in supply-side subsidies and rent regulations and by an attempt to improve the financial performance of housing associations over the past two decades. In 1995, several directors of housing associations and academic experts in social housing suggested making housing associations financially independent (Elsinga and Lind 2013; Haffner et al. 2009). This proposal was attractive to associations since it would allow them to operate with fewer constraints; it was attractive to the government, because it meant reduced subsidies and, thus, a relief for the government budget. The proposal was important since reducing the financial burden of housing loans was a smart way to meet the financial criteria of the EU in 1992 in preparation for the introduction of the common currency of the euro a decade later.

This proposal implied a huge departure from a rather government-dependent sector to an independent social housing sector with little government influence. Once this proposal was implemented, the main form of national public financial support left for the housing associations was access to the Dutch Social Housing Guarantee Fund (WSW), a private revolving fund backed by the government and funded by fees paid by housing associations. At the national level, this fund provides a guarantee to lenders that provide low-interest-rate loans to housing associations, but *not* to their for-profit competitors. At the local level, governments supported social housing by subsidizing the price of land for social housing associations.

For decades, coalition governments have come and gone. Parties that lean to the right, for example, liberal parties or Christian Democrats, typically want to reduce the size of the social housing sector or prefer little government interference, although they support housing associations, while the Social Democratic party typically sees a stronger role for the government. Although much research on social housing has been published, there is no consensus on a vision for the social housing sector.

The reason for the debate about social housing is the ambiguous position of housing associations, i.e., "private" organizations with a public mission. A key issue is the question of whether housing associations operate efficiently enough and whether there is adequate public supervision of the social rental sector. Interestingly, the debate about the Dutch social rental sector started long before the EC opened the discussion about the link between social and commercial housing providers.

Social Housing: State Aid and the Competition Discussion

The EC discussion on State aid and fair competition started in July 2005 with a letter from Loretta Dormal Marino, Director of State Aid at the Directorate General Competition of the EC, to the Dutch Minister of Housing. She considered the practice of renting social housing units to higher-income groups as unfair to unsubsidized for-profit housing associations (Elsinga et al. 2008; Gruis and Priemus 2008). In 2007, the umbrella organizations of institutional investors in real estate submitted a complaint to the EC about unfair competition by housing associations, arguing that housing associations did receive State aid for units provided to middle-income households, whereas institutional investors did not.

Related to the discussion on competition, there were two other ongoing debates in the EU Parliament. The first debate was about the question of whether the EC should be involved at all in the way member states oversee the targeting of housing association units to particular types of households. The second debate was about mixed-income neighborhoods and the question of whether public funds should be targeted to the neediest groups or whether some subsidies should go directly or indirectly to middle-income households in order to achieve social mixing.

On December 15, 2009, the EC concluded that State aid had indeed been used for social housing and that government support should be targeted to social housing that benefited lower-income groups (European Commission 2009). This decision became effective on January 1, 2012. There was, however, some compromise allowing middle-income groups to live in social housing to a limited extent. "Social housing" was defined as a rental unit with a rent up to €665 per month. Specifically, 90 percent of these social housing units should be allocated to households with an annual income below €34,000, leaving 10 percent of social housing units for households with a higher income with some kind of urgent need, such as health problems. If a housing association does not meet the 90 percent criterion, it is not eligible to receive a loan guarantee from the WSW.

The EC accepted this compromise, because its policymakers considered socially mixed neighborhoods a valid policy aim, while at the same time, the 90 percent requirement would approximate and ensure an equitable level playing field; i.e., housing associations would focus on low-income people, while the private market would concentrate on the middle-income. In January 2012, the Netherlands began implementing the EC decision of temporary regulations for services of general economic interest (SGEI). Dutch experts who had been critical of housing associations for appealing to all income groups welcomed this new regulation.

The Impact of the Temporary Regulations for Services of General Economic Interest (SGEI): Targeting and Market Prices

Interestingly, after the EC decision the Dutch Parliament investigated Dutch social housing associations in 2013 and 2014 in relation to corruption allegations. Also, the Dutch Parliament concluded that the Dutch social housing system needed more regulation, more targeting toward low-income groups, and more market rents. Together, the tensions between these directives posed a tremendous challenge to housing providers.

Moreover, in the aftermath of the financial and economic crisis, the Dutch government searched for solutions to reduce the national deficit. Since there was broad support for reducing social housing across the political spectrum, perhaps surprisingly, a majority of members in the Dutch Parliament supported a tax on affordable housing paid by the housing associations. This tax was considered a way for the national government to obtain revenue while at the same time reducing the volume of social housing.

Thus, as a result of this decision housing associations were forced to target lower-income groups in order to attain a fair playing field. The tax encourages housing associations to increase rents and to sell units at market prices in order to remain financially viable, a regular activity that has occurred since the 1990s, resulting in about 10,000 sold units each year (Elsinga and Wassenberg 2014).

The State aid debate has continued to fuel the Dutch discussion on social housing, i.e., whether to emphasize social mixing versus market efficiency, including the parliamentary inquiry, which has led to additional policies making social housing more efficient. Ironically, the social rental stock is now smaller, and lower-income groups pay higher rents than before the discussion, thereby making the affordability problem much more severe than ten years ago.

The German Case

History of Subsidized Housing

The EU has not affected the structure of the supply-side subsidy system in Germany, since its system was not intended to distort competition between social and private landlords (Bundesrepublik Deutschland n.d., Cornelius and Rzeznik n.d.), an outcome achieved by West Germany opting for a social market economy philosophy after World War II. According to this philosophy, social welfare is achieved by a strong role of the market, but with government intervention designed to support market forces (Busch-Geertsema 2000, 2004; Haffner et al. 2009; Haffner 2011, Kofner 2017, based on Oxley 2000). Social market policies, therefore, were temporary as well as actor neutral in order to prevent distortions in the behavior of the actors.

For housing policies these principles were implemented by designing a supply-side subsidy system for investors in rental and owner-occupied dwellings. The system of supply subsidies was introduced in 1956 and has existed in various forms since then, based on the needs identified by the national government for those subsidies (Haffner et al. 2009; Oxley et al. 2015). Supply subsidies are available to any organization or individual regardless of whether they are commercially oriented (i.e., the private sector) or aim for social goals (i.e., the municipal housing companies).[1] Regardless of whether they are profit oriented and/or subsidized or not, all investors in rental dwellings have a private legal status, not a public one (Kofner 2017, 63; see also Haffner et al. 2009).

Eligible recipients of supply subsidies must guarantee a certain rent or price level for subsidized rental or owner-occupied housing, respectively, must follow unit allocation rules, and must agree on a period that the subsidy scheme will run, which may last for several decades, but shortened more recently (see also the chapter authored by Rachel Bratt). As long as the unit is subsidized, it is regarded as a social housing unit, as the subsidy regulates setting rent levels below the market rate. Social rental units, however, owned by a special landlord type common in many other European countries, including the Netherlands, do not exist in Germany. When the subsidy period ends, the unit is considered as part of the regular private rental housing stock.

Supply-side subsidization was introduced based on the assumption that eventually subsidies could be withdrawn. Germany was well on its way as it moved into "a long period with predominantly relaxed urban housing markets" (Kofner 2017, 67; see also Deilman and Effenberger 2017; Haffner et al. 2009). The building boom of the 1990s following the reunification of former East and West Germany, in combination with slow population and household growth, resulted in a national oversupply of units. Thus, in the 1990s total dwelling production decreased to the lowest point since World War II, indicating that supply-side subsidies had been successful (Bundesamt für Bauwesen und Raumordnung 2010; Bundesregierung 2009).

National, state, and local governments[2] influence criteria for subsidization of social housing units. These criteria have changed over time, however. In 2006, the national government delegated the responsibility for the design of the supply-side subsidization to the (federal) states, allowing more flexibility, since some states have grown while other states have lost population (Bundesministerium für Verkehr, Bau- und Wohnungswesen 2001; Bundesregierung 2009; Haffner et al. 2009; Haffner 2011; Kofner 2017; Kholodilin 2017; Oxley et al. 2010, 2015; Westerheide 2011; Whitehead et al. 2016).

Kofner (2017, 66) regards this evolution as part of the change from "classical" to "modern" policies where the program design and regulation are now in the hands of the (federal) states instead of the national government. Decreased subsidies are now narrowly targeting low-income and other vulnerable groups, while in the past, they were much less focused on these groups. Subsidized rent levels used to be lower than market levels, but are now close to market levels. In the past, most participating developments were large-scale and located outside of city centers, while nowadays most developments are smaller and scattered throughout cities. In the early days, the negotiated subsidy periods were about 45–50 years, while now they range from 10 to 25 years. Finally, municipalities typically prefer subsidizing an investor/landlord who provides an existing unit rather than a developer who builds a new unit (Kofner 2017).

The desired flexibility across (federal) states was achieved. Whereas some states had taken advantage of subsidized loans from the central state-owned bank (Kreditanstalt für Wiederaufbau) in combination with their own financing, others had used federal government funds for other allowable purposes, such as building modernization (Kofner 2014). However, new construction failed to counterbalance subsidized units dropping out of the subsidized system, which resulted in a still relatively large private rental sector with 44 percent of units. The proportion of subsidized rental units decreased from about 20 percent of the housing stock in the 1980s to less than 5 percent in 2010 (Haffner 2011; based on Bundesamt für Bauwesen und Raumordnung 2007; see also Cornelius and Rzeznik n.d.; Kofner 2014). Kofner asserts that 3.3 percent of dwellings were subsidized rental dwellings, but fails to provide a date (2017, 62). Because of the oversupply of units, the decline in the number of subsidized rental units resulted in affordability problems in many large growing metropolitan areas, although not elsewhere (Bundesregierung 2009; Kofner 2017).

No Impact of EU Regulations, but Targeting and Market Forces Play a Role

Even though EU State aid regulations did not affect Germany's supply-side subsidy system, as was the case in the Netherlands with its social housing, a decline in subsidized housing did take place (Cornelius and Rzeznik n.d., Kofner 2017). Policymakers designed the subsidy system as a temporary intervention. In the late 2000s/early 2010s, high vacancy rates in some parts of Germany led to cutbacks in subsidies, causing a decrease in new construction. Since around 2010, the larger growing German cities like Berlin have experienced housing shortages and affordability issues (Bundesministerium für Umwelt, Naturschutz, Bau und Reaktorsicherheit 2015; Federal Institute for Research on Building, Urban Affairs, and Spatial Development 2017). Thus, the national government extended the program that ran from 2007 to 2013, contributing €518.2 million per annum to (federal) states from 2014 to 2019, in order to deal with the shortage of affordable housing (Kholodilin 2017; Kofner 2017; Oxley et al. 2015). This contributing amount was increased to €500 million after 2016 (Bundesministerium für Umwelt, Naturschutz, Bau und Reaktorsicherheit 2017). In addition, the national government increased annual funding for housing aid to asylum seekers and immigrants by another €500 million to reach a total of €1.5 billion after 2017 (Kofner 2017). Reversing the decision to stop subsidizing social rental housing in 2014, and increasing funding for asylum seekers and immigrants in 2016 and 2017, respectively, shows that a temporary supply-side system needs to be flexible in response to changing needs.

Conclusion

Implementing EU legislation on fair competition between providers of social rental housing and products on the housing market has had a huge impact on social housing in the Netherlands, while it has had no impact in Germany. These different impacts can be attributed to the different housing policy frameworks in these two EU member states. In the chapter, we discussed the differences between the two national housing policy frameworks and the larger question of how the European Commission interacts with its member states.

The Dutch housing system was originally set up to intervene in the rental market in order to relieve the housing shortage. The Dutch government created a subsidized social rental housing stock owned by social housing associations and municipalities. In contrast to Germany, there were two rental sectors: a social and a commercial one. Over time, the Dutch social rental sector increased much more than the commercial rental sector. Since social housing providers became financially independent in 1995, they have been increasingly considered as competitors of the commercial rental sector. A European court decreed that Dutch social rental housing constituted unfair competition with respect to commercial providers of rental housing. The Dutch ministries of finance and economics offered the same critique on social housing. As a result, social rental housing aimed at a broad target group lost the broad support it once had and was traded in for broad political support for the new Housing Act that aims to decrease and target the social stock to lower-income households.

The result was the end of the broad social housing model—as it had been generally known—in the Netherlands. Since 2011 social rental housing has only been available for lower-income groups—not for all income groups as had been the case in the past. The question is whether the EC regulation or the debate in the Netherlands, or some combination of the two, was the main driver for the switch toward social rental housing targeted to low-income households.

Germany's federal postwar supply-side subsidy system is based on the social market economy philosophy that advocates for government intervention supporting the efficient operation of market forces, allowing the government to intervene in the rental market through temporary subsidies, thus, creating social rental housing units. In this case, a landlord or an investor signs a contract with a municipality in order to provide subsidized rental housing for a limited period. This system does not result in the creation of a social rental sector as in the Netherlands. During the subsidy contract period, landlords or investors agree to provide units with rents below the market rate to households subject to an income limit, and in return receive a subsidy. Therefore, these subsidized rental dwellings fulfill the same role as social rental housing in other countries. Once the subsidy contract ends, landlords or investors no longer receive a subsidy, and the dwellings shift back to the private market.

The German social housing system has been consistent with EU regulation as it succeded in creating a level playing field where social housing is provided by a landlord or investor willing to accept the conditions of the subsidy programs. In sum, Germany has had a sufficient number of landlords and investors willing to provide subsidized rental housing in order to establish and maintain its own type of social housing system, until recently.

We conclude that interaction between the EC and member states has two sides. The EU imposes new regulation on member states, but at the same time member states or advocacy groups, such as landlords, can use EU regulations to create a breakthrough in national discussions.

The result of this competition debate initiated by the European Commission is that both the Dutch and the German case demonstrate that a level playing field between social and commercial rental housing has not prevented, but rather increased housing affordability problems in both countries.

Notes

1 The German model resembles the U.S. approach to subsidized housing in the 1970s and 1980s.

2 The German governmental system consists of the national or federal government, the (federal) state governments of 16 (federal) states, and local governments, also called municipalities (Haffner et al. 2009).

References

Bundesamt für Bauwesen und Raumordnung. 2007. *Wohnungs- und Immobilienmärkte in Deutschland 2006.* Bonn: Bundesamt für Bauwesen und Raumordnung.

Bundesamt für Bauwesen und Raumordnung. 2010. *Wohnungsmärkte im Wandel: Zentrale Ergebnisse der Wohnungsmarktprognose 2025.* Bonn: Bundesamt für Bauwesen und Raumordnung.

Bundesministerium für Umwelt, Naturschutz, Bau und Reaktorsicherheit. 2015. *Wohngeld-und Mietenbericht 2014.* Berlin: Bundesministerium für Umwelt, Naturschutz, Bau und Reaktorsicherheit, 2015.

Bundesministerium für Umwelt, Naturschutz, Bau und Reaktorsicherheit. 2017. *Kurzinfo Wohnraumförderung. Die Wohnraumförderung der Bundesregierung.* Berlin: Bundesministerium für Umwelt, Naturschutz, Bau und Reaktorsicherheit.

Bundesministerium für Verkehr, Bau- und Wohnungswesen. 2001. *Der soziale Wohnungsbau wird reformiert.* Bonn: Bundesministerium für Verkehr, Bau- und Wohnungswesen.

Bundesregierung. 2009. *Bericht über die Wohnungs- und Immobilienwirtschaft in Deutschland.* Berlin: Bundesregierung.

Bundesrepublik Deutschland. n.d. *Bericht der Bundesrepublik Deutschland nach Art. 8 der Entscheidung der Europäischen Kommission zur Freistellung von Diensten von allg. wirtschaftlichem Interesse.* Berlin: Bundesrepublik Deutschland. http://ec.europa.eu/competition/state_aid/public_services/2009_2011/germany_de.pdf

Busch-Geertsema, Volker. 2000. *Housing Policy in Germany.* Bremen, Universität Bremen.

Busch-Geertsema, Volker. 2004. "The Changing Role of the State in German Housing and Social Policy." *European Journal of Housing Policy,* 4(3): 303–321.

Committee on Employment and Social Affairs. 2013. "On Social Housing in the European Union (2012/2293(INI) A7–0155/2013." *European Parliament.* https://www.eesc.europa.eu/sites/default/files/resources/docs/ep-report-on-social-housing-in-the-european-union.pdf.

Cornelius, Julia, and Joanna Rzeznik. n.d. *National Report for GERMANY.* Bremen: Universität Bremen. https://www.tenlaw.uni-bremen.de/reports/GermanyReport_09052014.pdf

Deilman, Clemens, and Karl-Heinz Effenberger. 2017. "Germany." In *Sustainable Communities and Urban Housing. A Comparative European Perspective,* edited by Montserrat Pareja-Eastaway and Nessa Winston, 225–244. New York: Routledge.

Elsinga, Marja, Marietta Haffner, and Harry Van der Heijden. 2008. "Threats to the Dutch Unitary Rental Market." *European Journal of Housing Policy,* 8(1): 21–37.

Elsinga, Marja, and Hans Lind. 2013. "The Effect of EU-Legislation on Rental Systems in Sweden and the Netherlands." *Housing Studies,* 28(7): 960–971.

Elsinga, Marja G., and Frank A. G. Wassenberg. 2014. "Social Housing in the Netherlands." In *Social Housing in Europe,* edited by Kathleen Scanlon, Christine Whitehead, and Melissa Fernandez Arrigoitia, 25–42. West Sussex: Wiley-Blackwell.

European Commission. 2009. *State Aid No. E 2/2005 and N 642/2009: The Netherlands Existing and Special Project Aid to Housing Corporations.* Brussels: European Commission.

European Commission. 2011. *A Quality Framework for Services of General Interest in Europe*. Brussels: European Commission.

European Economic and Social Committee. 2012. *Opinion of the European Economic and Social Committee on Issues with Defining Social Housing as a Service of General Economic Interest (Own-Initiative Opinion)*. Brussels: European Economic and Social Committee.

Federal Institute for Research on Building, Urban Affairs and Spatial Development. 2017. "Housing Deficits Causing Considerable Rent and Price Increases in Many German Cities and Regions." *Research News*, 1(June): 7.

Gruis, Vincent, and Hugo Priemus. 2008. "European Competition Policy and National Housing Policies: International Implications for the Dutch Case." *Housing Studies*, 23(3): 485–505.

Haffner, Marietta E.A. 2011. *Secure Occupancy in Rental Housing: A Comparative Analysis. Country Case Study: Germany*. Delft: Delft University of Technology, OTB Research Institute for the Built Environment.

Haffner, Marietta, Joris Hoekstra, Michael Oxley, and Harry Van der Heijden. 2009. *Bridging the Gap between Market and Social Rented Housing in Six European Countries*. Amsterdam: IOS Press BV.

Kholodilin, Konstantin A. 2017. "Quantifying a Century of State Intervention in Rental Housing in Germany." *Urban Research & Practice*, 10(3): 267–328.

Kofner, Stefan. 2014. *The Private Rental Sector in Germany. OECD Research on Private Rental Sector: Consultancy Report*. Germany: N.P.

Kofner, Stefan. 2017. "Social Housing Germany: An Inevitably Shrinking Sector?" *Critical Housing Analysis*, 4(1): 61–71.

König, Christian. 2015. "National Housing Policy: An Issue for Prohibited State Aid." *Housing Finance International*, 29: 24–26.

Oxley, Michael. 2000. *The Future of Social Housing: Learning from Europe*. London: IPPR.

Oxley, Michael, Ros Lishman, Tim Brown, Marietta Haffner, and Joris Hoekstra. 2010. *Promoting Investment in Private Rented Housing Supply: International Policy Comparison*. London: Department for Communities and Local Government.

Oxley, Michael, Connie Tang, Colin Lizieri, Nick Mansley, Daniel Makic, Marietta Haffner, and Joris Hoekstra. 2015. *Prospects for Institutional Investment in Social Housing*. London: IPF.

Westerheide, Peter. 2011. "The Private Rented Sector in Germany." In *Towards a Sustainable Private Rented Sector: The Lessons from Other Countries*, edited by Kath Scanlon and Ben Kochan, 45–59. London: London School of Economics.

Whitehead, Christine, Kath Scanlon, Sarah Monk, and Connie Tang, with Marietta Haffner, Jens Lunde, Marc Lund Andersen, and Michael Voigtländer. 2016. *Understanding the Role of Private Renting—A Four-Country Case Study*. Cambridge and Copenhagen: University of Cambridge and Boligokonomisk Videncenter.

SECTION 6

Senior Housing

17

CONNECTIVITY AS AN INDICATOR OF OLDER PEOPLE'S HOUSING QUALITY

Stephen M. Golant

Introduction

Why do some older people, typically age 60 or age 65 and older, have happier, healthier, more independent, and more engaged lives, that is, age more successfully than others? Their different life experiences, demographics, socioeconomics, and personalities obviously are crucial, but a large body of research emphasizes that the quality of their residential environments also helps to explain these variations (Golant 2015a). Older people age optimally when they occupy dwellings, neighborhoods, and communities with physical and social environments that are more compatible with their lifestyles and supportive of their mental and physical functioning (Golant 2015a). In these places, older people achieve "residential normalcy"; that is, they report having pleasurable, appealing, hassle-free, stimulating, and memorable housing experiences, and they also feel competent, in control, and empowered (Golant 2011, 2015b).

The majority of older people report having such favorable experiences in their current homes and apartments. Consequently, they prefer to age in these places and are "reluctant to move from familiar settings where they have strong emotional attachments and social ties" (Golant 2016, 43). This is the case even when they experience life-changing events such as retirement, the death of a spouse, or declines in their health.

Their residential relocation patterns appear consistent with these preferences. In the United States, older people, especially the 79 percent who are homeowners, move less frequently than any other age group. Over a one-year period (2015–2016), only 3.2 percent moved (1.6 percent of homeowners and 10 percent of renters). Moreover, almost 62 percent of these residential moves were within the same county (U.S. Bureau of the Census 2016).

Older people in the United States are especially reluctant to transition to any residential arrangement they perceive as having institutional qualities, such as nursing homes and other long-term care facilities. Consequently, 93 percent of older people (age 65 and older) now age in place in ordinary homes and apartments. Even 77 percent of those in their mid-80s or older remain in their communities (Federal Interagency Forum on Aging-Related Statistics 2016).

But these residential inertia preferences and behaviors do not always produce positive outcomes (Golant 2008). Based on professional assessments, a significant proportion of older people occupy inappropriate housing arrangements. Their dwellings are unaffordable, require multiple repairs, and have physical features (e.g., stairs, slippery and cluttered floors, high cupboard shelves, and absent grab bars) that are incompatible with their mobility limitations. They occupy physically inhospitable and unsafe neighborhoods that are plagued with social disorders, such as crime, drug use, vandalism, and squatters (Kerr et al. 2012; Joint Center for Housing Studies of Harvard University 2013). These indicators suggest that when older people age in place, they do not necessarily age in the right places (Golant 2015a).

Unmet Connectivity Needs

This chapter focuses on older adults who have yet another category of housing-related problems. They have trouble accessing or connecting to activities, people, goods, services, and care in their communities. Consequently, they are not satisfying their obligatory needs for consumer items such as food, health, and long-term services and supports, or their discretionary wants for leisure and recreation, social relationships, religious attendance, and volunteering (Caro et al. 2010; Hayutin et al. 2010).

Residential normalcy is illusive when older people confront these connectivity barriers because they feel that their housing arrangements have split personalities. Although they still derive pleasure and enjoyment from their familiar physical and social surroundings, they feel incompetent and out of control because they cannot easily realize their everyday needs and wants (Golant 2015b).

Older people throughout the world confront these connectivity challenges. In its mission to transform cities globally into "age-friendly communities," the World Health Organization (WHO) identifies connectivity as a key strategy to support the "active aging" of older people, enabling them to optimize their health, independence, social engagement, and security (Golant 2014). To promote and implement its age-friendly principles, the WHO offered interested communities an extensive set of planning guidelines (World Health Organization 2007).

The literature makes it clear that older people can rely on multiple connectivity strategies (Figure 17.1), each with advantages and disadvantages. This chapter reviews the relative merits of these alternatives, an analysis complicated by older people's diverse lifestyles and capabilities and their occupancy of places with different connectivity opportunities.

Driving Everywhere: Is Connecting Really a Problem?

Auto Driving Dominates Despite Risks

The majority of older people rely on auto transportation. In the United States, about 82 percent of those age 65 and over had driver licenses in 2013 (National Center for Statistics and Analysis 2015), and they made more than 90 percent of their trips by car (Rosenbloom 2013).

However, auto transportation is less available to some groups of older people. Women are less likely to drive than men mainly because they never learned earlier in their lives, although this gender gap will narrow as the later-born baby boomer women enter old age. Women who live alone, especially in their 80s, however, are particularly dependent on others for their travel needs, and they typically have lower incomes, often making the operation of a car financially prohibitive (Rosenbloom 2013).

Racial and ethnic older minorities (e.g., Blacks, Asians, and Hispanics), especially recent immigrants to the U.S. and those with lower incomes, are also more likely to be carless (Rosenbloom 2013). This is especially troublesome because the older immigrant population (mainly Hispanics and Asians)—both current residents and new arrivals—will constitute a higher share of the future U.S. old (Federal Interagency Forum on Aging-Related Statistics 2016).

Persons at higher chronological ages (80s and older) make fewer trips by car and are more likely to travel as passengers (Turcotte 2012), because they have physical and cognitive impairments that make driving impossible, difficult, or more dangerous (American Association of State Highway and

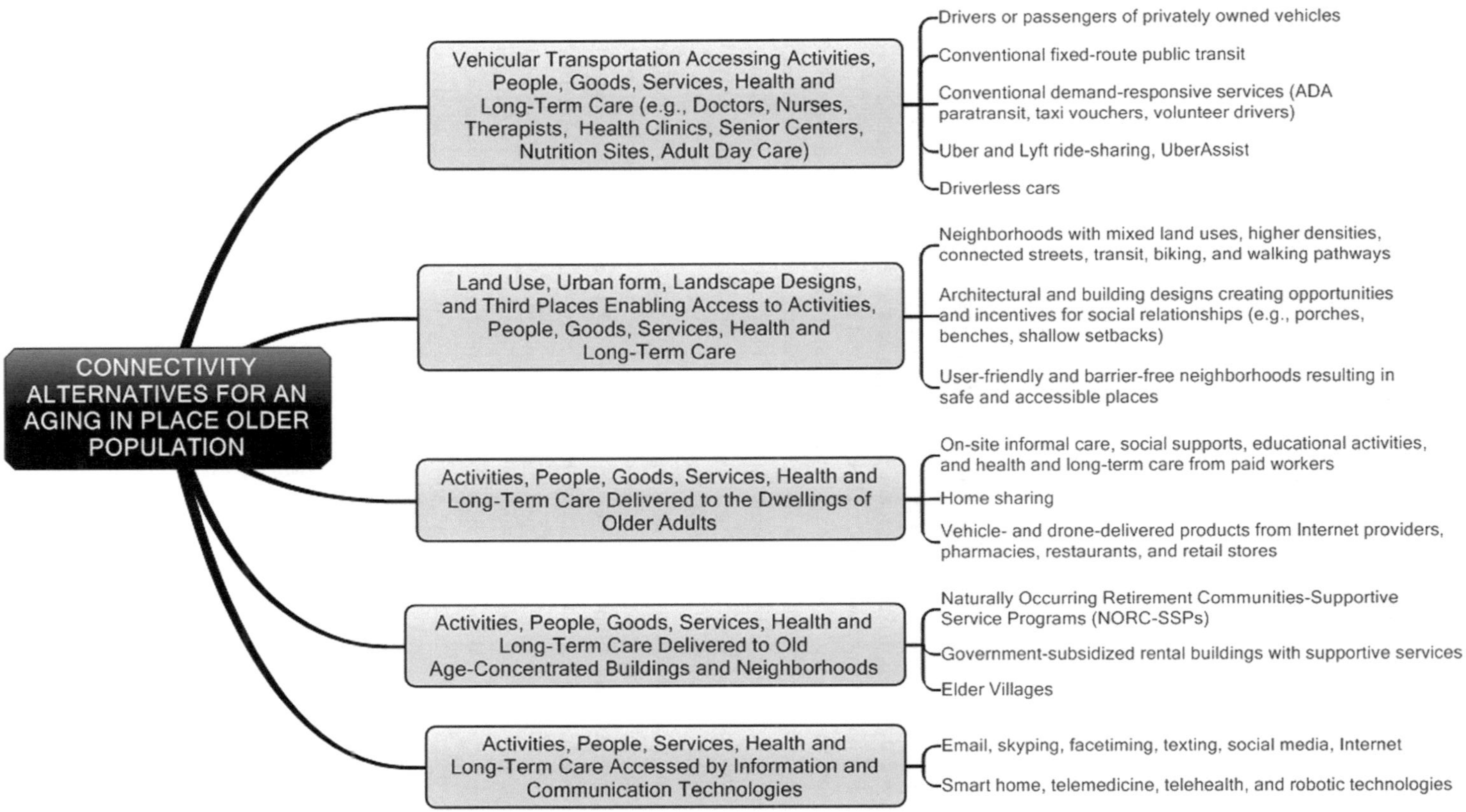

Figure 17.1 Connectivity Options for the Older Population Aging in Place.

Source: author.

Transportation Officials 2012). Although 80 percent of those ages 85 and older make over 80 percent of their trips by car, over half of their trips are as passengers (Turcotte 2012).

However, older people with declines in their cognitive, sensory, and motor capacities do not necessarily stop driving. Very frail older people can operate a car even though they have difficulties walking or using public transit (Gwyther and Holland 2012; Haustein and Siren 2014). They cope by self-regulating their driving behaviors (Kostyniuk and Molnar 2008). They go only to familiar destinations, during the daytime, at slower speeds, at less congested times, during favorable weather conditions, and on lower traffic volume routes (Alsnih and Hensher 2003; Dumbaugh 2008; Kim 2011).

Those who resist giving up driving—especially those over 80—are more involved in fatal crashes (controlling for miles driven) and generally are more seriously injured than younger populations (American Association of State Highway and Transportation Officials 2012; National Center for Statistics and Analysis 2015).

A very small share of older persons copes with these navigation issues by occupying active adult leisure-oriented communities where they reach their destinations by golf carts. Some states, like Florida, do not require a driver license to operate these vehicles (Simpson 2010).

The Disadvantaged Carless

Older people who cannot rely on auto transportation are typically more disadvantaged (Baily 2004). One study reported that they make "15 percent fewer trips to the doctor, 59 percent fewer trips to shop or eat out, and 65 percent fewer trips to visit friends and family, than drivers of the same age" (Transportation for America 2011, 3). Moreover, they feel less independent, more depressed, and less in control of their lives. They particularly feel guilty about having to rely on friends or family members to take them to their destinations (Dumbaugh 2008). That is why it is so debilitating for some women when their husbands die. They abruptly lose a person whom they depended on to take them everywhere (Haustein and Siren 2014; Lord et al. 2011).

The inability to reach destinations by driving is especially traumatic for older people, because the majority occupy places with relatively low population densities without public transit. In 2014, only 27 percent occupied the higher density central or core urban areas of U.S. metropolitan areas, while 53 percent lived outside of these locales, roughly referred to as suburban America. Another 20 percent lived in rural communities (U.S. Department of Health and Human Services 2015). Demographers predict that the share of older persons in these low-density locations will increase because of the aging in place of younger baby boomers already living in the suburbs and because more middle-aged and older residents move from urban cores to the suburbs than in the opposite direction. Only a relatively small share of older adults moves from suburban to downtown or nearby locations (Kotkin et al. 2015; Golant 2009).

More Optimistically: Safer and Driverless Cars

Yet, there are reasons for optimism. New auto technologies, such as lane departure warnings, collision warnings, and parking assist systems, are making driving easier, safer, and less limiting for older people with cognitive and physical deficits and at higher accident risk. As one report emphasizes, "Driver error is believed to be the main reason behind over 90% of all crashes" (Fagnant and Kockelman 2015, 169).

A more dramatic impact is promised by the advent of driverless cars—unquestionably a new travel paradigm. Private vehicle usage would become feasible even for the chronologically very old with questionable driver skills, greatly expanding their destination options without increasing their risk for accidents or injuries. Older persons would also benefit by the projected increase in more affordable (driverless) taxi transportation and on-demand ride-sharing programs. However, as true for any innovation, some older persons will only be late adopters (Peek et al. 2014). They will be intimidated by the technological design features of driverless cars and fear giving up personal control of their driving experience. Others with more severe physical or cognitive limitations will be reluctant users because they will still require assistance getting in and out of these vehicles and fear traveling alone to their destinations.

Alternative Vehicle Alternatives

Conventional Fixed-Route Public Transit

When auto transportation is not feasible, public transportation is an obvious alternative. However, older people rely on transit for just over 1 percent of their trips, mainly in places with higher population densities. Most transit use is concentrated in the urban cores of a very few metropolitan areas—known as legacy cities—that include New York, Chicago, Philadelphia, San Francisco, Boston, and Washington, DC (Cox 2014). In other urban communities, older people generally have few transit routes within walking distance of their homes (Transportation for America 2011). Similarly, only about 14 percent of rural older Americans report they have transit service available within a half mile of their dwellings (American Association of State Highway and Transportation Officials 2012).

The same health problems and impairments that limit the driving of older people also restrict their use of public transit. The physically vulnerable especially have difficulties walking to a bus or subway stop. They will also restrict their usage because of route schedules that do not match their travel needs, long waiting times especially in bad weather, difficulties getting on and off vehicles and finding a seat, worries about being pushed and jostled, and concerns for their physical safety (e.g., from crime and disrespectful youth) (Golant 2015a).

Demand-Responsive Services

Demand-responsive vehicle alternatives (also known as dial-a-ride, supplemental transportation programs, paratransit) offering home-pickup services by van-like vehicles attempt to eliminate or at least minimize these transit problems. We can distinguish three categories of these services.

First, under the Americans with Disability Act (ADA), public transit operators must provide paratransit service to persons with disabilities—a substantial share are seniors—who would otherwise not be able to access the fixed route network (for example, walk to bus stops from their dwellings) (Rosenbloom 2003). But eligible older persons must be seriously disabled to use these services. These typically heavily subsidized programs are also extremely expensive for transit companies to operate (Rosenbloom 2013).

Second, home-to-destination van services offered by both private (including nonprofit) and public vendors are available for those elders who require access to specific destinations, such as medical appointments and community senior centers (e.g., offering congregate meal programs). Sometimes these services are made affordable to low-income older persons through federally funded programs such as the Older American's Act (Title III), Medicaid Waivers, the Department of Veteran Affairs, and the Department of Transportation's Federal Transit Administration (Kaye et al. 2010). However, the availability of these services differs widely because of the vagaries of state and local government funding and philanthropic support. Consequently, in many communities, especially in rural America, the demand for these services typically outstrips supply (Rosenbloom 2003).

Moreover, these alternatives are not without downsides. Older people must plan trips 24 hours in advance, and the alternatives often have limited geographic coverage (especially on nights and weekends) and are available only for certain categories of trips (e.g., to a doctor or a medical facility). Also, some older people attach a negative stigma to traveling in these vehicles because they "broadcast" their vulnerabilities (Golant 2015a). Nonetheless, these alternatives unquestionably do offer a niche service to a small share of older consumers.

Taxis and, more recently, ride-sharing alternatives such as Uber and Lyft represent a third category of demand-responsive services. Their relatively high costs primarily restrict their use, but they do not suffer—at least to the same degree—from the aforementioned downsides. But not all these vehicles are able to accommodate disabled passengers, and some advocates worry about the absence of background checks of their drivers. Furthermore, the services offered by Uber and Lyft are usually unavailable to older persons without smartphones (to request these services). However, an increasing share of the future old are likely to use smartphones, and some local governments are making Uber services affordable to their lower-income older residents (Watkins 2015). Moreover, these companies are increasingly making their services available to disabled older persons such as in San Francisco, where some Uber drivers are trained to assist physically disabled seniors (Lien 2016). Additionally, older persons are benefiting from new communication platforms that are easier for them to use (Liepelt 2016).

Walking and Bicycling to Access Destinations

Walking or even bicycling to their destinations is another option for older persons. However, older people make less than 9 percent of their trips by walking (Rosenbloom 2003), a much smaller share than their European counterparts (Lynott et al. 2009). That is why many planning experts are critical of urban settlement patterns dominated by low-density, cul-de-sac neighborhoods that require Americans to drive rather than walk (or bicycle) to their destinations (Kotkin et al. 2015).

These concerns explain the call for new or redesigned places with higher density residential and nonresidential mixed land uses, and better connected pedestrian- or bicycle-friendly streets. These built environments are variously referred to as new urbanism, complete streets, or transit-oriented communities (when transit is also accessible by walking) (Steuteville 2002).

These places unquestionably make walking more feasible, especially when the pathways are smooth, flat, and well lit, have benches, are safely separated from traffic, and are free of crime. Because walking is a form of exercise, advocates also argue that older pedestrians receive collateral physical health benefits (Chaudhury et al. 2012; Winters et al. 2015).

Although older people do walk more in these neighborhoods, the causal relationship is unclear. These places may selectively attract older residents who are more motivated and able to walk (Rosso et al. 2011; Kerr et al. 2012). Moreover, there are few comparable studies of whether walking in traffic-free cul-de-sac neighborhoods (e.g., with pets) also yields health advantages.

Furthermore, however well designed these communities are, walking by older people is far from guaranteed. Some simply do not have the physical stamina to walk other than very short distances (Dumbaugh 2008). Others worry about falling, heavy traffic, poor weather, darkness, and carrying heavy parcels, especially when resting places are absent (Banister and Bowling 2004; Chaudhury et al. 2012; Turcotte 2012). Moreover, destinations such as doctors' offices or health centers are infrequently within walking distance.

These neighborhoods have other drawbacks, paradoxically because of their attractiveness to working aged populations. They have relatively high dwelling values and rents, making these neighborhoods unaffordable to lower-income older persons (Haughey and Sherriff 2010).

Activities, People, Goods, Services, Health, and Long-Term Care Delivered to the Dwellings of Older Adults

More than at any other time in history, older people have the ability to access products, services, activities, and people without leaving their dwellings. Traditionally, a small percentage of older people have shared their households with family members who are willing to help them access groceries and other household items. These living arrangements also offer older people enjoyable *in-situ* social relationships that reduce their need to travel (Ward 2016). Although these living arrangements are less likely now than in the past, in 2015, 6 percent of the male population age 65 and over and 16 percent of older females lived with family members other than their spouses. However, the prevalence of home sharing with family members may well increase in the future, because of the aforementioned projected increase in the relative size of the Asian and Hispanic older populations. These ethnic groups have been far more likely than older non-Hispanic Whites to live in such shared households (Federal Interagency Forum on Aging-Related Statistics 2016).

Older persons can also share their dwellings with persons unrelated to them, sometimes unmarried partners, but also friends, who help each other out (e.g., the Golden Girls). Other times, they lease their accommodations to boarders and charge them a lower rent (and/or board) in return for receiving help to obtain their needed household items. In 2015, 3.7 percent of older men and 2.6 percent of older women were in these living arrangements (Federal Interagency Forum on Aging-Related Statistics 2016). Nonprofit charitable organizations in some U.S. cities assist with these home sharing arrangements (http://nationalsharedhousing.org.), but commercial companies have also entered the market to help match seniors with boarders.

The physical accommodations of these arranged home sharers can vary. Some occupy a spare bedroom, but older persons risk interpersonal conflicts with their occupants and infringements on their privacy and lifestyles. Alternatively, they can occupy an accessory dwelling unit, typically a renovated garage or basement in a single-family dwelling with its own entrance, kitchen, bathroom, and living area (Golant 2015a). A very small share occupies a small, self-contained living unit on the side or the rear of the lot of an adult child's single-family unit. These structures are variously known as ECHO units or granny flats. However, zoning ordinances or neighborhood covenants may prevent older people or their family members who occupy single-family zoned neighborhood dwellings from introducing such multiple units (Wegmann 2015).

Such household sharing strategies have become less necessary as American society evolves into a full-fledged on-demand economy (Yoquinto and Coughlin 2015). There are now very few goods or products—from pharmacy items and prepared meals to paper towels—that cannot be delivered from either box stores or Internet providers. Consequently, older persons have fewer reasons to leave their dwellings. Moreover, receiving these items no longer carries any negative stigma as was once the case when such home deliveries were mainly used by the vulnerable and homebound or because they were funded by the government.

Older people with health problems or physical limitations also have more options to receive their acute care and long-term services and supports (LTSS) in their own homes (Kaye et al. 2010). Most of this assistance is provided by family members, the preeminent caregivers of older people (referred to as informal care), but this presumes that they live relatively close-by and are motivated and able to help. Paid workers supply a much smaller share of home-based assistance (referred to as formal care). These include professionals who make house calls, such as physicians, physician assistants, nurses, social workers, and occupational, physical, and speech therapists. Additionally, older people may hire direct care workers such as home health aides to perform labor-intensive and hands-on tasks such as lifting, carrying, and self-care (e.g., bathing, feeding, and toileting) (Stone and Harahan 2010). Many continuing care retirement communities (CCRCs) (now referred to as Life Plan communities) and assisted living properties also will deliver these health-related and independence-supporting services (once only available to their own occupants) to older people who live outside their facilities (Spellman and Townsley 2012).

Strong economic incentives have motivated the private sector to serve this highly lucrative elder consumer market. Additionally, in the United States, state governments now view programs that deliver services to the homes of their older poor constituents as less costly to offer than accommodating them in the Medicaid beds of their nursing homes. And for health care administrators, these in-home services are effective preventive strategies that reduce hospitalization and emergency room costs, savings encouraged by the 2010 Affordable Care Act legislation in the United States.

However, these are not always perfect solutions. The quality of in-home care offered by both family members and paid workers can vary greatly. For example, studies report medication administration errors and even physical abuse by caregivers (National Consumer Voice for Quality Long-Term Care 2012). Furthermore and paradoxically, the introduction of equipment and services sometimes results in the dwellings of older people looking and functioning more like institutions than private residences—the very outcomes that aging-in-place solutions were designed to avoid (Gubrium 1973).

Additionally, middle-income older persons often are unable to benefit from in-home care. Their incomes make them ineligible for most public programs targeting the poor, but unlike higher-income groups, they cannot afford the home care offered by the more expensive private sector providers (Golant 2014).

Activities, People, Goods, Services, Health, and Long-Term Care Delivered to Old Age-Concentrated Buildings and Neighborhoods

The large building or neighborhood concentrations of older residents resulting from their collective aging-in-place behaviors also present important connectivity opportunities. These clusters or critical masses of older consumers offer private and public sector providers various organizational and financial incentives to deliver their services or products. For example, rather than delivering food services to older people dispersed across multiple neighborhoods, a vendor can make a single trip to a residential cluster of older occupants, thus saving both time and delivery costs. Similarly, a concentration of low-income vulnerable elders in a government-subsidized rental building is justification for management to hire an on-site social or health care worker who can see multiple clients in a relatively short time (Golant et al. 2010).

The assistance needs of the older occupants in some of these residential enclaves have motivated nonprofit organizations and housing providers to offer service-related programs sometimes supported by federal funding (such as from the Older Americans Act and Medicaid Waivers programs), charitable service organizations, corporations, and resident membership fees. One such community-based model initiated in the mid-1980s is the Naturally Occurring Retirement Community-Supportive Service Program (NORC-SSP) first developed within large apartment complexes in New York City, but now available throughout the United States (Greenfield et al. 2013; Vladeck 2008). Funded by a variety of governmental and nonprofit entities, the housing development and its moderate-income older residents partner with health and social service organizations to help them age in place (Scharlach and Lehning 2016). The NORC-SSP is staffed by social workers and nurses, and the residents are offered a coordinated set of on-site or delivered services that includes meals, health care, and long-term services and supports. The NORC-SSP is "based on the principles of senior empowerment and community building" (Scharlach and Lehning 2016, 174), and the older residents are encouraged (although not required) to participate in various recreation and socialization programs and to serve as volunteers. However, the long-term sustainability of this model is unclear because of the vagaries of federal funding and because the apartment residents are not always enthusiastic participants (Enguidanos et al. 2010). Some occupants are reluctant to receive assistance from any government program or to be stigmatized as vulnerable or needy older persons.

A second model is represented by Elder Villages. These resident-driven and self-governing neighborhood-based grassroots organizations are much less likely to depend on government funding or the participation of nonprofit social agencies. Rather, older residents in a neighborhood or building who opt to be members establish their own nonprofit association that hires professional staff and recruits volunteers. Initiated 2001 in Boston (Beacon Hill Elder Village), such villages are growing in popularity throughout the United States. For an annual membership fee, they offer their predominantly middle-income occupants help with their grocery shopping, meal delivery, transportation, and preventive health needs. They screen possible vendors and providers and arrange for them to offer their products and services at a discounted rate. They also organize educational and recreational events for their residents.

However, to realize these connectivity benefits, older residents must subscribe to a "takes a village" service delivery model. That is, they must be willing to give up some of their self-reliant values and accept organized assistance in return for guarantees that they will have their unmet needs satisfied. The continued operation of these programs also depends on securing grants from private philanthropy and recruiting a sufficient number of volunteers. Also, to the disappointment of their members, these programs usually cannot guarantee the hands-on assistance required by physically or cognitively impaired older residents trying to age in place (Scharlach and Lehning 2016). Consequently, despite the growing popularity of these Elder Villages, their viability also remains uncertain (Scharlach and Lehning 2016).

Activities, People, Services, Health, and Long-Term Care Accessed by Information and Communication Technologies

Internet technologies have given older people seeking to age in place the ability to access and share "all types of interactive multimedia services with the capacity to transmit audio, video, and data" (Baker 2013, 1). In 2015, 58 percent of persons age 65 and older (and 50 percent of those 75 and older) reported using the Internet, up from 14 percent in 2000 (Perrin and Duggan 2015). These users can benefit from telecommuting employment and distance learning opportunities without leaving their dwellings or can serve as volunteers—such as helping younger children to read—without having to visit them in their homes or schools (Baker 2013). Those who feel comfortable using the Internet feel more secure knowing that their Social Security checks have been deposited directly into their bank accounts, and they can conduct most of their banking affairs online (Perrin and Duggan 2015), thereby avoiding potentially difficult or unsafe trips. It is also more likely that they can retrieve books from libraries, see movies, or listen to live symphonies in the comfort of their homes. Similarly, skyping and facetiming make it easy for them to have video social interactions with family or friends.

Going forward, new information and communication technologies known as smart home, gerontechnology, telemedicine, and telehealth technologies (sometimes involving robotic devices taking on humanoid or pet-like appearances) hold tremendous promise as connectivity options. Wireless sensor technologies—installed in the home, worn or attached to individuals—can now monitor or track the physiological functions, health status, mobility, and activities of older persons that once required trips to a doctor or visits by social work or health care professionals. When observed measures, activities, or behaviors deviate from some designated norm, they trigger video-conferencing interventions, home visits by care providers or emergency responders (Lee and Coughlin 2015). Older people can now consult with care professionals located in a different county or state.

It is too soon to know how readily older people will accept these new technological strategies and how effectively they will substitute for traditional office-based approaches. What is certain, however, is that older people will have to make crucial trade-offs to receive their benefits. Loss of privacy (i.e., always being under surveillance) will be a pervasive concern. So too, will be the reduced probability of spontaneous social interactions in "third places" (that is, semipublic places) such as stores, banks, professional offices, and restaurants (Oldenburg 1999). Others will feel angst about absent face-to-face contacts or physical touching with family, friends, or professionals, all-too-frequent vulnerability reminders, and the need to give up personal information (Forlizzi et al. 2004). Five other factors will also discourage their adoption. First, compared with the young, older people are less comfortable using new technologies; second, high speed broadband Internet is less available in some locations, such as rural counties; third, the equipment or services may be unaffordable to lower-income elders (Baker 2013); fourth, providers and consumers will have liability concerns (for errors or malfunctioning); and fifth, government programs and private insurance companies may be unwilling to cover (at least initially) the cost of buying or servicing these technologies (Lee and Coughlin 2015).

Conclusion

Aging-in-place older people experience various housing problems that make it difficult for them to achieve residential normalcy, that is, to occupy residential arrangements where they feel comfortable, competent, and in control of their lives and environments (Golant 2011, 2015b). This chapter discussed one category of problems arising from their difficulty to connect with people, goods, services, care, and activities.

Currently, older car drivers or passengers are the most advantaged when it comes to satisfying their needs and wants. Many older people who are able to walk to many of their destinations, share their homes, or occupy old age-concentrated residential enclaves also confront fewer connectivity challenges. However, these traditional accessibility approaches may become less important in a future society that routinely offers demand-responsive and ride-sharing alternatives such as Uber and Lyft, more available home- and digitally delivered products and services, and the promise of driverless cars and smart home technologies.

However, as is often true with the introduction of any new products and practices, there may be unintended and unfavorable consequences, that is, collateral damages (Golant 2015b). Older people will increasingly not require the on-site physical proximity of family members, friends, vendors, and care workers to receive their information, products, services, care, and even emotional supports. Consequently, and perhaps paradoxically, even as older people experience new and more effective ways to connect with everyone and everything, they will have to lose the positive experiences they enjoyed from their earlier face-to-face interpersonal relationships. The fear is that the most connected older people will also become the most socially isolated.

What this discussion makes abundantly clear is that there is currently no one-size-fits-all connectivity option that will enable all older adults to achieve their aging-in-place goals. Moreover, while even the most innovative strategies have advantages, they also have potential downsides. The diverse lifestyles and capabilities of older people will require both the private and public sectors to offer a variety of connectivity solutions if older individuals can live comfortable and competent lives in their current homes—and achieve residential normalcy.

References

Alsnih, Rahaf, and David A Hensher. 2003. "The Mobility and Accessibility Expectations of Seniors in an Aging Population." *Transportation Research Part A: Policy and Practice*, 37(10): 903–916.

American Association of State Highway and Transportation Officials. 2012. *Keeping Baby Boomers Mobile: Preserving the Mobility and Safey of Older Americans*. Washington, DC: American Association of State Highway and Transportation Officials.

Baily, Linda. 2004. *Aging Americans: Stranded without Options*. Washington, DC: Surface Transportation Policy Project.

Baker, Christopher A. 2013. *A Connection for All Ages: Enabling the Benefits of High-Speed Internet Access for Older Adults*. Washington, DC: AARP Public Policy Institute.

Banister, David, and Ann Bowling. 2004. "Quality of Life for the Elderly: The Transport Dimension." *Transport Policy*, 11(2): 105–115.

Caro, Francis G., Jeffrey A. Burr, Eilon Caspi, and Jan E. Mutchler. 2010. "Motives that Bridge Diverse Activities of Older People." *Activities, Adaptation & Aging*, 34(2): 115–134.

Chaudhury, Habib, Atiya Mahmood, Yvonne L. Michael, Michael Campo, and Kara Hay. 2012. "The Influence of Neighborhood Residential Density, Physical and Social Environments on Older Adults' Physical Activity: An Exploratory Study in Two Metropolitan Areas." *Journal of Aging Studies*, 26(1): 35–43.

Cox, Wendell. 2014. "New York, Legacy Cities Dominate Transit Urban Core Gains." *Newgeography.com*.

Dumbaugh, Eric. 2008. "Designing Communities to Enhance the Safety and Mobility of Older Adults: A Universal Approach." *Journal of Planning Literature*, 23(1): 17–36.

Enguidanos, Susan, Jon Pynoos, Maria Siciliano, Laura Diepenbrock, and Susan Alexman. 2010. "Integrating Community Services within a Norc: The Park La Brea Experience." *Cityscape*, 12(2): 29–45.

Fagnant, Daniel J., and Kara Kockelman. 2015. "Preparing a Nation for Autonomous Vehicles: Opportunities, Barriers and Policy Recommendations." *Transportation Research Part A: Policy and Practice*, 77: 167–181.

Federal Interagency Forum on Aging-Related Statistics. 2016. *Older Americans 2016: Key Indicators of Well-Being, Federal Interagency Forum on Aging-Related Statistics*. Washington, DC: U.S. Government Printing Office.

Forlizzi, Jodi, Carl DiSalvo, and Francine Gemperle. 2004. "Assistive Robotics and an Ecology of Elders Living Independently in Their Homes." *Human—Computer Interaction*, 19(1–2): 25–59.

Golant, Stephen M. 2008. "Commentary: Irrational Exuberance for the Aging in Place of Vulnerable Low-Income Older Homeowners." *Journal of Aging and Social Policy*, 20(4): 379-97.

Golant, Stephen M. 2009. "Aging in the American Suburbs." *Aging Well Magazine*.

Golant, Stephen M. 2011. "The Quest for Residential Normalcy by Older Adults: Relocation but One Pathway." *Journal of Aging Studies*, 25(3): 193–205.

Golant, Stephen M. 2014. *Age-Friendly Communities: Are We Expecting Too Much?, IRPP Insight*. Montreal: Institute for Research on Public Policy.

Golant, Stephen M. 2015a. *Aging in the Right Place*. Baltimore, MD: Health Professions Press.

Golant, Stephen M. 2015b. "Residential Normalcy and the Enriched Coping Repertoires of Successfully Aging Older Adults." *The Gerontologist*, 55(1): 70–82.

Golant, Stephen M. 2016. "Should Older Americans Live in Places Segregated from the Young?" *CSA (Certified Senior Advisors) Journal*, 67(3): 42–45.

Golant, Stephen M., Pamela Parsons, and Peter A. Boling. 2010. "Assessing the Quality of Care Found in Affordable Clustered Housing-Care Arrangements: Key to Informing Public Policy." *Cityscape*, 12(2): 5–28.

Greenfield, Emily A., Andrew E. Scharlach, Amanda J. Lehning, Joan K. Davitt, and Carrie L. Graham. 2013. "A Tale of Two Community Initiatives for Promoting Aging in Place: Similarities and Differences in the National Implementation of NORC Programs and Villages." *The Gerontologist*, 53(6): 928–938.

Gubrium, Jaber F. 1973. *The Myth of the Golden Years*. Springfield, NJ: Charles Thomas.

Gwyther, Holly, and Carol Holland. 2012. "The Effect of Age, Gender and Attitudes on Self- Regulation in Driving." *Accident Analysis & Prevention*, 45: 19–28.

Haughey, Rick, and Ryan Sherriff. 2010. *Challenges and Policy Options for Creating and Preserving Affordable Housing Near Transit and in Other Location-Efficient Areas*. Washington, DC: Center for Housing Policy.

Haustein, Sonja, and Anu Siren. 2014. "Seniors' Unmet Mobility Needs—How Important Is a Driving Licence?" *Journal of Transport Geography*, 41: 45–52.

Hayutin, Adele M., Miranda Dietz, and Lillian Mitchell. 2010. *New Realities of an Older America*. Stanford, CA: Stanford Center on Longevity.

Joint Center for Housing Studies of Harvard University. 2013. *Housing America's Older Adults*. Cambridge, MA: Joint Center for Housing Studies of Harvard University.

Kaye, H. Stephen, Charlene Harrington, and Mitchell P. LaPlante. 2010. "Long-Term Care: Who Gets It, Who Provides It, Who Pays, and How Much?" *Health Affairs*, 29(1): 11–21.

Kerr, Jacqueline, Dori Rosenberg, and Lawrence Frank. 2012. "The Role of the Built Environment in Healthy Aging." *Journal of Planning Literature*, 27(1): 43–60.

Kim, Sungyop. 2011. "Assessing Mobility in an Aging Society: Personal and Built Environment Factors Associated with Older People's Subjective Transportation Deficiency in the US." *Transportation Research Part F: Traffic Psychology and Behaviour*, 14(5): 422–429.

Kostyniuk, Lidia P, and Lisa J Molnar. 2008. "Self-Regulatory Driving Practices among Older Adults: Health, Age and Sex Effects." *Accident Analysis & Prevention*, 40(4): 1576–1580.

Kotkin, J., W. Cox, M. Schill, and A. Modarres. 2015. *Building Cities for People*. Orange, CA: Chapman University Press.

Lee, Chaiwoo, and Joseph F. Coughlin. 2015. "Perspective: Older Adults' Adoption of Technology: An Integrated Approach to Identifying Determinants and Barriers." *Journal of Product Innovation Management*, 32(5): 747–759.

Lien, Tracy. 2016. "Uber App Will Come Pre-Installed on Grandpad Tablet for Seniors." *Los Angeles Times*, January 7.

Liepelt, Kourtney. 2016. "Uber Tests New Platform with Canada Senior Living Pilot." *Senior Housing News*, August 25.

Lord, Sébastien, Carole Després, and Thierry Ramadier. 2011. "When Mobility Makes Sense: A Qualitative and Longitudinal Study of the Daily Mobility of the Elderly." *Journal of Environmental Psychology*, 31(1): 52–61.

Lynott, Jana J., Amanda Taylor, Hannah Twaddell, Jessica Hasse, Kristin Nelson, Jared Ulmer, Barbara McCann, Edward R. Stollof. 2009. *Planning Complete Streets for an Aging America*. Washington, DC: AARP Public Policy Institute.

National Center for Statistics and Analysis. 2015. *Older Population: 2013 Data, Traffic Safety Facts, Report No. DOT HS 812 199*. Washington, DC: National Highway Traffic Safety Administration.

National Consumer Voice for Quality Long-Term Care. 2012. *Consumer Perspectives on Quality Home Care*. Washington, DC: National Consumer Voice for Quality Long-Term Care.

Oldenburg, Ray. 1999. The Great Good Place: Cafes, Coffee Shops, Bookstores, Bars, Hair Salons, and Other Hangouts at the Heart of a Community. Philadelphia, PA: Da Capo Press.

Peek, Sebastiaan T. M., Eveline J. M. Wouters, Joost van Hoof, Katrien G. Luijkx, Hennie R. Boeije, and Hubertus J. M. Vrijhoef. 2014. "Factors Influencing Acceptance of Technology for Aging in Place: A Systematic Review." *International Journal of Medical Informatics*, 83(4): 235–248.

Perrin, Andrew, and Maeve Duggan. 2015. *Americans' Internet Access: 2000–2015*. Washington, DC: Pew Research Center.

Rosenbloom, Sandra. 2003. *The Mobility Needs of Older Americans: Implications for Transportation Reauthorization*. Washington, DC: The Brookings Institution.

Rosenbloom, Sandra. 2013. *Roadblocks Ahead for Seniors Who Don't Drive, Innovation in Infrastructure*. Washington, DC: Urban Institute.

Rosso, Andrea L, Amy H Auchincloss, and Yvonne L Michael. 2011. "The Urban Built Environment and Mobility in Older Adults: A Comprehensive Review." *Journal of Aging Research*, 2011: 1–10.

Scharlach, Andrew E., and Amanda J. Lehning. 2016. *Creating Aging-Friendly Communities*. New York: Oxford University Press.

Simpson, Deane. 2010. Third Age Urbanism: Retirement Utopias of the Young-Old. PhD diss., Swiss Federal Institute of Technology in Zurich (ETH Zurich), Switzerland.

Spellman, Sarah, and Scott Townsley. 2012. *Continuing Care at Home: Evolution, Innovation, and Opportunity*. Minneapolis, MN: CliftonLarsonAllen.

Steuteville, Robert. 2002. "The New Urbanism: An Alternative to Modern, Automobile-Oriented Planning and Development." *New Urban News*, July 25.

Stone, Robyn, and Mary F. Harahan. 2010. "Improving the Long-Term Care Workforce Serving Older Adults." *Health Affairs*, 29(1): 109–115.

Transportation for America. 2011. *Aging in Place, Stuck without Options*. Washington, DC: Transportation for America.

Turcotte, Martin. 2012. "Profile of Seniors' Transportation Habits." *Canadian Social Trends*, 93: 1–16.

U.S. Bureau of the Census. 2016. *Geographical Mobility: 2015 to 2016: Table 6*. Washington, DC: U.S. Bureau of the Census.

U.S. Department of Health and Human Services, Administration on Aging. 2015. *A Profile of Older Americans, 2015*. Washington, DC: U.S. Department of Health and Human Services.

Vladeck, Fredda. 2008. "Naturally Occurring Retirement Communities." *Designer Builder*, January/February: 43–47.

Ward, Lisa. 2016. "Beyond Airbnb: Sharing Homes Offers Seniors Financial and Social Advantages." *Wall Street Journal*, September 11.

Watkins, Morgan. 2015. "City Hopes to Use Uber to Give Rides to Seniors." *The Gainesville Sun*, June 4.

Wegmann, Jake. 2015. *Financing Ancillary Apartments on Residential Properties: Challenges and Solutions*. Berkely, CA: Institute of Governmental Studies.

Winters, Meghan, Christine Voss, Maureen C. Ashe, Kaitlyn Gutteridge, Heather McKay, and Joanie Sims-Gould. 2015. "Where Do They Go and How Do They Get There? Older Adults' Travel Behaviour in a Highly Walkable Environment." *Social Science & Medicine*, 133: 304-312.

World Health Organization. 2007. *Global Age-Friendly Cities: A Guide*. Geneva, Switzerland: World Health Organization.

Yoquinto, Luke, and Joe F. Coughlin. 2015. "The On-Demand Economy: Changing the Way We Live as We Age." *The Washington Post*, December 14.

18

HOUSING MODELS FOR AGING IN COMMUNITY

Sherry Ahrentzen and Ruth L. Steiner

Introduction

Across the world, approximately one in six persons will be 65 years and older by 2050, about twice today's proportion (United Nations 2013). By 2050, over one-third of the population of Japan (36.5 percent), South Korea (34.9 percent), Spain (34.5 percent), and Italy (33 percent) will be age 65 and older (United Nations 2013). Globally, persons over the age of 59 are projected to exceed the number of children (15 years and younger) for the first time in 2047 (United Nations 2013). Seventy-seven million American baby boomers—those born between 1946 and 1964—are turning 65 approximately 1 every 7 seconds (AgeLab 2016). Nowadays, about 25 percent of U.S. households include someone aged 65 or older (Lipman et al. 2012).

Coined "the silver tsunami" in the public media, the term "transformational demographics" may more accurately describe the trends associated with aging populations in the next few decades. The world has never experienced the age shift of younger to older persons that will be witnessed soon. The global population of older adults (OAs)[1] and households and homes with aging occupants has been surging, brought on by the aging of the baby boomers, ongoing improvements in health care and medical technology, decreasing mortality, and declining fertility. While there are important variations among countries (e.g., India and several African countries will not experience these demographic shifts as much as Japan, China, and South Korea), globally the age shift is being felt.

Like the diversity of the population itself, residences and living conditions of OAs cover a spectrum. Most OAs do not live in age-qualified housing or communities that are restricted to residents who are 55 years (sometimes 62 years) or older, but in what are characterized as "community-dwelling residences" (MetLife Mature Market Institute and Tenenbaum 2010). In 2007, 10 million of a total 23 million (46 percent) of older households resided in the suburbs (Hayutin et al. 2010). While most of the suburban housing stock was originally targeted to families with children or working-age adults, some of it may be inadequate for older individuals with chronic health conditions or functional limitations (Joint Center for Housing Studies 2014; Lipman et al. 2012). Compared to young and working-age adults, older households are more likely to reside in older residences (Lipman et al. 2012). These homes may have hazardous, tight, and thus challenging bathrooms; steep staircases; narrow halls and doorways; dated and inefficient electrical, heating, cooling, and plumbing systems; drafts, dampness, mold, and mildew; dated appliances that are difficult to use and maintain; and lighting and temperature controls that are difficult to maneuver for those with ailing health and sensory conditions.

While age-restricted retirement and senior living communities are frequently profiled in the media—from the early days of Sun City and Leisure World in the 1960s to today's booming alternatives such as The Villages (Florida) and continuing care retirement communities (CCRC)—surveys find that most individuals on the cusp of 65 or those over 65 want to age in place; i.e., they want to continue to live in the home and community in which they currently reside and in most instances have for years or decades (AARP 2012). Yet for numerous reasons, synopsized by Golant (2015), aging in the home and community one lived in during their 40s, 50s, or 60s is not feasible or healthy for many older people—yet living in a CCRC or seniors-only retirement community is neither desirable nor practical either. Further, in most European, Asian, and North American countries, fewer family caregivers will be available to support an aging society. Home health care services have been in growing demand, but projections for the future workforce are frighteningly inadequate (Weisfeld and Lustig 2015).

What viable alternatives can be embedded, integrated, and sometimes re-created in our existing residential infrastructure, particularly in the suburbs where the challenges to aging are significant? We believe a middle ground of residential options exists between the age-restricted senior living community and the suburban home one has lived in for the past decades that is becoming incompatible with one's level of health and mobility in terms of both accessibility inside the home and to services within the community. (We use the term "suburban home," recognizing that in the South and West most suburban-type homes are found inside the city limits.) This middle ground reflects a pattern of aging in community that many OAs and senior living organizations are touting as housing models that allow people to live in age-integrated communities but not necessarily in their current home, or to live in their existing homes but provided with a layer of community-generated service networks to support OAs in maintaining their homes and lives (Blanchard and Bolton 2013; Moran and Rollins 2007; Thomas and Blanchard 2009). These models go beyond simply addressing inaccessible dwellings to also considering the necessary social connectivity (see also the chapter authored by Stephen Golant). These housing models occur in urban as well as suburban neighborhoods characterized by access to friends, families, and services.

Since the 1970s, a considerable body of social epidemiological research has documented the health benefits of social integration and social support, including better mental and physical health outcomes, even lower mortality risks (Berkman et al. 2014; Brown and Lombard 2014). Social support can promote adaptive behavior or neuroendocrine responses in the face of stress or other health hazards (House et al. 1988). Other research suggests that the social environment may even impact patterns of cognitive functioning (Seeman et al. 2001).

To the extent that OAs increasingly live in suburban neighborhoods characterized by low density with separated land uses, and thus depend on driving to get around, it can be a challenge to maintain connections while aging in place. In such settings and circumstances, OAs may face a choice of making changes to their home—by adding an accessory dwelling unit (ADU), for example—or joining a network that provides social support and access to goods and services—the Village Movement—or moving to an intentional community within a larger urban or suburban context that fosters on-site connections, cooperation, and services as set and governed by the community—such as multigenerational or senior cohousing. In all cases, these models of aging in community draw on reserves of social capital to foster the well-being of older residents who choose not to live in conventional retirement communities or are unable to remain in their current homes without a meaningful level of assistance (Greenfield et al. 2013; Ivery 2014).

The remainder of this chapter profiles three of these residential options that typify an approach to aging in community: the Village Movement and naturally occurring retirement communities (NORCs); multigenerational and senior cohousing; and ADUs. The concluding section addresses not only the challenges but also the advantages of planning for aging in community.

The Village Movement and Naturally Occurring Retirement Communities

OAs who choose to age in place can be assisted by a variety of arrangements that provide supportive services, for example, the Village Movement and naturally occurring retirement communities (NORC) with supported services programs (NORC-SSP). Both of these housing arrangements include a support system that provides for the essential needs of OAs while allowing them to age in place in familiar surroundings. The difference between these two options relates to who provides the services and who pays for them (Golant 2015; Greenfield et al. 2013).

The Village Movement was founded in the Beacon Hill area of Boston in the early 2000s by a group of neighborhood residents who wanted to remain in their neighborhood. They wanted a model that focused on all aspects of their lives including housing, medical care, and social services. The involved residents created a grassroots, nonprofit membership organization that governs the Village and receives services with help from a staff, volunteers, and other strategic partners, such as service providers. They enrolled their first members in 2002. Since then this model has expanded to 230 Villages throughout the United States, the Netherlands, and Australia, all of which are connected through the Village-to-Village Network (Village-to-Village Network n.d.). Villages have materialized in a diversity of contexts, with 36 percent occurring primarily in an urban setting, 35 percent primarily in a suburban setting, 13 percent primarily in a rural setting, and 16 percent in a mixed geographic setting (Village-to-Village Network n.d.). Members pay an annual fee for the services provided by the local Village including concierge services, social and cultural programs, transportation to and from medical, retail, and entertainment services (possibly with driver or taxi coupons), free escorts after medical procedures, discounted home care services, and wellness programs. Low-income households and individuals may qualify for a reduced annual fee.

NORCs were first discussed in the literature by Hunt and Gunter-Hunt (1986). They initially defined a NORC as "a housing development that is not planned or designed for older people, but which over time comes to house largely older people" (Hunt 1992, 3; Hunt and Gunter-Hunt 1986; Hunt and Ross 1990). They found that in some stable neighborhoods, young families bought homes to raise families but never moved after the children had left the home. These communities could be seen as an "unintentional concentration of elders" who could turn into an "intentional community of elders" (Bookman 2008, 424). As originally defined, NORCs included housing that was not designed for aging. Thus, often there was a lack of neighborhood services (Bookman 2008). Today, NORCs can be classified into three types, based upon their geography: (1) housing-based, which are located in a single apartment building, housing complex, or cluster of apartments; (2) neighborhood-based, primarily comprising single-family homes or duplexes; or (3) community-based, which are similar to neighborhood-based except that their catchment area is larger (Golant 2015). High-rise apartment buildings are also called "vertical NORCs," while neighborhood-based NORCs are called "horizontal" (Enguidanos et al. 2010).

NORCs with supportive service programs (NORC-SSPs) were started at the Penn South Houses Program in New York City by the United Hospital Fund in 1985 and 1986 with funding from the United Jewish Appeal (UJA) Federation. NORC-SSPs include the following services: community engagement including educational and recreational activities, opportunities for seniors to take new roles, and community action initiatives; social work services to individuals, caregivers, and groups of seniors; nursing and health care-related services for individuals and the community; and ancillary services, such as transportation, housekeeping, and social adult day programs (NYOA n.d.). Like the Village model, NORC-SSPs organize health care, social services, and group activities on site; unite housing with their residents, health, and social services; promote independence by engaging seniors; provide OAs with vital roles in the development and operation of the program both in governance and through volunteer activities; and fill the gaps in Medicare, Medicaid, and the Older American Act, where most services are oftentimes insufficient or inadequately coordinated.

Greenfield et al. (2012) developed a framework for understanding the activities and services provided by the Villages and the NORC-SSPs for individuals with the long-term goal of aging in place (Figure 18.1). The internal resources include staffing, the use of volunteers, and an organizational mission, while the external resources include partnerships with external service delivery and planning organizations. The activities directed at individuals include civic engagement and empowerment through volunteer activities, social relationship building, and services to enhance access to resources. These activities and services are designed to provide individual benefits such as better physical and psychosocial well-being due to increased self-efficacy, social support and reduced isolation, and greater access to support and reduced unmet needs. The individuals who participate in Villages and NORC-SSPs also support broader goals for the community and support organizations that participate in the program.

Indeed, previous studies of Villages and NORC-SSPs suggest that some of these outcomes have been achieved. According to Anetzberger (2009), participants in the Community Options NORC program in Cleveland reported that the program enhanced their confidence that they could continue to live in their current residence and have more contact with neighbors. The program also enabled them "to feel better about themselves and more in control of their lives" (Anetzberger 2009, 10). A comparison of two Los Angeles NORCs—Park La Brea, a closed vertical NORC, and West Hollywood, an open horizontal NORC—and the participation of members in the Living Independently in a Friendly Environment (LIFE) program showed differences related to the type of community. In Park La Brea a larger number of members joined the program, more activities were planned for participants, and the participants volunteered for more hours. In West Hollywood, fewer OAs actually joined the LIFE program, fewer activities were scheduled, and participants volunteered for fewer hours, although OAs had joined the LIFE program more quickly. When funding ran out for the LIFE program, Park La Brea instituted membership fees, while West Hollywood folded its services into the existing senior center (Enguidanos et al. 2010). These results reinforce the importance of understanding the context of NORC-SSP programs and coordinating with existing programs targeted at OAs.

Villages and NORC-SSPs serve different markets and use different funding models that may be related to their history. While Villages derive significant funding from their members who volunteer or pay for their services, most of the services and funding of NORC-SSPs comes from government sources (Greenfield et al. 2013). NORC programs are "more likely to include individuals who were older, more functionally impaired, less economically secure, and living in communities predominantly middle class or lower" (Greenfield et al. 2013, 8). In a typical NORC-SSP community the services are coordinated by the staff of a lead agency, such as the Area Agency on Aging or another nonprofit social service provider (Golant 2015). Research on the impact of Villages and NORC-SSPs has not adequately evaluated whether these programs accomplish their goals of allowing OAs to be more independent (Golant 2015; Vladeck and Altman 2015). Yet, research on NORC-SSPs shows that "residents are more aware of the community services to help them, access them more easily, are more involved in their apartment projects' activities, are less likely to feel alone, isolated or depressed, feel as healthy as or healthier than they did in the past, feel safer because of their home modifications and express more confidence in their ability to live independently and remain in their communities" (Golant 2015, 229). Residents of these programs reported that they made more friends and interacted more frequently with their neighbors (Anetzberger 2009). Davitt et al. (2015) reached a similar conclusion but based on slightly different reasons.

> [O]lder adults who retire may be looking for other useful roles in their community, and volunteering in mutual support of other members may provide the social purpose needed.

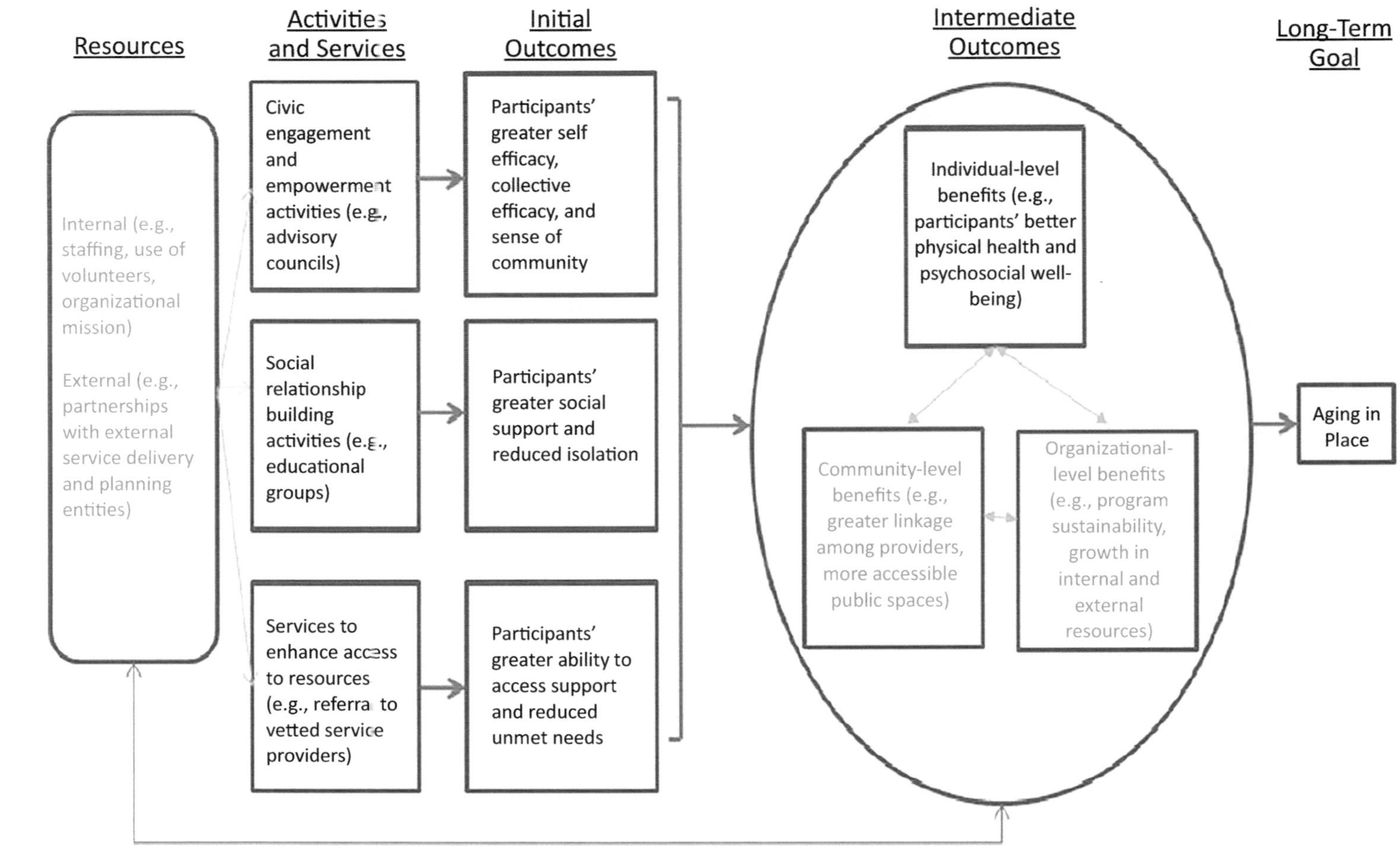

Figure 18.1 Conceptual Framework of Processes through which the NORC Program and Village Models Potentially Influence Aging in Place.

Source: Greenfield et al. (2012, 276). Used with permission.

> Likewise, older adults who are no longer able to drive are at risk of social isolation and limited access to community resources. By employing volunteers, older adults can remain connected to and socially engaged in their communities rather than isolated in their own homes.
>
> *(Davitt et al. 2015, 18)*

Multigenerational and Senior Cohousing

Multigenerational and senior cohousing and accessory dwelling units (ADUs) offer opportunities for OAs to move to a location where they can live close to or with family or friends, who can provide support services they may need for activities of daily living (ADLs) (Katz 1983).

The majority of cohousing communities consist of 15 to 35 private homes—each with its own kitchen, living area, bedrooms, and bathrooms—built around a common house or common area. The homes may be single-family attached or detached, townhomes, or multifamily residences. Unlike planned unit developments (PUDs) that may have common club rooms or swimming pools, cohousing is geared toward residents who want to not only know their neighbors but be engaged with them in community-oriented activities in social space, such as sharing meals on a regular (but not necessarily daily) basis, maintaining a community garden, or cleaning community spaces and grounds. Cohousing residents collaboratively plan community activities and frequently gather for parties, games, movies, or other events. Nonetheless, the private space of one's own home allows the resident to control the mix of privacy and community desired. With its proximity of residents and its imputed social connection, cohousing allows organized activities such as child care, elder care, and carpooling to happen with greater opportunity and ease. Nearly all cohousing communities have a common house with a kitchen and meeting space; others also have community gardens, bicycle repair or woodworking shops, outdoor walkways, open space, and other shared areas. When neighbors know each other well and mutually seek community sharing and decision-making, cohousing provides a level of support for those who need a helping hand, although not care for serious medical conditions (McCamant and Durrett 2011).

The legal structure is typically an Homeowners' Association (HOA), Condominium Association, or Housing Cooperative (Cohousing Association of the United States n.d.). Cohousing differs from traditional residential developments in that cohousing residents either develop the project themselves or co-develop it with an industry professional, and pool their equity to form a limited-liability corporation (LLC) (Durrett 2009).

Approximately 160 cohousing communities are currently being formed, established, or expanding in the U.S. (Cohousing Association of the United States n.d.). The numbers are much larger in Europe, where cohousing originated in 1967 (Killock 2012; Tummers 2016).[2] In the past 25 years, approximately 250 senior cohousing communities have been developed in Denmark with a population of approximately 5.7 million, nearly equivalent to the population of Wisconsin (Pedersen 2015). The Netherlands—with a population of 16.7 million, less than the population of the state of New York—has more than 300 cohousing communities, 73 being multigenerational and 231 being seniors only (Gemeenschappelijk Wonen n.d.). In comparison, the state of New York has currently eight cohousing communities (The Cohousing Association of the United States n.d.).

When cohousing communities were initially developed in the 1970s, they included households with a diversity of ages and family types in the community (e.g., singles, couples with or without children, or multigenerational families). In the 1980s, a group of seniors in Denmark began to discuss how they would handle aging in their community. In 1987, they opened the first senior cohousing community (Cohen 2005). Since that time, the number of multigenerational cohousing communities has grown, along with senior cohousing. In the U.S., Glacier Circle, the first senior cohousing development, was completed in Davis, California, in 2005 (Durrett 2009). Since then, 12 senior cohousing communities have been completed, and another 15 are forming (Cohousing Association of the United States n.d.).

One example of senior cohousing in the U.S. is Silver Sage Village (see Figure 18.2) in Boulder, Colorado (Durrett 2009; Silver Sage Village n.d.). The idea of senior cohousing in Boulder originated with a few residents of Nyland Cohousing, a multigenerational cohousing development in Boulder, who, while still appreciating the benefits of cohousing, also wanted a place specifically designed for the needs and preferences of senior residents (e.g., meal selections and meal times for older residents oftentimes do not correspond to those of families with young children). Silver Sage cohousing includes 16 duplexes and attached homes, ten of which are market-rate units and six designated as "permanently affordable" by the Boulder Housing Authority.[3] It also includes a community center and a common green, guest rooms, and arts and crafts and performance areas (Durrett 2009; Silver Sage Village n.d.).

While Silver Sage Village is an age-restricted development, it is located about 2 miles from downtown in the Holiday neighborhood of Boulder, which is replete with community gardens, parks, bike trails, pedestrian walkways, artists' studios, and a public library (Durrett 2009). Silver Sage Village reflects aging in community by being adjacent to a multigenerational cohousing community with 34 attached townhouses called Wild Sage, which was built and occupied a few years before Silver

Figure 18.2 Silver Sage Cohousing, a Senior Cohousing Development in Boulder, Colorado.

Source: architecture by McCamant & Durrett Architects. Used with permission.

Sage was established. Residents of the two communities occasionally attend joint social events (Wild Sage Cohousing n.d.).

Following a more urban scheme than Silver Sage is Durham Central Park Cohousing (Durham Coho) (Figure 18.3). This development is not marketed as senior but multigenerational cohousing as most residents are between the ages of 50 and 80, with a few having college-aged children living at home. Opened in 2014, Durham Coho is located in downtown Durham, North Carolina, within walking distance to the farmers' market, the public library, the YMCA, restaurants, shops, performing arts center, and theaters and within about 2 miles from the Duke University Medical Center, which residents can get to via the fare-free, hybrid electric bus service that connects the downtown to Duke University and medical facilities (Durham Central Park Cohousing Community n.d.).

The 24 private condominium units range in size from 850 to 1,700 square feet, and each unit has its own balcony or private porch. The architectural style of the building reflects its location in a reemerging neighborhood that once was an industrial part of downtown. Common spaces include parking, bike storage, two rooms for arts and crafts, a recycling room, a community kitchen/dining room, a laundry facility, a media room, and other amenities. The development also has a rooftop terrace and a backyard with vegetable gardens and ornamental plantings. Visitors may stay in the two private guest rooms with full baths (Durham Central Park Cohousing Community n.d.).

Emerging research on cohousing generally emphasizes its benefits for seniors. Cohousing offers the opportunity to engage with others regularly for both social and practical (e.g., housing maintenance) activities in one's immediate residential development, which may be an important point for those who do not have family support systems nearby (Glass 2009). In the case of Denmark and

Figure 18.3 Durham Central Park Cohousing in Downtown Durham, North Carolina.

Source: Fitch Architecture & Community Design. Used with permission.

Sweden, most senior cohousing residents are single, female, and generally satisfied with their choice of housing (Choi 2004). Among residents of Swedish senior cohousing, motivations to live there have changed over time, with increasing numbers of applicants seeking the social benefits involved—e.g., contact with neighbors, mutual support, shared common activities, reduced loneliness—rather than the practical reasons, such as housing and garden management or accessible housing features (Choi 2013). In the U.S., OA residents in cohousing were more likely to move to this type of housing because of a greater sense of community and expectations of mutual support (Glass and Choi 2015). Indeed, 93 percent of U.S. and 69 percent of European residents of cohousing stated that mutual support occurred in their developments (Glass and Choi 2015).

Accessory Dwelling Units

Accessory dwelling units (ADUs), granny flats, or in-law units are "additional living quarters on single-family lots that are independent of the primary dwelling unit" (Sage Computing, Inc. 2008, 1). ADUs can be added to the home of the OA, or they can be added to the home of a family member, friend, or other caretaker. The history of ADUs can be traced to the 1830s housing boom in London, when narrow service roads, called mews, were developed to serve as stables with upstairs living quarters for the workers in the main house (Antonineti 2008). The mews house was later exported to pre-Civil War American cities, such as Philadelphia and Washington, DC (Antonineti 2008). They were a common feature of many cities up to the World War II era, when they were used to address housing shortages (Sage Computing, Inc. 2008). The rapid growth of suburbs after the mid-1940s with the associated lower density housing eventually led most municipalities to prohibit ADUs. Yet in places where affordable housing was in short supply they continued to be built, often as illegally constructed units. Over time, the purpose of ADUs changed from being a source of housing for low-wage workers to becoming part of a family-run rental business to generate income. For other residents, ADUs became functionally independent housing units for elderly parents; hence, they have come to be called "granny flats" or "in-law units" (Antonineti 2008; Howe 1990). Some research suggests that health benefits may emanate from these arrangements of older parents living adjacent to or near their adult children. Using data from the large, nationally representative National Longitudinal Study of Aging dataset, Brinig (2015) found that living with one's adult children signaled a decline in health, but living close to but not with one's children positively affected the OA's health and longevity, with improvements in cognition as well.[4]

ADUs can be classified into three categories based upon their location in relationship to the primary dwelling unit and their differing levels of privacy for the ADU occupant: interior, attached, and detached (Sage Computing, Inc. 2008). Interior ADUs are typically built through the conversion of existing space, such as an attic or basement, and are located within the primary dwelling. Attached ADUs are living spaces added to the primary dwelling either at the rear or at the side, but they can also be constructed in an attached garage (Sage Computing, Inc. 2008). Detached ADUs are built separate from the primary dwelling; they can be constructed as a part of a detached structure, such as a detached garage, or they can be built separate from accessory and residential structures, such as small houses. A number of manufactured housing producers have prefabricated ADUs that can be built in a factory, trucked to the site, and set on a foundation via crane, thus minimizing the noise and disturbance to neighbors of site-built housing. Some of these ADUs have options for telehealth monitors as well as design features such as lighting, entry paths, and bathroom fixtures that are utilized by many aging adults (Petrowski 2010).

The U.S. Department of Housing and Urban Development initiated the Elder Cottage Housing Opportunity (ECHO) demonstration program in 1993, essentially consisting of a small prefabricated house in which an OA resides in a unit placed near the home of a host, which can be either relatives or close friends of the OA. The program was implemented with 80 small prefabricated homes in

Tennessee, New Jersey, Iowa, Kansas, and Missouri. An evaluation of ECHO noted benefits accruing from the daily-living support provided to OA residents by adjacent family members. Nevertheless, challenges for ECHO developers were numerous, including poor design criteria, unclear responsibilities, lack of guidelines for key groups, high costs, zoning constraints, vacancy, and difficulty in moving units from the factory site to the residential property as well as removing them from the property sites when no longer desired in those locations (Koebel et al. 2003).

ADUs have been adopted in many communities throughout Europe and the United States, including many cities in California (Antonineti 2008), Portland, Oregon, and Seattle, Washington. Florida enacted statewide legislation in 2004, allowing ADUs to be used in conjunction with rental rate limitations that would make the units affordable for low-income persons. An online survey found that 56 percent of public sector officials in Florida responded favorably about providing ADUs and supporting them in the future. About 64 percent of private sector builders and developers who participated in the survey supported ADUs to some degree (Florida Department of Community Affairs 2007). Yet many other states and communities have imposed a variety of restrictions, among them occupancy homeowner eligibility, location, or parking restrictions, making it difficult, if not impossible, for ADUs to be built (Cobb and Dvorak 2000; Liebig et al. 2006).

The City of Santa Cruz, California, overcame resistance to ADUs and passed its ADU ordinance in 2002, which established regulations for the location, permit process, deed restrictions, zoning incentives, and design and development standards of ADUs, which are permitted in designated residential zones on lots that are at least 5,000 square feet (no more than one ADU per lot allowed). But unlike many cities, Santa Cruz also provided technical assistance, including design suggestions developed by architects, a wage subsidy and apprentice program, and an ADU loan program (City of Santa Cruz 2003; Sage Computing, Inc. 2008). Homeowners who select one of these design recommendations may qualify for an expedited permitting process. Approximately 20 to 40 ADU permits are approved each year. This integrative ADU program (i.e., financing, permitting, design, technical assistance, ordinance changes, etc.) has received planning and environmental awards, and been used as a model by other communities (City of Santa Cruz 2003; Sage Computing, Inc. 2008). Similar ADU initiatives appear in rapidly growing, high-cost cities such as Portland, Oregon, Seattle, Washington, Vancouver, British Columbia, and the San Francisco Bay Area, among others (Chapple et al. 2017).

While the application process is typically onerous and deterring, many advocates see ADUs as a housing option that meets the desire for independence, affordability, social intimacy, and connection (Chapman and Howe 2001; Cobb and Dvorak 2000; Ross 2016). As of this writing, ADUs target a wide range of the population, not just older adults. The current interest in ADUs as an alternative housing model may in part reflect the growing tiny house movement (Pan 2015). Also, the growth and innovation of the modular/manufactured housing industry for infill homes can address some of the challenges, particularly the noise, traffic, and cost of on-site construction.

Conclusion

In the near future, housing and community planners will need to pay more attention to the challenges presented by the "silver tsunami" and the associated spiraling health care costs. The next generation of OAs may not be willing to simply age in place or go to retirement communities, 55+ communities, or other conventional senior living models (Golant 2015). While the existing models will continue to thrive, the new models of aging in community—Village, NORC-SSPs, senior cohousing, and ADUs—will offer greater choices for OAs than are currently available. These models offer options for age-integrated communities (but not necessarily resulting in OAs remaining in their current home), either with or without an added layer of community-generated services (Bookman 2008; Golant 2015; Greenfield et al. 2012; Vladeck and Altman 2015).

These new models of aging in community may pose challenges (including the affordability of social services), but ultimately they may offer OAs greater opportunities for control and connectivity with their community and neighbors. Some municipalities may need to make changes to land development regulations to allow cohousing and ADUs. Organizations that have traditionally provided support services to seniors may need to change their method of service delivery to accommodate the desire for OAs to age in their community. For example, they may need to provide in-home services instead of providing them at senior centers, or they may provide multiple services instead of a single service (e.g., a well-health visit instead of a blood-sugar check). Some organizations may require OAs to downsize or to move into smaller housing units than they might desire in order to obtain enhanced in-home services (a request that is not yet documented and would be controversial). A major challenge for these options is that they are often dependent upon unpaid support from family, friends, neighbors, or other volunteers, or that they do not have the public visibility or get the publicity that retirement communities receive. Thus, many OAs may not realize the possibilities for these alternative forms of aging in community. Planners and service providers to aging populations need to remove the barriers to creating these types of communities to make it easier for these options to be built into existing communities or to be marketed as viable residential options rather than as inferior or last-resort housing alternatives.

Notes

1 Most developed countries use the chronological age of 65 years as a definition of elderly, senior, or older adult. However, this marker does not adapt well to situations in all countries, for example, in Africa. Further, this chronological age marker suggests equivalence with physiological age, although these are not necessarily synonymous. The chronological marker of "older adult" is oftentimes associated with the age at which one can begin to receive pension benefits. Currently, there is no United Nations (UN) standard numerical criterion, but the cutoff suggested by the UN is 60+ years to refer to the older population. In this chapter, the marker for "older adult" varies depending upon the data source and conventional usage. In general, "near old" is 50–64 years; "older adult" is 65–84; and "oldest old" is 85 years and older.

2 We use the North American term "cohousing" to include similar housing types in other countries: *bofællesskeb* in Danish, *central wonen* in Dutch, and *kollektivhus* in Swedish.

3 Boulder's Permanently Affordable Homes Program includes homes sold at below-market-rate prices to income-eligible buyers who will occupy the home. An Affordability Covenant limits the resale price and places other restrictions on the home.

4 Because Brinig (2015) analyzed longitudinal data and is not based on an experiment with a randomized controlled trial, the findings should be interpreted cautiously.

References

AARP. 2012. *Beyond 50.05: A Report to the Nation on Livable Communities: Creating Environments for Successful Aging*. Washington, DC: AARP.

AgeLab. 2016. *Disruptive Demographics*. Cambridge, MA: MIT AgeLab.

Anetzberger, Georgia J. 2009. "Community Options of Greater Cleveland, Ohio: Preliminary Evaluation of a Naturally Occurring Retirement Community Program." *Clinical Gerontologist*, 33(1): 1–15.

Antonineti, Maurizio. 2008. "The Difficult History of Ancillary Units: The Obstacles and Potential Opportunities to Increase the Heterogeneity of Neighborhoods and the Flexibility of Households in the United States." *Journal of Housing for the Elderly*, *22*(4): 348–375.

Berkman, Lisa F., Ichiro Kawachi, and Maria Glymour, eds. 2014. *Social Epidemiology*. New York: Oxford University Press.

Blanchard, Janice M., and Anthony Bolton, eds. 2013. *Aging in Community*. North Charleston, SC: CreateSpace Independent Publishing Platform.

Bookman, Ann. 2008. "Innovative Models of Aging in Place: Transforming our Communities for an Aging Population." *Community, Work and Family*, *11*(4): 419–438.

Brinig, Margaret F. 2015. "Grandparents and Accessory Dwelling Units: Preserving Intimacy and Independence." *Elder Law Journal*, *22*(2): 381–398.

Brown, Scott C., and Joanna Lombard. 2014. "Neighborhoods and Social Interaction." In *Wellbeing and the Environment: Wellbeing: A Complete Reference Guide*, edited by Rachel Cooper, Elizabeth Burton, and C ary L. Cooper, 1–27. New York: John Wiley & Sons.
Chapman, Nancy J., and Deborah A. Howe. 2001. "Accessory Apartments: Are They a Realistic Alternative for Ageing in Place?" *Housing Studies*, 16(5): 637–650.
Chapple, Karen, Wegmann, Jake, Mashhood, Farzad, and Rebecca Coleman. 2017. *Jumpstarting the Market for Accessory Dwelling Units: Lessons Learned from Portland, Seattle and Vancouver*. Berkeley, CA: Terner Center for Housing Innovation.
Choi, Jung Shin. 2004. "Evaluation of Community Planning and Life of Senior Cohousing Projects in Northern European Countries." *European Planning Studies*, 12(8): 1189–1216.
Choi, Jung Shin. 2013. "Difference of Motivation of Move to Swedish Senior Cohousing by Longitudinal Analysis: Focus on Differences between the Years of 2001 and 2010." *Journal of the Korean Home Management Association*, 31: 81–92.
City of Santa Cruz. 2003. *Accessory Dwelling Unit Manual*. Santa Cruz, CA: City of Santa Cruz.
Cobb, Rodney L., and Scott Dvorak. 2000. *Accessory Dwelling Units: Model State Act and Local Ordinance*. Washington, DC: AARP.
Cohen, Raines. 2005. "The Next Phase: Senior Cohousing." *Communities*, 127: 33–39.
Cohousing Association of the United States. n.d. "Cohousing." https://www.cohousing.org. Accessed March 29, 2017.
Davitt, Joan K., Amanda J. Lehning, Andrew Scharlach, and Emily A. Greenfield. 2015. "Sociopolitical and Cultural Contexts of Community-Based Models in Aging: The Village Initiative." *Public Policy and Aging Report*, 25(1): 15–19.
Durham Central Park Cohousing Community. n.d. Durham Central Park Cohousing Community. https://www.durhamcoho.com/
Durrett, Charles. 2009. *The Senior Cohousing Handbook: A Community Approach to Independent Living*. Gabriola Island, BC: New Society Publishers.
Enguidanos, Susan, Jon Pynoos, Alexis Denton, Susan Alexman, and Laura Diepenbrock. 2010. "Comparison of Barriers and Facilitators in Developing NORC Programs: A Tale of Two Communities." *Journal of Housing for the Elderly*, 24(3–4): 291–303.
Florida Department of Community Affairs. 2007. *Accessory Dwelling Units: Report to the Florida Legislature*. Tallahassee, FL: State of Florida, Department of Community Affairs.
Gemeenschappelijk Wonen. n.d. Adressen van Woongemeenschappen (Community Addresses)
Glass, Anne P. 2009. "Aging in a Community of Mutual Support: The Emergence of an Elder Intentional Cohousing Community in the United States." *Journal of Housing for the Elderly*, 23(4): 283–303.
Glass, Anne P., and Jun Shin Choi. 2015. "A Comparison of Elder Cohousing in Europe and the United States." *The Gerontologist*, 55, Suppl 2: 94.
Golant, Stephen M. 2015. *Aging in the Right Place*. Baltimore. MD: Health Professions Press.
Greenfield, Emily A., Andrew Scharlach, Amanda J. Lehning, and Joan K. Davitt. 2012. "A Conceptual Framework for Examining the Promise of the NORC Program and Village Models to Promote Aging in Place." *Journal of Aging Studies*, 26(3): 273–284.
Greenfield, Emily A., Andrew E. Scharlach, Amanda J. Lehning, Joan K. Davitt, and Carrie L. Graham. 2013. "A Tale of Two Community Initiatives for Promoting Aging in Place: Similarities and Differences in the National Implementation of NORC Programs and Villages." *The Gerontologist*, 53(6): 928–938.
Hayutin, Adele M., Miranda Dietz, and Lilian Mitchell. 2010. *New Realities of an Older America: Challenges, Changes and Questions*. Palo Alto, CA: Stanford Center on Longevity.
House, James S., Karl R. Landis, and Debra Umberson. 1988. "Social Relationships and Health." *Science*, 241: 540–545.
Howe, Deborah A. 1990. "The Flexible House: Designing for Changing Needs." *Journal of the American Planning Association*, 56(1): 69–77.
Hunt, Michael E. 1992. "The Design of Supportive Environments for Older People." *Journal of Housing for the Elderly*, 9(1–2): 127–140.
Hunt, Michael E., and Gail Gunter-Hunt. 1986. "Naturally Occurring Retirement Communities." *Journal of Housing for the Elderly*, 3(3–4): 3–22.
Hunt, Michael E., and Leonard E. Ross. 1990. "Naturally Occurring Retirement Communities: A Multiattribute Examination of Desirability Factors." *The Gerontologist*, 30(5): 667–674.
Ivery, J. M. 2014. "The NORC Supportive Services Model: The Role of Social Capital in Community Aging Initiatives." *Journal of Community Practice*, 22(4), 451–471.

Joint Center for Housing Studies of Harvard University. 2014. *Housing America's Older Adults: Meeting the Needs of an Aging Population*. Cambridge, MA: Joint Center for Housing Studies of Harvard University.

Katz, Sidney. 1983. "Assessing Self-Maintenance: Activities of Daily Living, Mobility, and Instrumental Activities of Daily Living." *Journal of the American Geriatrics Society*, 31: 721–727.

Killock, John. 2012. *Is Cohousing a Suitable Housing Typology for an Ageing Population within the U.K.?* Self-Published.

Koebel, C. Theodore, Julia Beamish, and Karen Danielsen-Lang. 2003. *Evaluation of the HUD Elder Cottage Housing Opportunity (ECHO) Program*. Blacksburg, VA: Center for Housing Research.

Liebig, Phoebe S., Teresa Koenig, and Jon Pynoos. 2006. "Zoning, Accessory Dwelling Units, and Family Caregiving." *Journal of Aging and Social Policy*, 18(3–4): 155–172.

Lipman, Barbara, Jeffrey Lubell, and Emily Salomon. 2012. *Housing an Aging Population: Are We Prepared?* Washington, DC: Center for Housing Policy.

McCamant, Kathryn, and Charles Durrett. 2011. *Creating Cohousing: Building Sustainable Communities*. Gabriola Island, BC: New Society Publishers.

MetLife Mature Market Institute and Louis Tenenbaum. 2010. *The MetLife Report on Aging in Place 2.0: Rethinking Solutions to the Home Care Challenge*. New York: MetLife Mature Market Institute.

Moran, Jan, and Paul Rollins. 2007. "Aging in Community: How the Coming Baby Boom Eneration Will Transform Traditional Models of Independent Living." *Terrain.org: A Journal of the Built and Natural Environments*, 20(Summer/Fall): n.p.

Pan, Yuqing. 2015. "The Strange Allure of Tiny Homes—Explained." Realtor.com/news/trends/strange-allure-of-tiny-homes-explained/.

Pedersen, Max. 2015. "Senior Co-Housing Communities in Denmark." *Journal of Housing for the Elderly*, 29(1–2): 126–145.

Petrowski, Elaine. 2010. "Prefab 'In-law' Cottages Mix High-Tech Features, Comfort." https://www.aarp.org/home-garden/housing/info-10-2010/abcs_of_adus.html

Ross, Jaimie. 2016. "Accessory Dwelling Units: A Smart Growth Tool for Providing Affordable Housing." *Housing News Network*, 32(2): 17–20.

Sage Computing, Inc. 2008. *Accessory Dwelling Units: Case Study*. Washington, DC: U.S. Department of Housing and Urban Development.

Seeman, T.E., T.M. Lusignolo, M. Albert, and L. Berkman. 2001. "Social Relationships, Social Support, and Patterns of Cognitive Aging in Healthy, High-Functioning Older Adults: MacArthur Studies of Successful Aging." *Health Psychology*, 20: 243–255.

Silver Sage Village. n.d. Homepage. http://silversagevillage.com

Thomas, William H., and Janice M. Blanchard. 2009. "Moving Beyond Place: Aging in Community." *Generations*, 2: 12–17.

Tummers, Lidewij. 2016. *The Re-Emergence of Co-Housing in Europe*. London and New York, Routledge.

New York State Office for the Aging (NYSOA). n.d. *Naturally Occurring Retirement Community Support Service Program (NORC-SSP) and Neighborhood NORC (NNORC)*. https://aging.ny.gov/NYSOFA/Programs/CommunityBased/NORC-NNORC.cfm

United Nations. 2013. *World Population Ageing 2013*. New York: United Nations.

Village-to-Village Network. n.d. *About*. https://www.vtvnetwork.org/

Vladeck, Fredda, and Anita Altman. 2015. "The Future of the NORC-Supportive Service Program Model." *Public Policy and Aging Report*, 25(1): 20–22.

Weisfeld, Victoria, and Tracy A. Lustig. 2015. *The Future of Home Health Care: Workshop Summary*. Washington, DC: National Academy of Sciences.

Wild Sage Cohousing. n.d. *Wild Sage: A Cohousing Community*. http://www.wildsagecohousing.org

19

REALIZING INNOVATIVE SENIOR HOUSING PRACTICES IN THE U.S.

Deirdre Pfeiffer, Ashlee Tziganuk, Scott Cloutier, Julia Colbert, and Gracie Strasser

Introduction

Countries across the developed world have been in the throes of a dramatic demographic change—aging. In the U.S., the senior population aged 65 and older will likely double from 43 to 84 million from 2012 to 2050 (Ortman et al. 2014). By 2050, seniors will comprise an estimated one out of every five adults in the U.S. (Ortman et al. 2014). Seniors live differently than other age groups and use their housing and communities in unique ways. Older persons are more likely to live alone or in multigenerational households (Taylor et al. 2010). They also express a strong desire to age in place near their friends and family (Harrell et al. 2014). These qualities mean that distinct housing approaches are required to keep seniors healthy, socially integrated, and supported (Pfeiffer 2018).

Cohousing and accessory dwelling units (ADUs) are two innovative ways of housing seniors. Seniors living in cohousing have separate dwellings but shared communal spaces and often participate in the planning and maintenance of their communities. Seniors living in ADUs live in a secondary home on a single-family home property, or live in a primary home with a friend, family member, caretaker, or tenant living in the secondary home. U.S. housing planners have long expressed interest in adapting and scaling up these approaches. However, to date, few cohousing communities or legal and permitted ADUs have been built for seniors in the U.S. (Antoninetti 2008; Schafran 2012; The Cohousing Association of the United States 2017).

This chapter draws on semi-structured interviews with 22 U.S., Danish, and Canadian housing practitioners and experts on seniors to explore the challenges of meeting seniors' housing needs through cohousing and ADUs in the U.S., and to understand how these challenges could be overcome. We first introduce the concepts of cohousing and ADUs and describe our research method. We then draw on lessons from our interviews about the potential and challenges to house U.S. seniors through cohousing and ADUs, triangulating our findings with those from prior research. We conclude with a set of next steps for moving the conversation on cohousing and ADUs for seniors from visioning to production.

Cohousing

The concept of cohousing originated in Denmark, though cohousing was first coined and adapted to the U.S. context in 1988 by Charles Durrett and Kathryn McCamant, who founded The Cohousing Company around the same time (Durrett 2009). A cohousing community includes private and shared spaces. Typically, people living in a cohousing community occupy separate dwellings but use and maintain shared amenities, such as a kitchen, a garden, or a yoga studio. Dwellings may be individually or cooperatively owned or rented. A unique feature of cohousing is its participatory development and community governance (Tummers 2016). Residents take an active role in the ongoing maintenance of a cohousing community's shared spaces, which leads to strong social bonds among residents (Sanguinetti 2014; Sanguinetti and MacLane 2016). Some cohousing communities, like Harbourside in British Columbia, are even collaboratively conceived and constructed by their future residents. Other communities, like Takoma Village in Washington, DC, are built by private developers and marketed to people searching for greater community engagement and social interaction with neighbors (Critchlow 2015). Cohousing typically is not subsidized in the U.S. Thus, living in this type of community usually is out of reach for most low-income households (Sanguinetti 2015; Tummers 2016). However, there are a few examples of cohousing built by nonprofit affordable housing organizations, such as the ElderSpirit community in Abingdon, Virginia, which was financed through fundraisers by the Trailview Development Corporation (Glass 2009).

Senior cohousing communities in the U.S. are rare but becoming more popular. As of 2017, there were about 163 cohousing communities and another 134 in development (The Cohousing Association of the United States 2017). Most cohousing communities are intergenerational, but interest in senior-oriented communities has been increasing. There were 12 established senior cohousing communities and 13 in development as of 2017 (The Cohousing Association of the United States n.d.). An example of senior-oriented cohousing is Silver Sage Village in Boulder, Colorado, which has 16 homes and 24 residents. The community contains a common house, which hosts social gatherings, and is within walking distance to grocery stores, shops, and hiking trails (Durrett 2009; Silver Sage Village Cohousing 2017).

The Cohousing Company has been a leader in the promotion of senior-oriented cohousing communities like Silver Sage Village. This organization led a workshop called Getting It Built in the spring of 2017, which focused on three essential elements for successful cohousing projects: membership, development, and financing. The workshop helped members learn how to organize and make decisions in groups, plan, and design the communities with the help of professionals, and keep track of financing options and investment structures (McCamant and Durrett Architects 2017).

Accessory Dwelling Units

An accessory dwelling unit (ADU) is a technical term for a housing type known by many names, including a granny flat, backyard cottage, or in-law suite. An ADU is a secondary living quarter on a single-family home property. ADUs take diverse forms, from stand-alone cottages, to attached apartments, to basement conversions. The relationship between the resident of the primary unit and the ADU may vary, from landlord to renter, to guardian to dependent, to caretaker to person needing care (Litchfield 2011).

ADUs were more common in U.S. neighborhoods in the early twentieth century. The use and production of secondary units declined after the emergence of single-use or Euclidian zoning after 1926 (*Village of Euclid v. Ambler Realty Co.* 1926). Land use practices that privileged uniformity rather than diversity in the housing stock became the norm. Consequently, the single-family home became the American ideal (Liebig et al. 2006).

Estimates of ADUs in the U.S. are difficult to derive, given that most are built extralegally in jurisdictions that prohibit them outright or impose strict zoning and building code requirements. Gellen (1985) estimated that between 60,000 and 300,000 illegal ADUs were built each year in the early 1980s. It is likely that these numbers have increased in the subsequent years, as affordable housing has continued to be a pressing issue in many communities in the U.S. Only an estimated 330 jurisdictions in the U.S. allowed for ADUs in some form in 2012, a small fraction of the total number of localities nationwide (Schafran 2012). ADUs have been more prevalent in English speaking countries in the developed world, including Canada (Hare 1991). In the U.S., ADUs are more prevalent in major West Coast cities like Los Angeles and Seattle, due in part to state mandates that require localities to accommodate them (Chapman and Howe 2001; Liebig et al. 2006).

There are no estimates on the number of U.S. seniors living in ADUs. However, survey research suggests that seniors' interest in living in ADUs may be low (Chapman and Howe 2001; Varady 1990). In Seattle, for example, only 14 percent of ADU owners and 11 percent of ADU tenants were over age 65 in 1998 (Chapman and Howe 2001). Interest in ADUs tends to be higher among higher-income seniors, as well as those who are experiencing financial or physical distress (Chapman and Howe 2001; Varady 1990).

Moving From Visioning to Implementation

Cohousing and ADUs have the potential to meet some seniors' needs. Yet, most planners have not fully realized these approaches in the U.S. This is discouraging given that housing and senior advocates have long rallied around these approaches (e.g., Cobb and Dvorak 2000). Thus, planners should move the conversation from envisioning the potential of these approaches to realizing them. In the remainder of this chapter, we help to bridge the gap between envisioning and realizing cohousing and ADUs for U.S. seniors by drawing insights from semi-structured interviews with housing practitioners, academics, and experts on seniors in the U.S., Denmark, and Canada, and building on lessons from prior research.

Data and Method

The case studies of Denmark and Canada were chosen for investigation because both countries have experienced greater success in implementing cohousing and ADUs than the U.S. Cohousing, or *bofælleskab*, was originally developed in Denmark by architects Gudmand-Hoyer and Bodil Graae in the late 1960s (Egerö 2012). The architects' designs attracted significant interest, and two cohousing projects were developed by 1973. By 1981, the Danish government passed legislation for financing cohousing that still aids development today. Cohousing is quite common in Denmark, being found in most master plans (Egerö 2012). Senior-oriented cohousing has also been on the rise in Denmark, with an estimated 250 senior cohousing projects developed since 1987 (Pedersen 2015). Cohousing is quite prevalent in the Danish urban fabric, often in the form of retrofitted older buildings, and many seniors are able to move from their current home to a cohousing development within the same city or, in some cases, community. Amenities like grocery stores, health clinics, and even retail are commonly located on the bottom floors of cohousing complexes, and public transportation is typically nearby, which increases seniors' accessibility and mobility and enables them to age in place.

Almost all major cities in Canada allow ADU development in some capacity, an outcome of persistent population growth and housing affordability challenges—the same factors that are currently lending attention to ADUs in the U.S. (Mendez 2011; Touati 2013). Extralegal, unpermitted ADUs were common in Canadian cities in the early 1990s (Touati 2013). Ontario made ADUs legal in 1994 but shortly after revised its law to allow municipalities to ban ADUs after pushback from residents about potential adverse neighborhood effects, such as increased traffic and fears about an influx of student housing (Touati 2013). Interest in ADUs as a low-cost form of housing resurfaced in the wake of housing affordability challenges in many Canadian urban areas (Touati 2013). In 2016, the Canada Mortgage and Housing Corporation (CMHC) reported that 77 percent of 650 municipalities in Canadian metropolitan areas permitted ADUs—a 26 percent increase from 2006 (Canadian

Mortgage and Housing Corporation 2016). The City of Vancouver alone was estimated to have 26,600 ADUs in 2014, comprising about a fifth of the total city housing stock (Canadian Mortgage and Housing Corporation 2016). However, some Canadian cities, like Calgary, still struggle with implementation despite gains in legalization (Bell 2016).

We conducted semi-structured interviews with housing practitioners, academics, and senior advocates in the U.S., Denmark, and Canada to better understand the challenges and opportunities of cohousing and ADUs for U.S. seniors. We first developed a list of potential participants involved in cohousing and/or ADU activities in the three countries, which was derived from industry reports, scholarly articles, and media reports. We invited these actors to participate in the research over either email or a phone call. We followed up twice with those who did not respond to our invitation, then removed them from the list of potential participants. Additional informants were added through snowball sampling until themes began to repeat in the interviews (Small 2009; Yin 2014). In total, we asked 47 people to participate in an interview.

Our final sample had 22 participants (47 percent response rate). Fourteen of the participants had expertise on the U.S., four had expertise on Canada, and four had expertise on Denmark. The interviewees represented a wide range of perspectives, including those of developers, planners, academics, politicians, housing organization leaders, and journalists. All participants had direct experience in developing or regulating cohousing or ADUs, or had observed these processes as a resident, academic, or journalist. Twelve of the participants had expertise on cohousing. Seven had expertise on ADUs. Three of the participants (all from the U.S.) had expertise on cohousing and ADUs.

The interviews were conducted remotely (over the phone or through a video conference) from July 2016 to January 2017. We asked the participants about their perceptions of seniors' housing needs and the benefits and limitations of ADUs and cohousing in meeting these needs. We then asked the participants about their experiences with cohousing, ADUs, or both cohousing and ADUs, if applicable (e.g., involvement in policymaking or construction, etc.), and their perception about the factors affecting the outcomes of these experiences (e.g., implementation of a policy, completion of a project, etc.). We asked U.S. participants to reflect on the barriers to implementing cohousing and/or ADUs in the U.S.; Danish and Canadian participants were asked to reflect on why cohousing or ADUs were more prominent in their countries than the U.S. We also asked the Danish and Canadian participants to give advice to people who wanted to scale up these approaches in the U.S.

The content of the interviews was summarized, with illustrative quotes transcribed in full. The transcripts were coded using a deductive and inductive method to discern differences and commonalities in the opportunities and constraints of scaling up the two approaches in the U.S. We triangulated the interview data with findings from the existing literature on cohousing and ADUs in drawing conclusions (e.g., Antoninetti 2008; Chapman and Howe 2001; Durrett 2009; Liebig et al. 2006; Pfeiffer 2015; Sanguinetti 2014, 2015; Tummers 2016; Varady 1990).

Our approach to the research has strengths and limitations. One strength of our approach is our comparison of the experiences of diverse actors in three countries to distill the context-specific and overarching opportunities and constraints of cohousing and ADUs for seniors. One limitation of our approach is that our findings do not fully represent the experiences or perspectives of the population of actors involved in cohousing or ADUs in the three countries. We faced particular difficulty recruiting participants with expertise on Danish cohousing. Only four of the 16 people with expertise on Danish cohousing contacted responded to our request for an interview after two follow ups, compared to four of the seven Canadians and 14 of the 24 Americans. We found that language barriers and other possible differences in communication, such as Danes' tendency to check their email less frequently and take more time in responding to emails or phone calls, may have contributed to this outcome.

The Opportunities and Constraints of Cohousing and Accessory Dwelling Units for Seniors

We explore the opportunities and constraints of cohousing and ADUs for seniors in this section, triangulating themes from our interviews with those from existing literature. First, we discuss the barriers and limitations to meeting seniors' housing needs through cohousing and ADUs. Then, we address factors important to the success or failure of these innovative housing practices. We conclude by discussing unique challenges facing implementation in the U.S context.

Designing for Independence: "Privacy With Proximity" in Cohousing and ADUs

Cohousing and ADUs expand and protect seniors' independence, particularly their capacity to age in place as their needs and abilities change. These alternative housing choices offer seniors greater physical, social, and emotional accessibility than more common housing types available, while preserving privacy (Hare 1991).

Physical Accessibility

Seniors need homes that promote physical accessibility, within both the home and the surrounding community. Physical accessibility pertains to the physical structure of the home and its connections to the larger community and entails a match between the seniors' physical abilities and the layout of their home and community.

Good housing design can make aging in place easier for seniors, including abiding by principles of universal design, which offers seniors a safe environment through features like wide doorways and hallways, wheelchair accessibility, and first floor entry (Null and Cherry 1996). Other physical features internal to the home that participants felt were critical to seniors' well-being are the ability to communicate with those outside of the living space in case of accidents, the ability to easily maintain the home, and a sense of comfort and safety. Further, many participants mentioned that including physical features to enable seniors' aging in place was not enough; a home also must appeal to aesthetics to encourage aging people to stay long term. One Danish cohousing chairman advised that a home "should be able to accommodate wheelchair moving, but should not look like a special home for elderly people." Seniors want homes that are attractive, have plenty of light and space, and, most importantly, allow autonomy, according to our participants and the existing literature (Glass and Vander Plaats 2013; Liebig et al. 2006).

Seniors' livelihoods also depend on whether they can access their communities from their homes as their needs and abilities change (Harrell et al. 2014). Losing one's ability to drive can isolate seniors from their communities, particularly in the suburban U.S., where 46 percent of elderly people live (see also the chapter authored by Sherry Ahrentzen and Ruth Steiner). One U.S. cohousing founder and resident exclaimed, "You can't walk anywhere or do anything without a car. The minute they take your keys away from you, you're stuck." Many participants stressed that housing developers and planners need to consider designs that allow seniors to age in place not only in their homes but also in their communities, particularly by placing senior-oriented housing near public transit (Harrell et al. 2014).

Cohousing's unique membership and design-driven process have the potential to meet the physical needs of more community-minded seniors. Cohousing oftentimes starts with a core group of highly motivated and organized people who come together to build and maintain a community specifically tailored to meet their needs (Critchlow 2015). Seniors involved in building cohousing can incorporate attributes to foster their long-term accessibility within their homes and communities early in the planning process. One U.S planner stressed, "People in cohousing can often design their own units, which you can't do if you are just buying or renting some place."

ADUs built by senior homeowners may provide seniors financial or caretaking support so that they can remain in their familiar homes and neighborhoods (Antoninetti 2008). ADUs built for seniors by non-senior homeowners (e.g., seniors' relatives) allow seniors to downsize to a smaller home that is easier to maintain (Antoninetti 2008). Non-senior homeowners can retrofit ADUs to meet the physical needs of seniors, if they can afford it. Improving the accessibility of ADUs to community amenities is more difficult, as it depends on the location of the primary house (see also the chapter authored by Stephen Golant; Harrell et al. 2014).

Social Accessibility

Physical accessibility is only one important dimension of housing for seniors; social accessibility is another important dimension. Social accessibility pertains to housing and neighborhood characteristics that help to connect seniors to one another and build a sense of community.

Cohousing and ADUs provide greater social accessibility for seniors in different ways. Social interaction is integral to the planning and design of a cohousing community. Indeed, friends or relatives sometimes form cohousing communities and intentionally place their homes on the site to allow for social interaction. Cohousing also cultivates social interaction through communal governance and the design of communal spaces, such as community centers, dining halls, or community gardens (Critchlow 2015; Glass 2009; Sanguinetti 2014). A Danish cohousing chairman explained, "It is a great advantage that you get to know your neighbors well and meet them frequently in the common house or in the open space of the [cohousing] community. You very easily find other people to join you in new activities. If you need practical assistance you will always find someone who can help you." A U.S. architect with a focus on cohousing added that the ability "to have people to talk with through the course of the day . . . isn't necessarily available in today's standard residential environments or even . . . Sun City type for profit developments for older people." A Danish architect and cohousing resident provided a glimpse of what daily life is like for a senior living in cohousing:

> I see [my neighbors] regularly, because when I am home, I always have dinner in our common spaces. We have dinner 5 days a week, and we are taking turns cooking dinner, so every 6 weeks I am part of the group that is preparing, serving, washing up. And [for] 5 weeks I can go to a good and not expensive dinner, and this is good, because doing things together is the best way to get to know people. And to get to know people when you are over 50 years old is not easy, but when I moved in here, I got like 50 new people served on a silver platter.

Cohousing's frequent opportunities for social interaction through communal governance and shared spaces may forge deep social bonds and improve caretaking among residents (Critchlow 2015; Glass 2009; Sanguinetti 2014). Nearby neighbors provide emotional support or help in times of crisis (Critchlow 2015). The close proximity to neighbors afforded through cohousing's clustering of housing and communal spaces may keep seniors from becoming socially isolated and lonely and lead them to feel safe (see also the chapter authored by Sherry Ahrentzen and Ruth Steiner; Glass and Vander Plaats 2013).

Residents in one participant's cohousing community call these kinds of neighbors "proximate neighbors." Proximate neighbors provide much more immediate help and fulfilling social contact than the "check ins with professionals" who over time compose the social network of some seniors not living in cohousing. Proximate neighbors are common in neighborhoods without cohousing. However, residents have more proximate neighbors in a cohousing community, because cohousing's communal governance and spaces lead all residents to know one another (Glass and Vander Plaats 2013; Sanguinetti 2014). A U.S. national cohousing member recalled a time when "one of our neighbors was out of town, and their sprinkler system broke, and everyone pitched in and helped them collect their house and call the fire department." Having proximate neighbors is critical to the emotional health of seniors who have recently lost a spouse. Women are at particular risk of experiencing loneliness, as they tend to live longer than men. A Danish cohousing chairman explained: "Even when you lose your spouse you will still have the network around you and can continue an active life." Considering the rich social opportunities integrated into life in cohousing, participants described cohousing as "a college dorm for seniors" and "a village."

Unlike cohousing, ADUs are not intentionally designed to foster social interaction. However, ADUs do offer seniors the ability to live close to family and stay in the communities where they are rooted, while also maintaining some autonomy (Hare 1991). "With regard to ADUs," a U.S. ADU consultant explained, "I will say that many homeowners or many old people appreciate being able to live nearby their kids, and ADUs are one way to achieve that while also providing privacy." ADUs built by senior homeowners also enable seniors to house a caretaker on-site, avoiding the need to leave the community to live near a caretaker or in a supportive facility. As a Canadian councilman stated, ADUs are "a non-disruptive way of supporting seniors in place."

Emotional Well-Being

The social and physical accessibility of seniors' homes contributes to their emotional well-being, which is highest when a senior's home and community provide them with a sense of purpose and control. Evidence indicates that feeling that one's life has meaning is strongly associated with happiness (Lyubomirsky et al. 2005).

Studies indicate that living in cohousing can endow seniors who desire greater community engagement with a sense of purpose or meaning due to the participatory nature of decision-making and volunteering requirements for maintaining communal spaces (Critchlow 2015; Glass 2009). A U.S community land trust secretary reflected, "I think the democratic way of making decisions [in cohousing] makes [seniors] feel included. That decision-making power allows them to . . . produce something that feels meaningful . . . leave a legacy behind." Seniors desiring to contribute to upcoming generations may derive a sense of purpose from building an ADU on their property, as this approach allows them to house others needing a place to live. The U.S community land trust secretary felt that building an ADU leads seniors to feel like "a productive member of society."

The mix of private and common spaces in a cohousing community enables residents to interact with one another as they like, and retreat back to their private homes when they want to be alone, giving them a sense of privacy and control over their lives (Chapman and Howe 2001; Liebig et al. 2006). As one U.S. planning commissioner stated, cohousing and ADUs enable a senior "to take control of the next stage of your life." Cohousing gives seniors control by allowing them the ability to retreat back into a private home, which protects their privacy and independence. "You also need to be able to shut your own door and to be absolutely private," the Danish architect emphasized. "To be social also means you have to be private sometimes." Seniors living in cohousing also have control through self-governance, unlike seniors living in institutional settings. One Canadian cohousing resident explained, "We organize our own events, they aren't organized by a care committee. We play games and bridge, have a pub night, and it is all organized by ourselves." Two U.S. cohousing experts stressed that cohousing allowed seniors to create an institution that "fit more like a glove" rather than "a floppy old boot" or "a straightjacket," as institutions designed by experts and other outsiders were commonly characterized.

ADUs offer seniors control by providing "privacy with proximity" to family, friends, and others who can help with daily chores, transportation, and caretaking in times of illness (Hare 1991, 60).

ADUs also provide seniors greater control through financial security. Renting out or living in an ADU may offer seniors additional income or lower living costs, which may help to prolong their financial independence and may leave them with more money left over to pay for other expenses, such as health care.

The participants acknowledged that features of ADUs also threaten seniors' physical and social accessibility and emotional well-being. For instance, some types of ADUs are more suitable for seniors than others. A Canadian academic worried that seniors living in ADUs in converted basements might "feel like second class citizens" and that senior couples may find that ADUs are too small to meet their needs. A Canadian cohousing resident reflected that "isolations" might arise when seniors live near younger family members through ADUs, especially if their relatives perceive them as "being a burden on them."

Social Capacity, Regulations, and Financing: Challenges and Opportunities in Developing Cohousing and ADUs

Realizing the aforementioned benefits of senior cohousing and ADUs requires overcoming diverse challenges. Fostering and sustaining social capacity among grassroots and professional actors are key challenges to developing cohousing (Critchlow 2015; Glass 2009). Overcoming regulatory barriers to development is a barrier to developing ADUs (Antoninetti 2008; Liebig et al. 2006; Wegmann 2014). Cohousing and ADUs share the challenge of obtaining financing, which is difficult given banks' and lenders' lack of awareness of these housing types (Antoninetti 2008; Critchlow 2015). Our participants had various ideas on how to overcome these diverse challenges.

Building Social Capacity to Support Cohousing

Diverse forms of human and social capital are needed to support the successful development of cohousing (Labit 2015; Tummers 2016). Human capital pertains to the skills that people develop that enable them to accomplish goals (e.g., learning a skill like carpentry or studying law, accounting, or the arts). Social capital arises when people form relationships built on trust and shared values and can rely on these relationships to gain resources or knowledge, drawing on their human capital (Coleman 1988). Human and social capital are critically important to foster and harness in the process of developing cohousing; however, there are barriers to sustaining these forms of capital in projects involving seniors.

Opportunities to develop social capital arise early in the cohousing process. Cohousing projects begin with an interested group of potential residents, who may or may not know one another. Oftentimes a sense of community is created at the beginning of the development process, which can grow into social capital overtime (Labit 2015). However, participating in collaborative decision-making over a long time period may be challenging for seniors who lack self-motivation and determination (Glass 2009). The frequent and time-consuming nature of initial cohousing development meetings can strain seniors' capacities, slowing down progress (Glass 2009; Labit 2015). The development process may include many setbacks, such as financial barriers or issues with site selection (Critchlow 2015). Further, personalities and values may clash during consensus building, which can leave seniors unhappy or feeling compromised (Glass 2009; Labit 2015). One Danish architect and cohousing resident reflected on the early stages of cohousing development, "It is so much work you have to do. It is much easier to live in a prefabricated home, so it is difficult." Passion and tremendous reserves of stamina are required to continuously contribute throughout the planning and development process. A cohousing researcher reflected, "Ideally [cohousing participants] start building a sense of community with each other even before the physical cohousing is built, but that takes energy also to keep people coming together and build the relationships that will see you through the long haul."

Strong leadership can help to alleviate some of these early difficulties in planning cohousing (Critchlow 2015). Cohousing leaders play a vital role in working with professionals, like developers and architects, as well as in fostering connections among residents and keeping them involved (Critchlow 2015; Tummers 2016). A U.S. planner and cohousing architect mused on the role that leaders play in cohousing development:

> They learn how to partner, they are curious, they investigate, and there is a core group of go-getters that are willing to work every day in developing the community. After that, it's how do you succeed in maintaining the vitality and connections of people. You need a variety of strengths and personalities and being able to work and appreciate.

Some participants also stressed the importance of evolving leadership or nurturing future leaders to succeed the current ones. An unresolved challenge of cohousing is that it is designed to meet specific needs held by a group of people, but these needs may change as the community experiences turnover.

Professional partnerships complement strong leadership in aiding the success of a cohousing development process (Labit 2015). At the start of a cohousing project many seniors oftentimes have no idea how to begin the project or what planning, building, legal, and financial standards must be followed. People who have "never built anything in their lives" often initiate cohousing, according to a leader of a U.S. cohousing professional organization. Involving external help from architects, developers, and construction managers is critical to ensuring that the project follows professional development standards and is completed in a timely manner. As a Canadian cohousing resident explained:

> The complexity of building a multi-family residential community today is beyond the grassroots ability to organize. So expert project management is required. We like to say there are two wings to the bird of cohousing, and one is the potential group and the others are the professionals who support and bring the project through, on budget and on time and with a design that works for everybody. And without [both wings] you don't get to take off.

A challenge with involving external professionals is a lack of awareness within these professions of what cohousing is, along with its participatory planning process and communal governance and spaces (Critchlow 2015; Glass 2009; Tummers 2016). Some private developers contracted by prospective cohousing residents lack experience with participatory planning and may struggle to work with a large group of seniors and meet their needs. A Danish architect and cohousing resident stressed that "construction companies . . . must . . . learn to plan with others, not the way that they usually do." Another cohousing researcher reflected, "Being able to find a developer who understands the real concept of sense of community and is willing to work with you to build it into that model [is challenging]." Another challenge with working with developers in particular is that more lucrative projects might compete for their time. One solution is to involve these professionals early in the planning process and to work collaboratively with them throughout the process.

Overcoming Regulatory Barriers to ADUs

Regulatory barriers are the primary hurdles that prospective developers of ADUs must overcome (Antoninetti 2008; Liebig et al. 2006; Pfeiffer 2015). Regulations are laws that guide housing development. A particularly challenging issue is complying with local zoning ordinances. Rigid Euclidian zoning and homeowners' association rules prohibit families from adapting their single-family home properties as their needs change, including integrating more creative housing types that can benefit seniors, like ADUs (Howe 1990).

The crux of the controversy about ADUs is that they entail adding density to single-family home properties. ADUs often receive pushback from residents and local government when they are proposed (Antoninetti 2008; Liebig et al. 2006 Mendez 2011; Touati 2013). Acquiring a permit often requires homeowners to demonstrate owner occupancy and meet difficult site requirements, such as adequate setbacks and parking and wide passageways (Pfeiffer 2015; Wegmann and Chapple 2014). A U.S. planner used the example of the City of Los Angeles' requirement that ADUs have a 10-foot wide passageway to provide access, "which excludes about 95 per cent of lots" from having permitted ADUs. Many homeowners give up on obtaining a formal permit and instead opt to build an illegal or covert ADU, which can pose hazards to occupants if the ADU is not in compliance with standard building codes. Concerns that ADUs will degrade quality of life by increasing congestion or noise deter planners from advocating for ADUs in jurisdictions that prohibit them (Liebig et al. 2006; Pfeiffer 2015). A U.S. planning commissioner stated that local officials "are focused about keeping the cars off the streets, all parking on-site, keeping it covered, and then when you step back you won't end up with many ADUs."

Developing cohousing also requires overcoming some regulatory issues with zoning (Critchlow 2015; Tummers 2016). However, these challenges typically stem from acquiring a large enough site and meeting site requirements (Critchlow 2015). A leader of a U.S. cohousing professional organization reflected: "Cohousing communities typically want to focus on having a common house that is surrounded by clustered homes. That goes against most of the zoning in the suburbs in the entire country." Overcoming grassroots local NIMBY (not-in-my-backyard) issues that are not codified in the zoning regulations is less of a problem.

Reform of local zoning codes is needed to overcome regulatory barriers to cohousing and ADUs. Although there are occasionally national proposals to reform zoning, such as a White House initiative to challenge growth-stifling local zoning laws during the Obama presidency, the impetus to overcome regulatory barriers to ADU development lies with local actors. Each city has its own unique zoning and geographic differences, making it difficult to generalize recommendations for large scales (Chapple et al. 2011). As a U.S. ADU consultant explained, "Unfortunately, this type of development is on the local level, and there is nothing really nationally that one can do regarding zoning changes . . . that is the primary issue when it comes to ADU development."

Our participants stressed that local planners and politicians play an important role in helping to advocate for changes to zoning and other regulations to support ADU development. A Danish architect and cohousing resident mentioned the important role that a mayor in Stockholm, Sweden, played in spurring cohousing development in that city in the early 1980s. A promising trend in the U.S. is that concerns about housing affordability and aging in place are leading local regulators to advocate for or consider ADUs as an approach in jurisdictions that had formerly prohibited them (Antoninetti 2008; Liebig et al. 2006; Moore and Palleroni 2008; Pfeiffer 2015; Liebig et al. 2006). Examples of U.S. localities that have approved ADU regulations abound, including Montgomery County, Maryland; Arvada, Colorado; and Austin, Texas (City of Arvada n.d., Montgomery County n.d., Schafran 2012). Our participants also felt that grassroots initiatives and organizations were needed to push back against homeowners' associations and other NIMBY actors such as the local board of realtors that often combat and eventually stymie rezoning changes needed to legally build ADUs (Schafran 2012). Examples of such initiatives include the "friends of granny flats" in Denver, though it is one of the few (loosely) organized groups in the city (Schafran 2012).

Financing Cohousing and ADUs

A lack of financial capital is one of the most burdensome barriers to innovative senior housing development. We define financial capital as the monetary constraints or opportunities that exist for cohousing and ADU development. Our participants touched on themes of affordability and a lack of public subsidies and knowledge and support from mainstream financial institutions.

Cohousing and ADU projects require substantial financial outlays from residents at the start of the project, which can constrain access to low-income seniors and potentially limit the economic diversity of cohousing and ADU communities (Glass 2009; Tummers 2016). Personally investing in these projects allows residents the freedom to design their own living spaces, but it also creates issues of affordability (Critchlow 2015). Building a permitted ADU is costly. For example, constructing a 600-foot two-bedroom ADU in the San Francisco Bay Area cost an estimated $250,000 in the early 2010s (New Avenue 2014). This status quo puts ADUs out of reach for all but the most affluent homeowners, or leads homeowners to construct unpermitted and potentially temporary units that may expose residents to safety hazards. However, tract housing developers have recently started to incorporate ADUs in their communities, mainly to attract multigenerational families (Spivak 2012). Economies of scale in these developments may make ADUs more affordable to seniors (see also the chapter authored by Sebastiaan Gerards, Erik Nuyts, and Jan Vanrie).

Seniors also often must make large financial contributions to live in cohousing (Glass 2009). The required personal investment includes not only a share of the cost of building and maintaining the residences and communal buildings but also the cost of the land, which is expensive in high-demand coastal states, like California. A member of a U.S. national cohousing association explained, "For seniors who are on a fixed income, [cohousing is] pretty expensive unless they have a nest egg from all of the years spent working and equity from their previous homes. A lot of people don't have that. . . . Affordability is a big issue with getting [cohousing] spread around." Taking part in a cohousing community also requires a substantial time commitment from seniors, as discussed earlier. Low-income seniors who are not yet retired from their jobs may not have much time available to participate in these meetings. A U.S. planner and cohousing board member noted, "Right now cohousing is the purview of highly educated people with assets, and by that I mean people who figured out a way to sit in endless meetings with each other while things get worked out."

The upfront financial burden of cohousing and ADUs is further compounded by a lack of financial support and awareness in the public and private sectors. There is currently no public financial support in the U.S. for cohousing or ADUs other than the Mortgage Interest Deduction claimed by many homeowners, unlike in Scandinavia (Egerö 2012). One option could be to offer tax incentives for residents or developers who wish to build these projects in exchange for some affordability restrictions. However, integrating an affordability requirement may threaten the grassroots participatory design process that is at the heart of cohousing. A U.S. planning commissioner gave the example of a U.S. city that secured affordable rental housing concessions by donating land to a cohousing project. A lottery system was used to select residents for the affordable units after the design process was complete, which led the community to have lower social cohesion when the community was occupied, given that the lottery recipients did not necessarily know one another.

Cohousing and ADU developers also typically face challenges securing private financing for their projects, as lenders are hesitant to finance both housing types. The partially cooperative ownership structure of cohousing, where residents own units individually but share a stake in common spaces, is one concern (Critchlow 2015). The fact that residents play the role of developers in these projects, which creates a need for construction and mortgage financing, is another concern (Critchlow 2015). A leader of a U.S. cohousing professional organization explained, "Financial institutions don't understand cohousing. They can say, 'I don't want to work with 20 people; I want a developer to walk in.'" ADUs also are challenging to finance using traditional loan products, because they are seen as riskier investments. A U.S. planner stressed, "It's easy to go to a bank and get a mortgage to get an existing house. It's more difficult to go and get $100,000 for a backyard house." One solution to the problem of lack of financial capital is to provide greater education to private lenders, their underwriters, and secondary mortgage market investors about these housing types, and to provide greater education to prospective developers about ways of building cohousing and ADUs to maximize their investment potential.

Broadening Notions of the American Dream: Developing Cohousing and ADUs in the U.S. Context

Cohousing and ADUs are communitarian, bottom-up approaches to housing seniors. The values expressed by cohousing and ADUs are more consistent with cultures in countries like Denmark and Canada than the U.S., which espouses a more individualistic ethos of self-sufficiency, though some ethnic subgroups within the U.S. (e.g., the Amish, Blacks/African Americans, recent immigrants, etc.) may be more amenable to communal housing approaches. Many U.S. participants identified the threat that these housing types posed to conceptions of the American Dream as an additional challenge to housing seniors through cohousing and ADUs and stressed the importance of using education to help overcome this fear.

The individualistic nature of U.S. culture has had cascading effects on Americans' perceptions of housing. Many Danes and Canadians are more cooperative and have higher levels of social trust than Americans, who value their ability to purchase individual single-family homes, a desire cultivated in part by public policies such as the Mortgage Interest Deduction (Liebig et al. 2006). A U.S. cohousing resident attributed the success of cohousing in Denmark and Canada to their "much more socialistic . . . way of living . . . homogeneous society." The Danish communitarian tradition in particular stems back to early cooperative dairies and shops in the mid-nineteenth century (Egerö 2012). A Danish cohousing chairman mused, "[I]t has always been part of our culture to go together if we want to do something or get something done." In contrast, difficulties with producing cohousing in the U.S. stem from its "rugged individualism, dog-eat-dog [culture]," as a U.S. cohousing resident put it, and "a culture of people who are about picking themselves up by their bootstraps rather than caring and sharing," according to a U.S. cohousing organization member.

U.S. consumerism also poses several challenges to housing seniors through cohousing and ADUs. First, Americans are used to buying preexisting or new housing that fully meets their current and projected future needs rather than creating new housing or adapting older housing to meet their needs. Second, the American life cycle is premised on the idea of social mobility through continual housing upgrades rather than downsizing as needs change or redistributing resources toward the end of life, though the recent migration of U.S. empty nester baby boomers to smaller homes in U.S. cities runs counter to this trend (Bahrampour 2013). Many U.S. seniors cling to notions of the American Dream held in their younger years and ignore the need to change these notions as they age (Antoninetti 2008; Glass and Vander Plaats 2013). One U.S. architect and former city council member observed, "The American Dream lingers in elderly people and making that adjustment is really difficult. We are an aging population, and we are all still pretending we are young and going to live forever."

Anxiety over threats that cohousing and ADUs pose to the American Dream bubbles up when these innovative housing types are proposed in communities. This anxiety often manifests in the misconceptions and negative stigmas about the impacts that this housing will have on the single family home character of neighborhoods and localities. Concerns include a fear of increased density and parking and the decline in property value. These concerns impact proposals to build ADUs more than cohousing communities (Chapple et al. 2011). However, in hot housing markets like New York City or Los Angeles, homeowners will still construct ADUs even if they are unable to attain their neighbors' approval and official permits (Wegmann 2014). A Canadian housing academic mused about the potential to house seniors in legal ADUs on a large scale in the U.S.:

> I think this is more likely to work in cities that have a long history of ADUs and high house prices, so that property owners can see for themselves that the presence of ADUs in their neighborhoods doesn't impact their property values. Municipal authorities have a hard time passing ordinances and other regulation to allow ADUs when homeowners worry about maintaining the single-family character of their neighborhoods.

ADUs also are seen to threaten the American Dream of homeownership. Neighborhood NIMBY campaigns against ADUs are fueled by the fear that renters will move into this housing. Americans link renting more strongly with social class than Danes or Canadians. In particular, expanding rental options in the U.S. is seen as opening the door to low-income people who may threaten a neighborhood's safety, schools, and property values. There is more social support in the U.S. for using ADUs to house seniors or keep seniors in their homes, but the fear remains that when the senior leaves the property, a lower-income, younger renter will move in (Schafran 2012). A Canadian councilman reflected, "People [in the U.S.] go, 'I am okay with a little old lady, but when she sells it, it could be anybody in there and I don't want that.' If the senior moves out, then it remains a[n] [ADU] and it could be anybody. I don't want a dirty renter next to me." That ADUs will lower the socioeconomic status of a community is a legitimate concern in transitioning neighborhoods. However, researchers studying municipalities' experiences with ADUs on Long Island in New York theorize that ADUs occupied by family members or caretakers tend to remain vacant after a senior transitions to assisted care or dies (Anacker and Niedt [in progress]).

Cohousing faces some concerns about impacts to neighborhood character, particularly with density, but the bigger concern is that cohousing will diminish American individualism more broadly, as it is seen as a type of commune or village development. A member of a U.S. national cohousing association noted the challenge of "educating the neighborhood around you that you aren't a bunch of crazy wild-eyed idealists, drug-crazed hippies or communists."

Most of our participants agreed that broadening awareness of cohousing and ADUs among the general public would help to diffuse perceived impacts to the American Dream and make their development more palatable in the U.S. Providing on-the-ground examples through architectural renderings and brochures would allow the general public to experience diverse and thriving cohousing communities. A U.S. cohousing architect stressed, "People need to see this on the ground to understand what it really means. To see it as a genuine option." A Danish architect and cohousing resident attributed some of the success of cohousing in Denmark to the fact that advocates were able "to make ordinary people interested and . . . engage[d] in planning and building."

Education is necessary not only for raising awareness and altering misconceptions but also for managing expectations for those who wish to undertake these projects. It is important for future ADU residents and landlords to know what to expect in dealing with renters and maintaining the property and what best practices are. A Canadian academic explained, "[O]nce regulations have been introduced to make ADUs possible, I think it's important for public authorities to provide information to homeowners about the potential challenges of being a landlord-at-home, so that homeowners can make an informed decision about the worthiness of building and renting out an ADU." Our participants were hopeful that more knowledge and resources would become available on cohousing and ADUs from local planning departments and national advocacy organizations and agencies over time as more projects are developed.

Taking Action

Cohousing and ADUs are promising approaches to house seniors. Both housing types are better able to meet seniors' desire for physical and social connections and supportive care as they age in place than single-family homes in traditional neighborhoods (see also the chapter authored by Stephen Golant; Chapman and Howe 2001; Glass 2009). However, advocates and developers face multiple challenges in realizing their vision of housing seniors in these ways. Changes include developing sufficient social and financial capacity to build and maintain cohousing and ADUs over time and overcoming regulatory barriers to development (see also the chapter authored by Sherry Ahrentzen and Ruth Steiner). Housing advocates in the U.S. face the additional challenge of demonstrating that these housing types help to expand, rather than diminish, notions of the American Dream.

The experts that we interviewed in the U.S., Denmark, and Canada stressed the importance of educating different actors about cohousing and ADUs to implement these approaches on a larger scale. Prospective developers of cohousing and ADUs need to be aware of common issues that arise during the process of development and be exposed to best practices. Cohousing and ADU advocates have the task of educating financial institutions and their underwriters about the investment potential of these housing types and public agencies about their capacity to further seniors' well-being and advance social equity, outcomes deserving of public subsidies. Such education is beneficial not only for senior housing, but also in times of general affordable housing crises (Anacker and Niedt [in progress]). Finally, cohousing and ADU residents, developers, and advocates have the task of helping to overcome more broadly held public misconceptions about what living in and with these housing types means for American culture and quality of life.

References

Anacker, Katrin B., and Christopher Niedt. in progress. "Classifying Regulatory Approaches of Jurisdictions for Accessory Dwelling Units: The Case of Long Island."

Antoninetti, Maurizio. 2008. "The Difficult History of Ancillary Units: The Obstacles and Potential Opportunities to Increase the Heterogeneity of Neighborhoods and the Flexibility of Households in the United States." *Journal of Housing for the Elderly*, 22(4): 348–375.

Bahrampour, Tara. 2013. "The Kids Gone, Aging Baby Boomers Opt for City Life." *Washington Post*, August 9.

Bell, Rick. 2016. "Secondary Suites: A Day in the Life of Calgary's Wretched, Inhumane' Approval Process." *The Calgary Sun*, September 12.

Canadian Mortgage and Housing Corporation. 2016. *How the Strategy Works: Permitting Secondary Suites*. Ottawa, Canada: Canadian Mortgage and Housing Corporation.

Chapman, Nancy J., and Deborah A. Howe. 2001. "Accessory Apartments: Are They a Realistic Alternative for Ageing in Place?" *Housing Studies*, 16(5): 637–650.

Chapple, Karen, Jake Wegmann, Alison Nemirow, and Colin Dentel-Post. 2011. *Yes in My Backyard: Mobilizing the Market for Secondary Units*. Berkeley, CA: Center for Community Innovation.

City of Arvada. n.d. *Accessory Dwelling Units*. Arvada, CO: City of Arvada.

Cobb, Rodney L., and Scott Dvorak. 2000. *Accessory Dwelling Units: Model State Act and Local Ordinance*. Washington, DC: American Association of Retired Persons.

Cohousing Association of the United States. 2017. *The Cohousing Directory*. Boulder, CO: The Cohousing Association of the United States.

Coleman, James S. 1988. "Social Capital in the Creation of Human Capital." *American Journal of Sociology*, 94(Supplement): S95—S120.

Critchlow, Margaret. 2015. "Senior Cohousing in Canada: How Baby Boomers Can Build Social Portfolios for Aging Well." *Communities*, 166: 28–33.

Durrett, Charles. 2009. *The Senior Cohousing Handbook: A Community Approach to Independent Living*. Gabriola Island, Canada: New Society Publishers.

Egerö, Bertil. 2012. "Four Decades of Swedish Cohousing—What Chances of a Real Take-Off?" *Lecture at Self-Managed Cohousing: Born out of Need or New Ways of Life? Le Studium Conference*, Tours, France.

Gellen, Martin. 1985. *Accessory Apartments in Single-Family Housing*. New Brunswick, NJ: Transaction Publishers.

Glass, Anne P. 2009. "Aging in a Community of Mutual Support: The Emergence of an Elder Intentional Cohousing Community in the United States." *Journal of Housing for the Elderly*, 23(4): 283–303.

Glass, Anne P., and Rebecca S. Vander Plaats. 2013. "A Conceptual Model for Aging Better Together Intentionally." *Journal of Aging Studies*, 27(4): 428–442.

Hare, Patrick H. 1991. "The Echo Housing/Granny Flat Experience in the US." *Journal of Housing for the Elderly*, 7(2): 57–70.

Harrell, Rodney, Jana Lynott, Shannon Guzman, and Cheryl Lampkin. 2014. *What Is Livable? Community Preferences of Older Adults*. Washington, DC: AARP.

Howe, Deborah A. 1990. "The Flexible House: Designing for Changing Needs." *Journal of the American Planning Association*, 56(1): 69–77.

Labit, Anne. 2015. "Self-Managed Co-Housing in the Context of an Ageing Population in Europe." *Urban Research and Practice*, 8(1): 32–45.

Liebig, Phoebe S., Teresa Koenig, and Jon Pynoos. 2006. "Zoning, Accessory Dwelling Units, and Family Caregiving." *Journal of Aging and Social Policy*, 18(3–4): 155–172.

Litchfield, Michael. 2011. *In-Laws, Outlaws, and Granny Flats: Your Guide for Turning One House into Two Homes.* Newtown, CT: Taunton Press.

Lyubomirsky, Sonja, Laura King, and Ed Diener. 2005. "The Benefits of Frequent Positive Affect: Does Happiness Lead to Success?" *Psychological Bulletin*, 131(6): 803-855.

McCamant and Durrett Architects. 2017. *Getting-It-Built Workshop in Berkeley, CA.* Boulder, CO: The Cohousing Association of the United States.

Mendez, Pablo. 2011. *Ambiguity at Home: Unauthorized Geographies of Housing in Vancouver.* Unpublished Ph.D. diss., University of British Columbia in Vancouver, Canada.

Montgomery County. n.d. *Accessory Apartments.* Rockville, MD: Montgomery County.

Moore, Steven A., and Sergio Palleroni, eds. 2008. *The Alley Flat Initiative.* Austin, TX: The University of Texas Center for Sustainable Development.

New Avenue. 2014. "What Are the True Costs of an Accessory Dwelling?" *New Avenue Blog*, September 23.

Null, Roberta L., and Kenneth F. Cherry. 1996. *Universal Design: Creative Solutions for ADA Compliance.* Belmont, CA: Professional Publications.

Ortman, Jennifer. M., Victoria A. Velkoff, and Howard Hogan. 2014. *An Aging Nation: The Older Population in the United States.* Washington, DC: U.S. Bureau of the Census.

Pedersen, Max. 2015. "Senior Co-Housing Communities in Denmark." *Journal of Housing for the Elderly*, 29(1–2): 126–145.

Pfeiffer, Deirdre. 2015. "Retrofitting Suburbia through Second Units: Lessons from the Phoenix Region." *Journal of Urbanism*, 8(3): 279–301.

Pfeiffer, Deirdre. 2018. "Housing and Aging." In *Introduction to Housing*, edited by Katrin B. Anacker, Andrew T. Carswell, Kenneth R. Tremblay, and Sarah D. Kirby, 294–315. Athens, GA: University of Georgia Press.

Sanguinetti, Angela. 2014. "Transformational Practices in Cohousing: Enhancing Residents' Connection to Community and Nature." *Journal of Environmental Psychology*, 40: 86–96.

Sanguinetti, Angela. 2015. "Diversifying Cohousing: The Retrofit Model." *Journal of Architectural and Planning Research*, 32(1): 68–90.

Sanguinetti, Angela, and Chuck MacLane. 2016. "Seniors in Intergenerational Cohousing: Findings from a National Survey." *Lecture at Aging Better Together." Cohousing Research Network Conference*, Salt Lake City, Utah.

Schafran, Alex. 2012. *The Politics of Accessory Dwelling Units in the United States.* Paris, France: BIMBY.

Silver Sage Village. n.d. *Cohousing.* Boulder, CA: Silver Sage Village.

Small, Mario Luis. 2009. "How Many Cases Do I Need?: On Science and the Logic of Case Selection in Field-Based Research." *Ethnography*, 10(1): 5–38.

Spivak, Jeffrey. 2012. "Making Room for Mom and Dad." *Planning*, October: 8–13.

Taylor, Paul, Jeffrey Passel, Richard Fry, Richard Morin, Wendy Wang, Gabriel Velasco, and Daniel Dockterman. 2010. *The Return of the Multi-Generational Family Household.* Washington, DC: Pew Research Center.

Touati, Anastasia. 2013. "'Soft Densification' in Canada: The Example of 'Accessory Apartments' in Ontario." *Metropolitics*, 5: n.p.

Tummers, Lidewij. 2016. "The Re-Emergence of Self-Managed Co-Housing in Europe: A Critical Review of Co-Housing Research." *Urban Studies*, 53(10): 2023–2040.

Varady, David P. 1990. "Which Elderly Home Owners Are Interested in Accessory Apartment Conversion and Home-Sharing?" *Journal of Housing for the Elderly*, 6(1–2): 87–100.

Village of Euclid v. Ambler Realty Co., 272 U.S. 365 (1926)

Wegmann, Jake. 2014. *"We Just Built It:" Code Enforcement, Local Politics, and the Informal Housing Market in Southeast Los Angeles County.* Unpublished Ph.D. Dissertation. Department of City and Regional Planning, University of California, Berkeley, CA.

Wegmann, Jake, and Karen Chapple. 2014. "Hidden Density in Single-Family Neighborhoods: Backyard Cottages As an Equitable Smart Growth Strategy." *Journal of Urbanism: International Research on Placemaking and Urban Sustainability*, 7(3): 307–329.

Yin, Robert K. 2014. *Case Study Research: Design and Methods.* Thousand Oaks, CA: Sage Publications.

20

DESIGNED FOR ALL AGES

Multigenerational Housing as a Potential Housing Option in Flanders/Belgium

Sebastiaan Gerards, Erik Nuyts, and Jan Vanrie

The Growing Need for New Housing Concepts

In Flanders, the northern region in Belgium, people strongly prefer to age in place, which is supported by regional and national policy, similar to many other regions and countries in Europe as well as the United States (Golant 2011; Means 2007; Smetcoren et al. 2014, see also the chapters authored by Stephen Golant; by Deirdre Pfeiffer, Ashlee Tziganuk, Scott Cloutier, Julia Colbert, and Gracie Strasser; and by Sherry Ahrentzen and Ruth Steiner). Most Flemish people over 60 age in place in owner-occupied single-family homes, but these homes may have unused space and may not satisfy the current and future needs of their residents (Myncke and Vandekerckhove 2007). Thus, various researchers have started to focus on the need for a more diversified range of future housing options. For example, Smetcoren et al. (2014, 14) argued that "staying put in traditional family houses should not be considered as the standard for ageing well" and found that seniors, i.e., people over 65, are receptive to alternative housing options, such as communal housing.

Communal housing covers a broad range of housing types and subgroups, such as cohousing, small-scale assisted living, or large-scale projects where single-family houses are combined with a large community house, but typically consists of several housing units, where at least two households (1) share at least one residential space on a voluntary basis, (2) have at least one private residential space, and (3) co-manage residential space (Van den Houte et al. 2015). In the past decade or so, communal housing in general and for seniors in particular has received some attention in Flanders (Smetcoren et al. 2014; Van Herck and De Meulder 2009). However, practitioners, public policy analysts and makers, and academics have focused almost exclusively on shared outdoor spaces (e.g., gardens), but not on shared interior spaces (van de Weijer and Bervoets 2012). Therefore, we focus on a particular subgroup of communal housing, which does involve cohabitation under one roof, by investigating a new interpretation of multigenerational dwellings.[1]

In this chapter we define multigenerational housing as a self-managed communal dwelling where at least three different households cohabit in one housing unit. Although many scholars focus on multigenerational housing in the context of family members (Fuchs and Orth 2003), we focus on non-family members who may be in different age cohorts. Over the past few years multigenerational housing itself has been making a comeback in some countries in Europe (Gerards 2015; Labit 2015). In Flanders multigenerational housing is an interesting option to explore. Thirty-five percent of Flemings like the idea of living among other generations (ILIV 2013). Multigeneral housing is embedded in Flemish history due to its roots in historic farm housing (De Bleeckere and Gerards 2013).

A Design Assignment to Investigate the Potential of the Multigenerational Dwelling

In this chapter, we study the potential of multigenerational housing both from the viewpoint of possible future users (via a survey) and from an architectural perspective (via a design assignment). The data collection was integrated in a six-week long design assignment in 2014–2015 in the third year of the architecture bachelor's program at Hasselt University in Flanders. Each student was asked to first survey people with regard to their attitudes toward multigenerational housing and then to produce a tailored redesign of an existing single-family townhouse as a multigenerational home for them.

The following research questions are addressed in the chapter. First, do the survey respondents consider multigenerational housing a potential housing option? Second, how do the survey respondents perceive multigenerational housing in terms of expected costs versus benefits? Third, which interior spaces, if any, would they be willing to share? Fourth, is it architecturally feasible to redesign existing single-family townhouses as multigenerational dwellings that meet the different preferences of potential residents?

Survey

Forty-nine architecture students in their third year in the bachelor's program were asked to conduct five to seven structured interviews about people's attitudes toward multigenerational housing and then to design a multigenerational dwelling.[2] The dwelling had to serve five to seven people of at least three different adult age cohorts. In total, the students interviewed 405 people, on average eight per student, mostly from their social networks, i.e., family members, neighbors, and friends, ranging from 17 to 89 years. In other words, the students drew a convenience sample.

The authors designed a questionnaire with three parts. First, the students asked the respondents about their sociodemographic characteristics, their current housing situation, and their residential satisfaction. Second, the students introduced the concept of multigenerational housing. Third, the students examined perceptions of and attitudes toward multigenerational housing in three ways: (1) the students presented a list of interior and exterior spaces (i.e., living room, bathroom, kitchen, garden, etc.) and asked which of these spaces the respondents would want to use privately and which spaces they would want to share, (2) the students presented a list of eight possible advantages (increased social contact, younger generations may support older generations, older generations may support younger generations, reduced energy consumption, continued independence of the older generation, reduced housing costs by sharing spaces, reduced expenses by sharing household supplies, reduced expenses by sharing a car) and seven possible disadvantages of multigenerational housing, drawn from the literature (children causing nuisance, loss of privacy, forced contact with other residents, loss of independence, managing logistics, general maintenance obligations, the oldest generation being a burden for the others). For each enumerated advantage and disadvantage respondents indicated how important the respective advantage and disadvantage was for them on a 5-point Likert scale, and (3) the respondents were asked whether they would be willing to consider multigenerational housing as a future housing option.

Design Assignment

The students were then instructed to redesign an existing single-family townhouse with about 220 square meters (about 2,400 square feet, which can be considered a large dwelling but which is also not exceptional for this type) for a multigenerational dwelling. Interestingly, the Flemish townhouse accounts for about 29 percent of the Flemish housing stock but has received little attention from academics, practicioners, and policymakers (Canfyn 2014). The students had to base their multigenerational housing designs on the template and the information collected from the survey. In sum, the students had to translate the answers and preferences of their respondents into a multigenerational dwelling by redesigning the single-family townhouse.

In total, 49 multigenerational dwellings—one per student—were designed. In the following text we discuss the characteristics of the respondents that were interviewed and their preferences regarding multigenerational housing. We also take a closer look at the factors that may hint at the respondents' willingness to consider multigenerational housing and their willingness to share different interior and exterior spaces of the home. Subsequently, we compare the results of our survey with the design results and conclude, discussing some of the important challenges related to the redesign of multigenerational housing in Flanders.

Attitudes Toward Multigenerational Dwelling

Sociodemographic Description of the Sample

We use classes with a 12-year age span, starting from the age of 18, which is the official age of adulthood in Belgium. Hence, the age classes were 18–29, 30–41, 42–53, 54–65, 66–77, and 78–89. Respondents younger than 18 (n = 9) were excluded from the analysis. Thus, 396 respondents remained in the study. Fifty-six percent of the respondents were female, while 44 percent were male. The distribution of the different age cohorts was as follows: 18–29 years (36 percent), 30–41 years (14 percent), 42–53 years (25 percent), 54–65 years (10 percent), 66–77 years (10 percent), 78–89 years (6 percent). Thirty-seven percent of the respondents lived with a spouse or partner and children, and 32 percent with a spouse or partner without children. Eighty-six percent of respondents lived in owner-occupied homes, which is a little higher than the average of 70 percent in Flanders. Sixty-one percent lived in villages.[3] Seventy percent of the respondents rated their residential satisfaction as 8 or higher on a 10-point Likert scale.

Overall Perception and Willingness to Live in a Multigenerational Dwelling

In regard to the perceived advantages of living in a multigenerational dwelling, respondents indicated that reduced energy consumption was the most important advantage (average score 4.1/5), followed by older generations supporting younger generations (4.0/5) and younger generations supporting older generations (4.0/5). The most important disadvantages of living in a multigenerational dwelling were the loss of privacy (3.9/5), managing logistics (3.6/5), and general maintenance obligations (3.3/5).

Forty-eight percent of the respondents in our convenience sample stated that they would consider becoming a co-inhabitant of a multigenerational dwelling. This proportion is relatively high compared with other research in Flanders, which found that 32 percent of Flemish people in a survey would consider communal housing as a future residential option (ILIV 2013). Indeed, our results are more consistent with results obtained in Germany, where multigenerational housing has had a relatively long history and where in a survey 51 percent of 50–80 year-olds were open to this type of housing (Zok and Schwinger 2015). However, because of our reliance on a convenience sample our results should be used cautiously.

Factors Predicting the Willingness to Live in Multigenerational Housing

We utilized a regression analysis to explore the factors that predict whether respondents would consider residing in multigenerational housing. We focused on three different age groups, i.e., those age 18–29 (n = 141), 30–65 (n = 192), and 66 years and over (n = 63). We did not find statistically significant differences among these three groups in terms of their average willingness to consider living in multigenerational dwelling (age 18–29: 52 percent, 30–65: 45 percent, 66 and over: 46 percent; *Chi*²-value = 1.8, *df* = 2, *p* = 0.40). However, the variables that predict their willingness might differ because the needs and wants of these three age groups presumably also differ. As this difference would be highly relevant for implementing a housing option that requires the involvement of different generations with different needs, we decided to also look at each age group separately.

Table 20.1 shows the significant coefficients for four logistic regressions, one for each of the three age groups (18–29, 30–65, 66 and over) and one for all respondents combined. In each model, the dependent binary variable was considering (value: 1) versus not considering (value: 0) living in a multigenerational dwelling in the future. The following independent variables were utilized: sociodemographic variables (gender, age, etc.), residential satisfaction, preferences to use certain spaces

Table 20.1 Coefficients of the Variables Predicting the Willingness to Participate in a Multigenerational Dwelling.

	Type of Variable	*Age 18–29*	*Age 30–65*	*Age 66+*	*Total Sample*
Number of respondents		141	192	63	396
Nagelkerke R²		0.49	0.48	0.58	0.39
		Coefficients in the logistic regression			
Intercept		−6.5**	2.6	−6.6*	−4.2**
Age	Number				−0.02**
Cohort 54–65	Yes/no		−1.3*		
Education: primary degree	Yes/no		1.6		1.9***
Education: lower secondary degree	Yes/no		−1.6*		−0.9*
Education: higher secondary degree	Yes/no		−1.2**		0
Education: higher non-university degree	Yes/no		0		0.7*
Education: university degree	Yes/no		0.5		0.8*
Education: other degree	Yes/no		0		−2.7**
Household with children	Yes/no			4.3*	
Preference to use certain spaces privately					
Garden	Yes/no		−2.9**		−1.7**
Study room	Yes/no	1.4**			
Guest room	Yes/no	−1.7*			
Living room	Yes/no			−1.9*	
Perceived advantages of multigenerational housing					
Older generations may support younger generations	Number				0.5**
Less energy consumption	Number	1.5***			0.7***
Increased social contact	Number			1.5**	0.4**
Continued independence of the older generation	Number	0.9**			0.6*
Younger generations may support older generations	Number		1.3***		
Reduced housing costs	Number			1.5*	
Perceived disadvantages of multigenerational housing					
Loss of privacy	Number	−0.9**	−1.3***		−0.7***
Managing logistics	Number		−0.7***		−0.3*
General maintenance obligations	Number			−1.4**	

Source: authors.

* $p < 0.05$; ** $p < 0.01$; *** $p < 0.001$

privately, and the perceived advantages and disadvantages of multigenerational housing. Stepwise regressions were used, where insignificant variables were removed by the software at each iteration until the best fitting model was produced.

Our findings are consistent with research on other forms of communal housing, e.g., cohousing (Williams 2005). Respondents with a higher level of education are more likely to consider living in multigenerational housing. In regard to certain exterior and interior spaces, results show that respondents who prefer a private garden are also less open to consider living in multigenerational housing. Finally, with regard to the perceived advantages, respondents consider the following advantages, ranked by statistical significance: (1) older generations may support younger generations, (2) reduced energy consumption, (3) increased social contact, and (4) continued independence of the older generation. Conversely, in terms of the disadvantages, the only two factors that are statistically significant are loss of privacy and managing logistics. In sum, those who feel these particular disadvantages are important are less likely to consider living in multigenerational housing.

Columns 3 to 5 in Table 20.1 show the coefficients of the three age groups separately. The different models show many differences: Only loss of privacy appears with the same (negative) sign in multiple models. However, we do not find variables with different signs in multiple models, as this would suggest incompatible, i.e., mutually exclusive, preferences among the different age groups. Our main takeaway is that the preferences for and expectations of living in a multigenerational dwelling appear to be quite different for the different age groups but are, fortunately, not mutually exclusive.

Preferences for Private Use of Different Interior and Exterior Spaces

Our survey also addressed the propensity to use certain interior and exterior spaces privately. For example, all respondents preferred having their own bedroom, 90 percent wanted to have their own bathroom, and 61 percent (48, 34, 23, 18, 14, and 13, respectively) preferred having their own living room (study room, kitchen, dining room, storage room, laundry room, guest room). Only 8 percent of the respondents wanted a private garden.

For each of the interior and exterior spaces a logistic regression was performed to explore the variables that predicted whether or not respondents wanted this space to be private. In each model, the dependent binary variable was wanting this space private (value: 1) versus not wanting this space private (value: 0). As independent predictors, sociodemographic variables were used (gender, age, etc.), in addition to variables describing the household (living alone, living with a partner, etc.), the level of residential satisfaction, perceived advantages (older generations supporting younger generations, reduced energy consumption, etc.), and perceived disadvantages (loss of privacy, managing logistics, etc.). Stepwise regressions were used, where insignificant variables were removed until the best fitting model was produced. The results of these regressions show underlying patterns that could be useful for localities, nonprofit organizations, or for-profit developers seeking to design, implement, and market multigenerational housing. Table 20.2 shows the coefficients of these logistic regressions.

Four major results emerged from the analysis. First, gender has an impact on the preference for private use for three spaces: On average, men want a private study room but show less of an interest in a private bathroom or a private kitchen. Consequently, designing a multigenerational dwelling primarily (or only) for men might be quite different from designing one primarily (or only) for women. Second, as one might expect, the more importance people attach to the disadvantage of losing privacy, the more they want a space to be private. However, this does seem limited to the more intimate spaces, such as the bathroom. This does not apply to spaces such as guest, storage or laundry rooms. Third, strongly valuing the advantage of cost-saving seems to be a good indicator that rooms can be shared. Finally, the kitchen and living room are the two spaces that have the highest number of significant predictors of perceived advantages and disadvantages (i.e., four times). In other words, one might expect that differences in how respondents perceive the concept of multigenerational

Table 20.2 Coefficients of the Variables Predicting the Preference to Use Different Spaces Privately.

	Type of Variable	*Bathroom*	*Kitchen*	*Living Room*	*Dining Room*	*Study Room*	*Guest Room*	*Storage Room*	*Laundry*	*Garden*
Nagelkerke R^2		0.18	0.19	0.14	0.14	0.18	0.06	0.10	0.28	0.15
Intercept		0.4	−1.0	0.6	−1.5	−1.5*	1.0	1.8	3.2**	−3.2**
Gender: male	Yes/no	−0.5*	−0.3**			0.4***				
Age	Number		0.02**		0.02*			0.02*		
18–29	Yes/no					1.5***			0.8*	
42–53	Yes/no					1.0***				
54–65	Yes/no					1.0*				
66–77	Yes/no								1.9***	
78–89	Yes/no	−2.05***							2.0**	
Education: higher non-university degree	Yes/no								−0.7*	
Education: university degree	Yes/no								−1.2*	
Education: lower secondary degree	Yes/no								−0.2	
Education: primary degree	Yes/no								0.8	
Education: other degree	Yes/no								2.1**	
Partnered household	Yes/no		0.7*	0.8***						
Household with brothers/sisters	Yes/no	−1.5***								
Household with residing parents	Yes/no									−2.4*
Household with children	Yes/no									−0.8*
General housing satisfaction	Number			−0.1*						−0.2**
Perceived advantages of multigenerational housing										
Reduced expenses by sharing household supplies	Number		−0.5**	−0.3*	−0.7***	−0.4**			−0.6**	
Reduced housing costs	Number							−0.8***		
Less energy consumption	Number			−0.4*		0.5**				
Continued independence of the older generation	Number			0.3*					−0.8***	
Older generations may support younger generations	Number		−0.4*				−0.8***			
Perceived disadvantages of multigenerational housing										
Managing logistics	Number							−0.3*		
loss of privacy	Number	0.6***	0.3*	0.4***	0.6***					0.7**
Loss of independence	Number		0.3*							

Source: authors.

* $p < 0.05$; ** $p < 0.01$; *** $p < 0.001$

dwelling will be related to differences in their attitudes toward having a private or shared kitchen or living room. This is relevant information when discussing the possibilities of a multigenerational home with potential residents.

Comparison of Survey and Design Results

According to Cross (2006), design research yields a particular kind of knowledge, and brings scientific inquiry into the sphere of what is practically possible. In this sense, it occupies "a dialectical space between the world that is and the world that could be" (Margolin 2007, 4). We incorporated design into our study in the sense that we investigated whether the preferences of respondents could be translated into reality. We were particularly interested in whether the various preferences and concerns of the different age groups vis-à-vis multigenerational living were spatially compatible and feasible with the Flemish townhouse.

The spatial translation of the preferences into an actual design is not straightforward. First, although students had some flexibility in terms of selecting the group of residents for their design, these residents may have mutually exclusive preferences for private use of different spaces. For example, while one age group may prefer a shared kitchen, another age group may prefer a private kitchen in the housing unit. Second, students had to work within certain building constraints, which impose additional limitations on what is spatially achievable. Finally, architectural design always involves synthesizing different types of preferences, requirements, and constraints (from technical installations, to style issues, to building codes), which implies that any preference of any client always needs to be integrated in a broader synthesis. In sum, our study is not about searching for the "perfect" multigenerational housing design, but about 49 design suggestions, along with the aggregated survey results as a way to explore the possibilities and constraints.

General Observations Concerning the Different Locations of Multigenerational Housing

Consistent with the survey results, the students designed a private *bedroom* per resident (or per couple), although most bedrooms had more than a default bed and closet. Indeed, the students demonstrated that combining a standard bedroom with a seating group, a study room, or an outdoor area is feasible in a typical townhouse. With the exception of one student, who included an elevator in her design, all students provided a bedroom for the oldest inhabitant on the first (U.S.) or ground (Europe) floor, arguing that this design decision guarantees accessibility in the long run.

Most students included two *bathrooms* in their design, which makes sense in a multigenerational building. Eighty percent designed a private and wheelchair-accessible bathroom for the oldest resident on the first or ground level and an additional shared bathroom. Although the survey shows that the oldest respondents are willing to have a shared bathroom, most designs did not reflect this preference (see Table 20.2).

As mentioned earlier, 61 percent of the respondents wanted a private *living room*, but only a few students had designs that honored this preference. Given the dimensions of the existing building, designing private living rooms for each household turned out to be practically impossible. Instead, most students opted for an extended bedroom with a small sitting corner next to a main shared living room. A recurring concession was the design of a second shared living room by 47 percent of the students. While the main living room was designed as a central component of the shared space on the first or ground level, the junior living room was often a transition or buffer zone between private and shared space. We believe that this design solution constitutes one of the most prominent features of the design assignment, as even shared space can provide a certain degree of privacy in case it is used privately. Obviously, an important condition is a certain willingness to make arrangements concerning the private use of the shared parts of the building (e.g., when retired inhabitants can have private use of the junior living room during working hours).

On average, the older the respondent, the more likely he/she was to want a private *kitchen* (see Table 20.2). Nonetheless, all students designed shared kitchens on the first or ground level of the building. Only a few designs included a junior kitchen, which then appeared close to the private quarters of age group 30–65.

Just like the kitchen, the *dining room* designs were mostly shared, which is consistent with the survey results, indicating that only 23 percent wanted a private dining room. The fact that a higher proportion of respondents 66 years and older wanted a private dining room was only rarely incorporated in the designs. Unlike the living room, a junior dining room did not seem necessary for the students, as less than 10 percent of them provided this feature.

With regard to the *study room*, 70 percent of the students designed this space as shared space, mostly intended for school and college students. As with the junior living room, we consider its placement as a link between private and shared spaces. Consistent with the general preferences of the respondents, all designs included a shared *laundry* and *storeroom*. In most designs, these spaces were located in the basement of the building.

Finally, only 8 percent of the respondents wanted a private *garden*. Thus, all students indeed designed a shared garden, which was typically directly connected to the interior shared space on the ground level. Additionally, about one-third of the students designed complementary private outdoor spaces in the form of balconies or small roof terraces, mostly directly connected to private bedrooms.

Architectural Interventions for Achieving a Balanced Spatial Synthesis

Ideally, the spatial organization of a multigenerational dwelling emphasizes the most valued advantages—i.e., reduced energy consumption, the older generation supporting the younger generation, and the younger generation supporting the older—and minimizes the expected disadvantages, most notably the loss of privacy. Based on the designs, we find some inspiring architectural interventions that may help to achieve a good balance between the private and communal use of spaces within the home.

We distinguish the following five design strategies. First, the design should maintain the physical and social unity of the house and not split up the building into separate, isolated housing units. Overall, respondents were willing to share at least some spaces (e.g., 66 percent would share a dining room and 77 percent a kitchen), indicating that they did not expect a household to have a fully equipped apartment that could function independently of the rest of the building. This is also apparent in the student designs: A multigenerational dwelling is a large partially shared house and not an apartment building with separate, self-contained private dwellings. The design results show that it is possible to preserve the structure of the existing building. Obviously, avoiding interventions that alter the structure of a building has many advantages in terms of the affordability of the renovation.

Second, the design should show that a bedroom can contain more than a bed and a closet. By not limiting the bedroom to a place to sleep and dress, designers are better able to meet the preferences for privacy for each age group without resorting to creating independently functioning apartments.

Third, the design should provide buffer zones to create a spatial link between exclusively private and shared spaces. The designs show that a junior living room, a bathroom, or a shared study room could be considered as such a buffer zone. A related spatial strategy is the use of circulation areas as meeting places for the residents. In the same vein, shared spaces can also be regarded as circulation zones, through which the residents move to their private units.

Fourth, the design should include shared garden space, which can help to achieve the lowest threshold to becoming a co-inhabitant of a multigenerational dwelling. In a typical townhouse, however, access to a garden is only possible from the building itself, so sharing a garden also has

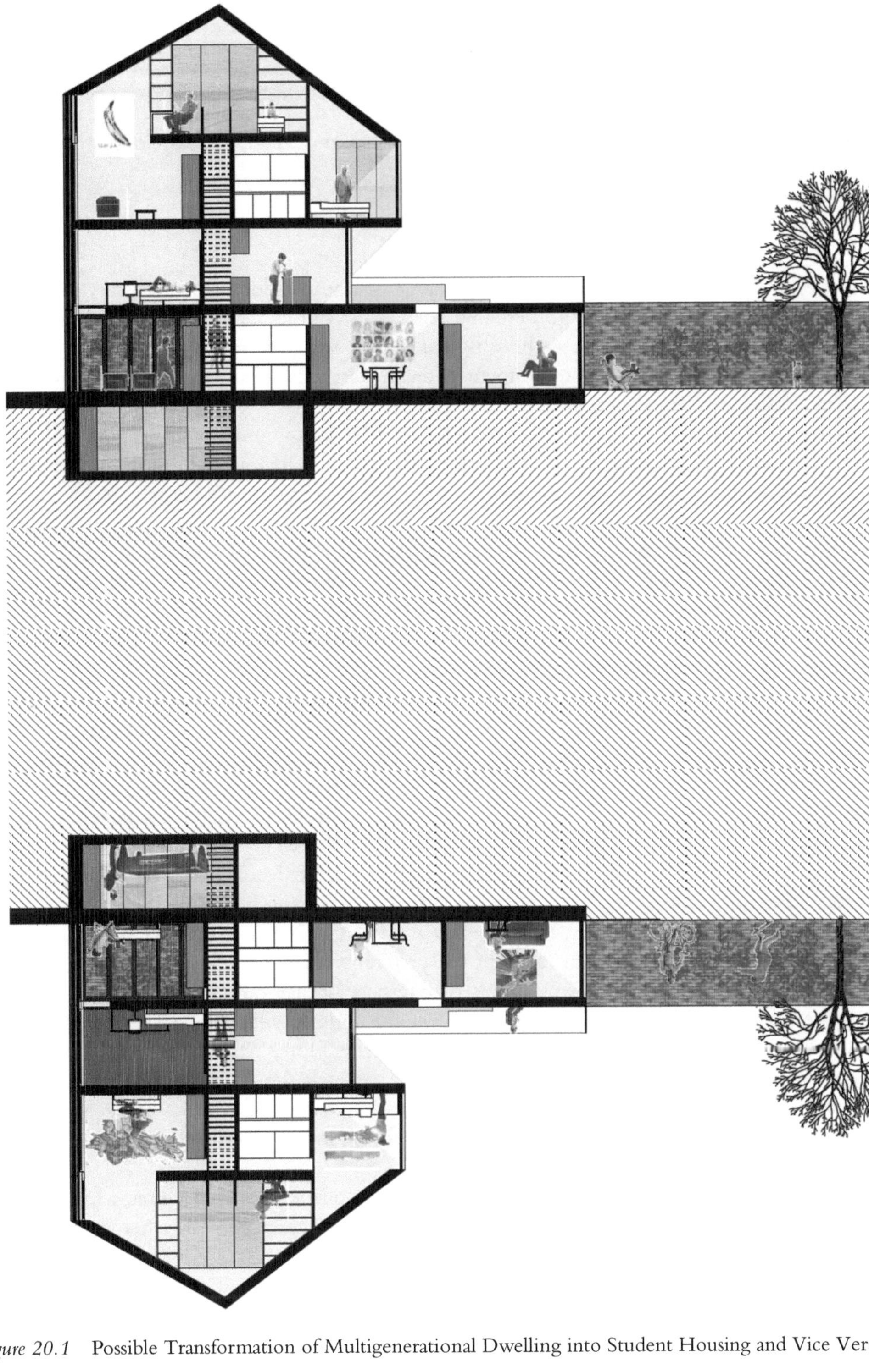

Figure 20.1 Possible Transformation of Multigenerational Dwelling into Student Housing and Vice Versa.
Studio credit: Ianthe Lindelauf. Used with permission.

implications for the design of the building (e.g., residents of the second floor should also be able to get to the garden). The student designs demonstrate that this is indeed feasible.

Fifth, the design should offer flexibility, which is necessary in multigenerational housing. Over time, there will be changes such as children leaving their parental home, the birth of children, or the death of a resident. Moreover, a flexible design also allows for the creation of new interpretations of the building, which may also increase the monetary value of the building (Schneider and Till 2005) (see Figure 20.1).

Conclusion

Previous research argues that a more diversified range of housing options is necessary in Flanders and elsewhere in developed societies (Smetcoren et al. 2014). In this chapter, we showed that a multigenerational dwelling may be a viable housing concept in an aging society. Our analysis confirms that a substantial proportion of our respondents were at least willing to consider multigenerational housing as a future residential option. The most important advantages people see in the concept are reduced energy consumption, the fact that older generations can support younger generations, and the fact that younger generations can support older generations. Obviously, the success of multigenerational housing stands or falls with the interest of different age groups to participate. Our analysis suggests that there is a general willingness to consider this housing option across age groups, although respondents had different motives and expectations. Based on previous research (Labit 2015), we expected that losing privacy would indeed be the most prominent obstacle to multigenerational housing. Our studio experience shows, however, that the architectural design of a dwelling can have a mediating role in this respect. Indeed, consistent with previous research (Gerards et al. 2015), several modes of sharing space are possible within a Flemish townhouse.

By linking our survey analysis to an actual student design project, we learned that multigenerational housing is a plausible option for many people and that it is feasible to implement the idea in a traditional Flemish townhouse. In sum, multigenerational dwellings are a viable and promising candidate in the search for alternative ways of housing. Obviously, several challenges still need to be addressed if multigenerational dwellings will be implemented in Flanders, along with other types of cohabitation.

There are at least three avenues in which to proceed. First, there is still a need for more applicable knowledge from the extensive gerontological literature on age mixing, particularly at the level of the individual building (Cook and Bailey 2013; Rowles and Bernard 2013). In this chapter we have only analyzed and discussed survey responses about the preference to use certain spaces privately. Thus, additional research into the actual experiences of intergenerational cohabitation seems warranted. Also, our results show that the factors predicting whether one is willing to consider multigenerational dwelling differs among age groups, and the multigenerational dwelling concept requires the involvement of different generations. Therefore, future research should investigate the differences and compatibilities among age groups. Gaining insights into the characteristics of people who are more likely to consider multigenerational housing could be useful to help identify possible future inhabitants.

Second, we have only focused on the concept of the multigenerational dwelling, which was also the starting point of the design assignment. A more extensive comparison with other forms of communal housing, such as cohousing or assisted living, would be interesting for future research, and would help to identify the profile of future multigenerational housing residents (see also the chapters authored by Stephen Golant; Deirdre Pfeiffer, Ashlee Tziganuk, Scott Cloutier, Julia Colbert, and Gracie Strasser; and Sherry Ahrentzen and Ruth Steiner).

Third, to further explore the possibilities for introducing the concept of multigenerational housing in the architectural landscape in Flanders, a constructive dialogue between designers and possible future inhabitants seems highly relevant. In such an endeavour, future inhabitants would become part of the design process, and actively modify and improve the final design (Qu and Hasselaar 2011). In reality, those processes would extend far beyond the framework of a six-week design assignment in architecture school, although student work can be a fertile breeding ground in this respect. It would also be interesting to investigate whether such a co-design process itself increases the interest in multigenerational housing. Most importantly, we believe that architecture, regardless of whether it is built or not, has an imaginative power, which can facilitate the necessary change in discourse and perspective concerning dwelling in Flanders and elsewhere.

Acknowledgements

We thank the 49 bachelor's architecture students at Hasselt University for their efforts and their inspiring design results. In addition, we thank our colleagues Sylvain De Bleeckere and Ludo Schouterden for their commitment and valuable input.

Notes

1 In this chapter we use the terms "housing" and "dwelling" interchangeably, although Gerards (2016) argues that there is a difference.
2 The instructors/authors did not provide any instructions on how to recruit respondents and did not obtain approval from their university's Institutional Review Board. The students did not provide any incentives for the interviewees.
3 The various geographic scales such as villages, suburbs, and cities were not operationally defined.

References

Canfyn, Filip. 2014. *Het syndroom van verkavelingsvlaanderen: een radicaal pleidooi voor stedelijk wonen*. Brussels: VUBPRESS.

Cook, Glenda, and Catherine Bailey. 2013. "Older Care Home Residents' Views of Intergenerational Practice." *Journal of Intergenerational Relationships*, 11(4): 410–424.

Cross, Nigel. 2006. *Designerly Ways of Knowing*. London: Springer.

De Bleeckere, Sylvain, and Sebastiaan Gerards. 2013. "Communal Housing: A Critical Review of Flemish Habitat." *MONU*, 18: 64–69.

Fuchs, Dörte, and Jutta Orth. 2003. *Umzug in ein neues Leuben: Wohnalternativen für die zweite Lebenshälfte*. München: Kösel Verlag.

Gerards, Sebastiaan. 2015. "Het Duitse Mehrgenerationenhaus." In *Wonen: Bouwstenen voor een habitologie*, edited by Sylvain De Bleeckere and Danny Windmolders, 97–105. Leuven: ACCO.

Gerards, Sebastiaan. 2016. *One Home, Three Generations: A Habitological Study on Multigenerational Dwelling as a Renewed Dwelling Concept for Flanders*. Unpublished Doctoral Dissertation. Hasselt University, Hasselt, Belgium.

Gerards, Sebastiaan, Sylvain De Bleeckere, and Roel De Ridder. 2015. "Designing Multigenerational Dwelling: A Workshop with Four Flemish Architecture Firms." *International Journal of Architectural Research*, 9(2): 20–30.

Golant, Stephen M. 2011. "The Quest for Residential Normalcy by Older Adults: Relocation But One Pathway." *Journal of Aging Studies*, 25(3): 193–205.

ILIV. 2013. *Samen thuis: het ILIV Thuis-Trendrapport 2013*. Antwerpen: ILIV.

Labit, Anne. 2015. "Self-Managed Co-Housing in the Context of an Ageing Population in Europe." *Urban Research and Practice*, 8(1): 32–45.

Margolin, Victor. 2007. "Design, the Future and the Human Spirit." *Design Issues*, 23(3): 4–15.

Means, Robin. 2007. "Safe as Houses? Ageing in Place and Vulnerable Older People in the UK." *Social Policy and Administration*, 41(1): 65–85.

Myncke, Raf, and Brecht Vandekerckhove. 2007. *Toekomstige ruimtebehoefte voor ouderen in Vlaanderen*. Brussels: Departement Ruimtelijke Ordening, Woonbeleid en Onroerende Erfgoed.

Qu, Lei, and Evert Hasselaar, eds. 2011. *Making Room for People: Choice, Voice and Liveability in Residential Places*. Amsterdam: Techne Press.

Rowles, Graham, and Miriam Bernard, eds. 2013. *Environmental Gerontology: Making Meaningful Places in Old Age*. New York: Springer.
Schneider, Tatjana, and Jeremy Till. 2005. "Flexible Housing: Opportunities and Limits." *Architectural Research Quarterly*, 9(2): 157–166.
Smetcoren, An-Sofie, Sarah Dury, Liesbeth De Donder, Nico De Witte, Sofie Van Regenmortel, Martinus Kardol, and Dominique Verté. 2014. "Denken over later? Attitudes van ouderen ten opzichte van verschillende woonvormen in Vlaanderen." *Ruimte and Maatschappij*, 6(2): 14–38.
Van den Houte, Kristien, Michael Ryckewaert, Bart Delbeke, and Stijn Oosterlynck. 2015. *Gemeenschappelijk Wonen*. Leuven: Steunpunt Ruimte en Wonen.
Van de Weijer, Marijn, and Wouter Bervoets. 2012."Een brede blik op hedendaagse woonpatronen." *Agora*, 28(5): 4–7.
Van Herck, Karina, and Bruno De Meulder, eds. 2009. *Wonen in meervoud: Groepswoningbouw in Vlaanderen*. Amsterdam: Sun.
Williams, Jo. 2005. "Designing Neighbourhoods for Social Interaction: The Case of Cohousing." *Journal of Urban Design*, 10(2): 195–227.
Zok, Klaus, and Antje Schwinger. 2015. "Pflege in neuen Wohn- und Versorgungsformen—die Wahrnehmung der älteren Bevölkerung." In *Pflege-Report 2015: Schwerpunkt: Pflege zwischen Heim und Häuslichkeit*, edited by Klaus Jacobs, Adelheid Kuhlmey, Stefan Greß, and Antje Schwinger, 27–54. Stuttgart: Schattauer.

SECTION 7
Gentrification

21
A MOVING TARGET
The Shifting Genealogy of Gentrification

Dennis E. Gale

Introduction

Over the past 40 years gentrification has become an increasingly critical issue in U.S. housing policy and planning. In the late 1970s and early 1980s, Congressional hearings and federally funded studies indicated growing national concerns (U.S. Congress 1978; U.S. Congress 1986; U.S. Department of Housing and Urban Development 1980). Similar signs of gentrification were being observed in certain cities in Canada, the UK and Western Europe, Australia, and New Zealand. Moreover, this issue has generated tremendous interest, media reportage, and scholarship. For example, a recent Google search of the term "gentrification" received 3,730,000 hits. A search of the term in the Stanford University Library catalog turned up 322 references, including monographs and digital sources.

Coined by sociologist Ruth Glass (1964), the noun "gentrification" referred to the renovation of older London neighborhoods by middle-class households, or the new gentry, who gradually replaced their working-class predecessors. The term took root in North American lexicons in the late 1970s. Today it bears physical, demographic, economic, social, and political implications (Lees et al. 2008). Previously, U.S. scholars and other observers had referred to gentrification as neighborhood resettlement, revitalization, regeneration, refurbishment, resurrection, and reclamation, among other terms (Laska and Spain 1980; Lees et al. 2008).[1]

In the following sections I first discuss the counterintuitive nature of gentrification, escalating as it did during the height of the nation's urban crisis. Second, I address the misunderstood origins of gentrification in the U.S. and the initial failure of federal urban renewal policies to exploit its potential. Third, I explore the elusive definitional challenges of gentrification and the difficulties this poses for empirical research. Fourth, I briefly examine the geographic range of gentrification with examples of scholarship in other nations. Fifth, I analyze two particularly bedeviling issues: the consumption versus production debate and the role of state sponsorship. Sixth, I discuss the issues of succession versus displacement and the differential outcomes of gentrification. And last, I probe the variable incidence of gentrification among U.S. cities and the implications it poses for housing affordability.

The Irony of Gentrification

Why was gentrification significant when it first gained substantial attention in the U.S.? Today many Americans do not fully appreciate the dire state of the nation's cities during the midcentury decades. Therefore, they may fail to understand the complex interplay of positive and negative outcomes resulting from this controversial socioeconomic phenomenon. By the 1950s and 1960s most American cities were experiencing physical deterioration and disinvestment in some neighborhoods. Officials feared that these conditions would spread to other areas of the city. Typically, these neighborhoods were inhabited by low- and moderate-income households, mostly renters. Vulnerable racial and ethnic minorities often predominated. Variously labeled "slums," "ghettos," or "blighted areas," the troubled sections contained overcrowded dwellings, insanitary facilities, poor property maintenance, and eroding public infrastructure (McDonald 2008). Not infrequently, crime rates were high, and gangs threatened residents. At the same time, cities were being drained of their middle-class households, as well as retail and office jobs, many of which were relocating to the outlying suburbs. Consequently, city revenues from property taxes and other sources were stagnant or declining.

As if such conditions were not enough, urban civil unrest in the 1960s, exacerbated by severe racial inequalities, political corruption, and other forms of official misconduct, decimated dozens of central city areas. Widespread opinion cast the American city in dystopian narratives, commonly subsumed under the rubric "urban crisis" (Beauregard 2003; Teaford 1990). Thus, when rising incidences of gentrification surfaced in the mid- and late 1970s, many observers were incredulous. Why would middle-class people buy or rent housing in what were then considered urban slums? And why were they not moving to the suburbs where such risks were much lower? House price differentials offered an incomplete answer, but survey evidence uncovered other factors, such as easy access to urban workplaces and a love of historic architecture (Gale 1982; Laska and Spain 1980). In a nutshell, gentrification was not only unanticipated, it was counterintuitive. By any measure this phenomenon—providing evidence that not all White middle-class Americans had given up on urban living—has contributed to momentous social, political, and economic impacts on American cities such as New York, Washington, Denver, Seattle, and Chicago.

Origins of Gentrification

Another enigma associated with gentrification is the fact that its existence in the U.S. long predated the 1970s, when it first gained significant media and scholarly attention. In fact it first appeared in a few Eastern and Southern coastal cities during and after World War I. Before the term "gentrification" took root, however, the phenomenon was most commonly referred to as "remodeling." From about 1915 to 1970 fully two generations or iterations of remodeling occurred. Boston, New York, Washington, DC, Charleston, and New Orleans were included in the first generation. Those cities, as well as Baltimore and Chicago, composed all or most of the second generation. When the third generation of gentrification appeared in the 1970s its momentum was driven mainly by the postwar baby boom population then reaching the household formation stage of adulthood.

Baby boomers' sheer numbers expanded demand for inner city housing, propelling gentrification into media, scholarly, and public policy realms. But the boomers' numbers accounted for only part of the gentrification momentum. Also, a weariness with modernist architecture and design, as well as with suburban living, and the maturation of a national historic preservation movement contributed to their interest in residing in an older neighborhood. Furthermore, enactment of the 1968 Fair Housing Act hastened the city-to-suburban outmigration of countless minority households; it seems likely that the units they left behind in some cases created new opportunities for urban gentrification. Meanwhile economic restructuring, involving deindustrialization, the growth of advanced services, and expansion of the global economy, increased employment opportunities in cities for white collar and creative class workers. Some of these households preferred a convenient commute to work and chose to buy or rent a home in the city (Florida 2005; Sassen 1991).

An associated paradox is the fact that when Congress considered policies to reverse urban decline in the late 1930s and postwar 1940s (Fogelson 2001), members failed to recognize the potential of the remodeling-cum-gentrification paradigm. By that time abundant evidence testified to the transformative powers of housing rehabilitation to revitalize older neighborhoods under primarily free market circumstances. Instead, the redevelopment paradigm was adopted in 1949. Complemented by public housing and interstate highway initiatives, urban redevelopment required that slums be cleared, residents and businesses be relocated, and new neighborhoods be constructed (Anderson 1964; Gans 1982; Jacobs 1962). Extensive public subsidies were necessary to offset what Congress perceived to be the failure of the urban marketplace.

Although Congress later supplemented redevelopment policies with housing renovation options, not until the late 1950s and 1960s were there signs that Capitol Hill was gradually accepting the gentrification paradigm. When urban renewal was terminated in the mid-1970s, a panoply of government programs followed. Most were supportive of housing rehabilitation and historic preservation and permitted only small-scale redevelopment. For better or worse, Washington policymakers had finally recognized the potency and popular appeal of the gentrification paradigm. Even though urban programs suffered drastic budget cuts in the 1980s, gentrification and related forms of urban development continued expanding in many cities, aided by tax incentives, public-private partnerships, economic restructuring, and government deregulation of the banking industry (Ryberg-Webster and Kinahan 2014).

Defining Gentrification

A key problem bedeviling gentrification discourse is that of definition (Lees et al. 2008 London et al. 1980). When formal recognition of the phenomenon appeared in the late 1970s and 1980s, gentrification denoted the renovation of older urban residential areas, such as the Park Slope area of Brooklyn (see Figure 21.1) and the Lafayette Park neighborhood of St. Louis (see Figure 21.2), by neighborhood newcomers. It was contrasted with "incumbent upgrading," which involved home

Figure 21.1 Park Slope, Brooklyn, Experienced Reinvestment in the 1970s.

Photo credit: author.

Figure 21.2 Newcomers to the Lafayette Square Neighborhood of St. Louis Were Attracted to Mansard-Style Houses Such as These.

Photo credit: author.

rehabilitation by long-term residents (Varady 1986). Gentrifiers—mostly young middle-class singles and couples—did much of the work themselves. Lesbian, gay, transgendered, and bisexual consumers, as well as students and those in creative occupations, were often in the vanguard. Once a trend was initiated, commercial real estate investors and firms often stepped in, actively marketing units for sale or rent. As the promise of inner city reinvestment became more apparent, realtors, for example, often led, rather than followed, trends by showing properties in "up-and-coming" and "undiscovered" areas.

While some gentrifiers rented an apartment, most purchased detached or attached single-family houses or condominiums in former rental apartment buildings. In some cities, factories, warehouses, or public schools were converted to loft or apartment units (Hamnett 2009a; Hudson 1987; Zukin 1982). Infill new construction created other opportunities. Once an area attracted early gentrification, the word spread, and demand increased for remaining properties. With prices and rents rising, would-be gentrifiers sought housing on the fringes of the most active streets, spreading gentrification to less costly sections (Abu-Lughod 1994). As housing turnover increased, commercial buildings underwent renovation, with pubs, coffee houses, and gourmet restaurants replacing bodegas, taverns, and dry cleaners. Not infrequently, national chain retail outlets replaced locally owned small businesses (Gotham 2005; Zukin et al. 2009). A case in point is the Mission District of San Francisco (see Figure 21.3).

The forgoing characterization might be thought of as "classic gentrification." "Modern gentrification," on the other hand, has evolved largely over the past 25 years. Its meaning is no longer confined to the renovation of older neighborhoods. Currently, it is used to denote any form of urban capital investment that draws middle- and upper-income people to a given area in which to reside, work, or pursue entertainment, shopping, dining, or cultural activities. Housing, offices, shops, malls, hotels, museums, galleries, concert halls, sports facilities, and parks are all examples. Modern gentrification

Figure 21.3 San Francisco Mission District Was Rapidly Gentrifying in the 1980s.
Photo credit: author.

has become synonymous with almost any form of urban development, including some examples subsidized by public authority (Wyly and Hammel 2001). Variations on the term include "new-build gentrification," "hyper-gentrification," "tourism gentrification," "re-gentrification," and "degentrification" (Gotham 2005; Lees et al. 2008). Put simply, Glass' original concept has been appropriated to embrace a variety of issues related to inter-class spatial conflict and the hegemony of capital over labor (Guterbock 1980). As a result, some authors point out, the meaning of "gentrification" suffers from "definitional overload" (Lees et al. 2008, 155). In my opinion, the term has mutated to a condition bordering on the amorphous. As a popular operetta states, "When everyone is somebody, then no one's anybody" (Gilbert and Sullivan 1889). If nearly all forms of urban reinvestment are gentrification, does gentrification retain any distinctiveness as a concept worthy of empirical inquiry?

In turn, confusion over definitions of gentrification has complicated research. Since gentrification was first widely recognized in the late 1970s and 1980s, defining it rigorously enough for persuasive empirical study has proven exasperating and elusive. As with pornography, we think we know gentrification when we see it. But gentrification has become immersed in a Sargasso Sea of theoretical and ideological turmoil, and there are many variables that capture facets of its occurrence but do not embrace its totality (Allen 1980; Glick 2008). Thus, researchers have employed metrics measuring shifts *inter alia* in householder educational or occupational attainment, household income, Black-to-White occupancy, and population mobility patterns (Lyons 1996). Such measures grasp at, but do not fully encompass, the ragged edges of the gentrification phenomenon.

Moreover there is much "leakage" in such studies (Atkinson 2000). For example, if two college students rent an apartment, replacing an electrician's family, their net income may be lower than that of their predecessor. Or, if an African-American accountant replaces a White truck driver's family, a racial transition has occurred. Note, however, that neither example conforms to popular assumptions about gentrification. Multivariate studies involving multiple household measures may offset some of these concerns, but because gentrifying areas are initially quite small, procedures such as U.S. Bureau of the Census suppression protocols may raise difficulties in achieving sufficient sample sizes. Thus, persuasive empirical research on gentrification is often troubled by analytical conundra such as these.

Gentrification in Other Nations

If Glass' London observations in the mid-1960s were among the first to document gentrification in scholarly circles, they would soon resonate in several other nations. On the European mainland, for example, similar patterns were unfolding in the 1970s, if not earlier (Couch et al. 2003). Most commonly, they appeared in or near the Altstadt (ancient city core) once encompassed by medieval city walls. The Gamla Stan (old city) island of Stockholm is an example. Termed "embourgeoisement" by the French and French-speaking Belgians (Van Criekengen 2009, 830), gentrification was especially prominent in the Marais section of Paris (see Figure 21.4) (Carpenter and Lees 1995). It also appeared in the Jordaan neighborhood of Amsterdam, the Nyhavn and Christianshavn sections of Copenhagen, the Kreuzberg and Neukölln areas of Berlin, and the Haidhausen sector of Munich (Appleyard 1979; Davidson 1979; Gale 1984). Variously composed of apartment blocks

Figure 21.4 Even Structurally Unsound Buildings in Paris' Marais Are Too Highly Desired to Demolish.
Photo credit: author.

(see Figure 21.5), three- to six-story row dwellings, or scattered preindustrial-age houses, most such neighborhoods were inhabited by working-class families when gentrification began.

However, gentrification in European cities typically progresses somewhat more slowly than in nations with less regulated market economies (Van Criekengen 2009). One reason is the fact that subsidized or social housing in European cities often composes a larger share of central city housing stocks than is the case, for example, with council housing in the UK or public housing in the US (Larsen and Hansen 2008). This tends to delay the more rapid reinvestment and household turnover characteristic of strong urban market economies (Carpenter and Lees 1995). Extensive European rent control policies over privately owned housing are an additional market retardant (Bernt and Holm 2005). Moreover, in Europe the national state generally maintains more authority over city development than is the case in the U.S. and other English speaking nations (Kantor et al. 2012).

While UK gentrification research has examined several cities such as Liverpool and Nottingham (Couch et al. 2003; Tallon 2013), London has received by far the greatest attention. Over the past half century vast areas of central and east London have undergone regeneration through private market gentrification or large-scale government sanctioned projects (Fainstein 1994; Hamnett 2003a). In the 1960s boroughs such as Camden, Kensington, and Westminster were experiencing gains in middle-class households, and in the 1970s so too were Islington, Hackney, and Lambeth (Butler 1996; Moran 2007). Once inhabited by working-class families, including immigrants, these areas are now home to legions of professional, technical, and managerial employees, as well as creative class workers (Florida 2005). Most have university educations and middle- to upper-level incomes (Butler and Robson 2003). Deindustrialization and the global economy strongly contributed to the demographic restructuring of London. One result is that the gentrification process is increasingly powered by large-scale development projects and corporate capital, rather than by individual households and small investors (Fainstein 1994; Jacobs 1996).

Figure 21.5 Example of an Apartment Block in Central Munich.

Photo credit: author.

But these gains have come at the price of growing income inequality among social classes, especially some immigrant groups (Imrie et al. 2009). Gentrification and other forces have increased the conversion of rental housing to ownership status, shrinking the supply of affordable apartments for working-class households (Hamnett 2003b). Hamnett (2003a) argues that the contraction of London's working-class population is not due to gentrification-induced displacement alone. More critical influences include the retirement, death, outmigration, and upward mobility of such workers. Thus life cycle factors may play a larger role than is generally recognized.

Canadian gentrification has been most extensive in Toronto and Vancouver, while Montreal, Ottawa, Halifax, and Edmonton have experienced more moderate levels. In general, the characteristics of "original" and gentrifier households compare with those in U.S. cities and London (Ley and Bourne 1993). For example, a study of Toronto found that from 1971 to 2001 the number of lower-income households in gentrifying areas declined at a higher rate than was the case citywide. Yet, perhaps somewhat more comparable to London and some European cities, there was little evidence of massive displacement, due in part to rent regulations and social housing supplies (Skaburskis 2012). Another Toronto study observed that in the 1960s and 1970s six of the seven early gentrifying areas were in close proximity to long-term upper-middle class neighborhoods (Ley and Bourne 1993). Similar patterns were found in Vancouver and Montreal, suggesting that stable bourgeois sectors in these three cities served to "germinate" additional middle-class resettlement in formerly working-class areas nearby.

While relatively few gentrification studies worldwide have analyzed the phenomenon's effects on civic affairs, Canadian researchers found that gentrifiers were frequently politically active in urban reform movements there (Ley 1996; Smith and Moore 1993). Slater (2004) noted that since the 1960s social mixing among middle and working classes in gentrifying areas declined in both Canada and the U.S. He cautioned against assumptions that Canadian and U.S. gentrification are essentially alike, noting differences such as the higher incidence of community organizing in gentrifying areas of Canada. Another scholar has warned against overestimating the extent and impacts of Canadian gentrification, noting that disinvestment and suburban outmigration were still the dominant trends even in the 1980s (Bourne 1993).

Australian cities such as Sydney, Melbourne, and Adelaide have experienced varying levels of gentrification. In Adelaide, for example, the more highly educated classes have been supplanting poor and working-class households at least since the mid-1960s. House price increases between 1986 and 1996 in the metropolitan area's inner core (where gentrification has been most extensive) substantially outpaced that in the overall metro area (Badcock 2001). In the larger cities of Australia and New Zealand the incidence of gentrification advanced through government intervention has risen over the past 40 years. As in other nations, conflict between classes in Sydney is underlain by racial and ethnic cleavages between, for example, Whites and Aboriginals (Shaw 2007).

Gentrification has been documented in a few Asian cities, as well. For example, a study of Shanghai notes that the phenomenon tends to be actively supported by state policies (He 2007). Similarly, Beijing authorities have promoted gentrification in the city center to advance tourism (Zhang 2013). Perhaps the single most consistent characteristic among all nations experiencing urban gentrification is what I term "upward socioeconomic transition," in which more affluent households succeed those of lesser circumstances. This is the inverse of traditional filter theory, in which households of progressively lower incomes inhabit aging and deteriorating urban housing.

Central Issues in Gentrification Research

Although scholarly discourse on gentrification has turned up numerous points of dissension or ambiguity, two issues have elicited considerable interest: the consumption versus production debate and the role of state sponsorship. The former issue seeks to isolate the seminal causal condition responsible for initiating the gentrification cycle. Consumption theorists stress the demand and supply axes of conventional housing market dynamics (Hamnett 2003a; Ley 1996; Slater 2004). They explain the rise of middle-class reinvestment in a given area as the product of increased demand for inner city housing due to a range of factors. Among them are increases in the number of adults at the household formation stage of life, the rising prices of suburban housing, the appeal of historic architecture and neighborhoods, and enhanced accessibility to urban employment, shopping, dining, culture, and entertainment. Hamnett (2003a), in particular, has observed that the global economy has increased demand for London's housing among the middle and upper classes. Ley found that growth in middle-class demand has resulted from an "emancipatory" shift in middle-class cultural values, which came to view older city housing not as slums but as charming living environments with cachet and prestige (Slater 2004). Generally, most consumptionists believe that gentrification is simply the outcome of shifting demand variables and demographic characteristics operating within a framework of economic restructuring.

Production theorists point to the producers of urban housing—investors, developers, and other real estate interests (Hamnett 2003b; Smith 1996). As neighborhoods age, production theorists observe that the difference between the present and future potential value of real estate increases. When the forces of decline cause present values to descend beyond a certain point, producers will exploit this difference or "rent gap." They acquire properties relatively cheaply, perhaps redeveloping or rehabilitating them, displacing their original occupants, and selling or leasing them to people of greater affluence (Smith 1979). Unlike consumption theory, production theory, shrouded in neo-Marxian garb, resounds with intimations of agency, imputing the deliberate exploitation of working-class families by malevolent capitalists and their bureaucratic handmaidens. Some observers have even referred to gentrification as revanchist, insisting that those who contribute to urban regeneration are vengeful in their intentions (Wyly and Hammel 2008; Smith 1996).[2] Finding worth in consumption and production theories, observers such as Lees et al. (2008) have called for efforts to resolve their differences.

A second key issue in gentrification research involves the extent to which governments have catalyzed urban gentrification. As pointed out earlier, the U.S. redevelopment paradigm gradually lost favor in the 1950s and 1960s. It was characterized by strong central governmental oversight, limited subnational government discretionary authority, and carefully circumscribed roles for capital. Clearance, rebuilding, and the replacement of most poor and working-class families by the middle class was its *modus operandi*. Over the late 1950s and 1960s the renewal amendments shifted the program's emphasis to rehabilitation. By the late 1970s Congress had adopted the gentrification paradigm, stimulating inner city private reinvestment through a variety of new mechanisms, including Urban Development Action Grants, Historic Preservation tax credits, and locally designed housing rehabilitation subsidies using federal Community Development Block Grant funding.

Over the next two decades federal authorities devolved more power from Washington to local governments and private sector actors. Funding for relocating displaced households was diminished, and loan and grant programs were augmented to stimulate housing renovation and infrastructure improvements. In addition, federal policy was shifting funding to subsidize eligible households rather than subsidizing affordable housing developments. Housing vouchers permitted eligible families to rent a privately owned dwelling instead of concentrating them in public housing (Varady and Walker 2007).

Through these mechanisms, as well as grants, loans, and tax incentives, federal agencies were orchestrating market forces to reverse urban decline. The gentrification paradigm had the added advantage of blunting community resistance by camouflaging public agency; unlike government-directed urban renewal, diffuse market-based reinvestment activity often lacks site offices, barren land tracts, and large project signs. Consequently, gentrification tends to proceed "under the radar," making organized resistance more difficult to achieve. Similar events have transpired in other nations undergoing urban gentrification. In the U.S., there seems to be little disagreement that state-sanctioned gentrification has expanded over the past 40 years. Thus, critics fault public authorities for supplanting social justice goals with gentrification-based household displacement and the destruction of poor and working-class communities (Hackworth and Smith 2001).

The consumption versus production debate and the issue of state-sponsored gentrification are encompassed by the theory of neoliberalism (Hackworth 2007; Künkel and Mayer 2012). At their heart neoliberal critics decry the disintegration of the safety net set by New Deal and War on Poverty programs, which helped protect the poor from unbridled market forces. Additionally, federal devolution of authority over urban programs has brought greater variability in the distribution of social benefits and social costs (Hackworth 2007). Public services have been contracted to private organizations, reducing the power of public employee unions over public budgets. Public-private partnerships in urban development have diminished government influence over affordable housing production and other social goals (Fainstein 1994). Critics of neoliberal policies argue that since the late 1970s state-sponsored programs have helped transform urban housing into a mere market commodity, thereby fueling the spread of gentrification and other forms of uneven development (Hackworth and Smith 2001). Others see gentrifiers as similar to a "colonial elite" and label gentrification "a meaner beast" (Atkinson and Bridge 2005, 3, 16).

Meanwhile, in the 1990s public housing production was reduced, and welfare reform under the Temporary Assistance for Needy Families program imposed time limits on the receipt of cash assistance. Poor people were encouraged to improve their status through education and employment. Despite increased federal housing voucher subsidies to permit eligible households to rent privately owned homes, the neoliberal critique concludes that the compromised social safety net has further exacerbated income inequality in the U.S. More generally, neoliberalism resides in the larger context of critical urban theory, which questions the very moral legitimacy of capitalist urban development (Brenner et al. 2012). After more than 80 years of trial-and-error public policy interventions, however, politically viable alternatives to former and current policies are difficult to envision.

This conundrum is especially so as cities and urban issues lose political saliency on the heels of Congressional reapportionment in response to metropolitan and Sunbelt population shifts. Moreover, as I will discuss further on, some cities (e.g., Boston, Washington, DC, New York, San Francisco, Seattle) have largely reversed urban decline, while others have made significant progress. This fact has undoubtedly contributed to a diminished sense of Congressional priority regarding urban policies in general. Hence, for better or worse, reliance on capitalism and market forces, in concert with public agency, is likely to remain the prevailing mode of urban revitalization, uneven though its benefits and costs may be. These circumstances leave critics with the challenge of pressuring officials to mitigate the negative impacts of gentrification and other forms of reinvestment through more expansive housing subsidies. Other policy responses involving excessive regulation of neighborhood housing markets show little promise of widespread acceptance in the current national political climate.

Displacement and Affordable Housing

A related research and policy issue involves the racial, ethnic, and socioeconomic composition of a given city area as gentrification progresses. The prevailing characterization portrays a shift from minority, poor, and working-class households to non-minority middle-income households. It assumes that most of the "original" residents will be forced from their homes by evictions, lease terminations, steep rent increases, or rising property tax burdens. At least two matters befuddle this impression. One conflates housing succession and displacement; the other cautions that gentrification has widely divergent impacts on existing renters and property owners (Freeman 2006).

Succession[3] versus Displacement

Normally, rates of household turnover in neighborhoods with large proportions of renters are higher than those in areas composed mostly of owner occupants. Without the ties of ownership, tenants are free to vacate their dwellings more readily. Household shifts that appear in all neighborhoods (e.g., a job change or loss, separation or divorce, death in the family, inheritance, retirement, or childbirth) are more likely to lead to voluntary turnover in renter-dominant areas. These types of changes underlie conventional succession patterns and are considered part of routine housing demand and supply cycles associated with virtually all capitalist economies (Hudson 1980). Gentrification-induced displacement, however, is most commonly perceived as the involuntary dislocation of an existing household due to property acquisition and/or occupation by a newcomer household of higher socioeconomic circumstances. Here, non-renewal of a lease, eviction, or unaffordable rent increases are the primary causes of turnover. Marcuse terms this "Direct Last-Resident Displacement" (Slater 2012, 185). I maintain that the larger phenomenon of household succession is often confused with the less prevalent one of Direct Last-Resident Displacement.

Marcuse's three other categories encompass factors that I will collectively term "indirect displacement." Indirect displacement not only occurs in a gentrifying neighborhood, but with time, its effects spread to other areas of a community. As capital reinvestment has accelerated in London, for example, it has diminished the citywide supply of affordable housing (Hamnett 2003a, 2009b). One element in this process is the well-known proclivity for gentrifiers to combine two or more smaller units in a given building into a single larger unit; this reduces the number of dwelling units in the structure while giving each new occupant more living space. Writ large, this process, while reducing overcrowding, also reduces the number of dwelling units in a neighborhood's housing stock. Together with housing price appreciation and the insecurities of rental tenure, indirect displacement diminishes a community's overall affordable housing supply.

Ironically, though, as time passes and gentrification advances, the market will attempt to cater to the heightened demand induced by "capital's voracious and clever appetite" (Jacobs 1996, 35). New infill construction, as well as the renovation of nonresidential structures for residential occupancy (e.g., industrial lofts or schools), will follow. Thus, more housing will be added to a community's supply, albeit almost completely priced for middle- and upper-income consumption. Furthermore, the risk increases that some units will revert from full-time to part-time occupancy as their wealthy owners utilize them for a pied-a-terre or second home. And as the sharing economy advances, conversion of apartments to short-term tourist rental units (e.g., Airbnb) will doubtless further impinge on local housing stocks. In both cases, former year-round occupancy dwellings are repurposed for temporary occupancy, diminishing shelter access for moderate-to- middle-income year-round residents.

Both direct and indirect displacement, along with the normal vicissitudes of succession, account for most household turnover in gentrifying areas. In cities such as Seattle, New York, Boston, and San Francisco, where gentrification has been rampant, these conditions have reached critical proportions. Yet, empirical research confronts ferocious challenges in determining the extent to which turnover rates in gentrifying neighborhoods are due to succession or displacement (Bradbury et al. 1982). Understandably, however, such distinctions fade into insignificance in the eyes of affordable housing activists and vulnerable families.

A related displacement issue raises the question, at what point does displacement cease to be operative in a given neighborhood? Typically, some early "pioneer" gentrifiers, often renters (e.g., artists, bartenders, baristas), will themselves later be displaced as rents, property prices, and real estate taxes continue to rise. Those who remain often lament rising rents and the neighborhood's shifting cultural character, as Ocejo (2014) points out in Lower Manhattan. Younger in-movers as well as executives, stockbrokers, tenured professors, and others will take up residence. Ironically, gentrifiers who perhaps once displaced a working-class family are now on the receiving end of this process. Thus, second- and third-order displacement may result over time. At the extreme, some particularly popular urban neighborhoods may even experience "super-gentrification" (Carpenter and Lees 1995). At this stage they become residential settings for the truly wealthy, including domestic and foreign investors seeking to place capital in secure commodities, such as real estate in Seattle or San Francisco. Should we consider middle-income households who move away due to rising rents or property tax burdens "displaced"? And how do we disentangle these households from others who relocate voluntarily due to common life cycle conditions? For example, affluent households with children may seek better schools for their children or safer neighborhoods with more recreational opportunities. Some householders will change jobs or retire, moving to the suburbs or to another community. Thus, the succession versus displacement dichotomy is somewhat muddied by these complexities.

Differential Outcomes From Gentrification

A further troubling characterization of gentrification-induced displacement is the fact that some incumbent residents are property owners. They may be working-class families who have inherited their homes and, due to gentrification, are later able to sell them or rent space to newcomers at a sizeable financial gain (Logan and Molotch 1987). Thus, even where a majority of a neighborhood's existing residents are displaced by gentrification, a smaller percentage can be expected to benefit from significant capital appreciation (Fainstein et al. 1983; Glick 2008). By no means can such beneficiaries be said to be disadvantaged by gentrification (although contingent outcomes such as the displacement of friendly neighbors may occur). On the other hand, in no way do such observations diminish the seriousness of the nation's affordable housing crisis, whether due to gentrification or disinvestment. Sadly, however, media and scholarly discourses too often omit reference to these realities. To be sure, the ranks of gentrification's beneficiaries are not limited to in-movers.

Incidence of Gentrification

Although gentrification has engendered enormous public attention since the 1970s, some cities have been substantially transformed by it, others have benefited significantly, and still others remain largely untouched. The same appears to be true for certain cities in the United Kingdom and Western Europe and perhaps in Canada, Australia, and New Zealand. Gentrification is most likely to appear in places undergoing extensive economic restructuring. It is also common in communities with attractive historic and environmental attributes and cultural venues. For discussion purposes, I will label these "Tier One" cities. Typically, such places draw large numbers of creative class workers, as well as retirees and tourists. Cities such as London, Paris, Berlin, Amsterdam, Toronto, Vancouver, Sydney, and Melbourne are examples. So, too, are U.S. cities such as Boston, New York, Washington, DC, Minneapolis-St. Paul, Columbus, Austin, Seattle, Denver, San Francisco, and Portland, Oregon. Smaller tourism-oriented cities such as Savannah, Charleston, Miami Beach, Annapolis, and Portland, Maine, as well as several university towns, may also be considered Tier One settings.

Tier Two cities compose the largest group. They include Chicago (Ehrenhalt 2012), Philadelphia (Adams et al. 1991), and Atlanta (Keating 2009). These communities, while experiencing significant levels of gentrification, mostly in their cores, still have substantial areas of disinvestment in areas of low market demand. In varying degrees the same may be true of Baltimore, St. Louis, Indianapolis, Kansas City, Pittsburgh, Cincinnati, Los Angeles, New Orleans, and Phoenix. Among the most challenged, Tier Three cities include Detroit, Newark, Buffalo, Oakland, and Memphis. Smaller examples include Camden, New Jersey; Gary, Indiana; Youngstown, Ohio; Reading, Pennsylvania; Lewiston, Maine; and Flint, Michigan. While small gentrifying areas may appear, Tier Three cities often compete poorly with others undergoing economic restructuring and possessing better locational advantages. Consequently, Tier Three central business districts, as well as several of their older residential sections, are under-capitalized. Unemployment, underemployment, and poverty rates are elevated, and public revenues are severely constrained (McDonald 2008).

Ironically, it is Tier Two, not Tier One, cities that find themselves in perhaps the most desirable circumstances in terms of overall human welfare. The very success of Tier One towns in attracting capital has led to the gentrification of most older neighborhoods possessing the most desirable attributes. Those with water or mountain views, nineteenth century architecture, or lively nightlife venues, for example, are often among the first to gentrify. Once this "low-hanging fruit" has been plucked, much of the formerly affordable housing stock has disappeared. Land prices in less desirable areas may accelerate as housing demand continues. Taking advantage of these conditions, unionized construction laborers insist on wage hikes. Developers in high-demand cities, such as New York and San Francisco, seek zoning concessions to construct higher density buildings, generating NIMBY opposition from neighbors. Such matters slow the pace of affordable dwelling construction, increase prices and rents, and limit the number of units produced by the public and nonprofit sectors. Meanwhile, waiting lists for such dwellings grow longer. Thus, the more gentrified a city becomes, the more difficult it is to produce enough affordable housing to satisfy existing demand levels (see also the chapter authored by W. Dennis Keating). Not only are the poor priced out, so also are moderate-income workers such as truck drivers, teachers, firefighters, and some employees who work in the nonprofit sector. In some neighborhoods, racially, ethnically, culturally, and economically diverse populations may not be able to stay.

Because Tier One housing market conditions cater mostly to upper-middle- and upper-income households, extraordinary expenditures are needed to increase affordable housing supplies. Tier Two cities, however, may have time to prepare preemptively. Measures such as land banking for affordable housing construction, housing trust funds, and inclusionary housing programs can help forestall some of the most pernicious impacts of runaway gentrification. Community development corporations, for example, can seek to develop socially mixed housing in areas somewhat distant from gentrifying neighborhoods. Yet, precisely because Tier Two cities are gentrifying more slowly, their revenue sources are more likely to be constrained; thus, it is more challenging for them to finance such measures.

Doubtless, some leaders in Tier Three cities would welcome more gentrification. But lacking many of the attributes found in Tier One and Two communities, they are unable to spur such growth. Fortunately, with fewer pressures on their housing markets, rents and sales prices remain comparatively moderate, at least in some less desirable areas. But property maintenance is more commonly deferred, leading to physical and economic decline. In addition, with extremely limited city revenues, maintenance and upgrading of public infrastructure are tightly constrained. Thus, Tier Three cities may not face the shortages of affordable housing found in Tier One and Two communities, but they lack the resources necessary to fully reverse their own overall downward trajectories. Clearly, the benefits and liabilities of gentrification and related forms of urban investment are differentially distributed nationally. And as the nation's inner suburbs age, many of the issues now facing cities will concern metropolitan residents, as well (Anacker 2015). It remains to be seen if twenty-first century social and economic conditions will free urbanized settings in the U.S. and other nations from the inequities and iniquities of uneven development.

Conclusion

Despite the enormous impact that gentrification has had on certain cities worldwide, this socio-economic phenomenon was initially unanticipated and underestimated in the U.S. and some other countries. Governments largely ignored the gentrification paradigm, instead implementing clearance and redevelopment, an alternative regeneration model. Over the last quarter of the twentieth century, however, the redevelopment paradigm in these places was supplanted by the gentrification paradigm as an approach to revitalizing declining city sections. As the public sector has embraced gentrification, the term now connotes almost any form of urban development, not just the rehabilitation of older buildings by owner occupants. Thus, any project increasing the socioeconomic status of people living, working, or pursuing leisure activities in its environs is now considered gentrification. In addition, the success of early or "classic gentrification" helped to embolden and indemnify commercial interests to undertake "modern gentrification" projects on a much larger scale with centralized management and financing.

Gentrification has been especially prominent not only in the U.S., but in certain cities in the UK and several other Western European nations, Canada, Australia, and New Zealand. Differences in national governance structures, urban development policies, market conditions, and cultural matters have yielded variations in the character, incidence, and progress of urban gentrification. In each case, however, the long-term effects usually include shifting demographic profiles, a decline in affordable housing, and the loss of religious and social organizations and small businesses. Researchers differ in their assessment of the initial factors spurring gentrification. Consumption theorists point to changing demand patterns, while production theorists insist that capitalist investment strategies trump other factors. Underlying these matters are deindustrialization, the privatization of government functions, the devolution of national government authority to subnational governments, the rise of the global economy, and neoliberal public policies, all of which have influenced the renewal of cities. While many researchers decry gentrification's displacement of working-class households, few have disentangled these forces from the larger dynamics of neighborhood succession. Similarly, few have acknowledged the financial benefits accruing to working-class property owners in gentrifying areas. Nonetheless, while federal urban redevelopment policies continued for about a quarter century in the U.S., the gentrification paradigm that followed has now persisted for more than 40 years. With or without state support, it remains the prevailing form of urban revitalization.

Notes

1 This preliminary confusion conspired with the later rise of the Internet to misdirect some scholars in subsequent generations who limited their search to the keyword "gentrification," thereby overlooking relevant research findings employing other terminology.

2 Careful explanation of the rationale for such claims appears to be lacking in the gentrification literature.

3 In this section I use the term "succession" in the broadest sense to refer to the turnover in occupancy in a given housing unit when one household succeeds another.

References

Abu-Lughod, Janet L., ed. 1994. *From Urban Village to East Village: The Battle for New York's Lower East Side*. Cambridge, MA: Blackwell.

Adams, Carolyn, David Bartelt, David Elesh, Ira Goldstein, Nancy Kleniewski, and William Yancey. 1991. *Philadelphia: Neighborhoods, Division and Conflict in a Postindustrial City*. Philadelphia, PA: Temple University Press.

Allen, Irving. 1980. "The Ideology of Dense Neighborhood Redevelopment: Cultural Diversity and Transcendent Community Experience." *Urban Affairs Review*, 15(4): 409–428.

Anacker, Katrin B., ed. 2015. *The New American Suburb: Poverty, Race and the Economic Crisis*. Burlington, VT: Ashgate.

Anderson, Martin. 1964. *The Federal Bulldozer: A Critical Analysis of Urban Renewal, 1942–1962*. Cambridge, MA: MIT Press.

Appleyard, Donald, ed. 1979. *The Conservation of European Cities*. Cambridge, MA: MIT Press.

Atkinson, Rowland. 2000. "Measuring Gentrification and Displacement in Greater London." *Urban Studies*, 37(1): 149–165.

Atkinson, Rowland, and Gary Bridge, eds. 2005. *Gentrification in a Global Context: The New Urban Colonialism*. New York: Routledge.

Badcock, Blair. 2001. "Thirty Years On: Gentrification and Class Changeover in Adelaide's Inner Suburbs, 1966–96." *Urban Studies*, 38(9): 1559–1572.

Beauregard, Robert A. 2003. *Voices of Decline: The Postwar Fate of U.S. Cities*. New York: Routledge.

Bernt, M. Matthias, and Andrej Holm. 2005. "Exploring the Substance and Style of Gentrification." In *Gentrification in a Global Context: The New Urban Colonialism*, edited by Rowland Atkinson and Garry Bridge, 106–120. New York: Routledge.

Bourne, Larry S. 1993. "The Myth and Reality of Gentrification: A Commentary on Emerging Urban Forms." *Urban Studies*, 30(1): 183–189.

Bradbury, Katharine L., Anthony Downs, and Kenneth A. Small. 1982. *Urban Decline and the Future of American Cities*. Washington, DC: The Brookings Institution.

Brenner, Neil, Peter Marcuse, P., and Margit Mayer, eds. 2012. *Cities for People, Not for Profit: Critical Urban Theory and the Right to the City*. New York: Routledge.

Butler, Tim, and Garry Robson. 2003. *London Calling: The Middle Class and the Remaking of London*. New York: Berg.

Butler, Timothy S. C. 1996. "People Like Us: The Gentrification of Hackney in the 1980s." In *Rising in the East?: The Regeneration of East London*, edited by Timothy S. C. Butler and Michael Rustin, 81–107. London: Lawrence and Wishart.

Carpenter, Juliet, and Loretta Lees. 1995. "Gentrification in New York, London and Paris: An International Comparison." *International Journal of Urban and Regional Research*, 19(2): 286–303.

Couch, Chris, Charles Fraser, and Susan Percy, eds. 2003. *Urban Regeneration in Europe*. Oxford: Blackwell.

Davidson, Hans F. 1979. "The Jordaan and Haarlem Neighborhoods in Amsterdam: Planning for the Future of a Historic Neighborhood." In *The Conservation of European Cities*, edited by Donald Appleyard, 221–235. Cambridge, MA: MIT Press.

Ehrenhalt, Alan. 2012. *The Great Inversion and the Future of the American City*. New York: Knopf.

Fainstein, Susan S. 1994. *The City Builders: Property, Politics and Planning in London and New York*. Cambridge, MA: Blackwell.

Fainstein, Susan S., Norman I. Fainstein, and P. Jefferson Armistead. 1983. "San Francisco: Urban Transformation and the Local State." In *Restructuring the City: The Political Economy of Urban Redevelopment*, edited by Susan S. Fainstein, Norman I. Fainstein, Richard Child Hill, Dennis Judd, and Michael Peter Smith, 202–244. New York: Longman.

Florida, Richard. 2005. *Cities and the Creative Class*. New York: Routledge.

Fogelson, Robert M. 2001. *Downtown: Its Rise and Fall, 1880–1950*. New Haven, CT: Yale University Press.

Freeman, Lance. 2006. *There Goes the 'Hood: Views of Gentrification from the Ground Up*. Philadelphia, PA: Temple University Press.

Gale, Dennis E. 1982. "Middle Class Resettlement in Older Urban Neighborhoods: The Evidence and the Implications." In *Internal Structure of the City: Readings on Urban Form, Growth, and Policy*, edited by Larry S. Bourne. New York: Oxford University Press.

Gale, Dennis E. 1984. *Neighborhood Revitalization and the Postindustrial City: A Multinational Perspective*. Lexington, MA: Lexington Books.

Gans, Herbert J. 1982. *The Urban Villagers: Group and Class in the Life of Italian- Americans*. New York: Free Press.

Gilbert, William S. 1889. *The Gondoliers*. http://en.wikipedia.org accessed January 10, 2019.

Glass, Ruth. 1964. *London: Aspects of Change*. London: MacGibbon and Kee.

Glick, Jonathan. 2008. "Gentrification and the Racialized Geography of Home Equity." *Urban Affairs Review*, 44(2): 280–295.

Gotham, Kevin Fox. 2005. "Tourism Gentrification: The Case of New Orleans' Vieux Carre." *Urban Studies*, 42(7): 1099–1121.

Guterbock, Thomas M. 1980. "The Political Economy of Urban Revitalization: Competing Theories." *Urban Affairs Review*, 15(4): 429–438.

Hackworth, Jason. 2007. *The Neoliberal City: Governance, Ideology and Development in American Urbanism*. Ithaca, NY: Cornell University Press.

Hackworth, Jason, and Neil Smith. 2001. "The Changing State of Gentrification." *Tijdschrift voor Economische en Sociale Geografie*, 92(4): 464–477.

Hamnett, Chris. 2003a. "Gentrification and the Middle-Class Remaking of Inner London, 1961–2001." *Urban Studies*, 40(12): 2401–2426.

Hamnett, Chris. 2003b. *Unequal City: London in the Global Arena*. New York: Routledge.

Hamnett, Chris. 2009a. "City Centre Gentrification: Loft Conversions in London's City Fringe." *Urban Policy and Research*, 27(3): 277–287.

Hamnett, Chris. 2009b. "Spatially Displaced Demand and the Changing Geography of House Prices in London, 1995–2006." *Housing Studies*, 24(3): 301–320.

He, Shenjing. 2007. "State-Sponsored Gentrification under Market Transition: The Case of Shanghai." *Urban Affairs Review*, 43(2): 171–198.

Hudson, James R. 1980. "Revitalization of Inner-City Neighborhoods: An Ecological Approach." *Urban Affairs Review*, 15(4): 397–408.

Hudson, James R. 1987. *The Unanticipated City: Loft Conversions in Lower Manhattan*. Amherst, MA: University of Massachusetts Press.

Imrie, Rob, Loretta Less, and Mike Raco, eds. 2009. *Regenerating London: Governance, Sustainability and Community in a Global City*. New York: Routledge.

Jacobs, Jane. 1962. *The Death and Life of Great American Cities*. New York: Random House.

Jacobs, Jane M. 1996. *Edge of Empire: Post-Colonialism and the City*. New York: Routledge.

Kantor, Paul, Christian Lefèvre, Asato Saito, H. V. Savitch, and Andy Thornley. 2012. *Struggling Giants: City-Region Governance in London, New York, Paris and Tokyo*. Minneapolis, MN: University of Minnesota Press.

Keating, Larry. 2009. "The Development, Redevelopment and Gentrification of Atlanta's Housing, 1980–2007." In *Past Trends and Future Prospects of the American City: The Dynamics of Atlanta*, edited by David L. Sjoquist, 251–289. Lanham, MD: Lexington Books.

Künkel, Jenny, and Margit Mayer, eds. 2012. *Neoliberal Urbanism and Its Contestations: Crossing Theoretical Boundaries*. New York: Palgrave.

Larsen, Henrik G., and Anders L. Hansen. 2008. "Gentrification—Gentle or Traumatic? Urban Renewal Policies and Socioeconomic Transformations in Copenhagen." *Urban Studies*, 45(12): 2429–2448.

Laska, Shirley B., and Daphne Spain, eds. 1980. *Back to the City*. New York: Pergamon Press.

Lees, Loretta, Tom Slater, and Elvin K. Wyly. 2008. *Gentrification*. New York: Taylor and Francis.

Ley, David F. 1996. *The New Middle Class and the Remaking of the Central City*. New York: Oxford University Press.

Ley, David F., and Bourne, Larry S. 1993. "Introduction: The Social Context and Diversity of Urban Canada." In *The Changing Social Geography of Canadian Cities*, edited by Larry S. Bourne and David F. Ley, 3–33. Kingston: McGill-Queen's University Press.

Logan, John R., and Harvey R. Molotch. 1987. *Urban Fortunes: The Political Economy of Place*. Berkeley, CA: University of California Press.

London, Bruce, Donald S. Bradley, and James R. Hudson. 1980. "Introduction: Approaches to Inner-City Revitalization." *Urban Affairs Review*, 15(4): 373–380.

Lyons, Michal. 1996. "Gentrification, Socioeconomic Change, and the Geography of Displacement." *Journal of Urban Affairs*, 18(1): 39–62.

McDonald, John. F. 2008. *Urban America: Growth, Crisis and Rebirth*. Armonk, NY: M.E. Sharpe.

Moran, Joe. 2007. "Early Cultures of Gentrification in London, 1955–1980." *Journal of Urban History*, 34(1): 101–121.

Ocejo, Richard E. 2014. *Upscaling Downtown: From Bowery Saloons to Cocktail Bars in New York City*. Princeton, NJ: Princeton University Press.

Ryberg-Webster, Stephanie, and Kelly L. Kinahan. 2014. "Historic Preservation and Urban Revitalization in the Twenty-First Century." *Journal of Planning Literature*, 29(2): 119–139.

Sassen, Saskia. 1991. *The Global City: New York, London, Tokyo*. Princeton, NJ: Princeton University Press.

Shaw, Wendy S. 2007. *Cities of Whiteness*. Malden, MA: Blackwell Publishing.

Skaburskis, Andrejs. 2012. "Gentrification and Toronto's Changing Household Characteristics and Income Distribution." *Journal of Planning Education and Research*, 32(2): 191–203.

Slater, Tom. 2004. "North American Gentrification? Revanchist and Emancipatory Perspectives Explored." *Environment and Planning A*, 36(7): 1191–1213.

Slater, Tom. 2012. "Missing Marcuse: On Gentrification and Displacement." In *Cities For People, Not For Profit: Critical Theory and the Right to the City*, edited by Neil Brenner, Peter Marcuse, and Margit Mayer, 171–196. New York: Routledge.

Smith, Neil. 1979. "Toward a Theory of Gentrification: A Back to the City Movement by Capital Not People." *Journal of the American Planning Association*, 45(4): 538–548.

Smith, Neil. 1996. *The New Urban Frontier: Gentrification and the Revanchist City*. New York: Routledge.

Smith, Patrick J., and P. W. Moore. 1993. "Cities as a Social Responsibility: Planning and Urban Form." In *The Changing Social Geography of Canadian Cities*, edited by Larry S. Bourne and David F. Ley, 343–366. Kingston: McGill-Queens University Press.

Tallon, Andrew. 2013. *Urban Regeneration in the UK*. New York: Routledge.

Teaford, Jon C. 1990. *The Rough Road to Renaissance: Urban Revitalization in America, 1940–1985*. Baltimore, MD: Johns Hopkins University Press.

U.S. Congress. House of Representatives. *Revitalization, Gentrification and the Low-Income Housing Crisis. Hearings before the Subcommittee of the Committee on Government Operations*. 99th Cong., 2nd sess., June 12, 1986.

U.S. Congress. House of Representatives. Task Force on Homeownership. *Hearings before the Subcommittee on Housing and Community Development of the Committee on Banking, Finance and Urban Affairs*. 95th Cong., 2nd sess., January 16–September 8, 1978. U.S. Department of Housing and Urban Development. 1980. *The Conversion of Rental Housing to Condominiums and Cooperatives: Volume of Conversion Activity in Selected Metropolitan Areas*. Washington, DC: U.S. Government Printing Office.

Van Criekengen, Mathieu. 2009. "Moving In and Out of Brussels' Historical Core in the Early 2000s: Migration and the Effects of Gentrification." *Urban Studies*, 46(4): 825–848.

Varady, David P. 1986. *Neighborhood Upgrading: A Realistic Assessment*. Albany, NY: State University of New York Press.

Varady, David P., and Carole Walker. 2007. *Neighborhood Choices: Section 8 Housing Vouchers and Residential Mobility*. New Brunswick, NJ: CUPR Press.

Wyly, Elvin K., and Dan J. Hammel. 2001. "Gentrification, Housing Policy and the New Context of Urban Development." In *Research in Urban Sociology, Volume 7: Critical Perspectives on Urban Redevelopment*, edited by Kevin Fox Gotham, 211–273. Stamford, CT: JAI Press.

Wyly, Elvin K., and Dan J. Hammel. 2008. "Mapping Neoliberal American Urbanism." In *Gentrification in a Global Context*, edited by Rowland Atkinson and Garry Bridge, 18–38. New York: Routledge.

Zhang, Yue. 2013. *The Fragmented Politics of Urban Preservation: Beijing, Chicago, and Paris*. Minneapolis, MN: University of Minnesota Press.

Zukin, Sharon. 1982. *Loft Living: Culture and Capital in Urban Change*. Baltimore, MD: Johns Hopkins Press.

Zukin, Sharon, Valerie Trujillo, Peter Frase, Danielle Jackson, Tim Recuber, and Abraham Walker. 2009. "New Retail Capital and Neighborhood Change: Boutiques and Gentrification in New York City." *City and Community*, 8(1): 47–64.

22

PREVENTING GENTRIFICATION-INDUCED DISPLACEMENT IN THE U.S.

A Review of the Literature and a Call for Evaluation Research

Miriam Zuk

Introduction

Despite over 50 years of scholarly research and debate (see the chapter authored by Dennis Gale), the processes of gentrification and displacement have continued to be highly contested in their definitions, causes, impacts, and potential policy responses. Gentrification is generally understood as the transformation of a working-class or vacant area of a neighborhood into middle-class residential or commercial use (Lees et al. 2013). Although in its origin the term "gentrification" inherently implied the displacement of working-class households, past and present discussions have centered on the influx of middle- and upper-class people into low-income neighborhoods (Glass 1964). Displacement can be a precondition or consequence of gentrification, and often occurs in neighborhoods where gentrification is nowhere in sight (e.g., due to a natural disaster). Due to the complex trajectories of households and neighborhoods alike, scholars continue to debate the scale and extent of the relationship between gentrification and displacement (Freeman 2005; Wyly et al. 2010).

This chapter reviews the academic and practitioner literatures on the causes, consequences, and policy responses to displacement in the U.S. I adopt the three-pillar framework of anti-displacement policy emerging from practitioners: stabilization, production, and preservation, while exploring the limited literature on the effectiveness of certain strategies. The chapter concludes with a call for more evaluation research as the demand for anti-displacement strategies continues to grow in U.S. cities.

Residential Displacement: Drivers, Scale, and Impacts

Residential displacement has become one of the key concerns of redevelopment and gentrification (Florida 2017). Yet displacement can also be driven by disinvestment in either housing (e.g., poorly maintained housing) or neighborhoods (e.g., decaying infrastructure) as well as regional housing market trends and exclusionary markets or practices. As characterized by early displacement scholars Grier and Grier (1978), disinvestment-related displacement occurs when the value of a property does not justify investing in its maintenance, thereby resulting in decay and abandonment. Reinvestment-related displacement, on the other hand, refers to cases where investments in a neighborhood result in increased rents to a point where it is profitable for landlords to sell the property or to raise the rent, forcing tenants to leave, and replacing them with higher-income tenants, thereby resulting in gentrification. Grier and Grier (1978) note, however, that disinvestment and reinvestment-related displacement are interrelated. Similarly, Marcuse (1985, 1986) argues that the disinvestment of urban neighborhoods creates vacant land ripe for investment through gentrification. From this perspective, gentrification may happen long after displacement has occurred.

The mechanisms of displacement can take on many forms. One of the most commonly recognized forms of displacement is landlord actions such as formal evictions. Desmond and Shollenberger (2015) argue that formal evictions processed through the court system are only part of a repertoire of landlord actions that can induce a tenant to leave. Informal evictions, which can include harassing, removing parking, cutting off utilities, or asking for untenable rent increases, are often "less expensive and more efficient than formal evictions" (2015, 1754). Although moves resulting from informal evictions may appear to be voluntary, such moves are in fact responsive to the conditions imposed by landlords. The authors argue that forced displacement is endemic to poor communities and not confined to gentrifying neighborhoods.

An overheated regional housing market is another driver of displacement that may or may not be related to investment. For instance, increased competition for housing may drive up rents independent of any investment in an area, displacing households that cannot afford to pay the increased rent. Furthermore, and especially in jurisdictions that do not have rent regulations, rent hikes can be a simple strategy used by landlords to remove a tenant that may require little paperwork or legal repercussions. Finally, Marcuse (1985) introduced the concept of exclusionary displacement to encompass situations where a household cannot move into a neighborhood based on conditions that are beyond their control (e.g., cost or discriminatory practices). When considering displacement from an exclusionary lens, its scale increases dramatically. When taking into account landlord discrimination against Housing Choice Voucher holders, neighborhood resistance to development, zoning restrictions on density, and the differing costs of neighborhoods.

A number of studies have attempted to quantify the scale of displacement in different neighborhoods and cities across the U.S. Yet given the challenge of "measuring the invisible" (Atkinson 2000, 163), or estimating how many people have either moved out of a neighborhood or been prevented from moving in, one can begin to understand the wide range of estimates that have been undertaken. For example, Zuk et al. (2018) reviewed ten studies on gentrification-induced displacement and found that between 5 and 30 percent of all moves resulted from displacement, which varies considerably among regions, cities, neighborhoods, and socioeconomic groups. Exclusionary displacement was consistently found in the form of increasing rents, resulting in a shift toward in-movers who were higher income, more likely to be non-Hispanic White, and with a higher degree of educational attainment than incumbent residents and out-movers. Less consistent was the evidence on direct displacement resulting from gentrification. The authors determined that the inconclusive evidence may be due to some definitional issues such as the omission of disinvestment-related displacement that may precede gentrification as described by Marcuse (1985), as well as methodological shortcomings in the reviewed studies. In addition to the evidence for exclusionary displacement, Ding et al. (2016) found that residents with low credit scores who moved from gentrifying neighborhoods were more likely to move to a lower income neighborhood compared to their origin neighborhoods than residents with low credit scores of non-gentrifying neighborhoods.

While the scale of displacement, especially associated with gentrification, is still unclear, the social impacts of displacement have received increasing attention. A number of studies document the detrimental effects of evictions on residents' mental and physical health (Burgard et al. 2012; Desmond and Kimbro 2015; Pevalin 2009). In their analysis of national survey data on low-income mothers, Desmond and Kimbro (2015) found that low-income mothers who had been evicted in the previous year experienced more material hardship, were more likely to suffer from depression, reported worse health for themselves and their children, and reported more parenting stress than similar mothers who were not evicted. Similarly, Marcus and Zuk (2017) found that displaced low-income households from the high-cost San Francisco Bay Area in California experienced a high rate of homelessness following forced moves. Displaced households moved to neighborhoods with fewer health care resources, greater environmental and safety concerns, and fewer job opportunities on average, leading to longer, more costly commutes.

Preventing and Mitigating Gentrification-Induced Displacement

A wide variety of affordable housing and community development interventions have been discussed in the practitioner and advocacy literature on anti-displacement. Numerous toolkits provide a wide range of anti-displacement strategies, yet there is little research on the effectiveness of different policies, especially as it pertains to reducing displacement (Greater Communities Collaborative 2007; Haughey and Sherriff 2010; Metropolitan Area Planning Council 2015; Pollack et al. 2010). In his overview of strategies to preserve and expand affordability in gentrifying neighborhoods, Lubell (2016) organizes a framework that includes six components: (1) preservation of affordable housing, (2) tenant protections, (3) social inclusion, (4) revenue generation for affordable housing, (5) incentives that increase the production of affordable housing, and (6) property acquisition for the preservation of affordability. Since gentrification and displacement can be traced to rising income inequality, many advocates and policymakers also include a series of strategies to help households build their wealth in their suite of anti-displacement strategies (Chapple 2017).

The following sections discuss anti-displacement strategies circulating in the practitioner literature grouped into three categories: first, neighborhood stabilization (including renter protection); second, production of affordable housing; and third, preservation of existing affordable rental housing.

Neighborhood Stabilization

As rents typically increase in gentrifying neighborhoods, advocates often point to the need for tenant protection, rent regulation, and other strategies to ensure that current residents are able to stay in the changing neighborhood. These can range from providing rental assistance and tenant counseling to proactive code enforcement and requiring landlords to have a just cause when trying to evict tenants. Neighborhood stabilization strategies can be loosely grouped into people-based versus place/housing unit-based strategies, and between those that are preventative versus those that are responsive (Table 22.1). In the following text I provide brief explanations and examples of these various

Table 22.1 Framework for Organizing Neighborhood Stabilization Strategies.

	Preventative	*Responsive*
People-based strategies	• Landlord anti-harassment protections • Just cause for evictions ordinances • Rental assistance • Tenant counseling	• Relocation benefits • Right to return policies • Evictee or neighborhood preference policies in housing subsidies • Tenant right to counsel
Place/housing unit-based strategies	• Condominium conversion restrictions • Rent regulation • Right of first refusal • Community land trusts • Proactive code enforcement	• Vacancy control in rent regulations • No net loss and one for one replacement

Source: author, based on Cash (2016).

strategies, followed by an in-depth review of the literature on rent control, one of the most commonly used and contentious anti-displacement strategies.

People-Based Strategies

Among preventative, people-based strategies, there are a number of policies and programs meant to either limit the ability of landlords to push tenants out of their homes, or help tenants meet requirements to stay. Just cause eviction ordinances limit the reasons for which landlords can evict tenants and are often adopted along with rent regulation to reinforce one another. Although some cities may pursue one without the other due to political feasibility, little is known about the effectiveness of either of these strategies in isolation (Hegarty 2017). In an effort to prevent landlords from pushing out tenants through informal means, a number of jurisdictions have also adopted anti-harassment protections. For example, New York City's anti-harassment policy was adopted in 2008 and prohibits landlords from interrupting utilities and other tactics used to push tenants out (*NYC Local Law 2008/007* 2008). On the tenant side, many nonprofits and some jurisdictions attempt to keep residents in their homes by providing short-term rental assistance to prevent evictions due to nonpayment of rent. Finally, there is counseling to help tenants both know their rights and seek resources.

People-based strategies that are responsive rather than preventative include relocation benefits and right of return policies. Relocation benefits, frequently used by government agencies when acquiring properties through eminent domain (*Uniform Relocation Assistance and Real Property Acquisition Policies for Federal and Federally Assisted Programs* 1970) are also beginning to be attached to certain evictions cases. For example, in San Francisco landlords are required to compensate tenants who receive no-fault evictions such as owner move-ins (*Tenants Rights to Relocation for No-Fault Evictions* 2006). In rapidly gentrifying, high-cost cities, relocation benefits may be insufficient to help tenants find homes in the same city or neighborhood. Right of return policies seek to guarantee displaced tenants the ability to move back into redeveloped properties and mixed-income developments (e.g., HOPE VI). Yet researchers have found limited rates of return due in part to numerous eligibility requirements (Chaskin and Joseph 2015; Joseph and Chaskin 2012). In a new expansion of the concept for the North/Northeast neighborhood of Portland, Oregon, that aims to attract displaced residents and to stabilize the existing African-Americanneighborhood, the right of return policy gives preference for public subsidy to residents who can demonstrate the risk of being displaced from or prior residence in the neighborhood of themselves, their parents, or even grandparents (Korn 2017).

Place/Housing Unit-Based Strategies

Place/housing unit-based strategies give tenants a chance to stay in their units even if a landlord wants to cease being the owner. Several jurisdictions have developed policies to give tenants the first right of refusal for purchasing the property. Perhaps the best known of these is Washington, DC's Tenant Opportunity to Purchase Act (TOPA), which has been the catalyst of preserving thousands of affordable homes (Gallaher 2016; Harrison Institute for Public Law 2006).

Other jurisdictions have limited landlords' ability to convert multifamily rental to ownership units through condominium conversion controls. Following waves of condominium conversions in the late 1970s and 1980s as well as the early 2000s, numerous jurisdictions have begun to regulate the process and now limit the number of possible conversions each year (Chambers 2006). Jurisdictions have adopted a diversity of criteria to limit the conversions, using vacancy rates or the proportion of rental housing units, or have limited the number of conversions to a percentage of total rentals or an absolute number. Condominium conversion controls are widespread in coastal regions of California, with over 67 percent of cities in the Bay Area and 27 percent of cities in Los Angeles County having some form of condominium conversion controls (Crispell et al. 2017). Crispell et al. (2017)

also documented a number of loopholes in policies found in the Bay Area that limit their ability to stabilize communities, although the effectiveness of this strategy has remained understudied.

Rent Control

Rent control or stabilization solutions may emerge when rents begin to increase (Minton 1997). A number of rent control models are discussed in the literature. For example, Lind (2001) identifies five main types of rent control along two dimensions. The first dimension distinguishes between rent control that covers rent changes for sitting tenants and rents generally. The second dimension is whether the aim is to protect the tenants against rents higher than market levels or against sudden large increases in rents, or to keep rents permanently below market levels in attractive areas. The first generation of rent controls sought to keep rents temporarily or permanently below market level (Arnott 1995). Such rent freezes or ceilings are still considered by jurisdictions today (Kinney 2016; Krisel 2017). Rent control models most common in U.S. cities, however, involve protecting sitting tenants only by de-controlling units upon vacancy, and allow annual increases due to inflation and/or cost increases. The primary goal of these models of rent control is to secure tenancy for current residents.

A number of scholars have looked at the ways in which rent control may impact the local housing market (Ambrosius et al. 2015) by affecting the number of new housing units that are produced (McFarlane 2003; Sims 2007), overall rental prices (Kutty 1996), housing maintenance (Kutty 1996), and even neighborhood racial, ethnic, and socioeconomic composition (Glaeser 2003; Sims 2011). Yet only a handful of studies have looked at the impact of rent control on mobility, turnover in units, and displacement in U.S. cities.

Clark and Heskin (1982) found that mobility rates of renters in the City of Los Angeles significantly declined by 14 percentage points in the two years after adoption of rent control in 1979. Gyourko and Linneman (1989) analyzed rental units in New York City, based on the Housing Vacancy Survey (HVC), and found significantly longer tenure length of residents in apartments that were rent-controlled versus not. Ault et al. (1994) arrived at similar findings, even after controlling for a diversity of factors at the individual, housing unit, and neighborhood levels.

Newman and Wyly (2006) analyzed displacement rates in New York City using five samples of the HVC between 1989 and 2002. They found that current residency in a rent-controlled unit did not significantly predict the rate of displacement from the previous place of residence; i.e., renters who were displaced from their previous home were not more likely to end up in a rent-controlled unit. In a follow-up study Wyly et al. (2010) found that residents in rent-controlled units were less likely to have been displaced from their previous residence in the 2005 and 2008 survey samples. The authors hypothesized that this finding was a result of the tight housing market where middle-class households aggressively compete for rent-stabilized units, thereby disadvantaging lower-income displaced residents. Finally, Diamond et al. (2017) analyzed the impacts of an expansion of rent control in the City of San Francisco in 1994. Utilizing a unique proprietary dataset of individual mobility, as well as parcel level assessor and transaction data, the authors found that the expansion of rent control increased the probability that renters stayed in their home by 20 percent.

In sum, these studies show that rent control ordinances significantly reduce household turnover, most likely due to both the cap on rent increases as well as just cause for eviction protections, which each of the cities also had in place. Yet, the combined effect, in addition to unintended consequences or an evaluation of a tenant's decision to move, has yet to be thoroughly analyzed.

Production of Affordable Housing

Developing new affordable housing is often included in the strategies considered for combating displacement to ensure that low-income residents are not priced out of neighborhoods. Cities have a number of tools at their disposal to influence the production of affordable housing, including fiscal strategies to generate resources for development, land use policies to incentivize or prioritize certain types of developments, and public investments that can be tied to housing affordability requirements. In general, affordable housing strategies can be categorized into those that generate funds to produce units or those that generate or incentivize the production of units. As shown in Figure 22.1, the other dimension of affordable housing strategies is whether units are funded by leveraging the market or through public investment. In the following text I discuss both dimensions and provide greater depth on recent debates about the relative benefits of developing subsidized versus market-rate housing and their impacts on displacement.

Generating Funds

Numerous jurisdictions across the U.S. have attempted to leverage market activity to generate funding for affordable housing through impact fees, as well as fees developers pay in lieu of developing units as part of inclusionary housing programs (discussed later). Impact fees, also known as linkage fees, on new market-rate housing developments have been traditionally used to fund infrastructure, but are now applied to fund new affordable housing development. These fees are especially common in Californian cities where the courts limited the ability of jurisdictions to impose inclusionary requirements (Palmer/Sixth Street Properties L.P. v. City of Los Angeles 2009). For example, over 20 percent of cities in the Bay Area have impact fees (Urban Displacement Project 2017). The nexus between new market-rate housing and the need for affordable housing is based on the idea that every resident who moves into a market-rate home will generate a need for services typically provided by employees paid less than the median income (Strategic Economics and Vernazza Wolfe Associates 2015).

Similar to housing impact fees, commercial impact/linkage fees, also known as jobs-housing balance fees, charge per square footage of commercial or retail space developed, with the logic that these developments create low-wage jobs for people who need affordable housing. One example of such

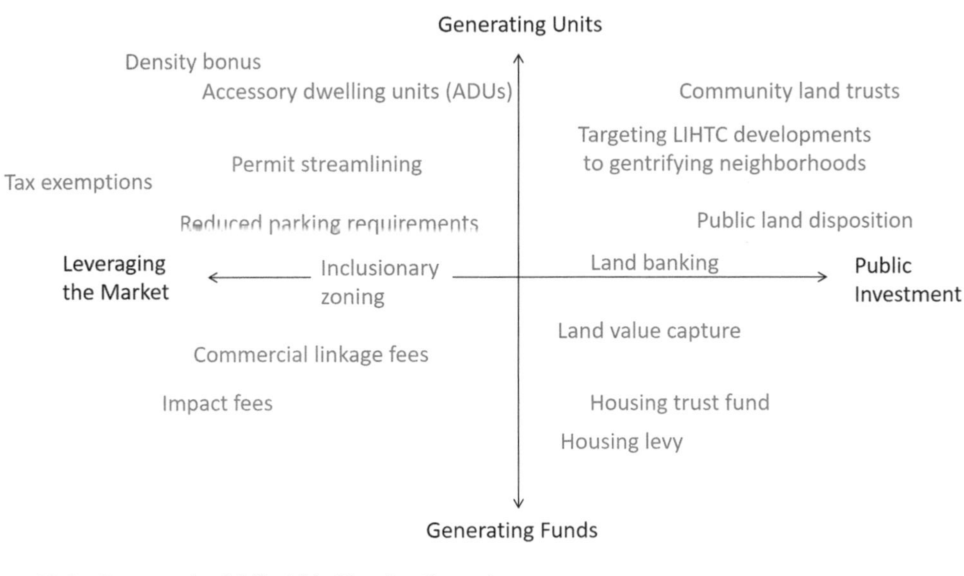

Figure 22.1 Framework of Affordable Housing Strategies.

Source: author, based on Cash and Dow (2017).

linkage fees is Boston's commercial linkage fee program. Established in 1983, the program generated over $51 million between 2006 and 2015, contributing to the creation of 2,181 affordable units (Boston Municipal Research Bureau 2015).

Jurisdictions also use their taxing powers to either generate funds or incentivize affordable housing development. There are a variety of tools, including property tax exemptions for owners of subsidized housing; property taxes to generate funding for affordable housing and the creation of a housing trust fund; and tax increment financing (TIF) districts to borrow against future improvements in land value. Examples of these tools include Portland's property tax exemptions for new developments that reserve a certain percentage of units for low- or very low-income tenants (City of Portland n.d.). Seattle's Housing Levy, on the other hand, has used property taxes to fund the development of over 12,500 affordable apartments, provide homeownership assistance to more than 900 first-time low-income homebuyers, and provide emergency rental assistance to more than 6,500 households since its passage in 1981 (City of Seattle 2017). Furthermore, the City of Portland has used its redevelopment authority to create TIF districts, which is the city's primary local funding source for affordable housing projects. In 2015 the City Council increased the TIF set aside for affordable housing from 30 percent to 45 percent (Templeton 2015).

Finally, cities have used their assets and investments to subsidize the development of affordable housing. Affordable housing advocates are beginning to push jurisdictions to dedicate a portion of their land for affordable housing (Hickey and Sturtevant 2015; Lane and Seifel 2015). The Fulton County/City of Atlanta Land Banking Authority, for instance, prioritized the transfer of land for affordable housing development, enabling affordable housing developers to acquire tax-delinquent properties at below-market prices for development of affordable housing (Sage Computing Inc. 2009). Between 1990 and 2009, the authority had facilitated the transfer of 50 to 100 properties per year. Cities may also engage in land banking, acquiring land that they later open up for affordable housing development. For instance, Portland's transit agency, TriMet, pursued land banking when trying to stabilize gentrifying neighborhoods around its light rail expansion (Zuk and Carlton 2015).

Generating Units

Perhaps the most commonly used and cited strategy for generating units by leveraging the real estate market is the creation of inclusionary zoning requirements on market-rate developments, also known as below-market-rate zoning (Freeman and Schuetz 2016; Hickey 2014). Inclusionary zoning programs are widespread—approximately 500 jurisdictions in 27 states and Washington, DC, have policies in place (Hickey et al. 2014). Inclusionary zoning often emerges in real estate markets that are experiencing significant growth and development, as evidenced by rising rents.

Advocates of inclusionary zoning often cite California as a success story because many cities have adopted ordinances. For example, over 70 percent of jurisdictions in the San Francisco Bay Area have inclusionary ordinances (Crispell et al. 2017). Inclusionary zoning policies have generated a large number of units of subsidized affordable housing. Nationally, Mallach and Calavita (2010) estimated that between 129,000 and 150,000 units had been produced through these programs, mostly in California, Massachusetts, and New Jersey. Yet research shows that the proportion of below-market units resulting from inclusionary policies is modest in comparison to the demand or even to the production of subsidized units from other funding sources, for example the Low-Income Housing Tax Credit (LIHTC) program. Powell and Stringham (2005) estimated that inclusionary housing policies in the Bay Area produced enough housing to meet only 5 percent of the affordable housing need. Freeman and Schuetz (2016) compared the production of subsidized housing from inclusionary policies to those funded by the LIHTC program in California, Boston, Washington, DC, and New York City, finding wide variability between jurisdictions, from a low of one inclusionary unit for every 25 LIHTC units produced in New York City to a high of four inclusionary units for every five LIHTC units developed in the Washington, DC, suburbs.

Some jurisdictions choose to incentivize affordable housing units through density bonuses, allowing developers to build at higher densities in exchange for affordable units. Density bonuses provide strong incentives in cities where there is interest in developing at greater densities or in taller buildings than what is permitted under the existing zoning code. In California, where the state passed a density bonus law in 1979, Garde (2016) found density bonuses to be the third most popularly used incentive among developers, behind the reduction of site development standards and direct public subsidies. The Grounded Solutions Network surveyed 187 incentive programs around the country and found that after density bonus incentives, 44 percent of programs provided zoning variances, 69 percent provided fee reductions, and 37 percent provided fee waivers to incentivize the development of affordable housing (Thaden and Wang 2017). However, without a strong housing market, incentives alone may not be sufficient to produce affordable units (Schwartz et al. 2012). For example, the City of Cambridge, Massachusetts, has had a voluntary inclusionary zoning program since the 1980s that offers density bonuses, but failed to produce any units in over a decade until it became mandatory in 1998. By 2010 this program had produced 385 affordable rental and for-sale homes (Schwartz et al. 2012).

The Debate Between Market-Rate and Subsidized Housing

The debate over the relative importance of subsidized and market-rate housing production in alleviating the current housing affordability crisis continues to preoccupy policymakers, developers, and advocates, especially in strong market cities. In their review of affordable housing strategies in strong markets, Freeman and Schuetz (2016) concluded that "most state and local programs have produced relatively small numbers of affordable units, and so are unlikely to substantially meet the demand for below-market-rate housing" (228). Similarly, the California Legislative Analyst's Office concluded that given the limited funding and high cost of development in California's coastal communities, perhaps the best approach to limiting displacement was to build more market-rate housing to reduce the cost of existing units and allow for the filtering process to meet demand for lower-income residents (Uhler 2016).

Yet, as Zuk and Chapple (2016) showed in their analysis of the relationship between housing production and displacement in the San Francisco Bay Area, the production of subsidized affordable housing was twice as effective at reducing displacement pressures when compared to the production of market-rate housing. Furthermore Zuk and Chapple found that the construction of market-rate housing in the 1990s led to both lower median rents and higher rent burden for low-income households. Other studies in non-US contexts have identified the spillover effects of market-rate housing developments that may add to increased housing costs (Brunes et al. 2016; Kurvinen and Vihola 2016; Ooi and Le 2013), indicating that further research is necessary to determine if and by how much the production of market-rate housing impacts housing costs and displacement rates of lower-income households.

Preservation of Existing Affordable Rental Housing

In many built-out neighborhoods experiencing gentrification pressures, there may be little room for new development. Furthermore, the cost of new development in strong housing markets can make the production of new subsidized housing at scale prohibitively costly. Therefore, strategies for preserving affordable rental units may be more cost-effective and feasible to counteract displacement forces in such communities. The preservation of affordable housing implies the loss of supply, which is caused by three factors: first, the expiration of affordability restrictions on federally subsidized affordable housing; second, increasing rents in non-subsidized affordable housing; and third, the loss of affordability in rent-restricted housing. In the following text I discuss each of these phenomena before exploring the programs emerging to preserve affordability.

In regard to the expiration of affordability restrictions on federal subsidized affordable housing, many federal subsidies, such as the U.S. Department of Housing and Urban Development's (HUD) project-based Section 8 program and the LIHTC program provide public subsidies on privately owned properties. Owners who receive these subsidies agree to restrict rents for low-income tenants for a fixed period of time. At the end of the period owners are offered the option to renew their contracts or exit the programs. Nearly 500,000 subsidized units are at risk of losing their affordability in the next five years, spanning a range of programs, though the majority receive project-based Section 8 subsidies (National Housing Preservation Database 2017; see Figure 22.2).

Lens and Reina (2016) found that project-based Section 8 properties whose affordability restrictions expired between 2000 and 2010 were located in low-opportunity neighborhoods that were experiencing improvements, indicating a higher tendency of property owners in gentrifying neighborhoods to exit the subsidy program. They further found that LIHTC properties set to expire between 2010 and 2020 were in higher opportunity neighborhoods than new and active LIHTC properties. Other researchers have also found that properties in areas with high price appreciation have higher odds of opting out of their rental subsidy (Ellen and Weselcouch 2015; Ray et al. 2015; Reina and Begley 2014). Independent of the changing nature of these neighborhoods, the loss of affordability restrictions can lead to displacement when tenants are unable to afford units after they lose their affordability restrictions.

Due to the scarcity of housing subsidies, most low-income households end up in unrestricted low-cost housing (also called naturally occurring affordable housing [NOAH]) that is poorly maintained by property owners and is located in low-opportunity areas (Schmidt et al. 2013). In regions with high and increasing costs, the availability of the NOAH stock has been rapidly decreasing as rents continue to increase and incomes have stagnated for many moderate- and low-income

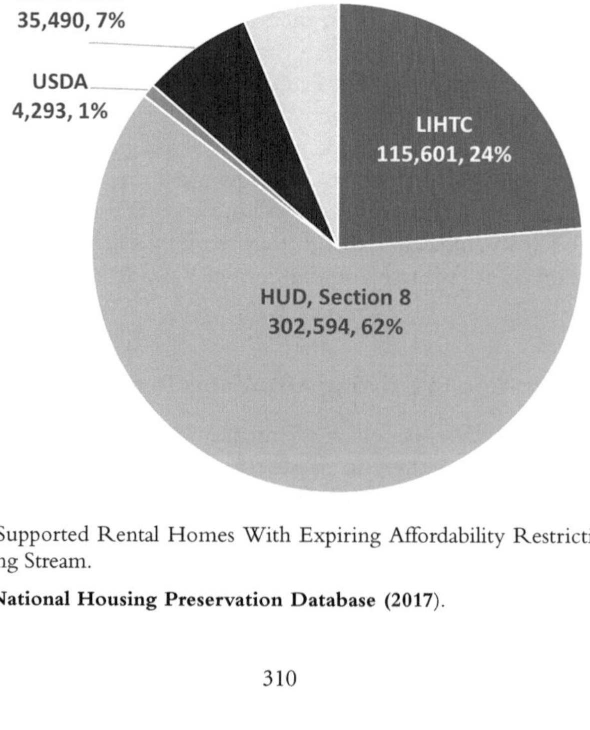

Figure 22.2 Publicly Supported Rental Homes With Expiring Affordability Restrictions Within Five Years by Funding Stream.

Source: author, based on **National Housing Preservation Database (2017)**.

households, resulting in non-subsidized housing that has become increasingly unaffordable. Nearly half of the nation's 100 largest metropolitan areas experienced absolute declines in their stocks of low-rent units between 2005 and 2015. Due to the growth in construction of high-rent multifamily units, 88 percent of metropolitan areas experienced decreases in the shares of low-rent units as well. Thus, there has been a severe and worsening mismatch of demand and supply, with the number of low-income renters far outstripping the number of available low-rent units, with real gross rent lower than $800 per month (Joint Center for Housing Studies of Harvard University 2017).

Finally, for rent-controlled units in jurisdictions with vacancy decontrol (i.e., where landlords can set rents at market rate once a unit is vacated) turnover can mean a significant loss of affordability, especially in neighborhoods with increasing rents. Although little research has documented this loss of affordability, in part due to limited data, Zuk et al. (2017) estimated that between 2010 and 2015, nearly 50 percent of the rent-controlled housing stock in the five Bay Area cities with rent control (i.e., Berkeley, East Palo Alto, Oakland, San Francisco, and San Jose) turned over, thereby reducing their affordability. The differential rates of eviction from rent-controlled units versus non-controlled units, however, has not been documented and requires additional research.

Finally, property owners may be motivated to exit the rental market when faced with regulations, resulting in a net loss of affordable units. In the overheated San Francisco market, the Eviction Defense Collaborative documented a 70 percent increase in owner move-in evictions between 2003 and 2016, with the greatest concentration in the gentrifying Mission District (Eviction Defense Collaborative 2016). Furthermore, Diamond et al. (2017) found that the expansion of rent control in 1994 resulted in a 15 percent reduction in supply due to owners converting to condos/tenancies in common (TICs) or otherwise removing units from the rental stock.

Preserving Affordable Housing

Preservation of affordable housing includes extending the affordability of either subsidized or unsubsidized rental units that are at risk of becoming unaffordable to low-income households. For subsidized rental housing, preservation refers to renewing a subsidy when they are due to expire. For unsubsidized rental housing, preservation usually refers to nonprofit developers purchasing rental units at risk of becoming unaffordable, investing to rehabilitate the units while also keeping rents at levels that are affordable to low-income residents (Schwartz et al. 2016).

Many preservation strategies fall into the same funding categories as production strategies, which also allow funds to be used for rehabilitation and subsidizing units. The LIHTC program, for instance, provides tax credits for acquisition and rehabilitation projects. Of the over 2.4 million low-income units in the LIHTC portfolio, tax credits provided to 900,000 units were used for acquisition and rehabilitation purposes (U.S. Department of Housing and Urban Development 2017). In addition to federal funding, a number of local jurisdictions have created funds specific to preservation. For instance, Alameda County in California's A1 bond measure, passed in 2017, will allocate $425 million for a Rental Housing Development Program that includes preservation. The City of Oakland passed Measure KK in 2016, which will allocate $100 million for anti-displacement and affordable housing preservation projects. As of this writing, over $16 million has been awarded for acquisition and rehabilitation of existing buildings for the preservation of 297 affordable housing units (Byrd 2017).

Williams (2016) highlights several non-governmental investment funds that seek to preserve the affordability of units that have an expiring subsidy (e.g., Avanath, New Generation Fund, and Housing Partnership Equity Trust). In 2000, the MacArthur Foundation started the Window of Opportunity: Preserving Affordable Rental Housing initiative, which aims to preserve 300,000 privately owned subsidized affordable units by 2020 (Schwartz et al. 2016). The initiative invests in preservation-themed loan funds, develops business practices around preservation, and works on federal policy, among other strategies. Although it achieved many of its goals it found that it was not on track to preserve the 300,000 subsidized affordable units by 2020 nor did it accomplish the creation of new federal incentives for preservation. One potential reason for the shortfall was that the initiative excluded for-profit owners, who it estimates control approximately 85 percent of the privately owned affordable housing stock.

Yet, the majority of low-income renter households live in non-subsidized units. Although many federal resources can be used to acquire, rehabilitate, and convert non-subsidized units into subsidized ones, the resources tend to be better suited for large buildings, whereas many NOAHs are in smaller buildings (Abdelgany 2017). To address this challenge, the City of San Francisco created its Small Sites program, an acquisition and rehabilitation loan program launched in 2014 that assists nonprofit and for-profit entities to buy small housing developments of five to 25 units and restrict their rents for the life of the project. Beginning with a $3 million pilot program, the city increased the budget to $75 million in 2017, with new funding sources coming primarily from the City's housing trust fund, in-lieu fees from the inclusionary zoning program, and bond funds. As of this writing, 137 units are on the path to preservation (Harris 2017). Other cities are experimenting with their inclusionary zoning programs to acquire existing units and stabilize them rather than producing new units. For example, Hickey (2015) discusses examples from New York City; Boulder, Colorado; Montgomery County, Maryland; Chapel Hill, North Carolina; and Edina, Minnesota, where developers are allowed to convert existing, market-rate units to deed-restricted, affordable units in lieu of developing new affordable units. Hickey found that these options may be attractive to developers that already own existing market-rate units in an area characterized by high development costs.

In addition to providing loans and funding mechanisms for acquisition and rehabilitation, policymakers and practitioners are considering a range of incentives, education, and other opportunities to preserve affordable rental units. For example, the Minnesota Preservation Plus Initiative identified a range of smaller interventions to meet the needs and challenges of landlords to maintain low rents (Schmidt et al. 2013). Among the interventions the authors of the initiative identified are local government rent subsidies such as Denver's new voucher equity program (Abello 2018), conditioning second mortgages or debts on landlord's commitment to keep rents affordable, and property tax discounts in exchange for property owners keeping rents below market rate, among others. In addition, the authors encouraged cities in Minnesota to use existing rent licensing programs as a potential entry point to more effectively communicate with property owners about existing programs and regulations. Due to the recent emergence of efforts around preservation, little research has evaluated the scale or effectiveness of these strategies.

Conclusions and a Call for More Evaluation Research

Despite significant data and methodological challenges in measuring the extent of displacement, most scholars agree that gentrification at minimum leads to exclusionary displacement and may involve direct displacement as well. To address the key concerns of urban redevelopment in the twenty-first century, policymakers, practitioners, and activists are adopting strategies to prevent or respond to displacement pressures. As discussed earlier, anti-displacement strategies can be largely categorized into three groups: first, neighborhood stabilization; second, production of affordable housing; and third, preservation of existing affordable rental units. There is a wide range of anti-displacement and affordable housing policies. Some policies, such as inclusionary zoning, have been adopted in many places; other policies, such as rent control, have been adopted in very few places.

Very little evaluation research has been conducted on the relative effectiveness of anti-displacement strategies. While some policies may have clear, trackable results, such as those that fund affordable housing developments, others such as education campaigns on tenants' rights may be less tangible and more difficult to track. Many scholars and activists have discussed the loopholes or lack of implementation of anti-displacement policies, making it important to analyze the effectiveness of the policies and their implementation rather than simply tracking whether a jurisdiction has them or not (Crispell et al. 2017). Given the nuances of policies and the growing concern about gentrification and displacement, assessing the effectiveness of anti-displacement strategies on a systematic basis is a difficult yet important direction for future research.

References

Abello, Oscar Perry. 2018. "Denver's New Local Rental Voucher Program Filling in Gaps—and Apartments." *Next City*, February 8.

Abdelgany, Somaya. 2017. *Catching Affordability Where It's at: Acquisition/Rehab of Oakland's Unsubsidized Affordable Housing*. Berkeley, CA:University of California at Berkeley.

Ambrosius, Joshua D., John I. Gilderbloom, William J. Steele, Wesley L. Meares, and Dennis Keating. 2015. "Forty Years of Rent Control: Reexamining New Jersey's Moderate Local Policies after the Great Recession." *Cities*, 49(December): 121–133.

Arnott, Richard. 1995. "Time for Revisionism on Rent Control?" *The Journal of Economic Perspectives*, 9(1): 99–120.

Atkinson, Rowland. 2000. "Measuring Gentrification and Displacement in Greater London." *Urban Studies*, 37(1): 149–165.

Ault, Richard, John Jackson, and Richard Saba. 1994. "The Effect of Long-Term Rent Control on Tenant Mobility." *Journal of Urban Economics*, 35: 140–158.

Boston Municipal Research Bureau. 2015. *City's Linkage Program Under Review*. Boston: Boston Municipal Research Bureau.

Brunes, Fredrik, Cecilia Hermansson, Han-Suck Song, and Mats Wilhelmsson. 2016. *NIMBYs for the Rich and YIMBYs for the Poor: Analyzing the Property Price Effects of Infill Development*. Stockholm, Sweden: KTH Royal Institute of Technology.

Burgard, Sarah A., Kristin S. Seefeldt, and Sarah Zelner. 2012. "Housing Instability and Health: Findings from the Michigan Recession and Recovery Study." *Social Science and Medicine, Part Special Issue: Place, Migration and Health*, 75(12): 2215–2224.

Byrd, Michelle. 2017. *NOFA Affordable Housing Fund*. Oakland, CA:. City of Oakland.

Cash, Anna. 2016. *Investment without Displacement: Neighborhood Stabilization*. Berkeley, CA: Urban Displacement Project.

Cash, Anna, and Paige Dow. 2017. *Beyond "Build, Baby, Build": Towards A Nuanced Conversation About Affordable Housing Development*. Berkeley, CA: Urban Displacement Project.

Chambers, Robert. 2006. "Pushed Out: A Call for Inclusionary Housing Programs in Local Condominium Conversion Legislation." *California Western Law Review*, 42(2): 355–376.

Chapple, Karen. 2017. "Income Inequality and Urban Displacement: The New Gentrification." *New Labor Forum*, 26: 84–93.

Chaskin, Robert J., and Mark L. Joseph. 2015. *Integrating the Inner City: The Promise and Perils of Mixed-Income Public Housing Transformation*. Chicago: University of Chicago Press.

City of Portland. n.d. *Chapter 3.103 Property Tax Exemption for Multiple-Unit Housing Development*. Portland, OR: City of Portland.

City of Seattle. 2017. *2016 Seattle Housing Levy*. Seattle: City of Seattle.

Clark, W.A.V, and Allan Heskin. 1982. "The Impact of Rent Control on Tenure Discounts and Mobility." *Land Economics*, 58(1): 109–117.

Crispell, Mitchell, Karolina Gorska, Miriam Zuk, Karen Chapple, Anastasia Loukaitou-Sideris, Gonzalez Silvia, and Abdelgany Somaya. 2017. "Anti-Displacement Policy Analysis." In *Developing a New Methodology for Analyzing Potential Displacement*, 181–261. Berkeley, CA: California Air Resources Board.

Desmond, Matthew, and Rachel Tolbert Kimbro. 2015. "Eviction's Fallout: Housing, Hardship, and Health." *Social Forces*, 94(1): 295–324.

Desmond, Matthew, and Tracey Shollenberger. 2015. "Forced Displacement From Rental Housing: Prevalence and Neighborhood Consequences." *Demography*, 52(5): 1751–1772.

Diamond, Rebecca, Tim McQuade, and Franklin Qian. 2017. "The Effects of Rent Control Expansion on Tenants, Landlords, and Inequality: Evidence from San Francisco." *Paper presented at the Urban Economics Association Conference*. Vancouver, BC, November 8–11.

Ding, Lei, Jackelyn Hwang, and Eileen Divringi. 2016. "Gentrification and Residential Mobility in Philadelphia." *Regional Science and Urban Economics*, 61(November): 38–51.

Ellen, Ingrid Gould, and Max Weselcouch. 2015. *High-Opportunity Neighborhoods in NYC Are Losing Affordable Housing.* New York: NYU Furman Center for Real Estate and Urban Policy.
Eviction Defense Collaborative. 2016. *City of Change: Fighting for San Francisco's Vanishing Communities.* San Francisco: Eviction Defense Collaborative.
Florida, Richard. 2017. *The New Urban Crisis: How Our Cities Are Increasing Inequality, Deepening Segregation, and Failing the Middle Class—and What We Can Do about It.* New York: Basic Books.
Freeman, Lance. 2005. "Displacement or Succession? Residential Mobility in Gentrifying Neighborhoods." *Urban Affairs Review*, 40(4): 463–491.
Freeman, Lance, and Jenny Schuetz. 2016. "Producing Affordable Housing in Rising Markets: What Works?" SSRN Scholarly Paper ID 2851175. Rochester, NY: Social Science Research Network.
Gallaher, Carolyn. 2016. *The Politics of Staying Put: Condo Conversion and Tenant Right-to-Buy in Washington, DC.* Philadelphia: Temple University Press.
Garde, Ajay. 2016. "Affordable by Design? Inclusionary Housing Insights from Southern California." *Journal of Planning Education and Research*, 36(1): 16–31.
Glaeser, Edward L. 2003. "Does Rent Control Reduce Segregation?" *Swedish Economic Policy Review*, 10: 179–202.
Glass, Ruth Lazarus. 1964. *London: Aspects of Change.* London: University College London: Centre for Urban Studies.
Greater Communities Collaborative. 2007. *Preventing Displacement Policy Fact Sheet.* San Francisco: Greater Communities Collaborative.
Grier, George, and Eunice Grier. 1978. *Urban Displacement: A Reconnaissance.* Bethesda, MD: The Grier Partnership.
Gyourko, Joseph, and Peter Linneman. 1989. "Equity and Efficiency Aspects of Rent Control: An Empirical Study of New York City." *Journal of Urban Economics*, 26: 54–74.
Harris, Ruby. 2017, January 25. *Small Sites Program: The Little Program That Could: Achievements, Challenges, and Next Steps.* Oakland, CA: East Bay Housing Organizations.
Harrison Institute for Public Law. 2006. *An Analysis of the Strengths and Deficiencies of Washington, D.C.'s Tenant Opportunity to Purchase Act.* Washington, DC: Georgetown University.
Haughey, Rick, and Ryan Sherriff. 2010. *Challenges and Policy Options for Creating and Preserving Affordable Housing Near Transit and in Other Location-Efficient Areas.* Washington, DC: Urban Institute.
Hegarty, Peter. 2017. "Alameda: Renters Win 'Just Cause' Eviction Protections." *East Bay Times*, May 17.
Hickey, Robert. 2014. *Inclusionary Upzoning: Tying Growth to Affordability.* Washington, DC: National Housing Conference.
Hickey, Robert. 2015. *Making Inclusionary Housing More Flexible: Four Ideas for Urban Settings.* Washington, DC: Center for Housing Policy.
Hickey, Robert, and Lisa Sturtevant. 2015. *Public Land & Affordable Housing in the Washington DC Region: Best Practices and Recommendations.* Washington, DC: Center for Housing Policy and the National Housing Conference.
Hickey, Robert, Lisa Sturtevant, and Emily Thaden. 2014. *Achieving Lasting Affordability through Inclusionary Housing.* Cambridge, MA: Lincoln Land Institute.
Joint Center for Housing Studies of Harvard University. 2017. *The State of the Nation's Housing.* Cambridge, MA: Joint Center for Housing Studies of Harvard University.
Joseph, Mark L., and Robert J. Chaskin. 2012. "Mixed-Income Developments and Low Rates of Return: Insights from Relocated Public Housing Residents in Chicago." *Housing Policy Debate*, 22(3): 377–405.
Kinney, Aaron. 2016. "San Mateo Landlords Jeer Recording of Sobbing Renter." *The Mercury News*, April 12.
Korn, Peter. 2017. "Payback, But How Many Will Return?" *Portland Tribune*, September 7.
Krisel, Brendan. 2017. "State Judge Upholds Rent Freeze For NYC Rent Stabilized Housing." *NY Patch*, March 28.
Kurvinen, Antti Tapio, and Jaakko Vihola. 2016. "The Impact of Residential Development on Nearby Housing Prices." *International Journal of Housing Markets and Analysis*, 9(4): 671–690.
Kutty, Nandinee K. 1996. "The Impact of Rent Control on Housing Maintenance: A Dynamic Analysis Incorporating European and North American Rent Regulations." *Housing Studies*, 11(1): 69–88.
Lane, Michael, and Libby Seifel. 2015. *Untapped Resources: Potential Bay Area Sites for Affordable Transit-Oriented Development.* San Francisco: Non Profit Housing Association of Northern California.
Lees, Loretta, Tom Slater, and Elvin Wyly. 2013. *Gentrification.* New York: Routledge.
Lens, Michael C., and Vincent Reina. 2016. "Preserving Neighborhood Opportunity: Where Federal Housing Subsidies Expire." *Housing Policy Debate*, 26(4–5): 714–732.
Lind, Hans. 2001. "Rent Regulation: A Conceptual and Comparative Analysis." *International Journal of Housing Policy*, 1(1): 41–57.

Lubell, Jeffrey. 2016. "Preserving and Expanding Affordability in Neighborhoods Experiencing Rising Rents and Property Values." *Cityscape*, 18(3): 131–150.

Mallach, Alan, and Nico Calavita. 2010. "United States: From Radical Innovation to Mainstream Housing Policy." In *Inclusionary Housing in International Perspective: Affordable Housing, Social Inclusion, and Land Value Recapture*, edited by Nico Calavita and Allan Mallach, 15–77. Cambridge, MA: Lincoln Institute of Land Policy.

Marcus, Justine, and Miriam Zuk. 2017. "Displacement in San Mateo County, California: Consequences for Housing, Neighborhoods, Quality of Life, and Health." Research Brief. Berkeley, CA: Institute of Government Studies, UC Berkeley.

Marcuse, Peter. 1985. "Gentrification, Abandonment, and Displacement: Connections, Causes, and Policy Responses in New York City." *Urban Law Annual: Journal of Urban and Contemporary Law*, 28(1): 195–240.

Marcuse, Peter. 1986. "Abandonment, Gentrification, and Displacement: The Linkages in New York City." In *Gentrification of the City*, edited by Neil Smith and Peter Williams, 153–177. New York: Routledge.

McFarlane, Alastair. 2003. "Rent Stabilization and the Long-Run Supply of Housing." *Regional Science and Urban Economics*, 33(3): 305–333.

Metropolitan Area Planning Council. 2015. *Managing Neighborhood Change: Anti-Displacement Strategies Toolkit*. Boston: Metropolitan Area Planning Council.

Minton, Jeffrey. 1997. "Rent Control: Can and Should It Be Used to Combat Gentrification?" *Ohio Northern University Law Review*, 23: 864–944.

National Housing Preservation Database. 2017. *2017 Preservation Profiles*. Washington, DC: National Housing Preservation Database.

Newman, Kathe, and Elvin K. Wyly. 2006. "The Right to Stay Put, Revisited: Gentrification and Resistance to Displacement in New York City." *Urban Studies*, 43(1): 23–57.

NYC Local Law 2008/007. 2008. NYC Administrative Code. Vol. 27–2004.

Ooi, Joseph T. L., and Thao T. T. Le. 2013. "The Spillover Effects of Infill Developments on Local Housing Prices." *Regional Science and Urban Economics*, 43(6): 850–861.

Palmer/Sixth Street Properties L.P. v. City of Los Angeles. 2009. Court of Appeal, Second District, Division 4.

Pevalin, David. 2009. "Housing Repossessions, Evictions and Common Mental Illness in the UK: Results from a Household Panel Study." *Journal of Epidemiology and Community Health*, 63(11): 949–951.

Pollack, Stephanie, Barry Bluestone, and Chase Billingham. 2010. *Maintaining Diversity in America's Transit-Rich Neighborhoods: Tools for Equitable Neighborhood Change*. Boston: Dukakis Center for Urban and Regional Policy, Northeastern University.

Powell, Benjamin, and Edward Peter Stringham. 2005. "The Economics of Inclusionary Zoning Reclaimed: How Effective Are Price Controls?" SSRN Scholarly Paper ID 1674474. Rochester, NY: Social Science Research Network.

Ray, Anne, Jeongseob Kim, Diep Nguyen, and Jongwon Choi. 2015. *Opting In, Opting Out a Decade Later*." Washington, DC: U.S. Department of Housing and Urban Development.

Reina, Vincent, and Jaclene Begley. 2014. "Will They Stay or Will They Go: Predicting Subsidized Housing Opt-Outs." *Journal of Housing Economics*, 23(Supplement C): 1–16.

Sage Computing, Inc. 2009. *Revitalizing Foreclosed Properties with Land Banks*. Washington, DC: U.S. Department of Housing and Urban Development.

Schmidt, Deidre, Sara Joy Proppe, Tim Thompson, and Cathy Bennett. 2013. *The Space Between: Realities and Possibilities in Preserving Unsubsidized Affordable Rental Housing*. Minneapolis: Minnesota Preservation Plus Initiative and One Roof Global Consulting.

Schwartz, Heather, Raphael Bostic, Richard Green, Vincent Reina, Lois Davis, and Catherine Augustine. 2016. *Preservation of Affordable Rental Housing: Evaluation of MacArthur Foundation's Window of Opportunity Initiative*. Santa Monica, CA: RAND.

Schwartz, Heather L., Liisa Ecola, Kristin J. Leuschner, and Aaron Kofner. 2012. *Is Inclusionary Zoning Inclusionary? A Guide for Practitioners*. Santa Monica, CA: RAND.

Sims, David. 2011. "Rent Control Rationing and Community Composition: Evidence from Massachusetts." *The B.E. Journal of Economic Analysis and Policy*, 11(1): 1–30.

Sims, David P. 2007. "Out of Control: What Can We Learn from the End of Massachusetts Rent Control?" *Journal of Urban Economics*, 61(1): 129–151.

Strategic Economics, and Vernazza Wolfe Associates. 2015. *Residential Impact Fee Nexus Study: Final Report*. Berkeley, CA: Strategic Economics and Vernazza Wolfe Associates.

Templeton, Amelia. 2015. Portland Council: Nearly Half of Urban Renewal Funding Will Go To Housing." *Oregon Public Broadcasting*, October 23.

Tenants Rights to Relocation for No-Fault Evictions. 2006. San Francisco Administrative Code.

Thaden, Emily, and Ruoniu Wang. 2017. *Inclusionary Housing in the United States*. Cambridge, MA: Lincoln Institute of Land Policy.
Uhler, Brian. 2016. *Perspectives on Helping Low-Income Californians Afford Housing*. Sacramento, CA: California Legislative Analyst Office.
Uniform Relocation Assistance and Real Property Acquisition Policies for Federal and Federally Assisted Programs. 1970. US Code. Vol. Title 42, Chapter 61.
Urban Displacement Project. 2017. *Policy Tools*. Berkeley, CA: Urban Displacement Project.
U.S. Department of Housing and Urban Development. 2017. *LIHTC Database*. Washington, DC: U.S. Department of Housing and Urban Development.
Williams, Stockton. 2016. "Innovative Investment Approaches for Preserving California's Existing Workforce Apartments." *Journal of Case Study Research: A Publication of the Center for California Real Estate*, I(I): 21–23.
Wyly, Elvin, Kathe Newman, Alex Schafran, and Elizabeth Lee. 2010. "Displacing New York." *Environment and Planning A*, 42(11): 2602–2623.
Zuk, Miriam, Ariel Bierbaum, Karen Chapple, Karolina Gorska, and Anastasia Loukaitou-Sideris. 2018. "Gentrification, Displacement, and the Role of Public Investment." *Journal of Planning Literature*, 33(1): 31–44.
Zuk, Miriam, and Ian Carlton. 2015. *Equitable Transit Oriented Development: Examining the Progress and Continued Challenges of Developing Affordable Housing in Opportunity and Transit-Rich Neighborhoods*. Washington, DC: Poverty and Race Research Action Council.
Zuk, Miriam, Anna Cash, Paige Dow, and Justine Marcus. 2017. *Investment without Displacement: Preserving Existing Affordable Housing*. Oakland, CA: Urban Displacement Project.
Zuk, Miriam, and Karen Chapple. 2016. *Housing Production, Filtering and Displacement: Untangling the Relationships*. Berkeley, CA: Institute of Government Studies.

23

URBAN RESTRUCTURING, DEMOLITION, AND DISPLACEMENT IN THE NETHERLANDS

Uncovering the Janus Head of Forced Residential Relocation

Reinout Kleinhans

Introduction

Across both sides of the Atlantic, continuous attention has been paid to improving the prospects of deprived neighborhoods and their residents since the 1970s. In the United States and many European countries, substantial housing and neighborhood restructuring programs have been implemented to fit this purpose. While these programs usually involve economic and social interventions, the emphasis has been on demolition of public or social housing and new construction of rental and owner-occupied housing on the original site. Renowned examples of such restructuring programs are HOPE VI (U.S.), Housing Market Renewal (HMR) Pathfinders in England, urban renewal in France (based on the *Loi de programmation sur la ville et la rénovation urbaine*), and urban restructuring in the Netherlands. Such urban renewal programs have several goals, ranging from the creation of mixed-income neighborhoods and poverty deconcentration to improving area reputations and social cohesion among residents (Dekker and Varady 2011; Kleinhans 2012).

Regardless of the differences in context, problems, and housing stock characteristics, all physical renewal programs usually require temporary or permanent relocation of significant proportions of the original residents from dwellings slated for demolition or upgrading. While large-scale relocation has often been contested, laws and regulations that provide the basis for housing and restructuring programs, and that are an indispensable part of a renewal operation, enable public housing authorities (U.S.) and housing associations (Europe) to relocate their tenants and sometimes also owner occupiers to other dwellings. In other words, the initial decision for relocation is usually not made by residents themselves, but by policymakers, housing authorities, and landlords. Hence, this type of residential mobility can be defined as *forced relocation*, as opposed to mobility as a "choice" made by households themselves.

Since the 1960s, forced relocation has fueled a large body of research connected to various urban policies, such as urban slum clearance in the 1960s and 1970s (e.g., Paris and Blackaby 1979; Thomas 1986), gentrification in the 1980s (up to the present), and urban renewal policies with a strong housing market-oriented approach since the early 1990s. Across many European countries, post-1990 urban renewal policies (as discussed earlier) share the general aim to lower concentrations of social rented or public housing in certain areas, in favor of building more owner-occupied housing and thus strengthening the "social mix." Forced relocation has been a recurring theme in these policies. However, the framing of this phenomenon in scientific debates shows similarities between countries. Research on relocation has been inspired by gentrification theory and research. A particular characteristic of this framework is a mostly negative displacement perspective on the outcomes of relocation processes (Kleinhans and Kearns 2013). The debate around gentrification has been fueled not only by policy and research findings, but to a certain extent also by neoliberal ideologies that, according to critics, increase the role of markets and capital at the expense of poor people and state-provided social security (e.g., Hackworth and Smith 2001; Lees et al. 2010; Uitermark et al. 2007). An example of a gentrification-based perspective is provided by Lees (2008), who argues that the "movement of middle-income groups into low income areas creates overwhelmingly negative effects, the most significant of which is the displacement of low-income groups" (Lees 2008, 2457). In sum, displacement of residents is one of the primary dangers mentioned by those concerned about the exclusionary effects of market-driven as well as state-driven gentrification (Newman and Wyly 2006, 27).

The discussion on neighborhood impacts of forced relocation has focused on concepts such as social mix and social cohesion (Uitermark 2003), whereas relocation outcomes for individual tenants have been framed in terms of displacement and (altered) place attachment (Kleinhans and Kearns 2013). This chapter focuses on the latter strand of outcomes. The scientific literature shows that scholars disagree about the extent to which the act of forced relocation is, by definition, a form of displacement that hurts households or if forced relocation may not necessarily be experienced as such, because it helps households to improve their living situation. This essay discusses the evidence from relocation research in the Netherlands and shows how it fits into the scholarly discussion on displacement. While pre-relocation expectations are usually negative, many households may actually be better off in terms of their housing and neighborhood situation after urban restructuring. Furthermore, the combination of counseling and legal compensations appears to have a mitigating influence regarding the negative impacts of both the process and the relocation outcome. The results give reason to counter-argue the predominantly negative discourse on displacement as an inevitable "dark side" of state-led gentrification (see also the chapter authored by Wouter van Gent, Willem Boterman, and Myrte Hoekstra) and urban renewal programs. The next section discusses theoretical perspectives on forced relocation and displacement. Subsequently, the Dutch policy context will be explained, providing a background for understanding relocation outcomes. The following section reviews the main outcomes, followed by conclusions and policy recommendations.

On Displacement, Relocation, and Gentrification

The first traces of the term "displacement" in the literature date back to the late 1970s. A particularly detailed definition can be found in a report from the U.S. Department of Housing and Urban Development, where displacement was defined as the situation in which:

> any household is forced to move from its residence by conditions which affect the dwelling or its immediate surroundings, and which: (1) are beyond the household's reasonable ability to control or prevent; (2) occur despite the household's having met all previously imposed conditions of occupancy and (3) make continued occupancy by that household impossible, hazardous or unaffordable.
>
> *(Grier and Grier 1978, 8; see also LeGates and Hartman 1986)*

In essence, this definition does not include any reference to housing policy, market forces, or any form of neoliberal policy. However, the term has become intricately connected to gentrification, which was coined as "the rehabilitation of working-class and derelict housing and the consequent transformation of an area into a middle-class neighborhood" (Smith and Williams 1986, 1). Marcuse used Grier and Grier's definition to specify displacement as a consequence of certain processes and through the way in which it is measured (Marcuse 1986, 156). He identified four forms for any particular property:

- Economic/physical displacement, in which residents are priced out of a dwelling through rent increases or by physical means such as demolition and upgrading
- Last-resident displacement, a measure in which only the last resident is displaced
- Chain displacement, which includes all residents who have been displaced from a property during a certain period
- Exclusionary displacement, which includes all people who have been unable to access property because it has been gentrified

In terms of the debate on implications of forced relocation, the importance of economic/physical displacement is obvious for renewal programs with a clear demolition component. However, exclusionary displacement has also gained recognition, in two forms. The first form of displacement, Marcuse's income-based exclusion (see preceding text), refers to the widespread phenomenon in gentrifying areas whereby previously affordable housing experiences rapid price increases and becomes unaffordable for the type of family that had lived in that area.

The second form of displacement is more subtle and does not require a move, because it relies on a different mechanism of exclusion. This type of exclusionary displacement often applies to long-term residents who witness gentrification processes in their neighborhood. Both the characteristics of newcomers (e.g., middle-class gentrifiers) and place-based changes to social and service environments can create much uneasiness among long-term residents who are confronted with rapid and fundamental changes in their neighborhood. This second form of displacement has been labeled "the hidden costs of gentrification" (Atkinson 2000), "indirect displacement" (Davidson 2008), and the "class-based transformation of place" (Davidson 2011). As mentioned earlier, the result is not necessarily a move out of the neighborhood, but an increasing sense of feeling "out of place" among long-term residents, which may be partly counterbalanced by decreases in local crime and the opening of new stores (Freeman 2006).

Over time, the notion of displacement has increasingly acquired a negative connotation. An important reason is that many gentrification studies have identified displacement as a significant problem (Atkinson 2004; Lees et al. 2010; see also the chapter authored by Dennis Gale), based on evidence or the premise that gentrification and displacement result in a degradation of the housing situation (broadly defined) of displacees. In fact, "[T]he fear of displacement has in the minds of many, however, come to dominate all other concerns regarding gentrification" (Freeman 2005, 487).

Another reason for the intricate relation between the concepts of gentrification and displacement relates to the roles of pure market forces vis-à-vis different forms of intervention by federal, state, or local authorities in urban restructuring programs that are not primarily driven by the market. Several scholars have argued that gentrification has become a well-considered strategy within urban regeneration policies (Atkinson 2004, 107; Hackworth and Smith 2001; Newman and Wyly 2006; Uitermark et al. 2007). In short, in this line of thinking, low-income groups are supposed to suffer most from displacement, especially in relation to having no access to a higher-quality dwelling and neighborhood, having to pay increased rents, and facing fewer housing opportunities in general. Schill and Nathan (1983) showed that those who experienced gentrification-induced displacement tended to experience improved housing and neighborhood conditions but also higher housing costs. Whether they were overall better off after relocation is unclear (Schill and Nathan 1983).

However, conflating forced relocation connected to state-led restructuring with displacement linked to gentrification ignores or downplays fundamental differences between these phenomena, especially in terms of institutional contexts (Kearns and Mason 2013; Kleinhans and Kearns 2013). First, the gentrification discourse tends to portray relocatees as rather powerless agents without any rights. However, in the Netherlands, national restructuring policies, based on Dutch Civil Law (*Burgerlijk Wetboek*), are implemented at the local level with legally established compensation mechanisms for residents facing a forced relocation (Kleinhans and Van der Laan Bouma-Doff 2008; Korthals Altes 2016). In the United States, types of compensation for public redevelopment projects are based on the Uniform Relocation Act (Manzo et al. 2008), while France's are based on Urban Renewal Law (*Loi de programmation sur la ville et la rénovation urbaine*). Such legal arrangements are rooted in a consensus that forced relocatees should not suffer negative outcomes from redevelopment programs without a proper compensation. Apart from rent protection based in national housing laws, such arrangements are either absent or much less elaborate in the context of pure, market-driven gentrification (at least in the Netherlands). The next section will revisit the importance of the Dutch institutional context with regard to forced relocation and urban renewal.

Urban Restructuring and Forced Relocation in the Netherlands

The Netherlands is internationally known for its relatively large social rented housing stock. In the autumn of 1997, the Dutch government issued a white paper entitled "Urban Renewal" (*Nota Stedelijke Vernieuwing*; MVROM 1997) in which the government analyzed the problem and stated the goals underlying a renewal policy that was geared primarily toward postwar neighborhoods with a large share of social rented housing (Musterd and Ostendorf 2008; Van Kempen and Priemus 2002). This renewal policy was called "urban restructuring."

Restructuring includes various measures: demolition, upgrading, or sale of social rented housing and the construction of new, more expensive owner-occupied or private rented housing. These interventions are intended to create more diversity in housing sizes and types, housing quality, prices, and tenure. Another characteristic of restructuring is the variety of goals, ranging from a stronger housing market position for restructuring areas, housing career opportunities, and poverty deconcentration to improved livability, reputation, social cohesion, and safety (Bolt et al. 2009; Kleinhans 2012; Uitermark 2003; Van Kempen and Priemus 2002). This policy has been extended over consecutive national administrations, which have emphasized specific goals, partly depending on the political agenda of the ruling parties. Scholars have studied policy outcomes from various (theoretical) viewpoints. With regard to in-movers and households moving within restructuring areas, many studies have reported positive outcomes of housing quality, livability, and housing career opportunities, while outcomes in terms of improved social cohesion, social capital, safety, and reputation are much more contested, ambivalent, or even questionable (Beckhoven and Van Kempen 2003; Dekker and Bolt 2005; Kleinhans 2005). The remainder of this chapter will focus on the context and outcomes just listed in relation to forced relocation.

As mentioned earlier, the institutional context matters, especially for the legal compensation arrangements for forced relocatees. Regardless of differences across cities in the Netherlands, forced relocatees moving from social housing are legally entitled to three forms of compensation, based on Dutch Civil Law (*Burgerlijk Wetboek*):

1. Another property comparable in size, type, and tenure, for which a priority status on the social rented housing market is issued, which gives relocatees priority over regular house applicants for eligible social housing
2. A reasonable allowance to cover relocation expenses, which is based on a national cost standard and indexed yearly
3. Additional assistance and counseling from housing associations, such as help with their search for suitable housing

Housing associations are allowed to relocate tenants only if these requirements are met (for full overviews, see Kleinhans and Van der Laan Bouma-Doff 2008; Posthumus and Kleinhans 2014; Tieskens and Musterd 2013). Regardless of the fact that the official restructuring policy was ended in 2015, this legal framework with these three compensation principles has remained intact, because it is established as a basic law in the Dutch Civil Law. Hence, any policy that employs forced relocation with regard to urban renewal (but not necessarily gentrification) must adhere to this overarching framework, which will continue to determine future relocation efforts of housing associations in cases of demolition of social housing.

Housing and Neighborhood Outcomes of Forced Relocation

This section reviews the outcomes of forced relocation in the Netherlands, based on studies predominantly conducted by scholars at universities and research institutions. Apart from summarizing outcomes in relation to the housing and neighborhood situation, other aspects are addressed, such as the role of pre-relocation attitudes, perceptions of counseling offered to relocatees, and the changes in housing costs. Methodological details of these studies were beyond the intended scope of this chapter; the interested reader should consult the referenced publications.

Housing Outcomes

By 2012, approximately 20 Dutch relocation studies had been conducted (Posthumus et al. 2012), and the number has steadily increased since then. While these studies have, by and large, a similar focus, the findings are highly mixed. A common element is that the majority of forced relocatees were (initially) very pessimistic about the impending move because they feared the stress and costs related to the move and were also concerned about their post-relocation housing situation. By and large, however, many relocatees moved from relatively unpopular social rented dwellings, regardless of their state of repair, to larger and higher-quality dwellings, either in the restructuring area, an adjacent neighborhood, or a different area elsewhere in the city.

Moreover, many studies show residents' subjective post-relocation evaluations matching changes in objective features of their new house compared to the old one, in terms of size, number of rooms, dwelling type, isolation, heating systems, and facilities (Bolt et al. 2009; Bolt and Van Kempen 2010; City of The Hague 2001, 2005, 2008; Doff and Kleinhans 2011; Kleinhans 2003; Leveling and Vos 2004; Oude Ophuis 2004; Posthumus et al. 2013a, 2013b, Tieskens and Musterd 2013). In some cases, improving one's situation was enabled by giving tenants priority to move to newly completed social rented housing. This had the favorable side effect of speeding up the demolition process (Kleinhans 2003, 2005).

However, several studies have found that significant numbers of relocatees to new neighborhoods would have preferred either not to have moved (City of The Hague 2001, 2005) or to have been able to relocate back to their "old" neighborhood (Wolf and Vriens 2006) or to a different area, i.e., not the one they were currently living in (Van der Zwaard and De Wilde 2008. These discontented respondents felt that the benefits that they had experienced did not make up for the uncertainty, trouble, and stress of the relocation process.

Traditionally, the Dutch social housing sector has been known for its quality and size, in Europe and beyond. However, high demolition rates and continuing restructuring have raised concerns about declines in the social rented stock for low-income households, the so-called target group of social housing. The decline in social housing is a problem because (1) the existing social stock must continue to be able to absorb the regular demand for social housing, (2) it must be available as a relocation option and as a buffering influence against displacement, and (3) it is needed for refugees. Continued privatization of social rented units in, for example, Amsterdam (see also the chapter authored by Wouter van Gent, Willem Boterman, and Myrte Hoekstra) and other large cities is a threat to the absorption capacity of the social rented sector. Overall, Dutch relocatees have a wider choice of available social housing and are therefore less constrained than, for example, American relocatees (Posthumus et al. 2013b). In general, Dutch cities are in danger of "going a step too far" in the further reduction of the social housing stock (Dol and Kleinhans 2012).

Neighborhood Outcomes

To a lesser extent, neighborhood improvement has also been reported. This is partly a reflection of the fact that urban restructuring areas are usually among the areas that are most disadvantaged. Consequently, relocation is very likely to move people into more prosperous neighborhoods (Bolt et al. 2009, 505; see also Wolf and Vriens 2006). But this type of mobility is different across different population groups. A crucial finding is that the share of non-Western immigrants (in the Netherlands, these are particular residents with at least one parent born in Turkey, Morocco, Surinam, or the Antilles) is a strong predictor of perceived post-relocation neighborhood improvement. Relocatees (whether they were native Dutch or immigrants) who managed to move to neighborhoods with lower concentrations of non-Western immigrants more often reported higher levels of perceived neighborhood improvement. There are two explanations for this finding. First, many native residents do not want to live in areas with high proportions of non-Western immigrants, so relocation enables them to separate themselves from immigrants. Second, neighborhood ethnic composition may be a (perceived) marker for the presence of concentrated poverty, crime and incivilities, and other problems leading to a poor image or reputation.

All else being equal, residents with a non-Western background appear to relocate less often to neighborhoods with a lower ethnic concentration than the "native" Dutch (Bolt et al. 2009; Doff and Kleinhans 2011; Posthumus et al. 2013b). There is disagreement on the extent to which this particular outcome is a result of non-Western immigrants who realize their preference to live in neighborhoods with clustered support networks (Doff and Kleinhans 2011; Tieskens and Musterd 2013), or whether this is mostly determined by socioeconomic factors such as low income, a low level of education, and large family size, which make it very difficult to find comparable housing in a better neighborhood (Bolt and Van Kempen 2010).

It is commonly assumed that forced relocation destroys social ties that are important for residents, in terms of either emotional or practical support and trust. In fact, in the context of the Netherlands, only a few studies have indeed reported the loss of supportive social ties (bonding capital) with former neighbors and fellow residents (Kleinhans 2003; Van der Zwaard and De Wilde 2008). Overall, the effects of forced relocation on the number and type of neighborhood-based social ties have been limited in the Netherlands. There are several explanations for this finding including the fact that many movers consider social ties and networks with neighbors relatively unimportant in comparison to social ties with family and friends. Social interaction with neighbors is usually limited and of a practical nature.

Respondents who reported improvement or degradation of social ties mostly referred to the extent to which they could get along with their new neighbors and other neighborhood residents in the post-relocation situation. A few movers experienced such difficulties, which negatively influenced their perception of their new housing situation. Older respondents were more likely to experience social stress. A few of them were homesick and idealized their former situation because they had lost their sense of belonging and social contacts in their former neighborhood (Kleinhans 2003). The extent to which meaningful ties are lost due to relocation is partly a function of moving distance. In some cases, residents relocated to dwellings only a few hundred meters from their previous location, enabling them to maintain ties with peers who had not relocated (Kleinhans 2005; Van der Zwaard and De Wilde 2008). People who experienced a loss of social ties and activities due to relocation also evaluated their post-relocation neighborhood change less positively (Doff and Kleinhans 2011). In the U.S., forced or unforced relocatees are sometimes happy to move from relatives and friends who are a burden for them (Briggs et al. 2010). However, this particular motivation has not been reported in Dutch research.

Pre-Relocation Attitudes and Search Strategies

Obviously, the difference between the pre- and post-relocation housing situation is strongly determined by objective and subjective differences between the two circumstances. In regard to the latter, there are initial attitudes toward relocation before the actual move, counseling provided by the housing association (as discussed later), and changes in housing costs (also discussed later). Several studies have looked at the issue of pre-relocation attitudes, inspired by the early example of Fried (1967), who found that "pre-relocation evidences of preparedness for change are the most important factors determining post-relocation adjustment-adaptation and tend to dwarf the importance of post-relocation situations and experiences" in Boston (Fried 1967, 100). Not surprisingly, many studies have found that anger, insecurity, stress, anxiety, and a range of other negative emotions are associated with impending forced relocation (Kleinhans 2003; Oude Ophuis 2004; Posthumus et al. 2012). On the one hand, many residents have reported highly mixed feelings, such as relief that they are enabled to leave disadvantageous dwellings, combined with a sense of insecurity in relation to their new housing and neighborhood situation (Kleinhans 2003; Van der Zwaard and De Wilde 2008; Wolf and Vriens 2006). On the other hand, several Dutch studies have found that residents who expressed a certain degree of understanding and approval before the relocation were more likely to report an improved housing situation afterward. These attitudes are based on several factors, such as (impending) changes in jobs and household composition, but also dispositional effects: Residents who were deliberately looking for opportunities to improve their housing situation often have succeeded in their attempts (see Kleinhans (2003) for a comprehensive discussion of this matter). The overall effect is that positive pre-relocation attitudes are associated with perceptions of better post-relocation outcomes (Doff and Kleinhans 2011; Kleinhans and Van der Laan Bouma-Doff 2008; Posthumus et al. 2013a; Wolf and Vriens 2006).

Negative discourses on forced relocation usually emphasize the resistance of tenants against forced relocation by stating that they do not want to leave. This assumption should be critically assessed in light of pre-relocation moving plans of tenants. A large body of research has shown that, on average, between 10 and 20 percent of the residents involved were already considering a move by the time they received the notice from the housing association. Because the priority status strongly improved their options on the housing market, many of them have been able to move upward in their housing career (Kleinhans 2003, 2005; Leveling and Vos 2004; Posthumus et al. 2012, 2013a, 2013b).

While there is an abundance of research about housing and neighborhood perceptions regarding post-relocation outcomes, only a few studies analyze the underlying mechanisms and the ways in which relocatees arrive at certain neighborhood outcomes. This perspective emphasizes coping behavior, in particular the search strategies of tenants facing relocation under different local conditions. The relatively scarce research on this matter has shown that relocatees adopt various strategies to deal with the impending relocation. Some behavior can be characterized by panicking and accepting the first relocation opportunity offered by housing associations. More better thought out strategies range from compromising on preferences and broadening the search area to resisting or negotiating with housing associations, refusing to accept standard alternatives, and maximizing benefits in terms of alternative dwellings offered by housing associations (Oude Ophuis 2004; Posthumus et al. 2012, 2013a; Posthumus and Kleinhans 2014). Many of these strategies are clear proof of relocatees' agency. These findings are in line with research in Seattle by Manzo and colleagues (2008), who state that "tenants are not passive victims" and that relocation research "reflects the human agency of tenants as they actively work to make sense of the redevelopment, plan for relocation and weigh their relocation options" (Manzo et al. 2008, 1872).

The Role of Counseling

As mentioned earlier, forced relocatees in the Dutch social housing sector are legally entitled to three main forms of compensation. One of them is the right to assistance and counseling from housing associations, such as help with their search for suitable housing, but also with organizing the actual move. While providing counseling is a legal obligation for housing associations, the nature and intensity are subject to strong local differences in contexts and with regard to the applied counseling methods. Hence, research on the effectiveness of and relocatees' satisfaction with the offered counseling shows a highly divergent picture.

Research has identified various shortcomings in applied counseling procedures, such as offering only standard and succinct information leaflets, incomplete or scattered information, and the lack of a personal approach, especially for seniors (e.g., Kleinhans 2003; Oude Ophuis 2004; Posthumus et al. 2012; Posthumus and Kleinhans 2014; Van der Zwaard and De Wilde 2008). A common reason for dissatisfaction among relocatees is the feeling that they are not being listened to by housing associations in general, and counselors in particular. Relatedly, many complaints refer to perceived promises of housing associations regarding specific information, the future availability of relocation options, or other issues. Another frequently observed problem is that counselors of housing associations are perceived to lack the necessary skills and/or proper resources from their own organizations in order to deliver well-tailored counseling to relocatees (Curley and Kleinhans 2010).

Apart from the content, the form of counseling also matters. Generally, the use of public information meetings in neighborhood centers, schools, or town halls is commonly rejected by respondents, because attendees feel that they are not listened to and cannot have a say. Others feel that the information offered at such meetings is too superficial to be of any use for one's private circumstances (e.g., Kleinhans 2003; Leveling and Vos 2004).

Housing associations have often underestimated the impact of relocation on residents' lives, especially on the lives of seniors or single-parent households. In some cases, substantial numbers of relocatees would have preferred the status quo, even if they benefited from the move (City of The Hague 2001, 2005). Partly as a result of failures in the early years of restructuring, Dutch housing associations have developed so-called "behind-the-front-door" approaches in collaboration with other renewal actors. Such proactive counseling approaches have been conducted to comprehensively tackle already existing social problems in relocation households (e.g., addiction, unemployment, debts) that might otherwise have been aggravated by relocation pressures (Curley and Kleinhans 2010; Posthumus et al. 2012).

Notwithstanding problems related to counseling, there is also evidence of positive impacts of counseling on post-relocation satisfaction, not just in response to the process, but also with respect to the new housing situation. In their survey of four large cities in the Netherlands, Posthumus et al. (2013a) found that residents who appreciated the counseling they had received were also significantly more satisfied with their new housing situation. Other research, both Dutch and American, has also found associations between the perceived quality of counseling and post-relocation satisfaction with the new housing situation (Kleinhans 2005; Posthumus et al. 2012; see also Varady and Walker 2007).

Changes in Housing Costs

A final determinant of post-relocation satisfaction with the new housing situation concerns the difference in housing costs (i.e., rent and utility costs) before and after relocation. Rents in the Dutch social housing sector are highly regulated and capped at a certain maximum. Relocation studies consistently show that almost all relocatees remain in the social housing sector (for an overview, see Posthumus et al. 2012). Changes in rent costs are subject to national regulations, and several factors play a role in determining the net change in housing costs. Thus the Dutch relocation model is successful because it ensures that housing costs do not go up too quickly. Several factors explain this finding.

First of all, urban restructuring usually targets the dwellings that have the lowest rents in the social housing stock. A move to another social rented unit often results in a higher rent for the new dwelling. But the net difference may not be paid if households are eligible for housing allowances (*huurtoeslag*), which are part of a general policy set of instruments to stimulate affordable housing in the Netherlands (see, e.g., Ronald 2013). In the context of relocation, housing allowances have a dampening effect on the net rent increase after relocation, with the result that eligible households can move to larger and better (sometimes even new) social rented dwellings without paying a much higher rent.

However, eligibility for housing allowances is capped in terms of household income. As a result, residents with incomes slightly above the eligibility criterion for housing allowances, who are nevertheless still entitled to social housing, pay the full net rent increase after relocation (Kleinhans 2003; Kleinhans and Van der Laan Bouma-Doff 2008). Consequently, it is more difficult for households with slightly higher incomes to attain a favorable price-quality ratio with relocation than for the lowest-income groups, who fully benefit from the compensation offered by their housing allowance.

In order to solve this problem, many housing associations have introduced the so-called rent habituation (*huurgewenning*). In this situation, relocatees do not pay the full rent price of their relocation dwelling immediately, but can use a transition period of several years, in which the rent is gradually increased toward the final level, to accommodate the rent increase in their budgets without their getting into financial problems immediately.

Conclusions

Since the early 1980s "displacement has been a central concept in the gentrification debate, and one which has migrated to substitute for the consideration of relocation within restructuring studies" (Kleinhans and Kearns 2013, 167). In the context of the negatively framed debate on displacement, gentrification, and urban renewal, this chapter has sought to review outcomes of forced relocation in the Netherlands, as part of a national urban restructuring policy, with demolition of social housing as one of its key elements. The review has shown that, while pre-relocation expectations are usually negative, many households may actually be better off in terms of their housing and neighborhood situation after urban restructuring and forced relocation. The combination of counseling and legal compensations appears to have a mitigating influence regarding the negative impacts of both the process and the relocation outcome. These results give reason to counter-argue the predominantly negative discourse on displacement as an inevitable "dark side" of state-led gentrification and urban renewal programs.

This chapter emphasized the need for a critical, but nuanced perspective on the outcomes of forced residential relocation, refraining from unilaterally negative or positive stances. In fact, the review has identified the Janus head of forced relocation, in the sense that the process and outcomes of Dutch forced relocation show different and sometimes highly contradictory faces regarding manifestations of the process and perceived outcomes.

The first manifestation of the Janus head concerns the housing and neighborhood outcomes of relocation. By and large, the majority of relocatees in all reviewed Dutch studies have reported benefits in their housing or neighborhood situation or both. This is partly a selection issue; as the most disadvantageous social housing and neighborhoods are targeted for urban restructuring, it is relatively easy to improve one's position. On the other side of the coin, however, significant shares of relocatees do not report improvements in their housing and/or neighborhood situation. Paradoxically, this outcome arises from the legal compensation mechanisms. While the priority status gives relocatees priority over regular house applicants for eligible social housing, this status is limited to comparable dwellings. This limited validity of the priority status limits relocatees' choice with regard to moving to different neighborhoods that do not suffer from similar problems as the restructuring areas.

Pre-relocation attitudes embody a second manifestation of the Janus head. Public discourse often emphasizes large-scale opposition from tenants against forced relocation. However, many residents often combine different perspectives in a nuanced perception of the impending relocation, such as relief to be enabled to leave disadvantageous dwellings, combined with a sense of insecurity in relation to their new housing and neighborhood situation. Many relocatees already had moving plans and/or perceived impending relocation and therefore saw relocation as an opportunity to move upward in their housing career. While they might be labeled as "displacees," they do not classify themselves as disadvantaged (Atkinson 2000, 310).

The third manifestation of the Janus head concerns the impact of relocation on social ties. An argument often stated against forced relocation is that it hampers important local social ties of residents in terms of emotional or practical support and trust. Indeed, some studies found broken ties, which specifically affected seniors who felt uprooted and lost a sense of belonging. However, the overall negative impacts have been limited in the Netherlands. Relocatees tend to explain the impact on their social networks in terms of the extent to which they can get along with their new neighbors.

Fourth, housing associations offer counseling to facilitate the relocation process and mitigate undesirable and unintended side effects. Several studies have found associations between the perceived quality of counseling and post-relocation satisfaction with the new housing situation (Kleinhans 2005; Posthumus et al. 2012, 2013a). On the other hand, studies have shown frequent and serious complaints of relocatees about this counseling, not the least because one-size-fits-all approaches often fail to adequately deal with the individual needs of relocatees and make them feel that they do not have a say in the process.

A final manifestation of the Janus head appears in how relocatees cope with the impending move and choice of new housing. Research has shown that relocatees adopt various strategies to deal with the process of relocation, once they have coped with the first phase of anger or frustration. While left-wing politicians or activists tend to portray forced relocatees as victims of a game that is played at their expense, the discussion provides proof of relocatees' agency, willingness, and ability to make the most out of it.

All in all, the Dutch experience with forced relocation provides a counter-narrative to the negative American and, to a lesser extent, European narrative that is predominantly rooted in a displacement discourse in the context of large-scale gentrification and negative experiences with urban renewal. The Dutch experience reveals the buffering or sometimes even empowering influence of a relatively large social housing stock, legal compensation rules, and a still-strong welfare regime, regardless of recent state retrenchment and policy reforms (Dol and Kleinhans 2012). While it may not be possible to provide relocatees in other developed countries with "similar opportunities—particularly in light of cutbacks in funding on both sides of the Atlantic—the Dutch example can be used to support a plea for a wider set of choices for displaced residents as well as the need for high-quality counseling" (Posthumus et al. 2013a, 29).

The findings in this chapter have implications for U.S. neighborhood redevelopment policies, in particular Choice Neighborhoods, the successor of the HOPE VI program. This initiative aims at improving housing and neighborhoods by replacing distressed public housing or distressed federally subsidized private housing with high-quality mixed-income housing, and improving neighborhood conditions. Local authorities need to clarify to residents much more explicitly how they can benefit from relocation, by providing a range of relocation choices (on-site, or to other neighborhoods) and further assisting them in the housing choice process. However, policymakers must be aware that increased knowledge may also lead to increased choices, such as relocatees moving to other homogenous low-income areas. Choice is the ultimate goal, and staying put is not by definition bad.

Counselors should make efforts to ensure that their clients understand the relocation process in general and the regional housing options in particular. This requires a proactive approach by counselors to ensure that all residents are aware of their choices and thus to avoid unequal outcomes for different population categories (such as non-Western immigrants versus native residents) that result from differences in knowledge and ability to deal with complex institutional arrangements of forced relocation.

Another implication relates to the increasing scarcity of social/public housing. Further reduction of this housing may lead to unacceptably long waiting lists and extremely limited opportunities for relocation. To prevent this, neighborhood revitalization can apply the principle of one-on-one replacement; for every demolished social dwelling, a new one is built in the same or in another neighborhood (Dol and Kleinhans 2012). This approach not only ensures stabilization of the social housing stock, but also opens up housing opportunities for low-income people in neighborhoods different from typical social/public housing areas.

References

Atkinson, Rowland. 2000. "The Hidden Costs of Gentrification: Displacement in Central London." *Journal of Housing and the Built Environment*, 15(4): 307–326.

Atkinson, Rowland. 2004. "The Evidence on the Impact of Gentrification: New Lessons for the Urban Renaissance?" *European Journal of Housing Policy*, 4(1): 107–131.

Beckhoven, Ellen van, and Ronald van Kempen. 2003. "Social Effects of Urban Restructuring: A Case Study in Amsterdam and Utrecht, The Netherlands." *Housing Studies*, 18(6): 853–875.

Bolt, Gideon, and Ronald van Kempen. 2010. "Dispersal Patterns of Households who are Forced to Move: Desegregation by Demolition: A Case Study of Dutch Cities." *Housing Studies*, 25(2): 159–180.

Bolt, Gideon, Ronald van Kempen, and Jan van Weesep. 2009. "After Urban Restructuring: Relocations and Segregation in Dutch Cities." *Tijdschrift voor Economische en Sociale Geografie*, 100(4): 502–518.

Briggs, Xavier de Souza., Susan Popkin, and John Goering. 2010. *Moving to Opportunity: The Story of an American Experiment to Fight Ghetto Poverty*. New York: Oxford Press.

City of The Hague. 2001. *Nota Herhuisvesting. Ervaringen van herhuisvestingskandidaten in diverse complexen [Memorandum Monitoring of Relocation. Experiences of Relocatees from Various Blocks]*. Den Haag: Dienst Stedelijke Ontwikkeling.

City of The Hague. 2005. *Onderzoek Herhuisvesting 2004. Feiten en ervaringen van herhuisvestingskandidaten uit diverse complexen [Monitor Relocation 2004. Facts and Experiences of Relocatees from Various Blocks]*. Den Haag: Dienst Stedelijke Ontwikkeling.

City of The Hague. 2008. *Barometer Herstructurering (Module 2). Ervaringen van herhuisves-tingskandidaten uit diverse complexen 2004–2006 [Barometer Restructuring (Module 2). Experiences of Relocatees from Various Blocks 2004–2006]*. Den Haag: Dienst Stedelijke Ontwikkeling.

Curley, Alexandra, and Reinout Kleinhans. 2010. "Combining Forced Residential Relocation with Supportive Services: What Can Dutch Housing Associations Learn from the American HOPE VI Program?" *Journal of Urban Regeneration and Renewal*, 3(4): 70–84.

Davidson, Mark. 2008. "Spoiled Mixture: Where Does State-Led 'positive' Gentrification End?" *Urban Studies*, 45(12): 2385–2405.

Davidson, Mark. 2011. "Critical Commentary: Gentrification in Crisis: Towards Consensus or Disagreement?" *Urban Studies*, 48(10): 1987–1996.

Dekker, Karien, and Gideon Bolt. 2005. "Social Cohesion in Post-War Estates in the Netherlands: Differences Between Socioeconomic and Ethnic Groups." *Urban Studies*, 42(13): 2447–2470.

Dekker, Karien, and David P. Varady. 2011. "A Comparison of Dutch and US Public Housing Regeneration Planning: The Similarity Grows?" *Urban Research & Practice*, 4(2): 123–152.

Doff, Wenda, and Reinout Kleinhans. 2011. "Residential Outcomes of Forced Relocation: Lifting a Corner of the Veil on Neighbourhood Selection." *Urban Studies*, 48(4): 661–680.

Dol, Kees, and Reinout Kleinhans. 2012. "Going Too Far in the Battle Against Concentration? On the Balance Between Supply and Demand of Social Housing in Dutch Cities." *Urban Research & Practice*, 5(2): 273–283.

Freeman, Lance. 2005. "Displacement or Succession? Residential Mobility in Gentrifying Neighborhoods." *Urban Affairs Review*, 4(40): 463–491.

Freeman, Lance. 2006. *There Goes the 'Hood': Views of the Neighborhood from the Ground Up*. Philadelphia: Temple University Press.

Fried, Marc. 1967. "Functions of the Working-Class Community in Modern Urban Society: Implications for Forced Relocation, Case Study West End, Boston." *Journal of the American Institute of Planners*, 33: 90–103.

Grier, George W., and Eunice S. Grier. 1978. *Urban Displacement: A Reconnaissance*. Washington, DC: U.S. Department of Housing and Urban Development.

Hackworth Jason, and Neil Smith. 2001. "The Changing State of Gentrification." *Tijdschrift voor Economische en Sociale Geografie*, 92(4): 464–477.

Kearns, Ade, and Phil Mason. 2013. "Defining and Measuring Displacement: Is Relocation from Restructured Neighbourhoods always Unwelcome and Disruptive?" *Housing Studies*, 28(2): 177–204.

Kempen, Ronald Van, Hugo Priemus. 2002. "Revolution in Social Housing in the Netherlands: Possible Effects of New Housing Policies." *Urban Studies*, 39(2): 237–253.

Kleinhans, Reinout. 2003. "Displaced but still Moving Upwards in the Housing Career? Implications of Forced Residential Relocation in the Netherlands." *Housing Studies*, 18(4): 473–499.

Kleinhans, Reinout. 2005. *Sociale implicaties van herstructurering en herhuisvesting [Social Implications of Urban Restructuring and Forced Residential Relocation]*. Amsterdam: IOS Press.

Kleinhans, Reinout. 2012. "A Glass Half Empty or Half Full? On the Perceived Gap between Urban Geography Research and Dutch Urban Restructuring Policy." *International Journal of Housing Policy*, 12(3): 299–314.

Kleinhans, Reinout, and Wenda van der Laan Bouma-Doff. 2008. "On Priority and Progress: Forced Residential Relocation and Housing Chances in Haaglanden, the Netherlands." *Housing Studies*, 23(4): 565–587.

Kleinhans, Reinout, and Ade Kearns. 2013. "Neighbourhood Restructuring and Residential Relocation: Towards a Balanced Perspective on Relocation Processes and Outcomes." *Housing Studies*, 28(2): 163–176.

Korthals Altes, Willem K. 2016. "Forced Relocation and Tenancy Law in Europe." *Cities*, 52: 79–85.

Lees, Loretta. 2008. "Gentrification and Social Mixing: Towards an Inclusive Urban Renaissance?" *Urban Studies*, 45(12): 2449–2470.

Lees, Loretta, Tom Slater, and Elvin K. Wyly, eds. 2010. *The Gentrification Reader*. New York: Routledge.

LeGates, Richard T., and Chester Hartman. 1986. "The Anatomy of Displacement in the United States." In *Gentrification of the City*, edited by N. Smith and P. Williams, 178–203. London: Unwin Hyman.

Leveling, Dieuwke., and Ellen Vos. 2004. *Gelukkig Verhuisd? [Happily Moved?]*. Amsterdam: Regioplan Beleidsonderzoek.

Manzo, Lynne C., Rachel G. Kleit, and Dawn Couch. 2008. "Moving Three Times Is Like Having Your House on Fire Once: The Experience of Place and Impending Displacement among Public Housing Residents." *Urban Studies*, 45(9): 1855–1878.

Marcuse, Peter. 1986. "Abandonment, Gentrification, and Displacement: The Linkages in New York." In *Gentrification of the City*, edited by N. Smith and P. Williams, 153–177. London: Unwin Hyman.

Musterd, Sako, and Wim Ostendorf. 2008. "Integrated Urban Renewal in the Netherlands." *Urban Research & Practice*, 1(1): 78–92.

MVROM. 1997. *Nota Stedelijke Vernieuwing [White Paper Urban Renewal]*. The Hague: Ministry of Housing, Spatial Planning and the Environment.

Newman, Kathe, and Elvin K. Wyly. 2006. "The Right to Stay Put: Gentrification and Resistance to Displacement in New York City." *Urban Studies*, 43(1): 23–57.

Oude Ophuis, R. 2004. *Tevreden verhuisd? [Moved Satisfactorily?]*. Amsterdam: Regioplan Beleidsonderzoek.

Paris, Chris, and Bob Blackaby. 1979. *Not Much Improvement. Urban Renewal Policy in Birmingham*. London: Heinemann Educational.

Posthumus, Hanneke, Gideon Bolt, and Ronald van Kempen. 2013a. "Victims or Victors? The Effects of Forced Relocations on Housing Satisfaction in Dutch Cities." *Journal of Urban Affairs*, 36(1): 13–32.

Posthumus, Hanneke, Gideon Bolt, and Ronald van Kempen. 2013b. "Why Do Displaced Residents Move to Socioeconomically Disadvantaged Neighbourhoods?" *Housing Studies*, 28(2): 272–293.

Posthumus, Hanneke, and Reinout Kleinhans. 2014. "Choice within Limits: How the Institutional Context of Forced Relocation Affects Tenants' Housing Searches and Choice Strategies." *Journal of Housing and the Built Environment*, 29(1): 105–122.

Posthumus, Hanneke, Reinout Kleinhans, and Gideon Bolt. 2012. *Bijwerkingen van herstructureringsoperaties. Verhuizingen, waterbedeffecten en veranderingen in de voningvoorraad [Side Effects of Restructuring Operations. Residential Moves, Spillover Effects, and Changes in the Housing Stock]*. Delft: Eburon.

Ronald, Richard. 2013. "Housing and Welfare in Western Europe: Transformations and Challenges for the Social Rented Sector." *LHI Journal of Land, Housing and Urban Affairs*, 4(1): 1–13.

Schill, Michael H., and Richard P. Nathan. 1983. *Revitalizing America's Cities: Neighborhood Reinvestment and Displacement*. Albany: State University of New York Press.

Smith, Neil, and Paul Williams (Eds). 1986. *Gentrification of the City*. London: Unwin Hyman.

Thomas, Andrew D. 1986. *Housing and Urban Renewal: Residential Decay and Revitalization in the Private Sector.* London: George Allen & Unwin.

Tieskens, Koen F., and Sako Musterd. 2013. "Displacement and Urban Restructuring in Amsterdam: Following Relocatees after Demolition of Social Housing." *Urban Research & Practice*, 6(2): 194–210.

Uitermark, Justus. 2003"'Social Mixing' and the Management of Disadvantaged Neighbourhoods: The Dutch Policy of Urban Restructuring Revisited." *Urban Studies*, 40(3): 531–549.

Uitermark, Justus, Jan Willem Duyvendak, and Reinout Kleinhans. 2007. "Gentrification as a Governmental Strategy. Social Control and Social Cohesion in Hoogvliet, Rotterdam." *Environment and Planning A*, 39(1): 125–141.

Varady, David P., and Carole Walker. 2007. *Neighborhood Choices: Section 8 Housing Vouchers and Residential Mobility*. New Brunswick, NJ: Center for Urban Policy Research.

Wolf, Susanne, and Gerard Vriens. 2006. Herhuisvesting na sloop. Ervaringen van bewoners uit herstructureringsgebieden *[Relocation After Demolition. Experiences of Residents from Urban Restructuring Areas]*. Utrecht, Gemeente Utrecht.

Zwaard, Joke van der, and Jokede Wilde. 2008. *Thuis in de nieuwe wijk? [At Home in the New Neighborhood?]* Rotterdam: CMO Stimulans.

24

STATE-SPONSORED GENTRIFICATION OR SOCIAL REGENERATION?

Symbolic Politics and Neighborhood Intervention in an Amsterdam Working-Class Neighborhood

Wouter van Gent, Willem Boterman, and Myrte Hoekstra

Introduction

Since the 1990s, urban policymakers in Western Europe have employed area-based strategies to regenerate deprived neighborhoods. Local authorities have sought to improve social and physical conditions in various ways, including initiatives in employment, health, safety, and education to improve the social situation of residents (termed *social* regeneration in policy and in this chapter), as well as redevelopment and renovation to improve housing conditions. These latter physical interventions are also often seen by planners and policymakers as an opportunity to transform neighborhoods socially. That is, by altering the housing stock and introducing new amenities, more affluent households can be attracted to poorer areas (Andersson and Musterd 2005; Goetz 2013). Various scholars in Europe and the U.S. have argued, however, that these policies favor middle-class residents to the detriment of the poor, who may be displaced or relocated from the neighborhood, or who may no longer feel "at home" because of the changed environment, new amenities, and social differences between them and newcomers, a new form of indirect displacement (Davidson and Lees 2005; see also Chaskin and Joseph 2013; Pinkster 2016; Van Gent et al. 2016). Consequently, some urban scholars have cast these regeneration strategies as a stealthy form of state-sponsored gentrification (Bridge et al. 2012). Not all regeneration efforts constitute gentrification, however, with *ipso facto* displacement and "loss of place." Relocation may be absent or negotiated, and rather than feeling displaced, some remaining residents may (cautiously) appreciate the arrival of new neighbors and amenities (Chaskin and Joseph 2013; Ernst and Doucet 2014; Freeman 2006; Kleinhans and Kearns 2013).

The aim of this chapter is to understand whether and how neighborhood interventions may constitute a form of state-sponsored gentrification, meaning that the state and its regeneration partners directly or indirectly promote this type of neighborhood change. Specifically, we focus on interventions and service provision in Van der Pekbuurt (VDP-buurt), a working-class area in Amsterdam. As large-scale redevelopment is not possible in this area, the City of Amsterdam and the Ymere housing association, the only association active in this neighborhood, rely on symbolic politics to make the area attractive for anticipated middle-class residents. "Symbolic politics" refers to the ways in which regeneration efforts are framed and experienced by politicians, policymakers, civil servants, administrators, and resident groups. Together with financial capital, organizational power, juridical power, and technical resources, the ability to dominate visions of reality (symbolic power) allows actors, in this case city government and partner nonprofits and for-profit companies, to impose that vision on reality (Bourdieu 1989).

Sakizlioglu and Uitermark (2014) outline two forms of symbolic politics: classification of the area and its population, and the politics of timing. The former refers to stigmatizing discourse, which may normalize the necessity of social change and divide residents between the deserving and undeserving (Goetz 2013; Nguyen et al. 2013; Paton et al. 2017). The latter refers to the strategic use of time by the state, in this case the City of Amsterdam, keeping residents in uncertainty for a long time by waiting out the protests, hoping that they will peter out (Gans 1962; Wallace 2016). We propose another—related—form of symbolic politics in regeneration: middle-class quarter making, which refers to policymakers catering to anticipated middle-class preferences through social policies, initiatives, and service provision. For instance, the City may fund or facilitate new upscale shops, specific cultural activities for new residents, or the temporary settlement of "pioneering" residents (i.e., students or artists). We see quarter making as a form of place making that is primarily done *in anticipation* of the arrival of middle-class residents.

The symbolic politics of middle-class quarter making may therefore constitute a "soft force" approach to state-sponsored gentrification, but this is not a foregone conclusion. Even when middle-class residents are catered to, efforts may also be directed at long-term residents. Kleinhans and Kearns (2013) argue that any appraisal of displacement requires an understanding of an area's population and housing market, policy aims and instruments, and the interface between the state and residents, including the level of community empowerment.

To understand how middle-class quarter making may amount to symbolic politics and how it is received by residents, this chapter discusses the recent history of VDP-buurt in Amsterdam. This relatively low-income area, located close to the city center, was scheduled for large-scale redevelopment to allow for more private forms of housing and a new social mix. Faced with relocation, tenants effectively resisted the redevelopment plans, which led to renegotiations. But while large-scale physical interventions have been stopped, the neighborhood remains subject to multiple policy initiatives for regeneration. Our case study focuses on how the City and the Ymere housing association are continuing their efforts to make the neighborhood and its nearby docklands more attractive to new middle-class residents. At the same time, however, the City has sought to provide social assistance to long-term residents, which may counterbalance middle-class quarter making efforts. This chapter highlights the tensions involved in efforts by the City to simultaneously empower residents and carry out quarter making.

The next section briefly discusses state-sponsored gentrification and the involvement of artists, culture, and other amenities, in order to introduce our concept of quarter making. After discussing the urban and policy context, we present our case study in three steps. First, we chart the organizational structure of neighborhood initiatives and identify organizations that are (partly) directed at a middle-class audience. Second, we discuss these middle-class symbolic markers in greater detail. Third, we present data on policymakers' and long-term residents' attitudes toward regeneration initiatives, with a focus on symbolic politics. We end with a discussion of state-sponsored gentrification and offer tentative explanations of our findings.

Gentrification, Policy, and Quarter Making

Originally, gentrification was conceptualized as a bottom-up process of small-scale investment in housing by in-migrating middle-class families. However, in their seminal paper, Hackworth and Smith (2001) shifted the scholarly focus to large-scale, capital-intensive redevelopment curated, and often also initiated, by the state. The restructuring of the housing market through liberalization, the sale of public land, and the facilitation of large-scale private investment in central city areas is now part and parcel of what is referred to as state-sponsored, or state-led, gentrification. The use of state power in the Atlantic Yards/Pacific Park project in Brooklyn, New York, is a recent example (Pason et al. 2017; Sze 2009). Yet state-sponsored urban redevelopment characterized by highly visible capital interests and the physical displacement of poor residents is more prominent in cities in the Global South where national and local governments implement urban renewal-type policies including widespread demolition (Lees et al. 2015). In Continental Europe, on the other hand, the social and state context has arguably made the process less relentless and more negotiated. Moreover, local urban and housing policies often play a decisive role in the process and outcomes. These policies have resulted in soft gentrification—i.e., neighborhoods with a stable mix of incomes—rather than complete gentrification.

In Amsterdam, for instance, despite liberalization trends in housing, direct displacement is rare as tenants' legal rights have remained strong. While the speed and depth of gentrification are limited by housing rights, and a large share of social housing (Van Gent 2013), policy reform, the sale of social rental housing, and the increasing demand for housing have all led to marked changes in Amsterdam's geography (Boterman and Van Gent 2014; Hochstenbach and Van Gent 2015). Although low-income households may not be directly displaced through forced evictions, the number of affordable housing units has been decreasing, leading to exclusionary displacement, and while some may welcome change, others may feel a loss of place because of changes in the built environment and the arrival of new amenities such as expensive shops and bars, and higher-income residents who disrupt long-term social relationships (Ernst and Doucet 2014; Pinkster 2016).

Western European accounts of the role of the state in promoting gentrification have generally focused on housing tenure as a key lever for engineering social change (Boterman and Van Gent 2014; Millard-Ball 2000; Murie 1991). Notwithstanding the importance of housing, the state may also promote neighborhood change via other means. In the wake of Richard Florida's creative class paradigm, policymakers have increasingly employed culture and creativity in regeneration strategies and neighborhood development (Florida 2002; Grodach et al. 2014; Lees and Melhuish 2015; Rota and Salone 2014). This may imply introducing artists and art to a neighborhood or changing the commercial landscape to suit middle-class tastes (Cameron and Coaffee 2005; Mathews 2010).

Artists can be key agents in the commodification of place due to what artistic production and consumption represent in wider society (i.e., art as exclusive and excluding) (Harris 2012). Importing such notions to places may, however, conflict with local tastes and interests and may be an instance of symbolic politics. Harris (2012) suggests that regeneration through arts may occur without displacement, yet he notes that "artists can be caught, intentionally or otherwise, in processes of gentrification with important social consequences" (237). Furthermore, as early gentrifiers, the pioneering artists may themselves be displaced or feel a loss of place when the gentrification process matures and high-income residents become dominant (Ocejo 2011).

Another form of cultural intervention is the modification of the commercial and amenities infrastructure of a neighborhood. Typically, gentrification manifests itself most visibly in the changing commercial landscape. High-end retail, restaurants, bars, and particularly coffee places are central to the iconography of gentrification (Bridge and Dowling 2001). While usually associated with market-driven processes, state actors may also facilitate or initiate the makeover of an area's commerce. Sharon Zukin argues that "in numerous cases state intervention has reinforced the cultural claims behind gentrification's market forces" (Zukin 1991, 194). Her account primarily refers to the realization of economic value through the symbolic reconstruction of areas into places of cultural consumption, partly through an emphasis on the authenticity and historicity of place. Local authorities may seek to capitalize on cultural activities through regulation, legal arrangements, and subsidies; e.g., the city may facilitate the creation of an arts and entertainment district. However, she states that "gentrification has received its greatest boost not from subsidies but from the state's substantive and symbolic legitimation of the cultural claim to space" (Zukin 1991, 194).

Introducing art and artists and favoring some commercial activities over others are part of the politics that symbolically reconstitute places so as to pave the way for their appropriation by new groups. This can be seen in terms of place making by state actors, together with the new middle class (Elwood et al. 2015). In situations where the commercial infrastructure is deliberately modified *in anticipation* of the arrival of desired groups—i.e., middle-class gentrifiers—we should perhaps rather talk of quarter making. We argue that state actors create the conditions for (anticipated) newcomers by making the area attractive and "safe" for middle-class consumption.

Study Area

VDP-buurt is a neighborhood in Amsterdam located on the northern bank of the river IJ directly opposite the historic city center. It is 3.1 square kilometers or about 1.2 square miles, and it is part of the North borough. Constructed in the 1920s for dockland workers, it was innovative for its working-class housing, red brick architecture, and garden village design. Canals, low-rise buildings, ample greenery, small squares, and crooked streets provide a village-like atmosphere. The eponymous Van der Pekstraat is its main thoroughfare.

The housing market in the VDP-buurt has been characterized by social rental housing since the 1920s. Of the neighborhood's 2,406 dwellings, 88.9 percent were social housing in 2014, yet this relatively high percentage may decline, as we will discuss. Dwellings are relatively small, with three rooms on average. Fifty-nine percent of the homes are less than 60 square meters. In 2013, assessed housing values averaged €158,000, well below the city average of €244,000. Recently, values have been decreasing, mainly due to the deteriorating housing stock. Many dwellings are poorly maintained, and a significant number lack central heating and adequate insulation, leading to low housing satisfaction among residents (Gemeente Amsterdam 2015).

Single-person households are in the majority, yet given the average housing size it is remarkable that about a quarter of households have children. Despite its reputation as a White working-class area, 50 percent of the residents of VDP-buurt are of non-Western descent (Gemeente Amsterdam 2015). Finally, about a third of households was subsisting on a minimum income[1] in 2013, a proportion that has increased since 2003.

Social Change

The location of VDP-buurt close to the ferry crossing to the city center and to nearby waterfront redevelopments has led to spillover demand for housing. Despite the large share of low-quality social housing, there has been an inflow of young residents of native Dutch and Western descent from outside Amsterdam.[2] This group constituted 20 percent of the neighborhood's population in 2013 and has been increasing at an above-average rate, compared both to the North borough and to the city. Thus, the area has seen modest increases in its share of single-person, higher educated, and native Dutch households. This is indicative of new demand, but is also the result of Ymere's policy of placing students and other short-term renters in vacant houses awaiting renovation. The neighborhood's tenure structure, with a high proportion of social housing, has largely forestalled rapid population change despite increasing housing market pressure.

Protest, Renovation, and New Amenities

Since the mid-1990s, the attractive location of VDP-buurt together with its challenged social and housing conditions have spurred renewal plans by the City and Ymere. These plans foresaw comprehensive renewal with social mixing, which was the *modus operandi* for regenerating deprived areas in Amsterdam at the time. Despite the area's cultural-historic value, Ymere maintained that renovation was neither worthwhile nor feasible. For Ymere, a heavily regulated private organization that owns nearly 78,000 dwellings, it made more financial sense to redevelop (i.e., demolish and build), to sell off some units of the newly developed dwellings as owner-occupied units, and to collect a higher rent for the remaining rental units. Fearing increasing rent levels, both short- and long-term residents organized into tenant and neighborhood organizations and mobilized to protect the neighborhood's cultural-historic value and its affordable housing.

In 2014, as part of an appreciation of early twentieth century garden villages, the Dutch government designated the neighborhood a "protected townscape," which meant that housing blocks could not be demolished but had to be renovated. After sustained resident protest, the City changed its position and struck a a deal with Ymere to renovate about 150 to 170 dwellings. Renovation would proceed piecemeal, allowing existing residents to move to a renovated house without experiencing a large rent increase. Of the renovated dwellings, 40 percent would be private rental or owner-occupied. As the renovation is costly and 60 percent will remain social rental units, the municipality is also contributing through a reduction in land development costs (Nul20 2014). This deal may be seen as a success for the activists, yet it is unclear what will be done with the more than 2,000 units.

Despite the thwarted redevelopment plans, the neighborhood has been undergoing a noticeable transition that is mostly visible through the emergence of new shops and cafés. Especially in the Van der Pekstraat, new businesses aimed at a middle-class clientele are starting to replace shops and cafés that had catered to long-term residents. Another major change is the renovation of the Van der Pekstraat, including the relocation of the street market, which also received a makeover and now includes stands offering organic and local produce as well as antiques.

The area surrounding VDP-buurt has also been changing rapidly: A new neighborhood called Overhoeks that consists of expensive, mostly owner-occupied housing is being built southwest of VDP-buurt. On the north side, where the local market used to be located, a new shopping center has been built. On the south side, a large hostel and entertainment facilities (e.g., bars, restaurants, clubs, and a movie theater) are being added to the redeveloped waterfront. The redevelopment of the northern IJ waterfront has arguably brought new attention to the North borough, as well as new residents, amenities, businesses, and visitors. As the neighborhood had long suffered from the closure of the shipping wharfs, these new developments are welcomed by many district politicians. The question is whether and how the long-term residents may benefit from these changes.

Policy Context

The pressure on Amsterdam's housing market offers Ymere and the municipality ample opportunities to draw more affluent residents to the neighborhood. Yet, in addition to social transformation, the interventions serve several policy objectives. Three types of policies are important in understanding the neighborhood's regeneration.

First, Dutch housing policies have been promoting homeownership since the early 1990s, and since the 2000s housing associations have been pushed to sell their dwellings as a financial strategy to invest in new development. After the 2008 Global Financial Crisis and the following budget crises in the Netherlands, new policies imposed taxes on housing associations to raise state income, forcing them to sell housing units at a premium in order to remain financially solvent (Boelhouwer and Priemus 2014).

Second, the municipality's growth agenda aims to secure Amsterdam's position as a competitive European metropolis through population growth and by accommodating highly skilled workers (Bontje and Lawton 2013). To attract more residents and visitors, the City seeks to expand the high density of cultural and commercial activities downtown to other residential neighborhoods adjacent to downtown. Government memoranda have stated that gentrification is desirable and that it should be promoted more (Van Gent 2013).

Furthermore, some local liberal politicians[3] and senior planners have publicly advocated for the social transformation of relatively low-income urban areas to accommodate new gentrifiers at the expense of social housing tenants (Capel 2016; Gadet 2015). In addition to developing high density housing, this pro-gentrification policy supports new amenities, such as cultural and entertainment venues, restaurants, cafés, and small specialty shops. Such municipal policies affect VDP-buurt, as well as the nearby waterfront revitalization (Savini and Dembski 2016). These local policies are linked to more general tendencies toward the sale and liberalization of social rental dwellings.

A stronger focus on economic growth has also transformed the municipality's cultural policies, including its best known initiative of *broedplaatsen* ("art factories"), an idea that originated in 1999 in response to concerns over the affordability of art spaces. *Broedplaatsen* policies entail the conversion of vacant industrial, housing, and office buildings into temporary facilities for starting artists and small creative-cultural companies. The municipality thereby subsidized exploitation costs for a limited period. Policy expectations, however, increased over time: Artists and start-ups were expected not only to produce art or commercial products and services, but also to contribute to local neighborhood development and social regeneration efforts (Bontje and Lawton 2013).

Third, area-based policies aim to improve the situation of poor residents. Since the 1990s, there have been several national programs to address urban social issues. The most recent one, the 40 Neighborhoods Program, also included the North borough and VDP-buurt. It funded programs for resident initiatives to improve livability and social contact among residents, alongside physical interventions. When the Dutch government announced the end of the 40 Neighborhoods Program in 2012, the City of Amsterdam stepped in and used the remaining two years of funding to develop the HSV Program, which included leadership training so that Amsterdam residents could effectively run and fund neighborhood initiatives through community-based projects or trusts. When a new liberal-socialist coalition came to power in Amsterdam in 2014, community-based initiatives were still seen as a primary agent of regeneration, yet the coalition initiated an overhaul of the HSV Program in connection with a restructuring of the City's governance. The mandate of the borough administration was reorganized, becoming more restricted, funding only grandfathered programs. In some cases City support was cut. Our interviews reveal that the overhaul meant that knowledgeable civil servants were transferred to other boroughs or departments.

Our Analysis

Amsterdam's growth agenda, the overall high demand for housing, and Ymere's financial priorities have put pressure on VDP-buurt, yet its current housing market structure has so far prevented a large influx of higher-income residents. Through sustained protest, tenants have effectively halted redevelopment and renegotiated renovations, at least for some parts of the neighborhood. However, even though redevelopment and renovation did not occur, the City and Ymere have been continuing their efforts through other means, for example, introducing and supporting new retail and cultural entrepreneurs, selecting students and artists to reside in temporary housing, as well as supporting and funding community-based organizations to work on neighborhood improvement projects. We ask the following questions to understand how these efforts may constitute middle-class quarter making, and to assess their social effects:

1. What organizations are involved in the transformation of VDP-buurt, and which of these are focused on middle-class residents?
2. How is the area and its regeneration represented, both by local organizations with a middle-class focus and by policymakers?
3. How is the regeneration viewed by residents?

Our analyses are based on various studies conducted by the authors. Between September 2015 and January 2016, we conducted an extensive social network analysis of the people involved in the regeneration of VDP-buurt, based on publicly available data in policy documents and on professional and personal websites. Individuals were identified and related to one another based on their organizational membership. This analysis was used to chart the overall organizational framework and to see which organizations are supported by the City and Ymere. Furthermore, in February 2016 we performed a content analysis of the public communication of organizations (e.g., websites, public newsletters, issued reports) to assess how the neighborhood's inhabitants and character were portrayed.

To assess the interface between policy and residents and how residents perceive the changes, we draw on interviews with five locally active civil servants and 27 residents.[4] These interviews, which were conducted between November 2014 and February 2015, focused on the VDP-buurt's regeneration. In addition, interviews were conducted with local representatives of six political parties, including members who were part of the borough administration in January 2016. We also asked these interviewees to comment on our network analysis. In general, their comments confirmed our analyses and helped to identify two additional interviewees.

Organizational Framework

Ymere and five municipal departments are the most important actors in terms of funding. Interestingly, our social network analysis and interviews identified 15 other organizations that play a role in the regeneration of VDP-buurt. These organizations are either commercial, nonprofit, or community-based. Of these, four are resident associations (one is mostly active in an adjacent neighborhood). Two are new community-based organizations (called trusts); one is focused on public space and social care in VDP-buurt, while the other seeks a broader audience and has been successful in obtaining multiple sponsors to organize cultural events in the adjacent park. Our network analysis shows that the latter organization—Noorderparkkamer—is strongly connected to the four cultural cooperatives and two of the three commercial associations. These six organizations form a cluster of artists and creative entrepreneurs, who live in Amsterdam North but not necessarily in VDP-buurt.

Interviewed borough administrators were generally enthusiastic about these cultural activities as they provide a boost to the area's image. Therefore, interviewees designated the cluster of cultural entrepreneurs as a "secondary circuit" of regeneration. While their activities differ from those of the municipality and Ymere (constituting the "first circuit"), these organizations and entrepreneurs are helping to introduce new residents to the neighborhood. Most of these organizations have received direct or indirect support from the municipality through cultural and social policies that seek to establish community-based organizations in vacant spaces, or are actively working in partnership with Ymere. Moreover, these seven organizations—including a group of urban designers—are the most active in producing symbolic markers that are typically middle class (see Table 24.1).

We defined the symbolic markers through an analysis of the discourse and content of the websites of the various organizations active in the neighborhood. Drawing on the work of Bourdieu (1984, 1989) and Zukin (1991) on how consumption and taste signify class position, we labeled the organizations that exhibit a clear focus on specific forms of consumption, which are linked to middle-class tastes (such as specialty coffee, "authentic" products, or yoga), as being "middle class symbolic markers in the literature and in the public debate." Furthermore, we also labeled those organizations that employ a discourse that stresses a pioneering attitude as middle class, as if VDP-buurt is "undiscovered" and exciting for potential pioneers, a *terra nullius*. Using such discourse does not necessarily mean that the organizations are trying to make the VDP-buurt a homogenous area. On the contrary, diversity is often celebrated as an interesting and exciting neighborhood trait. The next section further discusses these representations, which are reproduced, or actively produced, by eight organizations, most of which receive support from Ymere or the City of Amsterdam (see Table 24.1).

Table 24.1 Overview of Identified Organizations by Type and Association with the City of Amsterdam and Ymere, and Count of Organizations that (Re-)Produce Middle-Class Symbolic Markers.

Type of Organization	*Total Count*	*Receiving Material Support or Acting in Active Partnership With Housing Association or Municipality*			*No Partnership With Either HA or Municipality*	*Orgs. With Middle-Class Symbolic Markers*
		Ymere	*City of Amsterdam*	*Total*	*Total*	*Count*
Cultural cooperative	**4**	(1)	(2)	3	1	4
Residents association	**4**			0	4	0
Local retail and employment	**3**	(2)	(1)	3	0	2
Neighborhood trust	**2**	(1)	(2)	2	0	1
Forum for regeneration	**1**	(1)	(1)	1	0	0
Urban designers	**1**		(1)	1	0	1
Total	**15**	(5)	(7)	10	5	8

Source: authors.

Middle-Class Representation

Among the organizations, a substantial number exhibit middle-class symbolic markers. The most extreme, and highly prominent, example is the organization that publishes the weblog *I Love Noord*. The editorial board of about ten people posts text and media about "everything that makes you happy in Amsterdam North" (Founder, quoted in Virtueel Platform 2012), including performances, exhibitions, restaurant reviews, and community-based initiatives. It not only covers VDP-buurt, but explicitly aims to highlight all positive aspects of Amsterdam North as the forgotten borough. Its English and Dutch postings have been successful in raising attention in the old media and new social media (with 10,000 Facebook followers). A *New York Times* article that celebrated the borough's transformation from "blue collar to red hot" quoted an editor of the *I Love Noord* weblog as follows: "[The borough] is loaded with open spaces where wild entrepreneurs can start their own business. [. . .] There is a lot of raw energy here" (Williams 2012, n.p.). This quote fits well with the content and style of the weblog. In particular, the Eat and Drink section of the weblog features stylish locations geared toward middle-class tastes: craft beers, organic and Italian food, and cultural spaces "where interconnectedness is celebrated, restored and nurtured" (*I Love Noord*, March 4, 2016, n.p.).

Another interesting case is the Noorderparkkamer (NPK) trust, also mentioned in the *New York Times* article (Williams 2012). Like *I Love Noord*, NPK is a platform for local culture that expresses an inclusive discourse while mainly being managed by middle-class professionals. The website states: "What makes the Noorderparkkamer special is the quality programming made in close collaboration between artists, performers, volunteers, and local residents from all walks of life" (Noorderparkkamer 2016, n.p.). Yet our network analysis of affiliated participants shows that the professional and cultural middle classes constitute the vast majority. A short documentary about NPK confirms that it is primarily a site where artists and professionals organize cultural events (Ong-Alok et al. 2014).

I Love Noord and NPK are not the only active organizations in VDP-buurt, yet their activities and discourses exemplify what the neighborhood needs and ought to be in the eyes of the "secondary circuit" of regeneration. It is dominated by a rather specific group of middle-class professionals, who typically have a background in the creative and cultural industries and who have the skills and institutional capital to shape the representational frame to their tastes. In general, their framing is not conducted with malevolent intentions. Social and ethnic diversity is appreciated, and some middle-class newcomers worry that gentrification may be pushed too far and may lead to displacement (Tissot 2014). Yet they also dominate neighborhood activities. Their framing of what the neighborhood and the borough ought to be is very close to what municipal planners envision, which is why NPK and other cultural cooperatives receive subsidies and support from the municipality and Ymere.

Interviews with borough officials reveal that the middle-class view and aesthetic of these organizations are consistent with the borough's vision for the neighborhood. In its most ambitious form, this entails transforming VDP-buurt into an extension of the city center. A first step is to facilitate a different commercial infrastructure in the neighborhood, whose primary purpose is to draw visitors from wealthier neighborhoods. Thus, the borough hopes to create a virtuous cycle, spurring further investments in the area: "The renovation of Van der Pekstraat, the new shops and creative entrepreneurs are not for the neighborhood itself, but to draw more people from Overhoeks [the waterfront development] and to build [a new] shopping center, and thereby improve the neighborhood's image [. . .] also the organic market [the street market in Van der Pekstraat] is heavily subsidized [by the City, which hired a new manager who attracts new stalls and for marketing purposes], to draw a different clientele" (Interview, district civil servant and area manager of VDP-buurt, February 10, 2015). In short, quarter making through the support of cultural initiatives and the introduction of middle-class forms of consumption is believed to be necessary for "upgrading" VDP-buurt. As such, this constitutes a "soft power" approach to social transformation.

Not all interviewed local officials state that changing the social composition is necessary to achieve a neighborhood with a more middle-class flavor. Yet they do see that the neighborhood's changing image and new consumption patterns will have an effect. When asked about residents' views, local officials were very much aware that many long-term residents would prefer different kinds of amenities such as a community center or a street market. One official argued that most long-term residents do not understand the effects of middle-class amenities and renovations on their own ability to stay. The interviewed officials, however, believe it is inevitable that the population will change. They foresee that newly renovated housing will not have the same share of social housing and that the rents of any new social housing units will be too high for many low-income residents.[5]

Residents and Displacement

The previous sections showed that middle-class claims to space are legitimized symbolically by specific groups and organizations, which are in turn supported by the state. These efforts are to a certain degree counterbalanced by initiatives that focus on the social improvement of residents. While administrators and service providers tend to characterize long-term tenants as socially detached or too preoccupied with daily activities to become involved in neighborhood affairs, there are many active residents who seek to be involved in "their" neighborhood through small-scale resident organizations. Many local officials and residents agree that activities should benefit the entire neighborhood. Their efforts are often supported by the municipality and the housing association, in the belief that such residents who participate will have broader social networks, may acquire organizing skills, and will feel more attached to the neighborhood. This institutional support in direct (financial) or indirect (advice and training) form seems a key precondition for initiatives to succeed, as many new community-based organizations have ceased to exist after support was withdrawn after the reorganization in 2014 (Hoekstra and Pinkster 2019). In sum, the allocation of resources has shifted in recent years. As a result, different kinds of residents have been involved over time as a result of changes in resource allocation.

Active long-time residents mostly identified themselves as working class, either from Dutch or migrant origin, and have been active in informal networks for a long time. They are motivated by a sense of ownership of the neighborhood, which they argue was built and is meant for "people like them" (they also referred to themselves as "ordinary folks," or "the working class" in the interviews).

The social life of many White working-class residents used to revolve around neighborhood community centers (one of which was called "our home," while another was simply known as "the neighborhood center"). Due to municipal budget cuts, these have all closed after the GFC in 2008. As one long-term resident stated: "Many of the community centers are gone. It's a pity, though. Because, well, it affects things, a bit of social control. That's gone now." These residents described losing not only specific meeting places but neighborhood life in general: "This neighborhood has been completely shut down." Another argued that "It's no longer a working class neighborhood, now there's distance between people."

After the community center closed, some of these residents remained active in helping their neighbors or maintaining greenery and public gardens outside any organizational framework. A few residents have applied for subsidies offered by the HSV Program to organize activities in the neighborhood. However, they found that local institutions are not responsive to their needs and requests. Long-term active residents from migrant backgrounds have had similar experiences. For instance, a group of people of Moroccan descent, known by local officials and residents as "the men's group," has organized a variety of small events on topics such as debt prevention, domestic violence, and radicalization. Yet they claimed that their applications for financial support are routinely denied by the municipality—and this statement was verified by the social worker coaching the men's group. Long-term residents, regardless of whether they were native Dutch or non-native Dutch, felt abandoned by institutions that used to support them but now, according to them, favor middle-class residents.

Local officials counter-argued, stating that the activities of long-time residents lack a broad appeal, as they only involve some of the residents. Moreover, they expect that these non-middle-class, non-native groups will become less relevant once the regeneration efforts are completed. According to them, the neighborhood's changing population calls for different kinds of activities and a different kind of active resident (Baum 1998). These assessments of what kinds of participation are legitimate also reflect officials' own class identification. For example, the neighborhood's area manager (who lived in the neighborhood for a while) used her own experience to argue that the old working-class community centers were not "genuine" community centers but catered to "a specific group of long-term 'working class' residents who stopped being active in the center when the money stopped." To illustrate her point (and ours), she continued: "I used to live opposite [the center], I never went there, either." In contrast, she spoke positively about the newly established neighborhood trust, describing its participants as "active residents who take responsibility." The initiator of this trust (which seeks to provide social care and improve public space) agreed that his organization is better able to "take responsibility" for the neighborhood than the closed community centers, which he described as inefficient and marginal: "The ideas are nice, but if you look closer, it's too open-ended, they don't know how to make connections, how to reach their target group." Local officials mostly agreed with this assessment. One borough civil servant stated that "there are few people in the neighborhood who are willing and able [to participate]." She argued that residents have difficulties addressing funding requirements such as submitting a written budget and having to gain support from fellow residents.

In addition to NPK, the municipality and the housing association now support two organizations to improve the area: the local neighborhood trust that aims to become self-sufficient, as well as an organization for local entrepreneurs to develop the neighborhood's economic potential. These two organizations are the only surviving community-based organizations in the area itself, following the policy changes in 2014. They have survived because the residents that have been involved have the skills and the social capital to negotiate with the municipality and the housing association. These residents—mostly White middle-class Dutch—are able to rehearse the broader discourse of VDP-buurt's transformation from a closed working-class community into an "up-and-coming" area with a broad appeal. Active middle-class residents are better able to explain the relevance and timeliness of their organizations by referring to the current policy wisdom of bringing government closer to citizens and coping with budget cuts by relying on self-organization.

Furthermore, in contrast to other previous residents groups that often have a limited or no online presence, these residents are communicative and have established websites, making it easier for local officials and service providers to contact them and monitor their activities. In contrast, the lack of a contact email address among (active) poorer residents is seen as not only inconvenient but also strange. Thus, both symbolic and social advantages permit middle-class residents to raise money to organize things differently, allowing them to steer—and benefit from—regeneration at the same time. For example, the initiator of the neighborhood trust, even though he initially opposed regeneration, has lobbied the housing association to reserve some of the housing stock as "fixer-uppers" for young middle-class couples, after it became apparent that renovation would take place. This initiative allowed him to be involved in the negotiation process.

Discussion and Conclusion

For many years, urban and housing scholars have debated state-sponsored gentrification. Critical perspectives have been instrumental in highlighting the role of the state in urban change. Yet much of the critique seems to lose sight of the complexities of the state and the intricacies and context specificity of class relations (Brown-Sarracino 2016; Tissot 2014). To add to these debates, we have asked the question of whether we may view the state interventions and service provisions in VDP-buurt as beneficial social regeneration or as a form of state-sponsored gentrification.

The municipality and the housing association have launched and supported multiple initiatives in and around the neighborhood. By funding community-based initiatives, commercial change, and cultural cooperatives, the municipality and housing association are effectively paving the way for partial, if not full, gentrification. We found that social regeneration efforts are also biased toward middle-class residents and professionals working in the area. The activist residents have largely forestalled more radical interventions, which is quite an achievement. Yet despite ongoing residents' protests, the municipality and housing association have continued to engage in middle-class quarter making, mostly through what has been called a "secondary circuit" of regeneration. While some working-class residents may welcome the changes, our interviews reveal that most are dissatisfied about the changes in the allocation of funding, express uncertainty about the future, and experience a sense of loss of place. Consequently, we view state intervention in the VDP-buurt as a soft force approach to state-sponsored gentrification, leading to a partial gentrification process, at least for now.

There is a certain tragedy to this situation, because both middle-class residents and local civil servants and professionals show genuine social involvement and have expressed concerns over displacement. The middle-class residents, artists, and professionals living and working in the area may be subject to displacement at a later stage. Their own vulnerability is illustrated by a newspaper report on students who were selected by the housing association (through a local cultural entrepreneur) to temporarily move into vacant dwellings. Even though they knew it was a temporary arrangement, they grew fond of the area and complained that "first we had to make the area more attractive to outsiders, but now that there is sufficient spending power in the neighborhood, we can bugger off" (Segaar 2015).

State officials have demonstrated an awareness of the (future) negative effects for long-term residents, yet point to the municipality's growth agenda. As this agenda envisions a new population, policies and planning are perhaps biased toward accommodating these newcomers. In addition, middle-class residents are better positioned and have better access to public services in a number of ways (Baum 1998; Elwood et al. 2015; Matthews and Hastings 2013). Well educated and well connected individuals, either individually or organized in groups, are able to exert pressure on and influence service providers. This may be done through protest and "enraged response," or through their ability to navigate formal rules and regulations, but also informally as they have the social skills and knowledge to vocalize their needs and demand better services. The ability of middle-class residents to claim access to services points to the importance of social and cultural capital when individuals interact with state officials and civil servants. Service providers ("street level bureaucrats") are not detached agents but are embedded in society. Most officials, civil servants, and non-profit service providers hold a class position nearer to that of middle-class residents than to poor and working-class residents, often leading to uneven outcomes. When interactions between service providers and residents take place, class positions may produce "empathy from similarity."

Finally, many local policymakers view the neighborhood's transformation as inevitable, and, as noted earlier, the GFC, the economy, labor market change, increased housing demand, and national housing policies are all placing a strain on the area. Yet the transformation of areas like VDP-buurt is neither a natural occurrence nor accidental, but the outcome of political choices made at various levels of government at different points in time. The (neo-)liberalization of housing may inform some of these choices, but cultural, integration, and social policies and politics have also been relevant, resulting in an inconsistent and even contradictory set of initiatives. Recent public debates in Amsterdam and elsewhere in the Netherlands have led to political parties questioning the growth agenda and the continued demise of social rental housing. As politics may change, we may therefore see a shift in neighborhood interventions toward more focus on equity and social justice.

Acknowledgements

The authors would like to thank Patrick van Son for his assistance in data collection, and the editors for providing detailed comments and suggestions.

Notes

1 Minimum income refers to the municipality's poverty line, which is set at an income of up to 110 percent of the amount received in social benefits. Some households may earn this amount through part-time and/or low-wage employment. In 2016, these gross income thresholds were €1,264 for single-person households and €1,685 for couples per month.

2 In the Dutch ethno-national classification, when one or both parents are born abroad, a person is classified as non-native. A distinction is made between "Western" and "non-Western" non-natives: "Western" encompasses North America, Europe (excluding Turkey), Oceania (Australia, New Zealand, Melanesia, Micronesia, Polynesia), Japan, and Indonesia, and "non-Western" is the remainder.

3 We use the concepts "liberal" and "liberalism" in a Western European sense, referring to a political ideology centered around notions of liberty and freedom, but not social equality. The politicians are positioned between the center and the (non-radical) right end of the political spectrum.

4 Of the 27 residents, 16 had lived in VDP-buurt for more than 20 years. Twelve of these long-term residents may be classified as working class; 11 were native Dutch; and five were non-native Dutch.

5 Social rent may be up to about €750 per month, depending on housing characteristics and quality. There are individual rent subsidies, but social rental housing may be unaffordable for those earning the minimum income.

References

Andersson, Roger, and Sako Musterd. 2005. "Area-Based Policies: A Critical Appraisal." *Tijdschrift voor Economische en Sociale Geografie*, 96: 377–389.

Baum, Howell S. 1998. "Ethical Behavior Is Extraordinary Behavior; It's the Same as All Other Behavior: A Case Study in Community Planning." *Journal of the American Planning Association*, 64: 411–423.

Boelhouwer, Peter, and Hugo Priemus. 2014. "Demise of the Dutch Social Housing Tradition: Impact of Budget Cuts and Political Changes." *Journal of Housing and the Built Environment*, 29: 221–235.

Bontje, Marco, and Philip Lawton. 2013. "Mobile Policies and Shifting Contexts: City-Regional Competitiveness Strategies in Amsterdam and Dublin." *Tijdschrift voor Economische en Sociale Geografie*, 104: 397–409.

Boterman, Willem R., and Wouter P.C. van Gent. 2014. "Housing Liberalization and Gentrification; The Social Effects of Tenure Conversions in Amsterdam." *Tijdschrift voor Economische en Sociale Geografie*, 105: 140–160.

Bourdieu, Pierre. 1984. *Distinction. A Social Critiques of the Judgement of Taste*. London: Routledge.
Bourdieu, Pierre. 1989. "Social Space and Symbolic Power." *Sociological Theory*, 7: 14–25.
Brown-Sarracino, Japonica. 2016. "An Agenda for the Next Decade of Gentrification Scholarship." *City & Community*, 15: 220–225.
Bridge, Gary, Tim Butler, and Loretta Lees. eds. 2012. *Mixed Communities: Gentrification by Stealth?* Bristol: Policy Press.
Bridge, Gary, and Robyn Dowling. 2001. "Microgeographies of Retailing and Gentrification." *Australian Geographer*, 32: 93–107.
Cameron, Stuart, and Jon Coaffee. 2005. "Art, Gentrification and Regeneration—From Artist as Pioneer to Public Arts." *European Journal of Housing Policy*, 5: 39–58.
Capel, Sebastiaan. 2016. "De Ene Gentrification is de Andere Niet [Not All Types of Gentrification are the Same]" *D66 Amsterdam Zuid*, October 20.
Chaskin, Robert. J., and Mark L. Joseph. 2013. "'Positive' Gentrification, Social Control and the 'Right to the City' in Mixed-Income Communities: Uses and Expectations of Space and Place." *International Journal of Urban and Regional Research*, 37: 480–502.
Davidson Mark, and Loretta Lees. 2005. "New Build 'Gentrification' and London's Riverside Renaissance." *Environment and Planning A*, 37: 1165–1190.
Elwood, Sarah, Victoria Lawson, and Samuel Nowak. 2015. "Middle-Class Poverty Politics: Making Place, Making People." *Annals of the Association of American Geographers*, 105: 123–143.
Ernst, Olaf, and Brian Doucet. 2014. "A Window on the (Changing) Neighbourhood: The Role of Pubs in the Contested Spaces of Gentrification." *Tijdschrift voor Economische en Sociale Geografie*, 105: 189–205.
Florida, Richard. 2002. *The Rise of the Creative Class: And How It's Transforming Work, Leisure, and Community and Everyday Life*. New York: Basic Books.
Freeman, Lance. 2006. *There Goes The 'Hood: Views of Gentrification From The Ground Up*. Philadelphia: Temple University Press.
Gadet, Jos. 2015. "Laat die Gentrifiers maar Komen! [Let the Gentrifiers In!]" *Stadsleven*, June 5.
Gans, Herbert. J. 1962. *The Urban Villagers: Group and Class in the Life of Italians-Americans*. New York: Free Press.
Gemeente Amsterdam. 2015. *Gebiedsanalyse 2015, Oud Noord, Stadsdeel Noord*. Amsterdam: Gemeente Amsterdam.
Goetz, Edward G. 2013. "The Audacity of HOPE VI: Discourse and the Dismantling of Public Housing." *Cities*, 35: 342–348.
Grodach, Carl, Nicole Foster, and James Murdoch III. 2014. "Gentrification and the Artistic Dividend: The Role of the Arts in Neighborhood Change." *Journal of the American Planning Association*, 80: 21–35.
Hackworth, Jason, and Neil Smith. 2001. "The Changing State of Gentrification." *Tijdschrift voor Economische en Sociale Geografie*, 92: 464–477.
Harris, Andrew. 2012. "Art and Gentrification: Pursuing the Urban Pastoral in Hoxton, London." *Transactions of the Institute of British Geographers*, 37: 226–241.
Hochstenbach, Cody, and Wouter P.C. van Gent. 2015. "An Anatomy of Gentrification Processes: Variegating Causes of Neighbourhood Change." *Environment and Planning A*, 47: 1480–1501.
Hoekstra, Myrte S., and Fenne M. Pinkster. 2019. "'We Want to Be There for Everyone': Imagined Spaces of Encounter and the Politics of Place in a Super-Diverse Neighbourhood." *Social & Cultural Geography*, 20: 222–241.
Kleinhans, Reinout, and Ade Kearns. 2013. "Neighbourhood Restructuring and Residential Relocation: Towards a Balanced Perspective on Relocation Processes and Outcomes." *Housing Studies*, 28: 163–176.
Lees, Loretta, and Clare Melhuish. 2015. "Arts-Led Regeneration in the UK: The Rhetoric and the Evidence on Urban Social Inclusion." *European Urban and Regional Studies*, 22: 242–260.
Lees, Loretta, Hyun Bang Shin, and Ernesto López-Morales. 2015. *Planetary Gentrification*. Cambridge: Polity Press.
Matthews, Peter, and Annette Hastings. 2013. "Middle-Class Political Activism and Middle-Class Advantage in Relation to Public Services: A Realist Synthesis of the Evidence Base." *Social Policy & Administration*, 47: 72–92.
Mathews, Vanessa. 2010. "Aestheticizing Space: Art, Gentrification and the City." *Geography Compass*, 4: 660–675.
Millard-Ball, Adam. 2000. "Moving Beyond the Gentrification Gaps: Social Change, Tenure Change and Gap Theories in Stockholm." *Urban Studies*, 37: 1673–1693.
Murie, Alan. 1991. "Tenure Conversions and Social Change: New Elements in British Cities." In *Urban Housing for the Better-Off: Gentrification in Europe*, edited by Jan van Weesep and Sako Musterd, 137–148. Utrecht: Stedelijke Netwerken.
Nguyen, Mai Thi, Victoria Basolo, and Abhishek Tiwari. 2013. "Opposition to Affordable Housing: How the Debate is Framed and How Developers and Local Housing Actors Respond." *Housing, Theory and Society*, 30: 107–130.

Noorderparkkamer. 2016. *Over NPK.* Amsterdam: Noorderparkkamer.

Nul20. 2014. *Akkoord over Aanpak Van der Pekbuurt Amsterdam-Noord.* Amsterdam: Nul20.

Ocejo, Richard. E. 2011. "The Early Gentrifier: Weaving a Nostalgia Narrative on the Lower East Side." *City & Community*, 10: 285–310.

Ong-Alok, Gigi Pasco, Shelley Bontje, and Esther Huffener. 2014. *Goede koffie helpt!—De Noorderparkkamer.* Documentary. Amsterdam, Documentaire Instituut Amsterdam.

Paton, Kirsteen, Vikki McCall, and Gerry Mooney. 2017. "Place Revisited: Class, Stigma and Urban Restructuring in the Case of Glasgow's Commonwealth Games." *The Sociological Review*,65(4): 578–594.

Pinkster, Fenne M. 2016. "Narratives of Neighbourhood Change and Loss of Belonging in an Urban Garden Village." *Social & Cultural Geography*, 17: 871–891.

Rota, Francesca. S., and Carlo Salone. 2014. "Place-Making Processes in Unconventional Cultural Practices. The Case of Turin's Contemporary Art Festival Paratissima." *Cities*, 40: 90–98.

Sakizlioglu, N. Bahar, and Justus Uitermark. 2014. "The Symbolic Politics of Gentrification: The Restructuring of Stigmatized Neighborhoods in Amsterdam and Istanbul." *Environment and Planning A*, 46: 1369–1385.

Savini, Federico, and Sebastian Dembski. 2016. "Manufacturing the Creative City: Symbols and Politics of Amsterdam North." *Cities*, 55: 139–147.

Segaar, Eva. 2015. "Nu de Van der Pekbuurt is Opgeleukt, Kunnen We Weer Oprotten." *Het Parool*, April 7.

Sze, Julie. 2009. "Sports and Environmental Justice: 'Games' of Race, Place, Nostalgia, and Neoliberal New York City." *Journal of Sport & Social Issues*, 33: 111–129.

Tissot, Sylvie. 2014. "Loving Diversity/Controlling Diversity: Exploring the Ambivalent Mobilization of Upper-Middle-Class Gentrifiers, South End, Boston." *International Journal of Urban and Regional Research*, 38: 1181–1194.

Van Gent, Wouter P.C. 2013. "Neo-Liberalization, Housing Institutions and Variegated Gentrification; How the 'Third Wave' Broke in Amsterdam." *International Journal of Urban and Regional Research*, 37: 503–522.

Van Gent, Wouter P.C., Willem R. Boterman, and Marien W. van Grondelle. 2016. "Surveying the Fault Lines in Social Tectonics; Neighbourhood Boundaries in a Socially-Mixed Renewal Area." *Housing, Theory and Society*, 33: 247–267.

Virtueel Platform. 2012. "Blogging the City Bracht Bloggers uit de Hele Wereld bij Elkaar." *Virtueel Platform*, November 2.

Wallace, Andrew. 2016. *Remaking Community? New Labour and the Governance of Poor Neighbourhoods.* London: Routledge.

Williams, Gisela. 2012. "From Blue Collar to Red Hot." *New York Times*, November 4.

Zukin, Sharon. 1991. *Landscapes of Power: From Detroit to Disney World.* Berkeley, CA: University of California Press.

SECTION 8

Suburbs

25

HOUSING POLICY AND THE SUBURBAN METROPOLIS

A Focus on the United States and France

Bernadette Hanlon and John Rennie Short

Introduction

Housing policy after World War II was instrumental in the evolution and transformation of U.S. and French suburbs. We first focus our examination on the years from the end of World War II to the 1970s. Because U.S. suburbs and the French *banlieues* have changed significantly since the 1970s, we also examine more recent government actions in the realm of housing that stress the importance of social mixing in the suburban context. In the U.S., suburbs are perceived as places of opportunity in efforts to deconcentrate poverty in the inner city. In France, social mixing policy has been an integral element in the redevelopment of social housing projects in the *banlieues*.

Conceptually, there is a divide between the city and the suburb that we argue influences both the development and implementation of housing policy in both countries. We believe that efforts at social mixing and poverty dispersal are ultimately layered with a racialized vision of the metropolis that fails to account for broad structural problems that impact poor minority families and does little to help undo the deep fragmentation of social space in metropolitan regions. We begin our discussion by examining the influence of housing policy on postwar suburbanization in both countries, beginning with France.

Postwar Suburbanization in France

World War II had devastating impacts on France, with no fewer than one in five buildings across the country damaged (Cupers 2014). Wartime destruction was only part of the reason for the resulting severe housing shortage in the 1940s and 1950s. Certain "prewar precursors" (Rudolph 2015, 59) contributed to this shortage including a general lack of housing production before the war (Cupers 2014), and stringent rent control laws that resulted in rents so low that landlords had little incentive to build new rental units or maintain existing ones, especially for those at the bottom end of the housing market (Prost 1991). After World War II, almost half of French homes lacked running water, eight out of every ten homes did not have an indoor toilet, and some three million people were left homeless (Cupers 2014).

Early legislation in the 1920s attempted to deal with post-World War I housing problems. For instance, *Loi Loucheur* or *Loucheur Law* was introduced in 1928 with two effects. First, it led to suburban growth with the development of *pavillons*, or small, detached houses during the period between 1918 and 1930 (Castells 1977). Part of the law included subsidies to enable individuals to borrow at a reduced mortgage interest rate to buy land and build a house, which resulted in a haphazard collection of small suburban homes located outside Paris in particular. Second, the law did call for the development of some 200,000 *habitations à bon marché* (HBM) apartments and 60,000 *immeubles à loyer moyen* (ILM) or medium-sized rental buildings to help house families. HBMs included low-cost housing units specifically aimed at working-class and more destitute families; ILMs were reserved for upper-middle classes (Cupers 2014). However, not many units were built in the 1930s due to the lack of funding. Paris received an overwhelming majority of the units and ended up with around 50,000 HBMs and 20,000 ILMs (Rudolph 2015), not enough to deal with the large-scale housing shortage. In 1936, the *Office Public d'Habitations de la Ville de Paris* (OPHVP) published a report on social housing and found that most of the families living in these government-funded apartments were working class with three or four children who had previously lived in one room apartments (Rudolph 2015). The *Loucheur Law* did help set the stage for government involvement in the mass production of social housing that came in the 1950s.

Initially, at the very end of World War II, the focus was on reconstruction and rehabilitation of city and town centers rather than building anew in the suburbs (Cupers 2014). It was not until the late 1950s that the state became heavily involved in the construction of new housing, which came in the form of mass produced, modern housing in large estates or *grand ensembles*, mainly, although not exclusively, in the suburban periphery.

The need for decent housing had by then reached desperate levels, further exacerbated by population growth in the immediate postwar period (Cupers 2014). Poor housing conditions were most notable in the *bidonvilles* or shantytowns in the suburbs of Paris, built in large part by Algerian, Portuguese, and Spanish immigrants (Garbin and Millington 2012).

State-aided production of social housing was crucial to the migration of workers to new industrial plants built in peripheral locations, a necessary component of France's industrial growth (Cupers 2014). At that time, housing provision was viewed as a government responsibility, not only toward the poor, but also as part of a broader effort to embrace the needs of the industrial working and middle classes.

In the 1950s and 1960s, France, unlike the United States, never adopted the production of single-family homes as a viable strategy for large-scale housing provision, but rather viewed mass produced housing as both symbolic of French modernization and part of a larger project to aid industrial prosperity. During this time, the state dismissed the single-family home as "a thing of the past and not the future" (Cupers 2014, 10). For the French state, small individual homes were certainly not as striking nor as impactful as the modernist architecture of the *grand ensembles*. At the beginning of the 1960s, the French government was building half a million housing units each year in the form of large-scale housing estates. In the mid-1960s, only 21 percent of housing construction in France was in the form of single-family housing compared to 61 percent in the United States (Cupers 2014). The famous American developer William Levitt visited France to promote his tract-style residential development, but the French government was not interested. His firm created a few subdivisions, but they were perceived as too expensive for French working families.

It was not until the 1970s, when there was a shift away from the state-led production of housing estates, that the government turned its focus on the promotion of a mix of housing types, including some single-family housing units, especially in the developing new towns or *Les Villes Novelles* (Tuppen 1979). In 1977, state housing policies shifted from supply-side to demand-side subsidization in the form of subsidized mortgages that included low-interest loans called *Prêt d'Accession à la Propiérté* (PAP) and *Prêt Conventionné* (PC) aimed at first-time homebuyers. Between 1977 and 1984, about 60 percent of homebuyers benefited from one of these mortgages, and as a consequence, the homeownership rate increased from 40 to 54 percent between 1970 and 1984 (Gobillon and Le Blanc 2008). The French middle class in particular began to actively pursue ownership of the suburban home, as was the case in the U.S. As a consequence, large-scale social housing blocks became increasingly occupied by working and nonworking lower-income families and others unable to own a home. This trend intersected with the in-migration of North Africans, Turks, and sub-Saharan Africans during the mid-1970s, and, thus, social housing became increasingly occupied by immigrant groups and minorities.

Since the 1970s, the *banlieues*, despite the fact that they are varied in terms of neighborhood income, have, as Dikeç (2013, 28) notes, "become a shorthand term to denote certain groups of the population (and problems seen to be associated with them) who live in peripheral social housing estates." The *banlieues* embody fears about a lack of integration by ethnic and religious minorities into French society. As Dikeç (2013, 28) suggests, "In political and media discourses the problem of *banlieues* implies a threat to the values of the Republic, French identity and the authority of the state." Hargreaves (1996, 607) notes how mass media portrayal of the *banlieues* helped disseminate and reinforce "stereotypical ideas of people of immigrant origins as fundamentally menacing to the established order." As we discuss later, this characterization of the *banlieues* has greatly impacted contemporary policy and practice aimed at redeveloping these places.

Postwar Suburbanization in the U.S.

The postwar suburban story in the United States is quite different from that in France. In France as mentioned, the state subsized and produced large-scale apartment blocks in the periphery. In the case of the U.S, it was private developers who built low-density single-family homes for purchase in the suburbs. Beginning in about 1945, the White working and middle classes, precipated by government policies, moved on mass to this new suburban housing. The account of postwar suburbanization in the U.S. has been explored by a variety of urban scholars (e.g., Hayden 2004; Jackson 1987). Policies to promote suburban homeownership were instrumental. In 1934, the federal government established the Federal Housing Administration (FHA). This agency insured mortgages, and as a result, mortgages became much less financially risky for lenders who, after the Great Depression and World War II, were much more willing to lend money for home purchasing. In the 1940s, the Veterans Affairs (VA) mortgage program was similarly structured. Under the 1944 GI Bill, veterans returning from World War II could receive home loans insured by the government. Before the availability of FHA- and VA-insured loans, homebuyers were required to make at least a 50 percent down payment, and the mortgage had to be repaid within five years. As a result, only the wealthy could purchase a home at that time. With government-insured loans, homebuyers could put down 10 percent of the loan and pay it back in 30 years. This model soon became standard practice and opened up the possibility of home buying to the working and middle classes. By 1972, the FHA-VA programs had helped 11 million families purchase homes (Hanlon et al. 2009).

The strict guidelines of the federal government loan guarantee programs meant that the homes purchased were single-family and located in the suburbs. Government-insured loans were for the purchase of newly constructed rather than existing structures (which increased the likelihood of the loan being used in the suburbs). The FHA allowed personal and agency bias in favor of all White suburban subdivisions to influence the kinds of loans it guaranteed, as applicants in these subdivisions were generally deemed better credit risks. The effects were palpable. Between 1945 and 1959, less than 2 percent of all FHA-VA mortgages went to African Americans (Hanlon et al. 2009), and few if any government-backed loans were made to purchase homes in neighborhoods with large numbers of minority residents. During this time period, about 90 percent of loans were for homes in largely White suburbs (Hanlon et al. 2009).

While certain federal housing policies stimulated single-family housing construction for White families in the suburbs, other policies waged war against minority neighborhoods in the city (Short 2006). Most notable was the construction of the interstate highway system. The 1956 Federal-Aid Highway Act initiated the eventual construction of more than 41,000 miles of roads, one of the single largest infrastructure projects in the world at the time. Housing demolition in city neighborhoods occurred to make way for urban expressways. Between 1957 and 1968 alone, about 330,000 housing units were torn down as a result of the federal highway program (Short 2006). Most impacted were low-income minority neighborhoods.

Housing demolition was aided further by federal housing legislation. Initial plans for the Housing Act of 1949 were to eliminate substandard housing in the city and replace each unit torn down with a new unit. However, this did not happen. Title I of the 1949 legislation did result in the reauthorization of the public housing program established by the earlier Housing Act of 1937 and promised to replace demolished homes, but local practices and decisions ensured that much of the housing was replaced with tax-generating commercial enterprises (Short 2006). This trend continued further with the urban renewal program initiated by the 1954 Housing Act, which often resulted in the demolition of sound housing and the destruction of vibrant minority neighborhoods. Urban renewal came to be known as "Negro removal." With postwar housing legislation, American metropolitan areas were radically reorganized. Many White working- and middle-class families left for single-family homes in the suburbs; public housing, like in France, became increasingly occupied by low-income families; and cities, particularly those in the Midwest and Northeast, witnessed a concentration of more disadvantaged African-American residents.

Metropolitan Narratives, Social Mixing, and Contemporary Housing Policy

In both France and the United States, what emerged from postwar suburbanization is, in many respects, a rather simplified narrative about metropolitan regions. In France, this narrative emphasizes a sharp dichotomy between the *banlieues* (the sole location of social housing projects where the poor and immigrant groups live) and the city (the location of culture and the French elite). Newman (2015, xxviii) characterizes this dichotomy as follows: "the capital [Paris] is French and secular, the *banlieue* is immigrant (usually Arab or African) and Muslim. The capital is rich; the *banlieue* is poor.... The capital—and France itself—is the victim; the *banlieue* is the threat." Yet, as Newman (2015) notes, the *banlieues* are varied; they can be poor (e.g., Clichy-sous-Bois) and rich (e.g., Neuilly-sur-Seine), and cities, like Paris, have some poor neighborhoods where a diverse immigrant population lives. Within the *banlieues* themselves, the population is diverse in ways that include rich and poor, immigrant and French, old and young (Packer 2015). In France, the territorial stigmatization of the *banlieues* (Wacquant 2008) is juxtaposed against a much more positive characterization of the French city, particularly Paris, a global city that plays a significant role in the state's commitment to "elite urbanism" (Enright 2014, 386).

In the U.S., there was a very different metanarrative of city Black, suburb White; city poor, suburb rich; city heterogeneous, suburb homogenous (Nijman 2015). The depiction of challenges in inner cities in the United States is often held in pointed contrast with the perceived American suburban success story (Denton and Gibbons 2013). This, of course, has sharp racial undertones that, as we discuss later, have had important implications for low-income housing policy.

There was a solid basis for this dominant popular view. Certainly, federal policies during and before the postwar period greatly encouraged a precise racial division between the inner city and the suburbs. Mortgage lending, redlining, and real estate practices denied African Americans and other people of color access to certain suburbs and suburban homeownership. Exclusionary land use policies continued to further limit minority access (Pendall 2000). Yet it is important to recognize that suburbs today vary by socioeconomic status and racial and ethnic composition (Mikelbank 2004). As scholars of a new suburban history attest (Kruse and Sugrue 2006), suburbs have always been more heterogeneous than previously described. More recent studies highlight issues of suburban poverty (Kneebone and Berube 2013), refer to increasing suburban diversity (Anacker 2015), and note the gentrification of inner city neighborhoods (Lee et al. 2008). Any traditional division between city and suburb has been increasingly blurred with the intensified social and morphological complexity of metropolitan regions of the United States (Hanlon et al. 2006; Nijman 2015).

Portrayals of the morphology and social dimensions of suburb and *banlieues* alike have cultivated contemporary demands to re-envision social housing, create more socially mixed communities, disperse poverty, and more broadly revamp the suburban periphery. There are many deep assumptions about the cultural, class, and physical characteristics of the *banlieues* and the U.S. suburb that, we believe, have subsequent effects on policy (Wacquant et al. 2014, 1270). Much of this narrative is couched in a racialized vision of the metropolis and one that stresses the stigmatization of some communities over others. In the following text we explore this phenomenon in more detail by considering more recent housing policies related to the French *banlieue* and U.S. suburb.

Social Mixing and the French *Banlieue*

Beginning in 2003, the French government undertook an urban renewal program entitled the National Urban Regeneration Program (*Programme National de Rénovation Urbaine*, PNRU), which identified a total of 500 so-called vulnerable areas or *zones urbaines sensibles* (ZUSs) for rehabilitation and redevelopment. ZUSs are typically areas with large estates of social housing in the *banlieues*. With passage of *La loi d'orientation et de programmation pour la rénovation urbaine* in 2003, the state established an agency, *Agence Nationale de la Rénovation Urbaine* (ANRU), that acts as the central office for financing urban renewal projects with the goal of demolishing and reconstructing about 450,000 housing units (Lévy-Vroelant 2007). Urban renewal projects were built around a contract between municipalities and ANRU. PNRU was replaced in 2014 with a new ten-year national urban regeneration program that requires stakeholders and citizens at the local level to be part of the city contract process. In the contract, the list of measures needed to start the regeneration process of the targeted districts is included along with any financial commitments from local stakeholders. According to the OECD (2015), 145,000 social housing units were demolished and rebuilt, 300,000 units were rehabilitated, and 85,000 new units were built between 2003 and 2014 (OECD 2015).

Urban renewal in France is not something new. Demolition of working-class neighborhoods occurred first in Haussmann's Paris in the second half of the nineteenth century, and again in a number of cities in the 1960s and 1970s. Political justifications for these early demolitions focused on public health concerns (Gilbert 2009). The process initiated by the 2003 national urban renewal program involved the demolition of high-rise social housing to be replaced by mid-rise blocks of apartments, and with the added goals of controlling the behavior of individuals living in social housing as well as the inclusion of planning efforts to attract middle-class households (Gilbert 2009, 641). In the case of current urban renewal practices in the *banlieues*, motivations circle around deep concerns about the effects of the spatial concentration of poor and minority ethnic populations. Consequently, the focus is on boosting social mixing by including market-rate and ownership housing on redeveloped sites (Kipfer 2016). The objective of social mixing has become the cornerstone of much housing policy, not only in France, but in other parts of Europe (especially the UK, the Netherlands, and Sweden), and in the United States (Bolt and van Kempen 2013; Galster 2007).

In France, such social mixing has been a critical element of housing and urban policy since at least the 1990s. In 1990, the Besson Act created the right to housing, ensuring access to housing for the poorest households but also encouraging social equilibria in housing allocations (Bacqué et al. 2011) (see the chapter authored by W. Dennis Keating). In 1991, the Framework Act of Town Planning established a goal of 20 percent for municipalities with populations greater than 200,000, although the penalty for noncompliance was eliminated, leaving municipalities with freedom on how to implement this policy. Later in 2000, the Urban Solidarity and Renewal Law or *Loi Solidarité et de Renouvellement Urbain* (SRU) established obligations that every *commune* (with a population of

1,500 in the *Île de France* region and 3,500 in other regions) have at least 20 percent social housing units or face financial penalties. Some local municipalities preferred to accept fines than be subject to such requirements, generating much protest from those on the political right in France (Bacqué et al. 2011). In the context of France, it seems as though any willingness to promote social mixing by dispersing low-income housing associated with poor ethnic populations to a variety of locations is much more controversial than introducing a greater diversity of housing types and tenures to already existing social housing projects and neighborhoods. As we discuss later, dispersing poor populations, particularly poor minorities, is similarly problematic in the U.S. context.

One of the objectives of social mixing is to reduce the stigmatization of the *banlieues* (Arthurson 2013). Scholars document that social housing and the *banlieues* suffer from a negative public image, and, as Wacquant (1993, 369) states, living in a "low income [housing] estate means to be confined to a branded space," one that is marked as a place of poverty and disorder. In the 1980s and 1990s, and most recently in 2005, riots took place in the *banlieues*, further stigmatizing these places and, indirectly, certain ethnic groups (Dikeç 2007). In many ways, an intense security apparatus has emerged to monitor and control the behavior of inhabitants of social housing. Dikeç (2007) discusses in detail the ways in which urban policy toward the *banlieues* has, over time, become more authoritarian, with increased anxiety about immigration, Islam, and perceived threats to the French Republic's common culture and identity. Kipfer's (2016) interviews with planners and policymakers about 18 recent redevelopment projects in 13 municipalities in the Greater Paris region identify the ways in which the need for security is coded in racial terms. Of particular concern is the "specter of loitering youth" of African descent (Kipfer 2016, 612). Kipfer (2016) notes how the redesign of social housing projects often aims to prevent teenagers from gathering in entrance halls and public spaces. Based on his interviews, Kipfer (2016, 612) states that for planners a "key priority is to make projects 'open,' 'legible,' 'accessible,' and 'permeable' for the police, which by law must approve design plans."

What has evolved is a culture of policing, racial profiling (largely of young men of North African origin), and security in housing estates of the *banlieues*. The riots in 2005 started when Ziad Benna and Bouna Traore, two teenagers of Tunisian and Malian origin who lived in *Clichy-sous-Bois*, were electrocuted when hiding from police in a nearby power substation (Laurence and Vaisse 2005). Police practices and the "law and order" approach initiated in the 1990s have been characterized as counterproductive, perpetuating further social unrest (Jobard 2009). Some scholars claim that the demolition and reconstruction of social housing have had similar effects. For instance, Lagrange (2006) argues that designating certain areas in the *banlieues* for improvement raises expectations that, when they remain unfulfilled, lend themselves to further frustration. He states (2006, 122) that the 2005 riots heightened "claims aimed primarily at the state" and its failure to improve the lives of social housing residents after more than a decade of urban policy (Dikeç 2007). This raises doubts about the prospects for regenerating the *banlieues* given that structural problems (e.g., high rates of unemployment among young males of North African origin) have not been addressed.

Social Mixing and the U.S. Suburb

In the U.S., social mixing translates into the development of mixed-income housing and poverty dispersal. Goetz (2003) identified five key U.S. housing policy initiatives aimed at the deconcentration of poverty.

The first is the shift from project-based housing subsidies tied to particular neighborhoods to tenant-based subsidies (in the form of housing vouchers) to offer more locational choice to poor families. The idea is that poor residents will use these vouchers to live in low-poverty neighborhoods with better services, often referred to as "high-opportunity neighborhoods" (see the chapter authored by Kirk McClure). Second, there were efforts to refine tenant-based subsidies to further increase the dispersal of poor people. Third, beginning in the 1990s, an effort began to introduce different income groups into public housing, Fourth, there were policy efforts to disperse subsidized housing projects into places where there were traditionally few, if any, including the suburbs. Fifth, with the HOPE VI program, a major effort began to demolish public housing projects and replace them with mixed-income developments.

The social mix strategy is grounded in the neighborhood effects thesis. Based mostly on studies in the United States, this thesis suggests that living in neighborhoods of high poverty produces negative social effects, including social isolation and declining social capital, which, in turn, restrict social mobility (Chaskin and Joseph 2015). The idea that living in a deprived neighborhood is the most substantive driver of poor social and economic outcomes for local residents is not without critics (see Imbroscio 2008). The neighborhood effects argument centers, in part, on the notion that poor people living in high-poverty neighborhoods tend to have more localized social networks, that is, ties with people of similar low social status. These kinds of social networks might be useful for social support, but do little to connect poor people to resources and opportunities that would enable them to find employment and secure other types of economic benefits (Campbell et al. 1986). The policy rationale for dispersing poor people is that their new higher-income neighbors would provide them with the "right" connections, and would act as role models to help alter individual behaviors in ways that would encourage self-discipline and self-sufficiency (Chaskin and Joseph 2015). This approach to solving poverty has been criticized for its focus on individual "failings," with less attention paid to economic and social problems including, for instance, a lack of decent-paying job opportunities for low-skilled workers, and discrimination in the labor market (Wilson 1987). In addition, dispersed poor people use their neighbors less frequently for jobs than might otherwise be expected (Kleit 2001).

In the context of U.S. suburbanization, important policies address the dispersal of subsidized housing and, more specifically, the Housing Choice Voucher (HCV) program, which provides tenant-based subsidies in the form of vouchers for low-income households to rent in the private housing market (see the chapter authored by Andrew J. Greenlee). Recent studies suggest that HCV recipients are suburbanizing, albeit at a slow pace. According to Covington et al. (2011), in 2008 close to half of HCV recipients lived in the suburbs, an increase of 2.1 percent since 2000. They note that the metropolitan areas in the West of the United States experienced the largest increases in suburban poverty between 2000 and 2008, and also experienced the largest increase in the number of HVC recipients over this time period. The suburbanization of poverty has been attributed, in part, to policies that favor low-income housing dispersal (Kneebone and Berube 2013). It is important to note, however, that HCV recipients are more likely than the poor in general and the overall population to move to low-income suburbs where there is a lack of access to jobs (Covington et al. 2011). While there may be movement to the suburbs, it is not to prosperous suburban communities. Scholars have found that while vouchers do help reduce the housing cost burden of low-income families, these families have typically not used the vouchers to move to demonstratively better places (McClure 2010; see also the chapter authored by Kirk McClure).

This phenomenon is attributed, in part, to a lack of affordable units in desirable neighborhoods (Varady and Walker 2003). Also, the mobility of voucher holders is restricted by the availability of rental units, and obstacles to mobility disproportionately impact minority voucher holders (Basolo and Nguyen 2005). There are significant racial barriers for poor African Americans in particular to move into non-Hispanic White middle-class neighborhoods, especially those in the suburbs. In a recent article in *The New York Times*, Eligon (2015) describes racial barriers as a significant problem in the St. Louis metropolitan region, location of Ferguson, the suburb that experienced significant social unrest after the fatal shooting of Michael Brown, an 18-year-old Black man, by Darren Wilson, a non-Hispanic White police officer. Considering the effects of racial and ethnic segregation in the St. Louis region, Eligon (2015, n.p) explains that the east side of Ferguson—where Michael Brown lived—is the location of "a cluster of apartments that stretches into the census tract with the most

Section 8 renters in Missouri."Thus, Black voucher holders are unable to use their Housing Choice Vouchers to live anywhere they want in the region. They are largely confined to low-income neighborhoods in both the city and suburbs. Without addressing underlying structural racism and planning policies like exclusionary zoning, the re-concentration of Blacks in low-income suburban neighborhoods appears an inevitable outcome of contemporary poverty dispersal housing policy in the U.S.

Conclusion

During the postwar period, state-initiated housing policy was central to a widespread movement of people to the periphery in both France and the United States. In the case of France, the state actively produced collective social housing estates for families impacted by the devastation of the war, and to house workers during a period of industrial growth. The 1970s marked a shift away from social housing toward the development of new towns and the subsidization of single-family housing. The French middle class abandoned collective housing estates that became increasingly occupied by working and nonworking poor and immigrant groups. In the case of the United States, state production of social housing was focused in the city rather than suburbs. Aided by government-insured mortgages and real estate practices of the postwar period, the White working and middle classes left the city for the suburbs, buying single-family homes.

Sweeping changes in the global economy have greatly impacted the suburban periphery of both France and the United States during the post-World War II period. Beginning in the 1970s, the change from large-scale manufacturing to a more service-based economy in the U.S. significantly altered the lives of suburban and urban working-class families, leaving many behind as the new economy took shape. At the same time, the French suburban periphery has become a magnet for immigrants.

There are major differences between French and U.S. approaches to suburban housing issues. In France, the focus has been on the redevelopment of social housing in the *banlieues* to introduce middle-income residents by providing market-rate rental housing and ownership options. In the U.S., efforts have focused on (1) encouraging the spatial mobility of poor people, often to the suburbs, through the use of housing vouchers that can be used in the private rental market, and (2) encouraging middle-income people to move to mixed-income and mixed tenure developments in the central city (i.e., the HOPE VI program).

In the context of French suburbs, the territorial stigmatization of the French *banlieues* has greatly influenced the desire for a housing policy focused on social mixing, particularly in redeveloped social housing projects. An underlying assumption of this effort is that the problems of the poor people in the *banlieues* can be solved by improving the space and enticing middle-class people to renewed developments. In the U.S., suburbs are perceived as sites of success, and there is a belief that dispersing poor families to suburban places will help solve the social and economic problems associated with poverty. In many respects, poverty dispersal policy aimed at connecting people to employment and reducing segregation has proved challenging.

This focus on social mixing in France and America is in our minds problematic as a strategy to improve the lives of poor people. De Filippis (2013, 70) states that "poverty is not a function of the spatial configuration of poverty." He (2013, 70) goes on to say:

> The concentration of poor people in geographic space may well make the experience of poverty worse for some, and it certainly seems to lead to unjust outcomes such as poor quality services, schools, and amenities. But the root causes of poverty within capitalism are the organization of labor markets (and the structure and character of the demand for labor in those markets) and the centrality of wage labor to household income and wealth.

We agree with De Filippis and argue that efforts by the French and U.S. governments to solve the problems of poor people by moving them or others around are misguided. Dealing with structural problems such as unemployment and, we would add, racial and ethnic discrimination would better serve poor people wherever they live.

References

Anacker, Katrin B., Ed. 2015. *The New American Suburb: Poverty, Race and the Economic Crisis*. Farnham: Ashgate.

Arthurson, Katy. 2013. "Mixed Tenure Communities and the Effects on Neighbourhood Reputation and Stigma: Residents' Experiences from Within." *Cities*, 35: 432–438.

Bacqué, Marie-Hélène, Yankel Fijalkow, Lydie Launay, and Stéphanie Vermeersch. 2011. "Social Mix Policies in Paris: Discourses, Policies and Social Effects." *International Journal of Urban and Regional Research*, 35(2): 256–273.

Basolo, Victoria, and Mai Thi Nguyen. 2005. "Does Mobility Matter? The Neighborhood Conditions of Housing Voucher Holders by Race and Ethnicity." *Housing Policy Debate*, 16(3–4): 297–324.

Bolt, Gideon, and Ronald van Kempen. 2013. "Introduction Special Issue: Mixing Neighbourhoods: Success or Failure?" *Cities*, 35(December): 391–396.

Campbell, Karen E., Peter V. Marsden, and Jeanne S. Hurlbert. 1986. "Social Resources and Socioeconomic Status." *Social Networks*, 8: 97–117.

Castells, Manuel. 1977. *The Urban Question*. London: Edward Arnold.

Chaskin, Robert J., and Mark Joseph. 2015. *Integrating the Inner City: The Promise and Perils of Mixed-Income Public Housing Transformation*. Chicago: University of Chicago Press.

Covington, Kenya, Lance Freeman, and Michael Stoll. 2011. *The Suburbanization of Housing Choice Voucher Recipients*. Washington, DC: Brookings Institution.

Cupers, Kenny. 2014. *The Social Project: Housing Postwar France*. Minneapolis: University of Minnesota Press.

De Filippis, James. 2013. "On Spatial Solutions to Social Problems." *Cityscape: A Journal of Policy Development and Research*, 15(2): 69–72.

Denton, Nancy A., and Joseph R. Gibbons. 2013. "Twenty-First Century Suburban Demography: Increasing Diversity yet Lingering Exclusion." In *Social Justice in Diverse Suburbs: History, Politics and Prospects*, edited by Christopher Niedt, 13–30. Philadelphia: Temple University Press.

Dikeç, Mustafa. 2007. *Badlands of the Republic: Space, Politics and Urban Policy*. Malden, MA: Blackwell Publishing.

Dikeç, Mustafa. 2013. "Immigrants, *Banlieues*, and Dangerous Things: Ideology as an Aesthetic Affair." *Antipode*, 45(1): 23–42.

Eligon, John. 2015. "A Year after Ferguson, Housing Segregation Defies Tool to Erase it." *The New York Times*, August 8.

Enright, Theresa Erin. 2014. "Illuminating the Path to Grand Pari(s): Architecture and Urban Transformation in an Era of Neoliberalization." *Antipode*, 46(2): 382–403.

Galster, George. 2007. "Should Policy Makers Strive for Neighborhood Social Mix? An Analysis of the Western European Evidence Base." *Housing Studies*, 22(4): 523–545.

Garbin, David, and Gareth Millington. 2012. "Territorial Stigma and the Politics of Resistance in a Parisian *Banlieue*: La Courneuve and Beyond." *Urban Studies*, 49(10): 2067–2083.

Gilbert, Pierre. 2009. "Social Stakes of Urban Renewal: Recent French Housing Policy." *Building Research and Information*, 37(5–6): 638–648.

Gobillon, Laurent, and David Le Blanc. 2008. "Economic Effects of Upfront Subsidies to Ownership: The Case of the Prêt à Taux Zéro in France." *Journal of Housing Economics*, 17(1): 1–33.

Goetz, Edward. 2003. *Clearing the Way: Deconcentrating the Poor in America*. Lanham, MD: Rowan and Littlefield.

Hanlon, Bernadette, Short, John Rennie, and Thomas J. Vicino. 2009. *Cities and Suburbs: New Metropolitan Realities in the US*. New York: Routledge.

Hanlon, Bernadette, Thomas Vicino, and John Rennie Short. 2006. "The New Metropolitan Reality in the US: Rethinking the Traditional Model." *Urban Studies*, 43(12): 2129–2143.

Hargreaves, Alec G. 1996. "A Deviant Construction: The French Media and the 'Banlieues'." *Journal of Ethnic and Migration Studies*, 22(4): 607–618.

Hayden, Dolores. 2004. *Building Suburbia: Green Fields and Urban Growth, 1820–2000*. New Nork: Random House Vintage.

Imbroscio, David. 2008. "'[U]Nited and Actuated by Some Common Impulse of Passion' 1: Challenging the Dispersal Consensus in American Housing Policy Research." *Journal of Urban Affairs*, 30(2): 111–130.

Jackson, Kenneth. T. 1987. *Crabgrass Frontier: The Suburbanization of the United States*. Oxford: Oxford University Press.

Jobard, Fabien. 2009. "An Overview of French Riots: 1981–2004." In *Rioting in the UK and France. A Comparative Analysis*, edited by Dave Waddington, Fabien Jobard, and Mike King, 27–38. London, Willan Publishing.

Kipfer, Stefan. 2016. "Neocolonial Urbanism? *La Rénovation Urbaine* in Paris." *Antipode*, 48(3): 603–625.

Kleit, Rachel. G. 2001. "The Role of Neighborhood Social Networks in Scattered Site Public Housing Residents' Search for Jobs." *Housing Policy Debate*, 12(3): 541–573.

Kneebone, Elizabeth, and Alan Berube. 2013. *Confronting Suburban Poverty in America*. Washington, DC: Brookings Institution Press.

Kruse, Kevin M., and Thomas J. Sugrue, eds. 2006. *The New Suburban History*. Chicago: University of Chicago Press.

Lagrange, Hugues. 2006. "Autopsie d'une vague d'emeutes." In *Emeutes Urbaines et Protestations: Une Singularité Francaise*, edited by Hugues Lagrange and Marco Oberti, 37–58. Paris: Presses de Sciences-po.

Laurence, Jonathan, and Justin Vaisse. 2005. *Understanding Urban Riots in France*. Washington, DC: Brookings Institution.

Lee, Loretta, Tom Slater, and Elvin Wyly, eds. 2008. *Gentrification*. New York: Routledge.

Lévy-Vroelant, Claire. 2007. "Urban Renewal in France: What or Who is at Stake?" *Innovation*, 20(2): 109–118.

McClure, Kirk. 2010. "The Prospects for Guiding Housing Choice Voucher Households to High-Opportunity Neighborhoods." *Cityscape: A Journal of Policy Development and Research*, 12(3): 101–122.

Mikelbank, Brian A. 2004. "A Typology of U.S. Suburban Places." *Housing Policy Debate*, 15(4): 935–964.

Newman, Andrew. 2015. *Landscape of Discontent: Urban Sustainability in Immigrant Paris*. Minneapolis: University of Minnesota Press.

Nijman, Jan. 2015. "North American Suburbia in Flux." *Environment and Planning A*, 47(1): 3–9.

OECD. 2015. *OECD Territorial Reviews: Valle de México, Mexico*. Paris: OECD.

Packer, George. 2015. "The Other France: Are the Suburbs of Paris Incubators of Terrorism?: *The New Yorker*, August 31.

Pendall, Rolf. 2000. "Local Land Use Regulation and the Chain of Exclusion." *Journal of the American Planning Association*, 66(2): 125–142.

Prost, Antoine. 1991. "Public and Private Spheres in France." In *History of Private Life Volume V: Riddles of Identity in Modern Times*, edited by Antoine Prost and Gérard Vincent, 1–103. Cambridge, MA: Harvard University Press.

Rudolph, Nicole C. 2015. *At Home in Postwar France: Modern Mass Housing and the Right to Comfort*. New York: Berghahn Books.

Short, John Rennie. 2006. *Alabaster Cities: Urban US since 1950*. Syracuse, NY: Syracuse University Press.

Tuppen, John. 1979. "New Towns in the Paris Region: An Appraisal." *Town Planning Review*, 50(1): 55–70.

Varady, David P., and Carole C. Walker. 2003. "Housing Vouchers and Residential Mobility." *Journal of Planning Literature*, 18(1): 17–30.

Wacquant, Loic J. D. 2008. *Urban Outcasts: A Comparative Sociology of Advanced Marginality*. Cambridge, UK: Polity Press.

Wacquant, Loic J. D. 1993. "Urban Outcasts: Stigma and Division in the Black American Ghetto and the French Urban Periphery." *International Journal of Urban and Regional Research*, 17(3): 366–383.

Wacquant, Lois J. D., Tom Slater, and Virgilio Borges Pereira. 2014. "Territorial Stigmatization in Action." *Environment and Planning A*, 46(6): 1270–1280.

Wilson, William Julius. 1987. *The Truly Disadvantaged: The Inner City, the Underclass and Public Policy*. Chicago: University of Chicago Press.

26

A NEW GENERATION OF "SINGLE-FAMILY" HOMES

Multigenerational Homebuilding in the Suburbs of Phoenix, Arizona

Willow S. Lung-Amam

Introduction

In 2007, Taylor Morrison, a national homebuilder with properties in nine states across the U.S., was set to break ground on a 304-unit condominium project in Sunnyvale, a small suburb in California's Silicon Valley. However, rapidly declining home sales and prices in California and across the nation associated with the Great Recession forced the homebuilder to rethink how it was going to appeal to an ever-shrinking home buying market.[1] Four years later, Taylor Morrison came back with a new plan (Glantz 2011). Instead of building condominiums aimed at young professionals, it would build townhouses for multigenerational families, which were largely meant to appeal to the Valley's rapidly expanding market of new immigrant homebuyers, many of whom express distinct preferences for multigenerational living (Keene and Batson 2010). Nearly 3,000 miles away, Michael Cox and Aimee Norwood, a middle-aged couple, were about to sign a contract for a traditional single-family home in Aldie, Virginia, when Michael's father's health took a turn for the worse. The couple then changed their plans and instead signed a contract for a new home built by the Toll Brothers, a national homebuilder known for its high-end residential designs, with a separate suite for Michael's mother (Lerner 2015).

Such stories are not unique. Over the past 30 years, multigenerational living has become far more common in the U.S., particularly among Millennials hit hard by the Great Recession, baby boomers who are facing longer life expectancy and rising health care costs, and immigrants and racial and ethnic minorities for whom multigenerational living has long been commonplace (Fry and Passel 2014; Pfeiffer et al. 2016). But while various studies attest to the rising rate of multigenerational living in the U.S. and the factors that contribute to it, little is known about how the building industry is responding to these trends in places like Sunnyvale and Aldie.

This chapter investigates multigenerational homebuilding trends over the past several decades and how these homes are accommodating the diversity of multigenerational households. Utilizing real estate and homebuilding industry publications, developer data, and local, regional, and national newspaper articles, it shows that multigenerational building has been on the rise in the past several decades, particularly in the suburbs of the Sunbelt South and West, and after the Great Recession. I argue that these trends represent a remarkable shift in the design of suburban, single-family homes. As the birthplace of the first national multigenerational home model, the suburbs of Phoenix, Arizona, have proven to be an important testing ground for the multigenerational homebuilding industry. There, new home data and interviews with developers, real estate agents, and architects show the unique design qualities of these homes and why they have become popular among homebuyers and builders. However, they also reveal a host of design, planning, and policy challenges associated with homes and raise questions about developers' response to the wide-ranging needs and diversity of multigenerational households. The chapter concludes by addressing some of the challenges and opportunities that rising trends in multigenerational homebuilding represent for remaking suburbs into more sustainable, livable, and socially diverse and equitable places.

The Changing Form and Function of the Single-Family House

From the mid-nineteenth to the early twentieth century, the single-family home was widely considered the exclusive domain of the White, upper-class Americans. While urban centers were typically crowded with row homes and tenements, single-family homes were the defining feature of well-manicured and planned suburbs, like Riverside, Illinois, which served as an elite retreat from the social and environmental ills of the industrial city (Fishman 1987). No less important, but less well-documented, single-family homes also defined the landscape of a host of working-class, minority, and immigrant suburbs that oftentimes lacked even the most basic sanitary and sewer services (Nicolaides and Wiese 2006).

The mid-twentieth century, however, forever reshaped the American single-family home. The growing popularity of the Garden City movement in the early part of the century placed single-family homes at the center of an idealized neighborhood unit. This ideal was taken up and promoted by an increasingly influential group of large-scale developers, known as "community builders" (Weiss 2002). Aided by new building technologies and the federal government, which underwrote long-term Federal Housing Administration and Veteran Administration home loans to new and largely White homebuyers, these developers helped to drive an unprecedented suburban building boom in the post-World War II period (Jackson 1985). In iconic postwar suburbs, such as Levittown, New York, a more standardized version of the single-family home came within reach of millions of Americans, growing the middle class and a particular version of the American Dream (Jackson 1985; Kelly 1993). In new suburbs, the single-family home became the "modern minimum" of American middle-class identity, associated as it was with the nuclear family, privacy, consumption, and racial and ethnic exclusion (Archer 2005; Harris 2006).

Since the 1970s, however, while metropolitan areas have become more defined by political fragmentation and segregated development patterns (Soja 2000), and suburbs more socially diverse (Hanlon et al. 2009), single-family homes have also taken on new forms. Nationwide, they have expanded in size, producing vast tracts of what critics derisively refer to as McMansions (Duany et al. 2001). At the same time, they have become the focus of new design movements that are attempting to make suburbia more socially and environmentally sustainable. New Urbanists' and others' attempts to retrofit and rethink suburbs have helped to generate renewed interest in alley dwellings, garage apartments, in-law suites, and other design ideas that are catering to more varied uses, forms, and functions of the single-family home (Duany et al. 2001; Dunham-Jones and Williamson 2011). Among these is the need to accommodate an increasingly large number of multigenerational families.

The Rise of Multigenerational Living in the U.S.

Multigenerational living has been a significant feature of American households for much of the nation's history. In the early nineteenth century, about 70 percent of persons 65 years or older lived with their children or children-in-law, largely on rural family farms (Ruggles 2003). Among various other advantages, multigenerational living provided economic security that compensated for a weak public sector safety net. However, with the rise of wage labor and industrialization and the decline of agriculture and family-based professions at the turn of the century, many left their family farms for cities, and the economic advantages of multigenerational living waned. The strengthening of the social safety net with programs like Social Security and Medicare and the economic boon of the mid-twentieth century further diminished these advantages (Ruggles 2003). Between 1940 and 1980, the proportion of Americans living in multigenerational households declined by more than a half from 25 to 12 percent (Taylor et al. 2010). Starting in the 1990s, however, the tide began to shift. By 2000, the share of the American population living in multigenerational households climbed back to its 1960s levels, around 15 percent, before growing again to 16 percent by 2010 (Taylor et al. 2010).

Scholars have pointed to at least five main causes of the increasing prevalence of multigenerational households over the past several decades (Generations United 2011). First, the structure of the American family has changed. The number of unmarried adults has increased, and median age of first marriages has risen about six years between 1960 and 2010, to 29 for men and 27 for women (Taylor et al. 2010). This trend, coupled with later childbearing, lower fertility rates among women, and the rise in cohabitation and childbearing outside of marriage, has pushed alternative household structures and young adults to live with their parents longer (Bianchi et al. forthcoming; Weaver 2013).

Americans' increasing life expectancy and the rising costs associated with healthcare, childcare, and other household expenses have also contributed to increasing rates of multigenerational living. An increasing number of Americans have found themselves living with chronic health conditions and disabilities that require long-term care. As costs have risen, adult children have begun to play a more important role in providing this health care, especially in lower-income households (Bengtson 2001; Keene and Batson 2010). Young adults, on the other hand, have faced rising childcare costs and greater workplace demands and are more likely than their parents to be single parents who rely on other family members to take on financial, physical, and emotional responsibilities in the home (Weaver 2013).

Immigration, especially from Latin America and Asia, and America's increasing racial and ethnic diversity have also contributed to the rise in multigenerational living. Between 1980 and 2012, the share of immigrant-headed households in the U.S. doubled from 7 to 14 percent (Fry and Passel 2014). Both economic necessity and cultural values drive immigrants' higher rates of multigenerational households compared to native-born households (Keene and Batson 2010). In 2009, 16 percent of immigrant-headed households lived multigenerationally, compared with only 10 percent of native-born households (Fry and Passel 2014). Regardless of nationality, however, racial and ethnic minorities are far more likely than non-Hispanic Whites to live in multigenerational households. In 2012, roughly 27 percent of Asian Americans, 25 percent of African Americans, and 24 percent of Latinos lived in multigenerational households, compared to only 14 percent of non-Hispanic Whites (Fry and Passel 2014). Both Asian Americans' and Latinos' high rates of multigenerational households are typically associated with cultures of elder care, language barriers, and a lack of resources among some subgroups. Immigrants' uncertain legal status can also play a contributing role (Pfeiffer et al. 2016). African Americans' comparably high unemployment rates and proportion of single-parent households, low wages, and strong extended family networks contribute to their high rates of multigenerational living (Kamo 2000; Swartz 2009).

Finally, and perhaps most significantly, has been the impact of the Great Recession (2007–2009), which resulted in the highest unemployment rates since the mid-1980s (U.S. Bureau of Labor Statistics 2012). Between 2007 and 2009, the number of Americans living in multigenerational households grew to 51.4 million people, from 15.4 to 16.7 percent of the U.S. population (Taylor et al. 2011). Unable to locate or keep a job, many young adults moved back in with their parents (Pew Research Center 2010). According to the U.S. Bureau of the Census, between 2005 and 2011, the number of adults between 18 and 24 years of age living with their parents rose by about 5 percent to nearly 55 percent of young adults. Surprisingly, despite an improved labor market in recent years, multigenerational living among young adults has continued to increase. While unemployment rates dropped nearly five points between 2010 and 2015 for 18- to 34-year-olds, the percent of those living independently also decreased by 2 percent during the same period (Fry 2015). Commonly referred to as the "boomerang" or "sling" generation, young adults are far more likely than previous generations to live with their parents, in large part due to financial constraints (Parker 2012). Somewhat surprisingly, for seniors, multigenerational household formation and economic outcomes did not change much during the Great Recession (Pfeiffer et al. 2016).

While much research has examined how the changing lifestyles and circumstances of American households have given rise to the growth in multigenerational living, little has been said about how the iconic symbol of the American nuclear family—the single-family home—has been reshaped by these trends.

Methods

Tracing the building industry's response to rising rates of multigenerational living is challenging for a number of reasons. Perhaps most significantly, there is no standard definition of a multigenerational home. For some, it may be a house with two master suites. For others, it may be a house with a separate unit within, or detached from, the main home. Further, while the U.S. Bureau of the Census tracks multigenerational *households*, there is no database that tracks the development of multigenerational *housing*. In 2013, the National Association of REALTORS® (NAR) collected data on purchasers of multigenerational homes as defined by "a home that will house more than you and children under the age of 18."[2] Its report showed that in 2015, approximately 14 percent of all U.S. home purchases were made by multigenerational households. Among these, the largest share was made by those 49 to 58 years of age and was related to reasons of cost savings and children moving back into the home (National Association of REALTORS 2015). While useful, however, NAR's *Profile of Home Buyers and Sellers* says more about homebuyers and their intended use than about the form of multigenerational homes.

Because multigenerational homes are not identified by any particular housing typology or dataset, this analysis of multigenerational home development trends relied on a multi-step process that first identified homebuilders that marketed homes as multigenerational. To do so, my research assistants and I first looked for publicly available reports or data related to multigenerational homebuilding in real estate and homebuilding industry publications, such as those from the National Association of REALTORS® and National Association of Home Builders. NAR's *Profile of Home Buyers and Sellers* showed that the real estate and homebuilders industries were paying attention to trends in multigenerational living, but offered little data on multigenerational homes. We then queried local, regional, and national newspaper databases dating back to 1985 using the search terms "multigenerational" and "home."[3] This produced approximately 1,000 articles that we read to locate references to residential homes or home developments as being built for or marketed to multigenerational households and the developers of these homes. This review uncovered 34 independent, local, regional, and national developers involved in multigenerational homebuilding. To understand the scope of their efforts, we then investigated the lines of multigenerational homes built by regional and national developers. In places where these homes were being sold, we then tracked data on their locations and the number and price ranges of all homes within them.

The analysis revealed particular hot spots of multigenerational homebuilding in the suburbs of states in the Sunbelt South and Western U.S. To understand how the building trends were playing out on the ground, I conducted site visits of residential communities with large numbers of multigenerational homes in the Phoenix suburbs, which are considered by many developers as the birthplace of the multigenerational homebuilding trend. Lennar, a national homebuilder, launched its first multigenerational home model in the Phoenix suburbs, and other major homebuilders have since introduced their own multigenerational homes there as well. There, I toured several multigenerational homes and conducted in-depth interviews with nine developers, real estate agents, and architects about the design of these homes and their sales, use, and influence on the home buying market in Arizona and elsewhere. The interviewees were all key stakeholders who were selected based on their expertise in the multigenerational home marketplace. All interviews were transcribed and inductively and deductively coded to draw out important themes about the homes' development, design, use, and users.[4]

Multigenerational Homebuilding Trends in the U.S.

Multigenerational homebuilding appears to be a relatively recent endeavor. Our newspaper search of articles published between 1985 and 1999 found reference to only one development that was marketed to multigenerational households. This was an age-restricted master-planned community in Las Vegas, a type of housing that was booming at the time, which contained sections of the development for those 55 and older and others for families and younger couples (Heavens 1999).

Starting in 2000, however, reports of multigenerational home developments became more commonplace as local and independent developers began to experiment with the multigenerational home concept. In 2006, for instance, developers in the state of Washington marketed multigenerational homes that included both high-end homes and small planned unit developments (Bowden 2006).

By 2010, the multigenerational homes concept was gaining traction quickly among homebuilders. A significant milestone that marked the assent of the multigenerational home was Lennar's 2011 debut of its NextGen® home in the Phoenix suburbs. The NextGen® was a traditional single-family home that contained a separate attached unit, including a bedroom, a private and guest bathroom, a living room, laundry facilities, and a kitchenette. Five years later, NextGen® homes had been incorporated into more than 200 Lennar developments in 24 regional housing markets nationwide, comprising about 5 percent of Lennar's total home sales (Morrissey 2016).

Lennar's early successes caught the attention of other homebuilders that were struggling to find buyers in the late 2000s, when house prices were experiencing steep declines and foreclosure rates were increasing. This was particularly true in the Sunbelt South and West. California, Nevada, and Arizona were the three states that experienced the steepest house price declines in the late 2000s (Flanagan and Wilson 2013), and were among those in which, after 2010, developers constructed the most multigenerational homes. In Arizona, for instance, median home values dropped by nearly 48 percent between December 2006 and December 2011, compared to roughly 23 percent nationwide (Zillow n.d.). In 2011, Arizona's vacancy rate stood at 3.9 percent, compared to 2.4 percent nationally (U.S. Bureau of the Census 2011).

Following Lennar, Arizona-based developer Maracay Homes entered the multigenerational housing market by introducing the generation suite as a standard option in several of its signature Flex Design® homes in 2012. Like Lennar's NextGen®, the generation suite comes standard with a separate bedroom and living space, a guest bathroom, and an optional kitchenette. By 2016, the generation suite was available in 26 Maracay communities throughout the state. The New Home Company, a Southern California-based homebuilder, also launched its own multigenerational home line the same year. Its Life Space designs offer "compound estates, free-standing guest houses and private quarters for multi-generational families" (New Home Company 2012). By 2016, this option was available in nine of its developments, totaling over 550 units. Pardee Homes' launch of the GenSmart Suite® as one of its primary lines in communities throughout Nevada and California in 2013 further showed the increasing popularity of these homes among major regional developers. The GenSmart Suite® rebranded the basic multigenerational home design already used by several developers.

National homebuilders also adopted their own multigenerational home lines. The Toll Brothers, a homebuilder known for its high-end, luxury homes, began offering a Gen Suite option in select home lines and markets in 2012. By 2015, the option was available in all of its 50 markets in 19 states and accounted for roughly 5 percent of its sales (Morrissey 2016). Taylor Morrison sold multigenerational homes in communities in Arizona, Florida, and California, but as of mid-2016, had not yet introduced a standard multigenerational suite option in any of its home lines.

Since 2000, multigenerational homes have grown from an interesting idea to a proven, marketable concept. Combined, the three leading builders of multigenerational homes—Lennar, Maracay, and Pardee—have built multigenerational homes in more than 250 developments, 140 cities, and 14 states nationwide in only five years. When *The New York Times* first reported Taylor Morrison's entry into this growing marketplace in 2011, it did so under the headline "Multigenerational housing is a real estate growth niche" (Glantz 2011). Four years later, when Michael Cox and Aimee Norwood's story floated across a Saturday real estate section of *The Washington Post*, it did so under the headline "Multigenerational homes are in demand" (Lerner 2015). Taking a closer look at what has happened in the Phoenix suburbs during the past several years of the multigenerational home boom shows even more clearly their unique design qualities and why these homes have become popular among homebuyers and builders.

Multigenerational Homebuilding in the Phoenix Suburbs

On Black Friday 2011 in an IKEA parking lot in Tempe, Arizona, crowds gathered for the debut of the Evolution, Lennar's first NextGen® home model. Over the next four months, approximately 16,000 visitors walked through the home, which was then taken apart and donated to Habitat for Humanity (Thomas Watkins, interview with author, Gilbert, AZ, June 16, 2015). But while the house was deconstructed, the market for multigenerational homes was just beginning to take off. The event launched the NextGen® home line in Arizona.

The NextGen® was the brainchild of Howard Perlman, owner of the Perlman Design Group, an architecture firm based out of Las Vegas. In the mid-2000s, having watched the real estate market collapse and his own firm lay off about 200 people, Perlman was looking for new ideas to bolster the building industry and produce more work for his firm. Inspired by trends he had observed in his own neighborhood, Perlman began researching multigenerational living in the U.S. As he explained to me, he found himself both surprised by the numbers and intrigued about the possibilities for a lagging homebuilding industry:

> [H]omebuilders have had to use the same strategy for building since I can remember. Probably since Levittown, which is, they would get a subdivision; they'd build three models—a one-story, a small two-story, and a bigger two-story—all traditional family homes. I thought to myself, since a third of the population lives multigenerationally and almost a third work at home, that why shouldn't a model complex be a traditional home, a live-work home, and a multigenerational home?
>
> *(Howard Pearlman, telephone interview with author, June 17, 2015)*

Looking to inspire innovative home designs, Perlman created two models that he trademarked and shopped around to major homebuilders. After going down a list of about 40 that all rejected his ideas, he spoke with Alan Jones, President of Lennar's Phoenix branch. As Perlman explained, he had finally struck a chord:

> [Jones'] eyes glazed over as I was pitching him my idea and showing him the plan. He didn't say anything for 20 minutes, and at the end of that he said, "I like this idea and I [can] relate to [it] because my mother-in-law lives with us. [I] understand exactly what you are trying to do, and we're going to do this."
>
> *(Howard Pearlman, telephone interview with author, June 17, 2015)*

And "do this" they did. Five years after the NextGen® debut, Lennar was offering 14 multigenerational home models in 19 communities throughout the state, which made up about a third of the company's new home sales (Michael Dowell, interview with author, Gilbert, AZ, June 16, 2015).

Lennar's NextGen® homes include one- and two-story, four to six bedroom homes, all with private suites ranging from over 500 to 800 square feet. Lennar's original model, the Evolution, remains among its most popular (see Figure 26.1). It is a single-story home whose main structure contains up to four bedrooms and three bathrooms, a den, a great room, and a two-bay garage. Connected to the main structure by a dual access door is a 645-square-foot private suite, which has separate utilities, its own private entrance from the front porch, and a separate kitchenette, laundry area, living room, bathroom, bedroom, patio, and attached one-bay garage (see Figure 26.2). In accordance with local zoning ordinances, homeowners' association's covenants, codes, and restrictions, and master plan design regulations, many kitchenettes do not have full stoves because of provisions restricting the number of kitchens within single-family homes (Shin 2012). As Lennar's NextGen® Home Ambassador Tom Watkins and I walked through an Evolution model home, he explained that spatial separation was a key design concept. "It's completely divided, so you can completely live separate with each other but at the same time if you want to be together, of course, you can" (Thomas Watkins, interview with author, Gilbert, AZ, June 16, 2015).

In the Phoenix suburbs, two of the more popular developments in which these homes have been built are the Bridges, a 2,400-unit master-planned community in Gilbert, and Layton Lakes, an 832-acre project located in both Gilbert and the neighboring city of Chandler (see Figure 26.3). These master-planned communities feature amphitheaters, lakes, playgrounds, community parks, sports facilities, and walking trails. Both are located in cities that are among the fastest growing in the region. According to the U.S. Bureau of the Census, between 1980 and 2010, the town of Gilbert grew from just under 6,000 people to more than 200,000.

These suburbs' rapid population growth corresponded with a major building boom in the early 2000s. When the real estate market collapsed, they were also among the hardest hit. In the Phoenix-Mesa-Scottsdale Metro Area, building permits that had been steadily on the rise since 1995 dropped from a high of over 65,000 per year in 2004 to around 8,300 in 2010. Michael Dowell, Lennar Arizona's Senior Vice President of Marketing, credited the NextGen® with helping the division survive the harsh economic times (Michael Dowell, interview with author, Gilbert, AZ, June 16, 2015). As the Arizona division went from selling about 3,000 homes a year before the crash to about 400, Dowell explained that the NextGen® allowed families to pool their resources to meet the changing times. Many of the units were jointly purchased by a parent or parents and their adult children (Michael Dowell, interview with author, Gilbert, AZ, June 16, 2015). Maracay Homes' sales representative Parker Fairbanks added that the crises made multigenerational homes popular because they "tamed the market" into purchasing smaller homes that fit their basic needs (Parker Fairbanks, interview with author, Gilbert, AZ, June 15, 2015).

Whether the homes are billed as multigenerational or not, an increasing number of homebuilders in Gilbert and Chandler now offer flexible options for added privacy and internal separation, such as separate entrances, dual master suites, and custom doorways and walls. While walking me through Maracay's most popular model, the Cholla, Fairbanks explained that while the model did not include a generation suite option, its single-story layout, internal separations, and options for increased privacy make the home a popular choice for multigenerational households (Parker Fairbanks, interview with author, Gilbert, AZ, June 15, 2015). Catherine Kniss, a real estate agent whose primary market is Gilbert and Chandler, noted that the layouts and options common to multigenerational homes are increasingly common for all homes in Arizona—a trend that I observed while touring homes built by national homebuilders like Pulte Homes, Taylor Morrison, and Meritage (Catherine Kniss, interview with author, Gilbert, AZ, June 15, 2015). These trends are even more apparent among

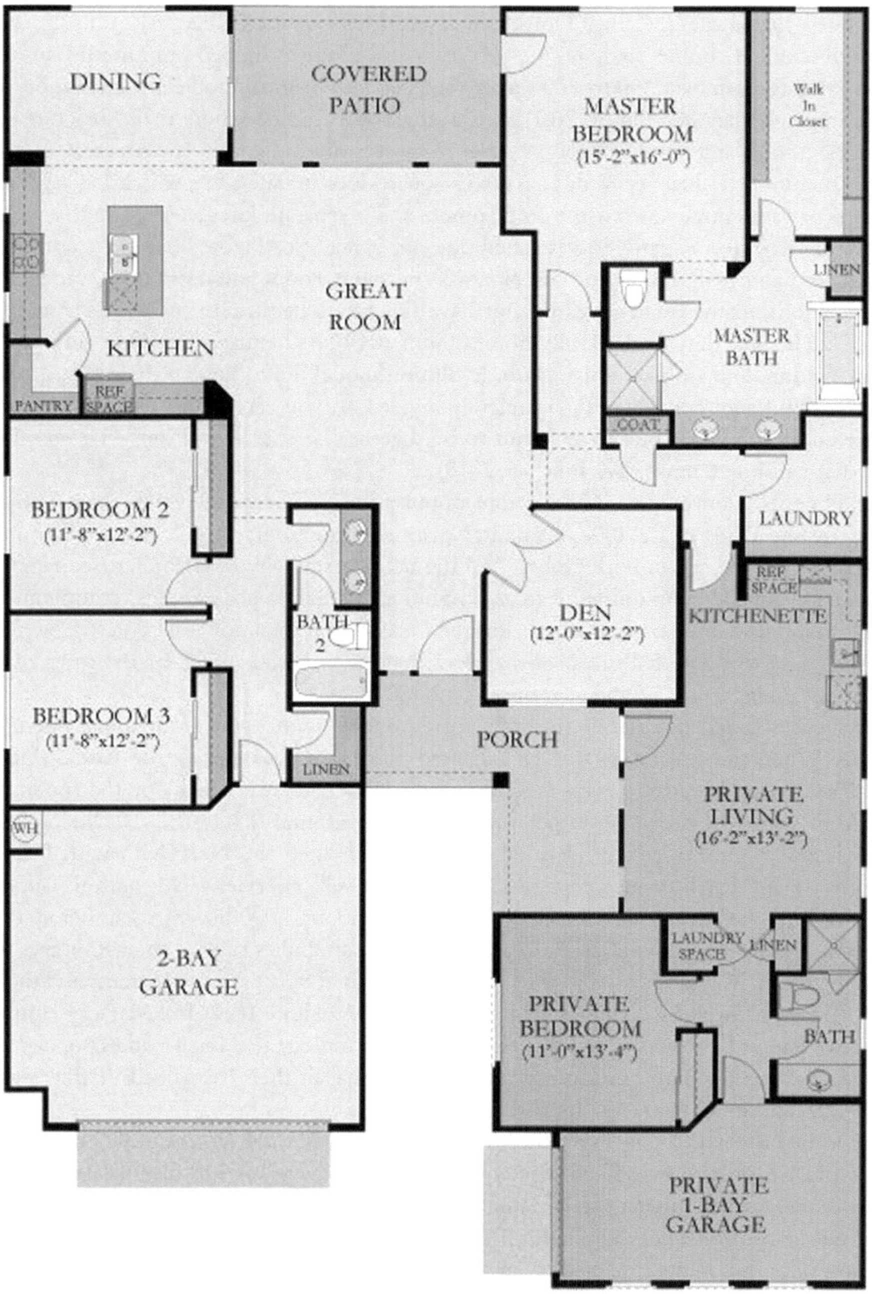

Figure 26.1 Lennar's NextGen® Evolution Floor Plan for the Layton Lakes community in Gilbert, AZ. The shaded area on the right indicates the separate suite that characterizes many multigenerational homes.

Source: Lennar Corporation. Used with permission.

Figure 26.2 Inside the Evolution NextGen® Suite in Layton Lakes, Gilbert, AZ.

Source: Lennar Corporation. Used with permission.

Figure 26.3 Layton Lakes, Gilbert, AZ, Where Multiple Developers Sell New Homes Marketed to Multigenerational Households.

Photo credit: author.

homebuilders that directly compete with Lennar in Arizona. Shea Homes, for instance, offers guest suites in many of its models, while Maracay Homes' generation suite has become a fierce regional competitor to NextGen® homes.

Another multigenerational home design is what are commonly referred to in Arizona as casitas or cabanas, which include both attached and detached units. Both Maracay and Shea Homes offer interior casita options (sometimes called guest suites) that have separate exterior entrances, kitchenettes, and bathrooms. Where local zoning, design restrictions, and homeowner's association regulations allow, detached buildings or accessory dwelling units (ADUs) are often connected to the main house via a patio (see Figure 26.4). Various municipal regulations around issues such as parking and setbacks affect the form and use of secondary uses and often severely restrict or prohibit their construction (Chapple et al. 2011). In the Phoenix metropolitan area, Pfeiffer (2015) identified three primary approaches used by suburban local governments to address ADUs, including fully embracing the units, limiting uses that preserved the single-family-home identity, and controlling or disallowing their construction. For example, the City of Gilbert allows ADUs within single-family residential districts, provided that they meet certain setbacks and other regulations, but does not restrict their occupancy. However, the City of Chandler allows ADUs, but does not permit them to be used for any dwelling purposes. According to Kniss, casitas originated in Arizona as detached units, but have increasingly been constructed as attached units as builders have become increasingly concerned with production costs after the foreclosure crises (Catherine Kniss, interview with author, Gilbert, AZ, June 15, 2015). The challenges of navigating the host of local regulations regarding ADUs have also likely played a role in their limited use as a multigenerational home design strategy.[5]

With a building trend as difficult to define as that of multigenerational homes, there is little reliable demographic data on who is living in these homes or why. When I posed this question to homebuilders and real estate agents in Arizona, they commonly responded with a range of different

Figure 26.4 In Chandler and Gilbert, Some Homes of the Artesian Ranch Development Include Detached Casitas that May Be Used to Accommodate Multigenerational Households.

Source: Arizona Multiple Listing Service. Used with permission.

buyer types, including work-at-home professionals who need a home office, young couples or empty nesters looking to rent out space to help pay the mortgage, and adults looking to accommodate other family members. However, they agreed that the bulk of homebuyers make use of their multigenerational unit to accommodate adult children returning from college, guests, aging parents, and disabled relatives. Most common among these homebuyers were households trying to accommodate older adults or adult children, who Mike Dowell estimates make up nearly 80 percent of NextGen® homebuyers in Arizona (Michael Dowell, interview with author, Gilbert, AZ, June 16, 2015).

Even more so than the single purpose that multigenerational homes may serve, most homebuilders and real estate agents with whom I spoke agreed that the multiple purposes that these homes serve over time are key to their popularity. Kniss described her typical buyer as a young couple trying to plan ahead for an uncertain future:

> They are selling their old home for reasons of space, family expanding, family changing, jobs changing. They're [thinking], "Our children are small right now, this is how we would like to use the space in the interim, and here is how we also see ourselves using the space for [our] parents as they are getting older."
>
> *(Catherine Kniss, interview with author, Gilbert, AZ, June 15, 2015)*

Multigenerational homes are adaptable to many households' changing needs and uses, including those that may be temporary. Arizona, for instance, has a number of "snowbirds"—residents, typically retirees, who live in the state only during the winter months. With multigenerational homes, spaces that might be occupied by snowbirds during the winter might be used as children's play spaces, home offices, guest suites, or temporary rental units during other seasons. Tom Watkins pointed out that this flexibility is particularly useful for seniors, who make up a significant portion of Arizona's population:

> You are senior and thinking, "Hey, I don't want to be shipped off somewhere, I'm going to plan ahead." You've got a space right over there, ready to go for a healthcare worker. But should you have that stroke or something happens that catches you off guard, a family member can rush right in, have a place, and stay with you. [You do] not feel [like] you are trying to figure it out.
>
> *(Thomas Watkins, interview with author, Gilbert, AZ, June 16, 2015)*

Multigenerational homes act as a kind of insurance to support seniors' desire to age-in-place and plan for an uncertain future.

These homes are also popular among some of Arizona's ethnic minority and religious groups. According to Kniss, this includes the region's large Mormon population, who traditionally have large and extended families and often require large homes that can provide separate spaces for common activities like homeschooling (Catherine Kniss, interview with author, Gilbert, AZ, June 15, 2015). Cultural preferences and needs for multigenerational housing may also affect the home buying decisions of immigrants, especially those from China and India, many of whom are employed in Chandler's robust high-tech sector (Lung-Amam 2013). According to the 2010 U.S. Census, the percentage of Asian Americans in Chandler grew from 4.3 percent in 2000 to 8.2 percent in 2010—the largest growth among any racial group in the city (Johnson 2011).

While demographics, the financial crisis, and changing life circumstances seem to play a part in the popularity of multigenerational homes in Arizona, Parker Fairbanks argued that the biggest driver of changes in the building industry is shifting cultural attitudes around multigenerational living: "All the builders are starting to get this demand. The people are asking for this stuff. The builders that were worth their salt came in and said, 'We have to satisfy this need in the market.'"

Multigenerational homes are not simply a necessity for households as they were only a few years ago, Fairbanks argued; today families are proudly choosing to live multigenerationally and planning accordingly. "This, in my opinion, is one of the biggest cultural shifts I have ever seen in the 20 years I have been [selling real estate]," he noted (Parker Fairbanks, interview with author, Gilbert, AZ, June 15, 2015).

Conclusion

Suburban, single-family homes in the U.S. have taken on different forms at different times to accommodate their many uses, users, and ideals. In recent decades, as the lifestyles, preferences, and needs of suburban Americans have changed, homebuilders have been challenged to adapt, especially during harsh economic times. The extent to which they have done so through the building of multigenerational homes remains a rich area for exploration. This chapter's initial exploration of this trend has been limited by a lack of standard definition of multigenerational homes, limited data to track them, and a case study within only one market area. As the building trend continues to take shape, additional research into different market areas can further show the scale of these practices as well as their prospects and possibilities for helping to make suburban communities more environmentally, economically, and socially sustainable and equitable.

This chapter has shown that the building industry has made major strides over the last three decades in adapting the form of single-family homes to meet the growing needs of multigenerational households. In the Phoenix suburbs, the clearest beneficiaries appear to be middle-class residents who are burdened by rising health care, childcare, home, and other costs, especially those who have been hard hit by the Great Recession. For many households, multigenerational homes appear to help weather hard times, be they associated with a stagnant economy, severe illness, or simply uncertainty. Some developers and real estate agents whom I spoke with argued that the benefits of these homes went even further, to include bringing families and communities together, increasing a household's quality of life, and promoting long-term residence and more stable neighborhoods. Such prospects should inspire further research into the possibilities of these homes in informing urban policy and planning solutions.

But while the developers, real estate agents, and architects whom I spoke with largely touted the benefits of multigenerational homes, there is also reason to question whether this trend is contributing to more sustainable suburban communities. For example, research has shown that multigenerational living can also impose economic costs on some residents, such as older residents who may bear the burden of the household expenses, and strain familial relationships over issues such as childcare or discipline (Oberlander et al. 2009; Pfeiffer et al. 2016).

Contemporary multigenerational home trends also appear to be contributing to urban sprawl. All of the subdivisions that I visited were located in relatively new, auto-dependent developments far from Phoenix's downtown. With the minimum square footage that many developers told me they wanted in order to build a multigenerational home, such sprawling development patterns, which have been shown to have myriad social and environment impacts, will likely continue (Duany et al. 2001). More multigenerational home designs and policies are needed to encourage compact development and infill strategies in already urbanized areas.

Yet another challenge is affordability and different cultural preferences for multigenerational living. Given the size and features of these homes, their cost is often out of reach of many working-class and lower-income families who make up a significant share of multigenerational households. Further, homebuilders have also developed relatively standard models that, as Howard Perlman described, largely appeal to "Americans who grew up watching 'Leave It to Beaver' and still want to live that way" (Howard Pearlman, telephone interview with author, June 17, 2015). However, research about the preferences and needs of racial and ethnic minorities and immigrants suggests that there are several features of these homes that may not work as well for these groups, such as their focus on privacy and separation (Crowe et al. 2014). Even if the benefits of multigenerational homes outweigh their costs, land use regulations will continue to challenge developers' desire and ability to build them. Local zoning and design regulations as well as private homeowners' restrictions adjudicate the very features that define multigenerational homes as such, like second kitchens, accessory dwelling units, and separate entrances and utilities (Shin 2012). In many communities, these requirements severely restrict, delay, or even derail the approval and building process. Over the next few years, the prospects and challenges associated with these homes will be debated in more and more communities around the nation. Indeed, the factors that spurred the rapid rise in multigenerational living in the first decade of the twenty-first century—from changing economic conditions and lifestyles, to growing life expectancy and immigration—do not appear to be changing anytime soon.

Notes

1 The author would like to thank Upendra Sapkota and Pranjali Rai for their generous research assistance.
2 Importantly, researchers have used various definitions of multigenerational households that have led to different estimates of their population. For a discussion of these different definitions see Generations United (2011) and Pfeiffer et al. (2016).
3 The term "multigenerational" is well-established in the academic literature in reference to multigenerational living trends and homes. The terms "multigenerational" and "housing" were also searched, but produced a substantive number of overlapping articles with "multigenerational" and "home." The earliest available year for articles through Lexis Nexus, a popular newspaper search engine, was 1985. The rise in rates of multigenerational living did not occur in the U.S. until the 1990s, and we found no mention of the trends in industry publications prior to 1985.
4 Inductive coding allows themes to emerge from the bottom-up about topics of interest, whereas deductive coding emphasizes a more top-down approach to test theories or a hypothesis about a topic. Interview transcripts were hand-coded using both approaches.
5 The single-family home appearance of many of the multigenerational homes and their common amenities such as secondary-unit garages have also likely helped to avoid common complaints about ADUs in other communities, especially with regard to parking (Chapple et al. 2011).

References

Archer, John. 2005. *Architecture and Suburbia: From English Villa to American Dream House, 1690–2000*. Minneapolis, MN: University of Minnesota Press.
Bengtson, Vern L. 2001. "Beyond the Nuclear Family: The Increasing Importance of Multigenerational Bonds." *Journal of Marriage and Family*, 63(1): 1–16.
Bianchi, Suzanne M., V. Joseph Hotz, Kathleen McGarry, and Judith A. Seltzer. Forthcoming. "Intergenerational Ties: Alternative Theories, Empirical Findings and Trends, and Remaining Challenges." In *Caring and Exchange Within and Across Generations*, edited by Alan Booth, Ann C. Crouter, Suzanne Bianchi, and Judith A. Seltzer. Washington, DC: Urban Institute.
Bowden, Marilyn. 2006. "Multigenerational Housing: The Family, Family, Family Home." *Naples Daily News*, September 24.
Chapple, Karen, Jake Wegman, Alison Nemirow, and Colin Dentel-Post. 2011. *Yes in My Backyard: Mobilizing the Market for Secondary Units*. Berkeley, CA: Center for Community Innovation.
Crowe, David, Paul Emrath, Rose Quint, Steve Melman, Ashok Chaluvadi, and Heather Taylor. 2014. *What Home Buyers Really Want: Ethnic Preferences*. Washington, DC: National Association of Home Builders.
Duany, Andres, Elizabeth Plater-Zyberk, and Jeff Speck. 2001. *Suburban Nation: The Rise of Sprawl and the Decline of the American Dream*. New York: North Point Press.
Dunham-Jones, Ellen, and June Williamson. 2011. *Retrofitting Suburbia: Urban Design Solutions for Redesigning Suburbs*. Hoboken, NJ: John Wiley & Sons.
Fishman, Robert. 1987. *Bourgeois Utopias: The Rise and Fall of Suburbia*. New York: Basic Books.
Flanagan, Christine, and Ellen Wilson. 2013. *Home Value and Homeownership Rates: Recession and Post-Recession Comparisons from 2007–2009 to 2010–2012*. Washington, DC: U.S. Bureau of the Census.

Fry, Richard. 2015. *More Millennials Living with Family Despite Improved Job Market*. Washington, DC: Pew Research Center.

Fry, Richard, and Jeffrey S. Passel. 2014. *In Post-Recession Era, Young Adults Drive Continuing Rise in Multi-Generational Living*. Washington, DC: Pew Research Center.

Generations United. 2011. *Family Matters: Multigenerational Families in a Volatile Economy*. Washington, DC: Generations United.

Glantz, Aaron. 2011. "Multigenerational Housing is a Real Estate Growth Niche." *The New York Times*, April 21.

Hanlon, Bernadette, John Rennie Short, and Thomas J. Vicino. 2009. *Cities and Suburbs: New Metropolitan Realities in the U.S.* New York: Routledge.

Harris, Dianne. 2006. "Race, Class and Privacy in the Ordinary Postwar House, 1945–60." In *Landscape and Race in the United States*, edited by Richard H. Schein, 127–156. New York: Routledge.

Heavens, Alan. J. 1999. "Housing for All Ages as Buyers Long for Simpler Times, Developers are Embracing a 'Multigenerational' Concept, with Homes for Older and Younger Families Together." *The Philadelphia Inquirer*, October 10.

Jackson, Kenneth T. 1985. *Crabgrass Frontier: The Suburbanization of the United States*. New York: Oxford University Press.

Johnson, Weldon B. 2011. "Chandler Proving to be Mecca for Asians in Arizona." *The Arizona Republic*, March 16.

Kamo, Yoshinori. 2000. "Racial and Ethnic Differences in Extended Family Households." *Sociological Perspectives*, 43(2): 211–229.

Keene, Jennifer R., and Christie D. Batson. 2010. "Under One Roof: A Review of Research on Intergenerational Coresidence and Multigenerational Households in the United States." *Sociology Compass*, 4(8): 642–657.

Kelly, Barbara M. 1993. *Expanding the American Dream: Building and Rebuilding Levittown*. New York: SUNY Press.

Lerner, Michele. 2015. "Demand Rises for Properties that Can House More than One Generation." *The Washington Post*, September 30.

Lung-Amam, Willow. 2013. "That 'Monster House' Is My Home: The Social and Cultural Politics of Design Reviews and Regulations." *Journal of Urban Design*, 18(2): 220–241.

Morrissey, Janet. 2016. "Multigenerational Homes that Fit Just Right." *The New York Times*, April 8.

National Association of Realtors. 2015. *2015 National Association of REALTORS® Home Buyer and Seller Generational Trends*. Washington, DC: National Association of REALTORS.

New Home Company. 2012. *The New Home Company Introduces "Life Space" Home Designs*. Aliso Viejo, CA: The New Home Company.

Nicolaides, Becky, and Andrew Wiese. 2006. *The Suburb Reader*. New York: Routledge.

Oberlander, Sarah E., Fatma M. Shebl, Laurence S. Magder, and Maureen M. Black. 2009. "Adolescent Mothers Leaving Multigenerational Households." *Journal of Clinical Child and Adolescent Psychology*, 38(1): 62–74.

Parker, Kim. 2012. *The Boomerang Generation: Feeling OK about Living with Mom and Dad*. Washington, DC: Pew Research Center.

Pew Research Center. 2010. *Millennials: Confident. Connected. Open to Change*. Washington, DC: Pew Research Center.

Pfeiffer, Deirdre. 2015. "Retrofitting Suburbia through Second Units: Lessons from the Phoenix Region." *Journal of Urbanism*, 8(3): 279–301.

Pfeiffer, Deirdre, Katrin B. Anacker, and Brooks Louton. 2016. *What Are the Effects of Doubling up on Retirement Income and Assets?* Chestnut Hill, MA: Center for Retirement Research at Boston College.

Ruggles, Steven. 2003. "Multigenerational Families in Nineteenth-Century America." *Continuity and Change*, 18(1): 139–165.

Shin, Stephanie H. 2012. *Planning for a Multi-Generational Future: Policies, Regulations, and Designs for Multi-Generational Housing in the United States*. Master's Thesis. Massachusetts Institute of Technology, Cambridge, MA.

Soja, Edward. 2000. *Postmetropolis: Critical Studies of Cities and Regions*. Oxford: Blackwell Publishers.

Swartz, Teresa T. 2009. "Intergenerational Family Relations in Adulthood: Patterns, Variations, and Implications in the Contemporary United States." *Annual Review of Sociology*, 35: 191–212.

Taylor, Paul, Rakesh Kochhar, D'Vera Cohn, Jeffrey S. Passel, Gabriel Velasco, Seth Motel, and Eileen Patten. 2011. *Fighting Poverty in a Tough Economy, Americans Move in with Their Relatives*. Washington, DC: Pew Research Center.

Taylor, Paul, Jeffrey Passel, Richard Fry, Richard Morin, Wendy Wang, Gabriel Velasco, and Daniel Dockterman. 2010. *The Return of the Multi-Generational Family Household*. Washington, DC: Pew Research Center.

U.S. Bureau of the Census. 2011. *American Community Survey: 5-Year Estimates*. Washington, DC: U.S. Bureau of the Census.

U.S. Bureau of Labor Statistics. 2012. *The Recession of 2007–2009*. Washington, DC: US Bureau Labor Statistics.

Weaver, Jessica Dixon. 2013. "Grandma in the White House: Legal Support for Intergenerational Caregiving." *Seton Hall Law Review*, 43(1): 1–74.

Weiss, Marc A. 2002. *The Rise of the Community Builders: The American Real Estate Industry and Urban Land Planning*. Washington, DC: Beard Books.

Zillow. n.d. *Arizona Home Prices and Values*. Seattle, WA: Zillow.

SECTION 9

The Future of Housing

27

ADDRESSING AFFORDABILITY CHALLENGES

The Role and Scope of Local Housing Plans

Ingrid Gould Ellen

Over the past several decades, renters across the United States have seen steadily rising rents, while their incomes have stagnated. As a result, the proportion of renter households paying more than 30 percent of their income on housing (a typical threshold used to measure cost burden) rose from 24 percent in 1960 to 49 percent in 2014. Meanwhile, the share of renters in the bottom quintile of incomes who pay more than 30 percent of their income on rent rose from 66 percent in 1960 to 83 percent in 2014 (Joint Center for Housing Studies 2015). The past few years have delivered an even crueler math, especially in high-cost, growing cities such as New York, where between 2005 and 2014, the median rent rose by 15 percent after controlling for inflation and the median income rose by less than 2 percent (NYU Furman Center 2016). As a result, households have found themselves with less disposable income left over for such critical needs as nutritious food, health care, and clothing.

Gyourko et al. (2013) dubbed high-cost, growing cities like New York as "superstar" cities.[1] On the demand side, an increasing number of people want to live in these cities, as a result of their labor market opportunities, amenities, and cultural vitality. Demand is rising especially among college-educated and high-income households who are willing and able to pay very high prices and rents. Consider that in 2015, the mean income of the 20 percent of households in New York City with the highest incomes was $252,200, up 50 percent from the mean income of this top group in 1990 of $168,400 (in 2015 dollars). Yet incomes at the low end of the city's distribution have failed to keep up. During this same time period, the mean income of the 20 percent of households with the lowest incomes actually fell in real terms, from $16,400 to $9,500.

On the supply side, these are cities in which it is extremely difficult to build because of both geographic and regulatory constraints. (In economic terms, the supply of housing is highly inelastic.) Regulatory constraints include restrictive zoning as well as complicated and prolonged approval processes. These regulations surely constrain supply; but many of these cities also have limited land and room for expansion.

The combination of robust demand and restricted supply has led to explosive increases in rents and house prices, crowding out lower-income households or forcing them to economize on space and pay higher shares of their incomes on rents. Affordability is the key challenge for housing policy in these high-cost cities, as residence in them appears to be increasingly becoming a luxury good, with lower-income households faced with a shrinking number of affordable places to live. For example, the income of in-movers to New York City rose significantly in real terms between 1990 and

2010, while the median income of out-movers actually declined (Lobo and Salvo 2013). Even in more affordable cities that have not seen the same overall growth in rents and house prices, there are neighborhoods that are experiencing these pressures.

Faced with these growing affordability challenges, many cities across the country, such as Boston, New York, and Seattle, have recently adopted local housing plans, which either establish new commitments or step up prior commitments to produce and preserve affordable housing. These plans are generally separate and distinct from mandated federal and state plans. Despite the prevalence of such local housing plans, there has been little examination of their purpose, scope, and effectiveness. What do they include? What goals do they set? What populations do they target? This chapter begins such an inquiry, drawing on recently released plans from ten selected cities, most of which are high-cost, growing cities, paying particular attention to the case of New York City. The chapter also draws on discussions with participants in a National Community of Practice facilitated by the NYU Furman Center and Abt Associates, which includes 14 leading public and private sector practitioners working to address affordability problems in cities around the country. While this group has greatly enhanced my understanding of the content and purpose of local housing plans, the arguments expressed here are my own.

Role of Local Governments in Housing Policy

Housing policy discussions and analyses tend to center on the federal government. A cursory review of the titles of the 788 articles published in the first 26 volumes of *Housing Policy Debate*, the leading housing policy journal in the United States, reveals that only 148, or less than a fifth, have focused on local housing policy. Despite this lack of attention, local governments play an increasingly critical role in housing policy. Indeed, the most important housing policy decisions in the United States are arguably made at the local level.

Before describing the role of local governments, it is worth stepping back to consider why an ample stock of high-quality, affordable housing is so important for cities. To begin with, affordable housing may be key to attracting and sustaining economic activity. Businesses need a diversity of workers in order to thrive and grow; and those workers in turn need a range of affordable housing options. (While workers can potentially commute from more affordable areas beyond city limits, long commutes are expensive and may not be sustainable.) Further, there is considerable research showing that firms in larger and denser cities with more diverse economic activity are more productive (Glaeser 2011; Kolko 2010; Quigley 1998), and an ample supply of housing, serving a range of incomes, is key to enabling that density and diversity. The economic diversity that affordable housing supports may help to attract workers and residents too, as many are drawn to cities for their diverse neighborhoods and cultural vitality.

Cities also have a strong fiscal interest in ensuring that their housing stock is well-maintained, though this will tend to be a less pressing motivation in high-cost, growing areas where affordability is the chief challenge. Because housing is fixed in space, its condition influences the value of neighboring properties. The existence of a dilapidated building with unkempt grounds may discourage other investment and reduce the value of surrounding homes. Conversely, an attractive, well-maintained building may boost confidence in the area and increase property values. Research has documented these spillovers in New York City, showing that investing in affordable housing can, in some instances, breathe new life into distressed communities and increase the value of surrounding properties, through repopulating streets, constructing attractive new buildings, and removing pockets of blight that can discourage other private investment (Schwartz et al. 2006).

In addition to these neighborhood externalities, higher-quality, affordable housing might enhance occupants' health and well-being, and support their human capital development (Collinson et al. 2016). Healthier and more educated residents could in turn benefit a city's economy by forming a more robust and productive workforce. Children may especially gain from living in stable, decent homes (Leventhal and Newman 2010). Recent research also highlights the long-run benefits that young children glean from growing up in lower-poverty neighborhoods (Chetty et al. 2016). Finally, city residents might support affordable housing for equity reasons as well.

In short, local governments have a strong interest in supporting an ample supply of affordable housing, but what should they do to support it? Basic public economics warns local governments against spending tax revenues on redistributive programs, as the mobility of both people and capital threatens to undermine any effort to redistribute income at the local level. While cities surely have some ability to engage in redistribution—especially large cities that offer unique concentrations of economic and cultural activity—local officials should avoid setting taxes that are far out of line with those of other cities (O'Flaherty 2005).

However, local governments can and do play a central role in housing policy even without spending large amounts of local tax dollars. First, together with local housing authorities, they are responsible for implementing federal and state programs, such as the Housing Choice Voucher program, the HOME Investments Partnership program, and the Community Development Block Grant program, and they have considerable discretion in deciding how to deploy and target these resources, particularly the block grants.

Second, local governments have many tools at their disposal that federal and state governments do not have (or do not use), such as enacting zoning ordinances that govern what type of buildings can be built in which locations, adopting inclusionary zoning programs to incentivize the inclusion of affordable housing in market-rate development (or to require such inclusion in the states where mandatory inclusionary zoning is allowed), establishing and enforcing building and housing codes, controlling the process of awarding building permits, and using property taxes to incentivize the creation or renovation of affordable housing.

Third, local governments are arguably better positioned than higher levels of government to understand their needs and market conditions and to tailor and combine programs and policies to achieve their locally defined goals. Local government leaders can weave together federal and state funding streams and complement them with local revenue sources, regulations, administrative reforms, and tax policies to build a broad-based and effective local housing strategy. Of course, local officials do not always act wisely or fairly. Still, it seems ever more critical for local officials to play a more central role as housing needs grow and the political environment in Washington makes any significant expansion of federal aid unlikely.

The Potential of Local Housing Plans

Given the complexity of funding streams and regulations, and the large number of policies that affect housing, comprehensive local housing plans can be enormously helpful for city leaders, at least in theory. Housing plans can provide a framework that local officials can use to draw together the many relevant policy strands into a single, coordinated strategy to tackle affordability. Housing plans can be stand-alone initiatives, launched by a mayor or a city council, or they can be a component of a mandated general or comprehensive plan.

In general, there is little research examining what makes housing plans effective. Varady (1996) is an exception, drawing lessons for U.S. cities from four exemplary plans in Great Britain. The article highlights the potential of good plans to engage the public and marshal resources. Community of Practice participants similarly endorsed the potential power of plans, but they expressed the importance of central leadership and commitment to garner resources and forge collaboration across multiple agencies. They argued that some mandated plans were viewed as "check-the-box" exercises that offered little in the way of strategic prioritization. Some research has questioned the degree to which mandated housing plans actually lead to more production (Lewis 2010). In a more recent analysis, Ramsey-Musolf (2016) finds that municipalities in compliance with California's housing plan requirement see greater low-income housing production but less overall housing production.

The Community of Practice participants stressed the need for different kinds of plans in different contexts. They agreed that aspirational plans that set out general goals can be highly useful but should be paired with more detailed plans that include more concrete actions and goals. The Community of Practice participants highlighted several key purposes of local housing plans.

Most fundamentally, they argued that housing plans allow local officials to lay out a set of over-arching objectives that they aim to accomplish through their plan, which can help to garner political support for housing initiatives and investments and allow city officials to make the case of why a healthy supply of high-quality, affordable housing is so important to a city. They can mobilize groups around a set of shared objectives and may even help to combat resistance in individual neighborhoods to specific affordable housing developments, as residents may better appreciate how the planned development will further the broader goals of the housing plan. Laying out a set of broad goals also grants policymakers some flexibility to make policy adjustments or even change course if shifting trends make alternative policies better suited to achieving them.

New York City's series of housing plans offers some examples for how plans can be used to establish a larger housing policy commitment. Ever since the release of Mayor Koch's Ten Year Capital Plan for Housing in 1986, every mayor in New York City has released a plan for housing, and these plans seem to have helped in garnering substantial resources for affordable housing investments. Over the past few decades, New York's spending on housing has dwarfed that of other large cities (Ellen and O'Flaherty 2013). The most recent plan, Mayor de Blasio's, lays out eight high-level "Guiding Principles," such as "Our housing policies must address the City's changing demographics and expand the range of those we serve," "Economic diversity must be a cornerstone of housing development," and "Our municipal tools and assets should be deployed more effectively" (City of New York 2014, 7). After articulating these principles, the plan then lays out four broad goals: (1) fostering diverse, livable neighborhoods; (2) preserving the affordability and quality of the existing housing stock; (3) building new affordable housing for all New Yorkers; and (4) promoting homeless, senior, supportive, and accessible housing. These are all general goals that most New Yorkers embrace.

Housing plans also typically include more specific plans for particular programs to achieve those goals, together with targeted objectives around those programs. Many recent city plans include such numerical goals, typically focused on production or preservation. Mayor de Blasio's plan aims to build or preserve 200,000 units of affordable housing over the next ten years, targeted to a range of specified affordability levels. In doing so, de Blasio follows in the footsteps of his predecessors who, since Mayor Koch, have also set a count of new or preserved affordable units as their administration's housing goal. Other cities typically set unit counts as goals as well. Mayor Ed Lee of San Francisco announced a plan in early 2014 to build or restore 30,000 affordable units by 2020 (City of San Francisco 2014). Mayor Martin Walsh of Boston released a plan in the fall of 2014 to build 53,000 new housing units in the city by 2030, targeted to a range of income levels.

The Community of Practice participants emphasized that such explicit, programmatic goals can help to ensure budgetary support for a plan over multiple years and encourage state legislatures to take needed actions. They might even provide political cover for public officials to take actions that are locally unpopular. Programmatic goals also help to provide transparency and accountability to the public. Both advocates and the media can track a city's progress toward meeting its goals and hold officials responsible for delivering on them.

Despite their popularity and potential benefits, focusing on a simple, numerical production or preservation goal has downsides as well. First, such a simple count of total units ignores important differences among them. It treats studio apartments the same as three-bedroom apartments, units affordable to middle-income households the same as those affordable to extremely low-income households, units located in the city's highest-poverty neighborhood the same as those located in the lowest-poverty neighborhood, and units that cost $500,000 to build the same as those that cost $100,000.

Second, a goal to produce or preserve a particular number of units is overly simplistic, as it implies that there is a quantifiable housing gap to address: Once we build 50,000 units, the need will be met. But it is clearly not so simple or static. More fundamentally, a single-minded focus on producing a certain number of units may lead cities to overlook potentially promising responses such as assistance to allow renters to avoid eviction, counseling to improve voucher lease-up rates, or efforts to repair public housing units.

Thus, while production count goals can be essential in tracking and monitoring plan implementation, officials might instead target the number of households assisted, which would encompass a broader set of initiatives. Further, they might try to adopt metrics to track progress on other goals, such as the share of voucher holders or other assisted households living in opportunity areas. Identifying metrics for some goals would require potentially controversial definitions. While we may all agree on how to count housing units, we might not agree on what constitutes a "diverse and livable neighborhood."

Finally, housing plans can also serve as useful management tools for city officials. Because housing efforts typically call for coordinated actions across multiple agencies, the roadmaps that plans provide can help to mobilize the city bureaucracy and align internal efforts. City officials can—and should—hold regular, cross-agency meetings to ensure that plan implementation is on track and to identify and address problem spots.

Of course, there is tremendous variation across local housing plans in their scope and their choice of goals. There is no one housing plan that is right for every city. Plans will differ depending on local market conditions, resources, the political environment, and other idiosyncratic factors, including whether the plan is initiated by the mayor, as in the case of New York City, or mandated by state legislation, as in California, for example. Plans often differ in their intended audiences, their level of detail, and the degree of accessibility (Healey 1996).

Cities need to refine goals to tailor them to individual circumstances. For example, stopping all displacement in some cities will not be viable, but promoting housing stability and helping residents find new affordable homes may be more achievable. Even after narrowing down to viable goals, city officials will still need to make difficult choices about which activities and goals to prioritize given limited resources. Should they target rental or owner-occupied housing? How much should they focus on preservation versus new production? Should they choose to build more units in lower-cost areas or fewer units in higher-cost areas? To what extent should they incorporate regional goals and initiatives? What income levels should they target? City officials should be explicit about their choices and clearly articulate their rationale for making them. A few of these trade-offs are worth discussing more fully.

Whom to Serve?

Perhaps the most fundamental question city officials must address in developing plans is whom they should aim to serve. Clearly, lower-income groups face greater needs than middle-income households. But local governments may be justified (and indeed more so than the federal government) in targeting at least some of their scarce subsidy dollars to the middle class, in order to bolster the city's economy (Ellen 2017). Middle-income workers fill critical administrative positions in private businesses, and they are often employed in jobs key to the city's day-to-day life such as teachers, police officers, and city employees. Middle-income workers educate our children, keep our streets safe, and take care of us in hospitals.

Further, middle-income households are facing growing affordability strains in high-cost, growing cities. Consider that between 2000 and 2012, the share of New York renters with incomes between

80 and 120 percent of area median income who paid more than 30 percent of their income on rent rose from 14 percent to 30 percent (NYU Furman Center 2013, 34). Rent burdens also rose for households earning between 120 and 200 percent of area median income. So cities may choose to provide housing assistance to middle-income households to address growing needs and promote longer-term economic development. And of course, cities can stretch their dollars further when focusing on middle-income households. Creating a unit affordable to a middle-income household requires less subsidy support than creating one affordable to a lower-income household.[2]

There are also good arguments for strategies to increase the overall supply of housing at all affordability levels (Jacobus 2016). Even luxury housing, although it may increase housing prices in the immediate vicinity, can help to alleviate affordability problems in the larger city through filtering, as it opens up the supply of older homes that the wealthy would have otherwise occupied (Hankinson 2018). To be sure, the filtering process will fail to deliver homes for low-income households who cannot afford to pay a rent that covers operating and maintenance costs, but more market-rate construction helps to alleviate general affordability pressures (Rosenthal 2014). Thus, growing cities should continue to push to increase the supply of housing and scrutinize their regulations, which can impede new development. As many economists argue, significant barriers to new construction will worsen affordability challenges for middle- and moderate-income households (Glaeser et al. 2005).

Demand-Side Versus Supply-Side Interventions

One of the most basic questions in housing policy is whether to provide tenant-based or place-based subsidies. When considering federal policy, most experts agree that tenant-based subsidies (also referred to as demand-side subsidies) are more flexible and cost-effective than supply-oriented programs that subsidize the production of housing.[3] But for local governments, long-term vouchers are extremely expensive and involve fairly substantial redistributive spending, which public economics textbooks warn could trigger middle-class flight. Further, in some housing markets, demand-side subsidies may not live up to their promise if the supply response to increased demand is limited or constrained. Still, local officials should consider adopting more modest demand-side subsidies, such as time-limited, eviction-prevention subsidies that help families stay in their homes when they face financial bumps in the road.

Identifying Dedicated Funding Streams

As noted earlier, one of the benefits of a housing plan is to ensure budgetary support over multiple years, even as political winds change. Among other benefits, this certainty can help both to draw in private developers and to allow nonprofit developers to build needed capacity. But city officials need to identify a source of funds. They have multiple choices, including one-time infusions (like housing bonds) or dedicated funding streams, such as real estate transfer taxes, city and state pension funds, or property tax incentive programs (Brooks 2016). There is likely to be political opposition to any new revenue generating mechanism, but identifying a funding source is essential.

Analysis of Selected Plans

To help illustrate the variation in local plans, and to examine the scope of their activities and how they have managed these trade-offs, I draw on an analysis of ten local housing plans that were released in 2014 or 2015.[4] In conducting that analysis, my colleagues and I aimed to select recently released plans from ten large cities that were facing affordability pressures. We also made sure we had some regional variation, though we ended up with disproportionate representation from the Western region. The ten include Boston, Chicago, Denver, Honolulu, Los Angeles, Minneapolis/St. Paul,

New York City, Raleigh, San Francisco, and Seattle. Although they are all large and generally high-cost cities, Table 27.1 shows that their characteristics and market conditions vary. New York City housed nearly 8.5 million people in 2014, while Raleigh, North Carolina housed just under 440,000. While all but one of the cities (Chicago) saw a growth in population between 2000 and 2014, their growth rates spanned from 6 percent in Los Angeles and New York City to a stunning 37 percent in Raleigh. The median income in these cities ranged from $48,000 in Chicago to $85,000 in San Francisco. And poverty rates varied, too—from 23 percent in Boston to under 10 percent in Honolulu County. (Note that the Honolulu and Minneapolis/St. Paul plans cover a county or a set of counties in the metropolitan area, not just the central cities, and the statistics in Table 27.1 describe conditions in the geographies covered by the plan.)

As for the housing market, the homeownership rate ranged from 31 percent in New York City to 70 percent in the Minneapolis/St. Paul metropolitan area. The high homeownership rate in the Minneapolis/St. Paul area is driven in part by the inclusion of suburban areas; if we restrict ourselves to cities, the city in the group with the highest homeownership rate was Raleigh, at 49 percent. (The homeownership rates in the individual cities of Minneapolis and St. Paul were 48 and 47 percent, respectively.) Finally, and perhaps most relevant, we see a large variation in housing vacancy rates. While the rental vacancy rate in 2014 was below 3 percent in Boston, San Francisco, and Seattle, it was above 6 percent in Chicago and Raleigh.

To provide a broad comparison of the plans for these ten cities, my colleagues and I developed a rubric to summarize their scopes and overall goals. Obviously, we had to make some judgment calls, but the rubric helped us to analyze them as consistently as possible across cities. For some plan features, we simply note if the plan includes the feature (e.g., did it identify revenue sources or set numerical targets). For programmatic efforts, we give a city's plan two stars if the plan dedicates significant attention to the effort and one star if it mentions the objective but does not give it significant or sustained attention. For example, a plan that identifies expanding the overall supply of housing as an important goal and includes recommendations that it identifies as designed to achieve it would get two stars under the "Expanding the overall supply of housing" category. By contrast, a plan that does not include a specific goal or discussion of expanding the overall supply of housing but includes

Table 27.1 City Characteristics, 2014.

City	*Total Population*	*Population Growth (2000–2014)*	*Homeownership Rate*	*Median Income ($2014)*	*Poverty Rate*	*Rental Vacancy Rate*
Boston	656,051	10.2%	34.9%	$56,902	22.6%	2.6%
Chicago	2,722,407	−6.4%	43.6%	$48,734	22.0%	7.3%
Denver	663,862	16.5%	48.1%	$54,941	15.7%	4.1%
Los Angeles	3,928,827	6.0%	36.7%	$50,544	22.4%	3.4%
Minneapolis-St. Paul[1]	3,495,176	13.3%	69.9%	$69,111	10.3%	3.7%
New York	8,491,079	5.7%	31.2%	$52,996	20.9%	3.4%
Oahu-Honolulu[2]	991,788	11.7%	54.4%	$74,634	9.7%	5.4%
Raleigh	439,884	37.2%	49.1%	$53,475	16.6%	6.3%
San Francisco	852,469	8.9%	37.6%	$85,070	12.0%	2.5%
Seattle	668,337	15.7%	45.5%	$70,975	14.4%	1.2%

Source: U.S. Bureau of the Census. n.d.

1 Based on the Metropolitan Statistical Area (MSA) for Minneapolis-St. Paul-Bloomington, MN-WI Metropolitan Area.
2 Based on Honolulu County, Hawaii.

(as one recommendation among many) a general recommendation to support higher density near transit stations would receive one star in this category. Two people reviewed each plan to ensure consistency. In reviewing the plans in the following text, I focus on the "two-star" cells, since the plans that only mention a specific activity may not truly incorporate or embrace it.

Table 27.2 highlights the characteristics of the plans. The first set of rows (Panel 1) shows the degree to which the ten different plans support activities related to the construction or preservation of subsidized housing. All ten of the plans include efforts to preserve the affordability of the existing subsidized stock, and nine of the ten aim to increase the number of dedicated, affordable rental units. While some plans do not clearly distinguish between new construction and rehabilitation goals, some lay out specific distributional targets. The New York City plan, for example, aims to preserve 120,000 units and build 80,000 new units.

As for targeted income levels, all the plans reference the need to serve a range of income levels. Some plans lay out specific numerical targets. Chicago, for example, aims virtually all of its planned 41,000 units (78 percent) at households earning less than 60 percent of area median income, and more than half (62 percent) are targeted to households earning less than 30 percent of area median income. Some plans make middle-income housing production a top priority, like Boston, which targets about 40 percent of its units to middle-income households (and another 42 percent to the market-rate, unassisted stock). Similarly, Los Angeles targets 48,000 of its planned 60,000 units to households earning above 120 percent of the area median income.

Panel 2 shows that a smaller share (though still a majority) emphasize the construction or preservation of housing more generally. Six of the plans prioritize efforts to expand the overall housing stock. The Seattle plan, for example, calls for 50,000 units over ten years, 60 percent of which would be market-rate units. Boston's plan calls for 53,000 units over 20 years, 42 percent of which would be market-rate units.

Six of the plans also dedicate attention to preserving the affordability of the naturally occurring affordable rental stock (the unsubsidized but still affordable stock). For instance, the Minneapolis/St. Paul plan estimates that nearly 60 percent of rental units affordable to households earning less than 50 percent area income in the region are unsubsidized rentals, and it states that helping to preserve these approximately 120,000 units is a critical part of the region's overall housing strategy. The New York City plan promises to develop new incentive programs to encourage landlords to agree to restrict rents in rapidly gentrifying areas.

As a whole, the plans dedicate considerable attention to increasing and preserving the supply of subsidized housing. As shown in Panel 3, they pay less attention to demand-side programs to supplement households' ability to pay for housing. While six of the ten include an effort to increase the homeownership rate through down payment assistance, low-interest mortgages, or counseling services, only two dedicate serious attention to tenant-based rental assistance. Chicago, for example, plans to set aside $75 million over five years in Trust Fund dollars to support rental assistance for roughly 15,000 extremely low-income renters. Perhaps more surprisingly, only one of the plans dedicates serious attention to providing emergency assistance to help renters at risk of eviction stay in their homes. Of course, it is likely that several of these cities actually offer eviction prevention assistance (for instance, New York City offers emergency assistance through its homeless prevention program, Homebase), and the housing authorities in all of these jurisdictions administer Housing Choice Voucher programs. But it is notable that only one of the cities appears to integrate such demand-side subsidies into its housing plan. With the exception of homeownership assistance programs, these plans are very focused on brick and mortar efforts to increase supply and generally do not integrate demand-side assistance into those efforts.

Panel 4 summarizes the tools that the various plans envision using. Perhaps surprisingly, all envision spending city, county, or regional funds on housing, which may be a reflection of the large size of these cities. Half of the plans also focus on some kind of property tax program to further incentivize

Table 27.2 Summary of Local Housing Plans.

	Boston	*Chicago*	*Denver*	*LA*	*Minn./St. Paul*	*New York*	*Oahu*	*Raleigh*	*San Francisco*	*Seattle*
1. Extent to which plan supports activities related to the preservation or construction of income-restricted housing:										
Preserving the affordability of existing dedicated affordable rental units	★★	★★	★★	★★	★★	★★	★★	★★	★★	★★
Increasing the number of dedicated affordable rental units	★★		★★	★★	★★	★★	★★	★★	★★	★★
Increasing the number of dedicated affordable homeownership units	★		★	★	★	★	★★	★★	★★	★★
2. Extent to which plan supports activities related to the preservation or construction of market-rate housing:										
Preserving the affordability of unsubsidized rental units		★★	★★	★★	★★	★★	★		★★	
Expanding the overall housing supply	★★	★★		★★	★★				★★	★★
3. Extent to which plan attempts to supplement household income/assets through the following activities:										
Increasing the ability of low-income households to afford the rents of unsubsidized units (e.g., through tenant-based rental assistance)		★★		★	★★		★★★			★
Increasing the rate of homeownership through down payment assistance, low-interest mortgages, or counseling services	★★	★★	★★	★	★★			★★	★★	
Providing emergency assistance to prevent an eviction or meet other short-term needs	★	★		★★	★	★				
4. Extent to which plan incorporates policy tools from the following domains:										
Expenditure of city, county, or regional housing resources	★★	★★	★★	★★	★★	★★	★★	★★	★★	★★
Tax policy	★★	★★	★	★		★★	★★	★	★	★★

(*Continued*)

Table 27.2 (Continued)

	Boston	*Chicago*	*Denver*	*LA*	*Minn./St. Paul*	*New York*	*Oahu*	*Raleigh*	*San Francisco*	*Seattle*
Land use regulation	★★	★★		★★	★	★★	★★		★★	★★
Use of publicly owned land/building	★	★★	★	★	★	★★	★★	★★	★★	★★
Housing authority programs or resources	★★	★★	★	★★	★	★★			★★	★
Building code reform				★★		★★	★★		★	★★
Building permitting processes	★★			★★		★	★		★★	★★
Housing finance agency programs or resources	★		★	★★	★	★★	★★	★★		
5. Extent to which plan addresses revenue:										
Identifies the revenue sources that will support the proposed activities	✔	✔	✔	✔	✔	✔	✔	✔	✔	✔
Includes provisions to generate new revenue to support proposed activities	✔	✔	✔	✔	✔	✔	✔	✔	✔	✔
6. Extent to which plan includes activities that further economic integration and promote access to opportunity:										
Expanding the affordable housing stock in "opportunity" areas (to promote AFFH goals)					★★				★★	
Preserving or expanding affordable housing stock in changing/gentrifying areas	★★		★		★★	★★			★	★★
Mixed-income development	★★	★★★★		★	★★	★★	★★	★★	★★	★
Mobility counseling	★				★★					
Reducing housing discrimination				★★	★★				★★	★★
7. Extent of inter-agency collaboration:										
Product of inter-agency dialogue	✔		✔	✔	✔	✔	✔	✔	✔	
Includes actions requiring inter-agency collaboration	✔	✔	✔	✔	✔	✔	✔		✔	✔
Assignments of responsibility to specific agencies	✔	✔	✔	✔	✔	✔	✔			
8. Extent to which plan addresses goals or initiatives that go beyond housing:										
Transit or transit-oriented development (TOD)		★★	★★	★★	★★	★★	★★	★★	★★	★★
Environmental sustainability	★★		★★	★★	★★	★★	★		★★	
Resilience to natural disasters	★★			★		★★			★★	

Community development		★★	★	★	★★	★★	★★	★★		
Workforce development				★		★★		★★		
Education		★		★	★★					★
Health	★	★		★	★				★	
Addresses needs of broader geographic regio:n	✔				✔				✔	
9. Extent to which plans focus on mo:nitoring and implementation:										
Numerical targets/goals	✔	✔	✔	✔		✔	✔		✔	✔
Discussion of how action items will be impl:emented	✔	✔		✔	✔	✔	✔		✔	✔
10. Audience:										
Plan is communicable to a public audience	easy	easy	easy	Technical	Moderate	easy	Moderate (jargon)	Moderate (jargon)	Technical	easy

Coding:
[No stars] = Not addressed.
★ = Addressed briefly but not the focus of significant or sustained attention in the plan.
★★ = Receives significant attention in the plan.
★★★ Focus is on homeless population.
★★★★ Affordable Requirements Ordinance requires residential developments that receive city financial assistance or involve city-owned land to provide a percentage of their units at affordable prices.

For Procedural Characteristics:
[no check] = The procedural characteristic does not apply.
✔ = The procedural characteristic applies.

Source: author.

owners to provide affordable housing. Only four programs rely on housing finance agency programs or resources. As for regulatory tools, seven give significant attention to land use regulations, but only four promise building code reforms, and only four call for reforms to the permitting processes to reduce costs. Six of the ten emphasize publicly owned land, and five include efforts to draw on housing authority resources or programs. (As noted earlier, however, even these plans identify a limited set of housing authority resources; for example, only one of the plans incorporates Housing Choice Vouchers.) As a whole, then, the plans appear to draw on a fairly wide variety of tools, though levels of engagement with these tools vary. The three areas that appear least used are building code reforms, permitting reforms, and housing finance agency programs.

As for revenues, Panel 5 shows that all of the plans identify revenue sources to support the proposed activities and all of them also include specific proposals for generating new revenue to support the plan's activities. For example, the city of Boston estimates that it will need to increase its annual housing budget from $31 million to $51 million in order to meet its housing goals. It calls for creating a new line item in city and state budgets to support senior housing programs, and proposes lowering the minimum size of developments required to pay linkage fees, and raising the price that developers must pay to avoid building units on site through the city's inclusionary zoning program. Seattle's plan calls for a new tax on real estate transfers, a doubling of the city's property tax levy for low-income housing, and the creation of a new partnership with local employers to support affordable housing. New York City's plan lays out multiple sources of funds, including the city capital budget, property tax incentives, and city and state pension fund investments in housing. This specificity on funding sources suggests these cities treat these plans as meaningful guides for action. That said, some of the identified funding sources extend beyond these cities' control, as they either are state and federal sources or require voter or state approval.

Panel 6 summarizes the degree to which the plans include activities that can help to further economic integration and promote access to opportunity. Seven of the ten plans include programs to encourage mixed-income housing. Only two include efforts to expand the affordable housing stock in opportunity areas. For example, the Raleigh plan calls for using local resources to provide gap financing to allow tax credit developers to build housing in areas that are near transit and jobs and that currently have little affordable housing. The Minneapolis/St. Paul plan calls for adding affordable housing in higher-income areas, for reducing concentrated poverty, and for developing indicators to measure their progress.

Interestingly, despite the gentrification pressures that virtually all of these cities are experiencing, only four of them focus attention specifically on preserving or expanding the affordable housing stock in changing/gentrifying areas. Meanwhile, only four state an aim to combat discrimination in housing markets, none offer any discussion of how they will accomplish that aim, and only one includes counseling or other efforts to help families move to higher opportunity areas. It is possible that these cities describe more robust efforts in their Consolidated Plan's Analysis of Impediments to Fair Housing Choice (now replaced by the Assessment of Fair Housing) or in local housing authority plans required by the U.S. Department of Housing and Urban Development (HUD). But even if so, it is telling that activities aimed at furthering integration are not incorporated into their broader housing plans.

Panel 7 considers inter-agency collaboration. It shows that eight of the plans emerged from an inter-agency dialogue. Nine of the plans include actions that would require collaboration across agencies, and seven of them assign responsibilities to specific agencies. This fairly robust level of cross-agency involvement suggests some centralized support from City Hall for the efforts. Indeed, five of the plans include introductory letters from the city's mayor: Boston, Chicago, Denver, New York, and Seattle.

Panel 8 addresses the degree to which the plans incorporate broader planning goals that go beyond housing. The results of this panel are mixed. On the one hand, all but one of the plans include a focus on transit investments or on transit-oriented development, six devote significant attention to environmental sustainability, and five embrace community development as an area of emphasis. But few plans dedicate attention to the other goals, such as resilience to natural disasters, workforce development, education, or health, despite clear connections between housing and these objectives.

As for regional collaboration, both the Honolulu County plan and the Minneapolis/St. Paul plan address regional issues, as they were prepared by regional, or at least county-wide, bodies and explicitly aim to address housing needs of the region. But Panel 8 shows that only two of the eight city plans dedicate attention to the needs of the broader region. Given the political challenges in forging regional collaborations, this is perhaps not surprising, but this omission may limit these plans' ultimate effectiveness. As for monitoring and implementation, Panel 9 shows that eight of the plans include specific numerical targets or goals that can be tracked over time, and eight include a discussion of how at least some of the policies or action items will be implemented.

Finally, and perhaps most fundamentally, Panel 10 shows that these ten plans differ in their intended audience. The California plans are clearly aimed at the state agency staff who will review them and judge whether they satisfy the housing element requirement. The Minneapolis/St. Paul, Raleigh, and Oahu plans seem intended to communicate to policymakers or agency staff. The plans released by Boston, Chicago, Denver, New York, and Seattle, by contrast, are clearly designed to be public documents that highlight the attention that the mayor is giving to these issues. As public documents, they also invite external monitoring of the degree to which the city agencies are making progress in meeting their stated goals.

Conclusions

What can we learn from these local housing plans? First and most fundamentally, the emergence of these plans suggests that cities are increasingly seeing a lack of affordable housing as a problem they need to address, and are not just waiting for the federal government to respond.[5] The fact that five of the plans seem so clearly aimed at a broad, public audience also suggests that many mayors see political benefits in staking out a commitment to affordable housing.

As for their content, these ten plans lay out ambitious goals, involve collaboration among multiple agencies, identify dedicated sources of revenue, and are fairly comprehensive in their scope of activities and policy tools. These plans may not be representative, of course, given that these ten cities are generally large and have high-capacity local governments and housing providers.

Despite these plans' evident strengths, there are some potential areas for improvement. For example, a narrow majority focus on increasing the size of the overall housing stock. Most are very focused on the subsidized stock. Further, few of the plans mention demand-side assistance; they are almost entirely focused on building new units and preserving existing ones. While cities—and these cities in particular—need more housing, demand-side subsidies can be a very useful complement to the suite of production and preservation programs. At the very least, cities should be thinking strategically about how to work with the local housing authority to use their allotment of federal Housing Choice Vouchers to bolster their housing plans (see also the chapters authored by Rachel G. Kleit; and Lawrence Vale and Yonah Freemark). For example, cities could do more to support efforts to help voucher holders reach opportunity neighborhoods through landlord outreach, mobility counseling, or project-basing of vouchers.

While funding their own long-term voucher programs to supplement the supply of federal vouchers may be deemed too expensive, cities could do more to provide short-term emergency assistance to households who hit a bump in the road and are struggling to make their rent payments. As noted, some might do this already through their departments of social services, but it would be useful for cities to incorporate these programs into their housing strategies and perhaps to expand them or complement them with housing counseling efforts to help families who are facing a longer-term struggle to find new homes that are more affordable.

The majority of these plans also overlook local housing and building codes, despite the reality that such codes can powerfully shape affordability by allowing more inexpensive forms of construction and preservation and/or permitting more people to occupy a given unit. Few of the plans involve any efforts at regional coordination. The politics are clearly challenging, but such coordination could go a long way toward supporting affordability in the broader region, benefiting both cities and their surrounding suburbs. Such coordination need not involve actual regional governmental structures. Finally, these cities generally leave fair housing goals out of their plans, and few pay attention to locking in affordability—and thus long-run diversity—in gentrifying areas.

Again, there is no one housing plan that is suitable for every city and every set of market conditions. And to be clear, this chapter only addresses the content of the written plans and offers little evidence about the extent of their implementation, but I hope it has provided a useful start. As an increasing number of cities adopt serious housing plans, even when not mandated to do so, more research should be done to explore their scope, implementation, and, most critically, effectiveness. In the coming years, local governments are likely to feel even more pressure to engage in housing policy, and yet we know far too little about the success of their efforts in marshaling resources to alleviate the rent burdens that challenge so many of their residents.

Acknowledgements

I am extremely grateful to Jeff Lubell, Mark Willis, and Traci Sanders for their help in shaping this paper and for their detailed comments on earlier drafts. (The errors are my own.) I also thank Kimberly Burnett, Rebecca Cohen, and Amy DeHuff for help in reviewing housing plans and Maxwell Austensen and Monica Griffith for other research assistance. I also benefited from discussions with participants in the National Community of Practice co-facilitated by the NYU Furman Center and Abt Associates as well as generous support from the Ford and MacArthur Foundations.

Notes

1 Specifically, Gyourko et al. (2013) categorize a city as a superstar if it is experiencing high demand, or the sum of its price and quantity growth over the prior two decades is above the sample median, *and* it has an inelastic housing supply, which they proxy by a price growth to housing unit growth in the top 10 percent.
2 One challenge in targeting middle-income households is that cities often rely on federal subsidies to support their housing subsidy and production programs, and few if any permit serving households with incomes above 60 percent or 80 percent of the local area median income. But to the extent that cities are spending their own resources, they may want to target at least some of those resources to middle-income households.
3 For an overview of the theoretical and empirical evidence on the relative cost-effectiveness of housing vouchers and certificates, see Schill (1993). See also U.S. Government Accountability Office (2001), which found the total per-unit costs for housing production programs ranged from 12 percent to 27 percent more than the cost of voucher programs.
4 In reviewing these plans, and developing the rubric described in the following text, I worked with Jeff Lubell from Abt Associates and Mark Willis and Traci Sanders from the Furman Center.
5 Since the passage of the Housing Element Law in 1969, the state of California has required that every local government adopt a housing element as part of its long-term, comprehensive physical plan, so both Los Angeles and San Francisco have issued plans for years (Ramsey-Musolf 2016).

References

Brooks, Mary. 2016. *State and Local Housing Trust Funds*. Washington, DC: National Low Income Housing Coalition.

Chetty, Raj, Nathaniel Hendren, and Lawrence Katz. 2016. "The Effects of Exposure to Better Neighborhoods on Children: New Evidence from the Moving to Opportunity Experiment." *The American Economic Review*, 106(4): 855–902.

City of New York. 2014. *Housing New York: A Five Borough, Ten-Year Plan.* New York: City of New York.

City of San Francisco. 2014. *Mayor Lee Outlines Bold 2014 Agenda in State of the City Address.* San Francisco, CA: City of San Francisco.

Collinson, Robert, Ingrid Gould Ellen, and Jens Ludwig. 2016. "Low Income Housing Policy." In *The Economics of Means-Tested Programs*, edited by Robert A. Moffitt, 59–126. Chicago, IL: University of Chicago Press.

Ellen, Ingrid Gould. 2017. "Fiscal Federalism and Mixed Income Housing." *Cityscape: A Journal of Policy Development and Research*, 19(1): 253–255.

Ellen, Ingrid Gould, and Brendan O'Flaherty. 2013. "How New York Housing Policies Are Different—and Maybe Why." In *New York City-Los Angeles: The Uncertain Future*, edited by Andrew Beveridge and David Halle, 286–309. New York: Oxford University Press.

Glaeser, Edward. 2011. *Triumph of the City: How Our Greatest Invention Makes Us Richer, Smarter, Greener, Healthier, and Happier.* New York: Penguin Press.

Glaeser, Edward L., Joseph Gyourko, and Raven Saks. 2005. "Why Have Housing Prices Gone Up?" *The American Economic Review*, 95(2): 329–333.

Gyourko, Joseph, Christopher Mayer, and Todd Sinai. 2013. "Superstar Cities." *American Economic Journal: Economic Policy*, 5(4): 167–199.

Hankinson, Michael. 2018. "When Do Renters Behave Like Homeowners? High Rent, Price Anxiety and NIMBYism." *American Political Science Review*, 112(3): 473–493.

Healey, Patsy. 1996. "The Communicative Work of Development Plans." In *Explorations in Planning Theory*, edited by Seymour J. Mandelbaum, Luigi Mazza, and Robert W. Burchell, 263–288. New Brunswick, NJ: Center for Urban Policy Research.

Jacobus, Rick. 2016. "Why We Must Build." *Shelterforce*, March 9.

Joint Center for Housing Studies. 2015. *America's Rental Housing: Expanding Options for Diverse and Growing Demand.* Cambridge, MA: Harvard University Press.

Kolko, Jed. 2010. "Urbanization, Agglomeration, and Coagglomeration of Service Industries." In *Agglomeration Economics*, edited by Edward Glaeser, 151–180. Chicago, IL: University of Chicago Press.

Leventhal, Tama, and Sandra Newman. 2010. "Housing and Child Development." *Children and Youth Services Review*, 32(9): 1165–1174.

Lewis, Paul. 2010. "Can State Review of Local Planning Increase Housing Production?" *Housing Policy Debate*, 16(2): 173–200.

Lobo, Arun Peter, and Joseph Salvo. 2013. "A Portrait of New York's Immigrant Melange." In *One Out of Three: Immigrant New York in the Twenty First Century*, edited by Nancy Foner, 35–63. New York: Columbia University Press.

NYU Furman Center. 2013. *State of New York City's Housing and Neighborhoods in 2013.* New York: NYU Furman Center.

NYU Furman Center. 2016. *State of New York City's Housing and Neighborhoods in 2015.* New York: NYU Furman Center.O'Flaherty, Brendan. 2005. *City Economics.* Cambridge, MA: Harvard University Press.

Quigley, John M. 1998. "Urban Diversity and Economic Growth." *The Journal of Economic Perspectives*, 12(2): 127–138.

Ramsey-Musolf, Darrel. 2016. "Evaluating California's Housing Element Law, Housing Equity and Housing Production (1990–2007)." *Housing Policy Debate*, 26(3): 488–516.

Rosenthal, Stuart. 2014. "Are Private Markets and Filtering a Viable Source of Affordable Housing? Estimates from a Repeat Income Model." *The American Economic Review*, 104(2): 687–706.

Schill, Michael H. 1993. "Distressed Public Housing. Where Do We Go from Here?" *The University of Chicago Law Review*, 60(2): 497–554.

Schwartz, Amy, Ingrid Gould Ellen, Ioan Voicu, and Michael Schill. 2006. "The External Effects of Subsidized Housing." *Regional Science and Urban Economics*, 36: 679–707.

U.S. Bureau of the Census. n.d. *American Community Survey*. Washington, DC: U.S. Bureau of the Census.

U.S. Government Accountability Office. 2001. *Federal Housing Programs: What They Cost and What They Provide.* Washington, DC: U.S. Government Accountability Office.

Varady, David P. 1996. "Local Housing Plans: Learning from Great Britain." *Housing Policy Debate*, 7(2): 253–292.

28

AFFORDABLE HOUSING

Program Financing and Policies in U.S. States

Victoria Basolo

State governments in the United States of America (U.S.) have played an increasingly vital role in the delivery of affordable housing.[1] The range of state housing functions and strategies has broadened over the past 40 years, and many state housing agencies have exhibited a notable level of innovation (Basolo and Scally 2008; Nenno 1996). In fact, state housing approaches follow many paths, including the requirement that localities take actions toward meeting housing goals, the administration and distribution of selected federal housing resources, the development of state funding mechanisms for affordable housing, and the formulation and management of housing programs (Davis 2006; Nenno 1996; Scally 2009). These general options are discussed in more detail in the following text.

State housing policy is uniquely situated between federal housing policy and programs and the activities of local governments, with localities often expressing a significant level of resistance to affordable housing development (Nguyen et al. 2013; Scally and Tighe 2015). Traditionally, state housing activity has tended to be indirect such as the administration of federal programs, participation in the establishment of building standards, and promotion of local housing planning through a comprehensive planning process. Over time, these roles have expanded along with more direct state housing efforts (Nenno 1996; Scally 2009).

States typically identify a range of housing needs and seek to create or administer programs to address these needs. For example, states may foster first-time homeownership, promote affordable rental housing, and improve housing conditions. However, state programs are influenced by their source of funding, and a significant proportion of the affordable housing support available to states comes from the federal government (Mueller and Schwartz 2008). Federal funds come with constraints in the form of regulations and administrative rules. As a result, the types of programs and populations served by these initiatives typically are shaped by the federal government. However, state governments do have some flexibility in program design in these cases. For example, despite the limits imposed on federal housing programs such as the Low-Income Housing Tax Credit (LIHTC), states have a level of discretion in implementation that allows them to express some of their policy preferences such as where LIHTC projects should be located (Ellen et al. 2015).

States have the most control over housing policies and programs when the funds are internally generated. Historically, however, states have dedicated relatively few internal resources to affordable housing, but this trend seems to be shifting. State general funds, general obligation bonds, and housing trust funds are all examples of state-based revenue that has been committed to affordable housing over the past few decades (Basolo and Scally 2008; Nenno 1996; Scally 2012).

This chapter examines state housing policies and programs with a particular focus on the financing of affordable housing activities. The chapter begins with a discussion of the rationale for state level efforts in the housing domain and the pivotal role states can play in affordable housing policy, followed by a description of typical housing policies, programs, and funding mechanisms at the state level in the U.S. After this overview of state housing activity, the chapter then narrows its focus, presenting a case study of California, a politically progressive state with a diverse population and complex housing issues. The purpose of the case is to provide a rich description of specific housing policy activities in a state with tremendous housing needs and a mix of housing funding sources. In the conclusion I discuss the future of housing policymaking at the state level.

States and Affordable Housing

The role of states in affordable housing programs can be conceptualized and contextualized in numerous ways. Constitutionally, states have always had the power to take a leading role in affordable housing policy. Two conditions have worked against strong state involvement in affordable housing. First, funding housing assistance programs requires the political will to do so and the identification of a source of revenue for this purpose. For most of this nation's history, states have lacked the political will necessary to take an active role in creating and funding affordable housing programs (Goetz 1993; Nenno 1996). Second, it was the federal government that took the lead in housing policy beginning in the 1930s and has provided the lion's share of the funding for affordable housing programs since then (Basolo 1999). These two conditions are likely related because states may have viewed housing assistance as a function of the federal government, thereby relieving the pressure on states to develop affordable housing initiatives.

Over the past 30 years, federal devolution, budget decisions, and administrative policies have impacted housing policies and programs (Goetz 1995; Mueller and Schwartz 2008; Scally 2009). The impacts on housing assistance are tangible and mostly negative. For example, despite a growing need for housing assistance, the Center on Budget and Policy Priorities (2017) reported that public housing funds decreased by over $1.6 billion from 2010 to 2014, and, between 2012 and 2014, the number of households using housing voucher assistance had decreased by 85,721. However, as the federal government has retreated from its strong leadership in housing policy, states have ramped up their efforts (Scally 2012).

The change in federal disposition toward housing policy is not the only rationale for states' increasing concern with affordable housing. There are at least two other reasons why states have promoted and supported affordable housing. First, many state leaders understand that their economy depends on a strong labor force and that labor force needs affordable housing. Recent trends in the U.S., however, reveal the challenges associated with availability of sufficient affordable housing: In 2013 the median household income was at 1993 levels (adjusted for inflation), and housing rental costs increased at twice the overall inflation rate in 2015. Combined with an undersupply of affordable housing in many areas, these circumstances have led to a significant number of cost burdened households (i.e., households paying more than 30 percent of their income toward housing) throughout the country; especially hard hit are low-income households (Joint Center for Housing Studies of Harvard University 2015). Second, localities often resist affordable housing development within their boundaries, and states can require, even mandate, that these jurisdictions address housing issues.

Research indicates, however, that many cities do not support affordable housing because it is redistributive in nature and perceived as detrimental to the local economy (Basolo 1999). Also, at the local level, the siting of affordable housing may be contentious as residents near proposed affordable housing locations may protest against it (Nguyen et al. 2013). The problem, however, is that housing exists in space and affordable developments must be located in someone's neighborhood. State governments, through regulations and funding mechanisms, can encourage the development and support of affordable housing in local communities, at least to some degree.

Housing policy is inherently political at the state level of government. The political climate or context of the policymaking environment can influence the policies adopted by government (Choi et al. 2010). Basolo and Scally (2008), for example, found that liberal ideology was associated with affordable housing policy innovation.

In general, the research indicates that states are "in the mix" when it comes to affordable housing. In fact, state involvement has grown from early partnership arrangements in a handful of states to proactive policies, coordination responsibilities, and new funding streams in many states (Nenno 1996; Scally 2012). Research also shows that the policy work is undertaken within, and influenced by, the state political environment. For these reasons, it would be expected that there would be policy variation across states due to differences in political context.

State Affordable Housing Policy and Programs

State governments historically have not been proactive in the affordable housing policy domain. While the federal government dramatically expanded its housing policies during the Great Depression era, except for a handful of Northeastern states involved in public housing, state level housing policy was nearly non-existent until the emergence of housing finance agencies (HFAs), the first appearing in New York in 1960 (Nenno 1996; Rosan 2014). Other states followed New York State and created their own HFA; by 1981, 44 states had an HFA (see Nenno 1996). Today, all 50 states have an HFA (National Council of State Housing Agencies 2016a).

The academic literature on HFAs, and state housing efforts in general, is very limited (Scally 2009). However, drawing on existing literature, especially more recent work, and documents from the practice literature, primarily reports, provides an understanding of state housing organizations and their past and present policies and programs. Given that HFAs were the first widespread housing organizations to appear across the states, they are a logical starting point for a discussion of state housing policies and programs.

Housing Finance Agencies: Housing Financers and Administrators of Federal Programs

In one of the most in-depth studies of housing finance agencies, and certainly one of the more recent, Scally (2009) identified an array of past and present roles assumed by HFAs. While HFAs began with a limited scope of responsibilities, Scally identifies five current roles of these organizations including financer, administrator/monitor, planner, convener/coordinator, and policy innovator. In this section, the roles of financer and administrator will be emphasized as they appear to be the dominant activities of HFAs. A state's role in housing planning and policy innovation may involve other state housing organizations and will be discussed later in the chapter. States as conveners/coordinators is a role embedded in state housing networks, and there is not much systematic research on the topic. For this reason, the convener/coordinator role is not discussed further in this chapter.

One of the early areas of state activity was the issuance of tax-exempt Private Activity Bonds (PABs) to support housing assistance, and this activity continues today (National Council of State Housing Agencies 2016b; Nenno 1996; Scally 2009). PABs are used to finance a range of activities including mortgages for lower-income homebuyers (mortgage revenue bonds or MRBs) and to develop or rehabilitate multifamily housing (multifamily housing bonds or MFHBs) to serve lower-income renters. HFAs typically administer the housing programs funded by these bonds.

Under federal rules, states are subject to a cap on the total annual amount of PABs.[2] Repayment of this type of bond comes from revenue generated by loans made from the bond proceeds; these loans are set at a below-market interest rate made possible by the tax-exempt status of the bonds. Therefore, the bonds are not a state or federal obligation, but they do result in lower tax revenue to the federal government.[3] In 2015, according to the Council of Development Finance Agencies (CDFA) (Council of Development Finance Agencies 2016), states reported issuing just over $6.6 billion in MFHBs (37 of 50 states reporting) and about $4.57 billion in MRBs (42 of 50 states reporting). As shown in Table 28.1, the average issue of MFHBs for the reporting states was about $178.5 million. However, there was considerable variation across states in the MFHB issuances. Thirteen of the reporting states did not issue any MFHBs, and California issued the largest total amount of these bonds at approximately $2.86 billion. California, however, did not have the largest issue when taking the size of the state into consideration; New York had this distinction with a per capita figure of $85.72.

The average issue of MRBs for the states reporting was approximately $108.7 million. Again, there was substantial variation across the states. Eleven states issued zero MRBs, while Minnesota had the largest total amount of MRBs at $739.9 million. On a per capita basis, North Dakota had the highest issue at $274.65. All states reporting to CDFA had PAB carry forward at the beginning of 2015, indicating some level of unused capacity for these bonds.

The PAB cap includes other types of bonds, and the distribution of PAB issuances across bond types suggests priorities for the states. In addition to MFHBs and MRBs, the CDFA reports PAB issuance for exempt facilities, Industrial Development Bonds, student loans, and agriculture. Of the states reporting on all PAB bond types and having a total issuance greater than zero (n = 35), on average, MFHB represented 36.7 percent of the total PAB issuance in 2015. MRBs made up an average of 41.5 percent of the total PAB issuance (see Table 28.1). However, again, these distributions are highly variable, with some states issuing none of one type of bond and others issuing 100 percent of their bonds on only one bond type. Taking into consideration PABs issued for all activities by states in 2015, MFHBs composed 55.7 percent of the total, and MRBs were 31.8 percent of the total PAB issues. Combining both MFHB and MRB totals, these housing-related bonds represented about 87.5 percent of the total PAB issues in 2015. Thus, at least in terms of PABs, states appear to strongly support affordable housing activities.

Mortgage Credit Certificates (MCCs) are another option for state governments. A state can convert some of its PAB cap authority, at a conversion rate set by law, to MCC authority. The MCC was created by Congress in 1984 to assist households to purchase a home by lowering their federal

Table 28.1 Private Activity Bond Issuance in the U.S., 2015.

	Multifamily Housing Bonds	*Mortgage Revenue Bonds*
Average (Total $)	$178,537,837.84	$108,726,190.48
Minimum	$0	$0
Maximum	$2,861,000,000	$739,900,000
Average (Per Capita $)	$14.35	$41.02
Minimum	$0	$0
Maximum	$85.72	$274.65
Avg. % of Total PAB	36.7%	41.7%
Minimum	0%	0%
Maximum	100%	100%

Note: The sample size varies for the indicators due to there being no reports from some states. For multifamily housing bonds (MFHBs), average total and average per capita dollars, n = 37. For mortgage revenue bonds (MRBs) on the same indicators, n = 42. For both MFHBs and MRBs, average % of total PAB issuance is based on 35 cases.

Source: author, based on Council of Development Finance Agencies (2016).

income tax liability. Slightly different program elements exist across states, and some states partner with lower levels of government to make MCCs more widely available. In a typical MCC program, an eligible household, usually a lower-income, first-time homebuyer, applies to the state HFA (or local jurisdiction) for the credit. When awarded, the credit reduces the income tax owed to the federal government by the amount of the credit awarded to the applicant. Participating lenders may count the amount of the credit as income, making it easier for a homebuyer to qualify for a mortgage. MCCs are underutilized currently. Beginning in 2007 with the start of the Great Recession, mortgage interest rates plummeted, and in 2015 they were at low levels, which may have made MCCs a less attractive financing option. According to CDFA (2016), with 45 states reporting, approximately $10.5 billion in MCC carryover was abandoned in 2015 due to nonuse within the allowed time frame.

States use other financing via a federal tax benefit (see also the chapter authored by Kirk McClure). For example, the Low-Income Housing Tax Credit (LIHTC) is a well-known and generally heavily utilized program supporting the development or rehabilitation of multifamily developments serving lower-income households. There are two levels of tax credit: (1) the 70 percent subsidy (commonly called the 9 percent credit), which is most coveted and cannot be combined with other federal subsidies; and (2) the 30 percent subsidy (called the 4 percent credit), which is combined with other subsidies (Novogradac and Company n.d.). Currently, the federal government provides about $8 billion in tax credits to the states based on a minimum level for smaller states and total state population for states exceeding the minimum (U.S. Department of Housing and Urban Development 2015a). For this reason, California, with the most state residents, received the largest amount of tax credits in 2015, just over $91 million in 9 percent credits (California Tax Credit Allocation Committee n.d.).

The LIHTC program has been very successful in producing affordable housing. According to the U.S. Department of Housing and Urban Development (2015b), for the nation as a whole, the LIHTC assisted 28,448 developments with over 2.5 million dwelling units during the period 1995–2014. The basics of the program are simple. States receive tax credits and distribute them to developers of lower-income housing, who then sell the tax credits to investors to raise capital for the housing project. Investors use the tax credits to reduce their federal tax liability with a one credit to one dollar value.[4] Because the tax credits are provided to states on a per capita basis, the funds available to states and thus the number of developments vary by state. The number of units produced overall by the program also varies by year. Figure 28.1 shows the number of units produced annually

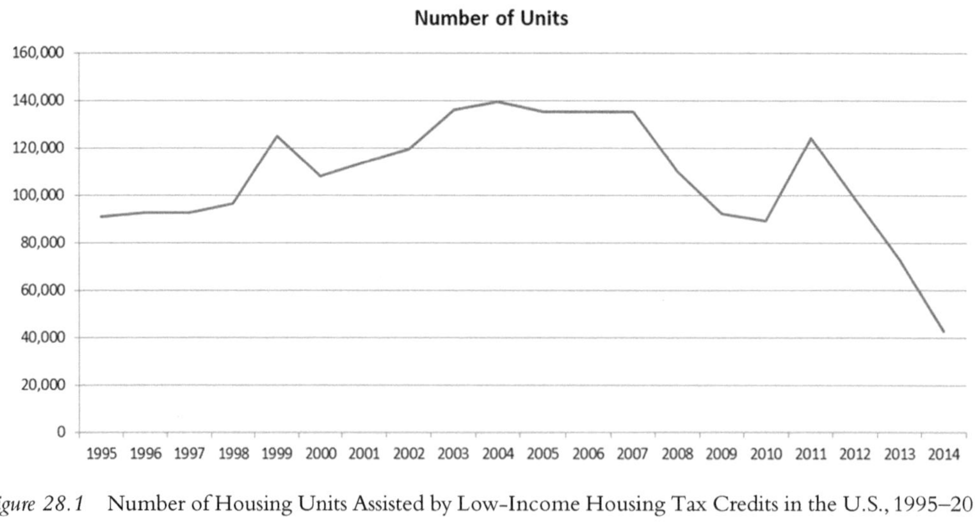

Figure 28.1 Number of Housing Units Assisted by Low-Income Housing Tax Credits in the U.S., 1995–2014.
Source: U.S. Department of Housing and Urban Development 2015a.

from 1995–2014. It is important to highlight the decline in LIHTC units during the Great Recession, beginning with the start of the Recession in 2007.[5] The demand for tax credits plunged during this period because investors who had experienced reduced revenue and losses during the economic slump did not need the tax reduction provided by the LIHTC. Federal action in response to the recession kept the program afloat for a time until it recovered.

The states also administer other federal resources including the Community Development Block Grant (CDBG), Home Investment Partnerships Program (HOME) funds, and several other programs including those aimed at special populations such as the Emergency Shelter Grant (for the homeless) and Housing Opportunities for Persons With AIDS (Scally 2009). The CDBG and HOME program funds are widely known and represent a notable portion of federal funds going to state and local governments. They are distributed to states based on a formula accounting for various factors (e.g., housing and demographic characteristics). States then provide these CDBG funds to non-entitlement localities for purposes allowed under the federal regulations, which, among other uses, can include assistance to support the delivery of lower-income housing opportunities. In 2015, HUD reported expenditures of approximately $145.2 million in CDBG funds through the state program on housing construction, rehabilitation, homeownership assistance, and other housing-related services (U.S. Department of Housing and Urban Development 2015b). The HOME program also involves a state distribution of federal funds to localities, developers, and eligible nonprofit organizations to produce or preserve affordable housing. In their assessment of the HOME program at the state and local levels, Mueller and Schwartz (2008) show that these federal funds tend to be used to develop, acquire, or rehabilitate for rent and ownership housing, and generally do not serve the very lowest-income groups. Nevertheless, HOME funds are an important resource from the federal government to states; between 1992 and 2014, HFAs have administered just under $8.96 billion in HOME funds (National Council of State Housing Agencies 2016b).

HFAs today remain primarily financers and housing program administrators. In fact, the National Council of State Housing Agencies (NCSHA; 2016b) identifies tax-exempt housing bonds, the LIHTC, and HOME "at the center of HFA activity within the states" (v). If these were the only state housing efforts underway, then it would be easy to conclude that not much has changed in the work of HFAs over the past few decades and that states are basically a pass through for federal funds. However, research shows that some HFAs have expanded their activities, and more than one agency involved in affordable housing policy and programs exists in many states (Basolo and Scally 2008; Scally 2009). In the next section, other state housing efforts will be discussed. These activities may be an HFA responsibility or fall under another state agency involved in housing.

State-Initiated (Non Federal) Housing Policies and Programs

State-initiated housing policies and programs have existed for over four decades. These efforts began slowly and developed over time (Nenno 1996). While state-based initiatives vary considerably from state to state, some of these activities are more common across the states, while others are relatively limited to a subset of states. These activities can be divided into revenue generation/sources to fund the creation and support of state affordable housing programs, actual state affordable housing programs, and state regulatory actions to promote affordable housing. In this section, some common and less common initiatives for these three activities are discussed.

One of the longest running efforts by states to establish funds for housing programs is the creation of a housing trust fund (HTF). According to the Center for Community Change (Housing Trust Fund Project n.d.), the definition of HTFs is:

> distinct funds established by city, county or state governments that receive ongoing dedicated sources of public funding to support the preservation and production of affordable housing and increase opportunities for families and individuals to access decent affordable homes.

The first HTF, the Delaware Housing Development Fund, appeared in 1968[6] (Nenno 1996; Scally 2012). By the close of 1992, 36 states had established an HTF; by 2011, this figure had increased to 47 states, and it remained at this level in 2016[7] (Center for Community Change [Housing Trust Fund Project] 2013, 2016). In fiscal year 2015, state HTF revenue was estimated at $790 million (Center for Community Change [Housing Trust Fund Project] 2016). In its 2016 survey of HTFs, the Center for Community Change identified four states with unfunded HTFs: Alabama, California, Idaho, and Rhode Island; in other words, states without "dedicated sources of public funding" for an HTF. In a survey of HTFs in 2016, the most frequently reported activities eligible for state HTFs were new affordable housing construction and preservation/rehabilitation of multifamily housing. Other types of activities funded by HTFs, to name a few, were tenant assistance rental programs, housing for the homeless, down payment assistance, and emergency repairs (Center for Community Change [Housing Trust Fund Project] 2016, Technical Assistance Collaborative 2014).

Some states use other forms of funding for affordable housing programs. For example, state housing tax credits, similar in concept to the LIHTC, are available in 17 (34 percent) of the states. Each state has its own rules, related to the use of the state credit; for example, about 40 percent of the states with a state housing tax credit program require the credits to be used with the federal tax credits (National Council of Housing Agencies 2016b). Another less common strategy used by states to raise affordable housing funds is the issuance of general obligation bonds (GO bonds). Typically, GO bonds are used to finance large infrastructure activities such as transportation projects, but some state and local governments have issued these bonds to be used for affordable housing efforts. GO bonds are tax-exempt, but repayment of these bonds is very different than it is for PABs. These bonds are backed by the government entity issuing them, and therefore, taxpayers through their government are "on the hook" for paying off the bond debt; thus, taxpayers must vote to issue GO bonds (Smith and Greenblatt 2017). For example, in 2016, the Rhode Island State Legislature placed a $50 million GO bond measure aimed at affordable housing needs and urban revitalization on the ballot; it passed, as had previous Rhode Island GO bond measures for affordable housing (Ballotpedia n.d.a).

GO bonds may be the source of revenue for an HTF in some states. For example, Washington State has taken this approach, and it also has funded the HTF with document recording fees (Center for Community Change [Housing Trust Fund Project] 2016, Washington State Finance Committee 2002). Irrespective of the source, GO bonds, HTFs, or tax credits, these state-generated revenue sources are used to fund affordable housing programs. The types of programs funded with these resources can vary widely because they are typically designed to address specific needs in the state. State-generated funds may go to senior housing, first-time homeownership, farmworker housing, homeless programs, and other housing-related activities. In general, these sources typically are used to create or rehabilitate affordable housing instead of providing direct subsidies to renters or owners. Eighty percent of the $50 million in GO bonds approved by Rhode Island voters in 2016, for example, was committed to affordable housing development.

States have taken pathways other than direct funding to promote affordable housing. These efforts have arisen for several reasons including a concern about the extent that publicly funded efforts such as HTFs at the state level can be expanded or at least be sustained (Scally 2012). Arguably, the two most cited state affordable housing approaches are regulatory rather than direct program funding. Specifically, they are state mandates for localities to plan for housing and to promote inclusionary housing.

The number of states requiring localities to plan for housing has been estimated to range between 25 and 50 percent (Pendall 2007). In a recent article, Ramsey-Musolf (2017) provides an exhaustive inventory of housing planning across the states. He is careful to qualify the housing plan requirement among the states based on contextual differences. Ramsey-Musolf shows that 23 states (46 percent) require local housing plans under various conditions and that three states make the housing plan optional for localities.[8]

Some research suggests that state-mandated housing planning can influence the support of localities for affordable housing (Basolo 1999). However, there is limited research on the effect of state-mandated planning, and existing results present a complicated picture. For example, a recent study from California, a state requiring local housing plans, shows that cities with compliant housing plans (reviewed and found compliant by the state) did produce more low-income housing over time than non-compliant cities, but overall housing production in the compliant cities over time declined (Ramsey-Musolf 2016). The strength of state requirements for housing plans is likely to influence results from state to state. Pendall (2007) found that only a handful of states, California, Florida, New Jersey, Oregon, and Washington, have strong planning requirements that compel localities to include more effective actions promoting affordable housing.

As mentioned earlier in this chapter, localities often resist the development of affordable housing in their jurisdictions. This propensity by localities has resulted in some states adopting inclusionary or anti-snob policies to address the issue. Probably the two most widely known examples of this type of policy are from New Jersey and Massachusetts. In the case of New Jersey, the state policy stemmed from the ruling of the New Jersey Supreme Court on the *Mt. Laurel I and II* cases (1975 and 1983) concerning a locality's economic discrimination through zoning practices. Taken as a whole, the Mt. Laurel decisions are known as the Mt. Laurel Doctrine and resulted in a requirement that all localities in the state plan, zone, and take "affirmative actions" toward meeting their "fair share" of the regional need for low- and moderate-income housing. The Council on Affordable Housing within the New Jersey Department of Community Affairs has the responsibility for administering activities associated with state policies created in response to decisions on the Mt. Laurel cases. These cases have been written about extensively (Basolo and Scally 2008; Calavita et al. 1997; Kirp et al. 1995).

A Massachusetts law known as Chapter 40B or the Comprehensive Permit Act began in 1969 as an effort to streamline affordable housing developments in that state. It addressed unreasonably burdensome conditions placed on developers seeking project approval. The appeals process is open only when a jurisdiction has less than 10 percent of its housing inventory affordable to low- and moderate-income households (Bratt and Vladeck 2014; Cowan 2006). Research on 40B has focused on its contentious history or politics (Krefetz 2016) and its effectiveness (Fisher 2008; Fisher and Marantz 2015). Generally, researchers and policy observers have mixed opinions on the effectiveness of inclusionary and anti-snob policies.

This section has provided a general overview of the work of states in the domain of affordable housing. Table 28.2 lists the sample of major governmental housing efforts by initiating level of

Table 28.2 Selected State Housing Efforts by Origin of Program and Usage Level.

	Most Common	*Less Common*
Federal (Authorization, Programs, Funding)	Housing bonds Mortgage revenue bonds Low-Income Housing Tax Credit CDGB HOME	Mortgage Credit Certificates
State-Initiated (Policies, Funding, Programs)	Housing trust funds	Required housing plan Inclusionary housing/fair-share General obligation bonds State tax credits

Source: author.

government discussed in this section divided by the author's assessment of their degree of use in states. In the next section, a brief case study of housing activities in California is presented.

Affordable Housing Efforts in California

States are actively engaged in affordable housing efforts, although the level and type of state housing activities vary from state to state. In this section, the focus is on California, a state that is considered liberal and, thus, more likely to have a higher level of support and innovation related to housing needs (Basolo and Scally 2008). In addition, this state was selected because the author has extensive experience in the affordable housing field in California. I draw upon experiential knowledge gained through many conversations, meetings, as well as presentations with housing professionals in the state.

Several characteristics of California are important to understand when discussing housing policies and programs in the state. First, California is considered a liberal or progressive state. This claim is supported by voting and party dominance in the state. Currently, the governor is a Democrat, the Democrats have a majority in the house and senate, and Hillary Clinton, the Democratic presidential candidate in 2016, received 61.6 percent of the vote within the state (Ballotpedia n.d.b). Second, the state essentially divides affordable housing activities across three entities: the California Housing Finance Agency (relatively narrow in scope), the Department of Housing and Community Development (a policy hub with a wide scope of programs), and the Tax Credit Allocation Committee[9] (also narrow in scope). Finally, the state's socio-demographics and significant housing needs distinguish the state from many others in the nation.

California is the largest state in the U.S. based on its population of over 38 million residents (see Table 28.3). According to Career Trends (n.d.), using 2015 data, California has the eighth highest cost of living index, driven largely by the cost of housing, which is 45 percent above the average for the U.S. as a whole. Consistent with this estimate, at $385,500, California's median housing value is second only to Hawaii; similarly, California, at $1,149 a month, comes in just behind Hawaii as the state with the highest median contract rent. California has a relatively high median household income at $61,818, well above the national figure (U.S. Bureau of the Census 2016), but it also has a very high housing wage of $28.59;[10] only Hawaii and the District of Columbia have a higher housing wage (National Low-Income Housing Coalition 2016). California's poverty rate of 16.3 percent is also above the national average of 15.5 percent. While homeowners comprise more than 50 percent of the households in the state, California's homeownership rate is almost 10 percentage points below the national rate. Overall, 45.1 percent of households are housing cost burdened (i.e., they are paying more than 30 percent of their income on housing), which is much higher than the national figure; a higher proportion of renter households are cost burdened (56.9 percent) than owner households (35.7 percent) in California.

Federal housing policy recognizes a range of special populations in its programs. As mentioned previously in the chapter, the homeless and persons with AIDS are two of these groups. While these special populations tend to be undercounted, it is still possible to compare available figures. Recent counts show that California has 115,738 homeless (0.30 percent of the total population), which is higher than the nation as a whole (.18 percent); and 4,728 people (0.01 percent of the total population) have HIV,[11] which is similar to the national percentage.

Clearly, California has serious housing needs related to affordability for renters and owners. The housing cost burdens and the low homeownership rate suggest a need for assistance to renters and potential owners. Furthermore, there are sufficient numbers in the special populations, specifically the homeless and people with AIDS, that deserve the attention of the state.

A substantial level of federal assistance comes to California due to its population size. Therefore, a good starting point to understand the state's orientation to affordable housing is a review of two documents prepared as part of receiving certain federal funds. First, the Consolidated Plan (CP), required

Table 28.3 Housing and Socio-Demographics, California and the U.S.

Characteristic	*California*	*United States*
Population	38,421,464	316,515,021
Cost of Living Rank[a]	8th	n.a.
Median Housing Value	$383,500	$178,600
Median Contract Rent	$1,149	$776
Housing Wage[b]	$28.59	$18.60
Median Household Income	$61,818	$53,889
Poverty Rate (Individuals)	16.3%	15.5%
Homeownership Rate	54.3%	63.9%
Housing Cost Burdened[c]	45.1%	35.1%
Owner-Occupied	35.7%	26.3%
Renter-Occupied	56.9%	51.8%
Homeless Persons[d]	115,738	564,708
Persons Diagnosed With HIV[e]	4,728	40,040

Note: The data are from the U.S. Census, American Community Survey, 2015 five-year estimates, unless noted otherwise.

a Career Trends (n.d.).

b National Low-Income Housing Coalition (2016); U.S. average calculated by author from 50 states plus Washington, DC.

c Housing cost burden was calculated from ACS, 2015 five-year estimates, by the author. This calculation (universe = occupied housing units) used paying 30 percent or more of income toward housing and included only households reporting at least some income and/or some cash rent in the denominator.

d National Alliance to End Homelessness (2016) (from statutory Point-in Time Counts).

e Centers for Disease Control (2016).

Source: author, based on sources listed earlier.

by HUD of jurisdictions receiving certain federal funds (CDBG, HOME, ESG, and HOPWA), states goals and provides projected funding allocations. Second, the Qualified Allocation Plan (QAP) is a requirement of the LIHTC program and is the framework for the selection of projects to receive tax credits. The QAP is usually revised annually and reveals the state priorities in any given year based on the discretionary portion of the scoring criteria.[12]

California's CP, 2015–2020, states six goals, which can be collapsed into five general categories: production and rehabilitation of affordable rental units, support for affordable homeownership and rehabilitation of owner-occupied housing, homeless assistance, economic development, and non-housing community development (see Table 28.4). The goals summary table in the CP shows the state's planned division of program funds over the period 2015–2019. In this table, the funds for CDBG, HOME, ESG, and HOPWA total more than $414 million. Over 80 percent of the funds during the period are directed toward affordable housing activities, 32.6 percent for rental housing, and 28.7 percent for ownership programs. Homeless activities are to receive 23.4 percent of the total funds. The remaining funds, about 15 percent, go to non-housing activities. These goals are consistent with needs, particularly the preference for rental assistance in this renter high cost burdened state, followed by homeownership assistance in a state with a relatively lower homeownership rate and a high rate of housing cost burdened owners.

The QAP as an element of the LIHTC program focuses only on affordable rental production and rehabilitation. Moreover, in California, distribution of the tax credits supports developments that provide nearly all units affordable to low-income households. As shown in Table 28.4, about 98 percent of the tax credit allocation of the 9 percent and 4 percent credits is going to producing low-income units. California also expresses preferences through the percentage of tax credits distributed to the

Table 28.4 California's Use of Federal Housing Programs/Authorization.

Selected Housing Efforts	*California*
Consolidated Plan, 2015–2020[a]	
Production & Rehab of Affordable Rental Units	32.6%
Affordable Homeownership & Rehabilitation	28.7%
Homeless	23.4%
Economic Development	3.8%
Non-Housing Community Development	11.5%
LIHTC Program, 2015[b]	
Housing Credits Awarded, 2015 (9%)	$91,101,325
State Tax Credits Awarded	$111,069,513
Number of Units Produced	4,903
Percentage Low-Income	97.8%
Housing Credits Awarded (4%) With State Credits	$137,554,828
Number of Units Produced	589
Percentage Low-Income	98.1%
Housing Credits Awarded (4%) Tax-Exempt Housing Bonds	$132,476,827
Number of Units Produced	13,012
Percentage Low-Income	97.9%
Private Activity Bonds[c]	
Tax-Exempt Housing Bonds Issued	$2,861,000,000
Mortgage Revenue Bonds Issued	$200,000,000
Mortgage Credit Certificates	Abandoned carry forward

a Total used for percentages includes CDBG, HOME, ESG, and HOPWA and excludes approximately $3.2 million from the Lead Hazard Control Program.
b Data on the LIHTC program come from the California Tax Credit Allocation Committee (n.d.).
c Data on Private Activity Bonds are from Council of Development Finance Agencies (2016).

Source: author, based on sources listed earlier.

type of development and other variables (e.g., location of proposed project), as well as special scoring criteria (e.g., access to transit and other services). In California, the QAP distributes almost a third of the tax credits to projects serving affordable, large, family developments. California also provides amenities points for locations that offer access to transit, parks, libraries, grocery stores, and health services, as well as provision of good Internet service and the use of green building techniques. This type of scoring reflects a preference for avoiding isolation of LIHTC developments from needed services and for supporting the state's aggressive agenda to reduce greenhouse gas emissions.

California uses PABs, tax-exempt housing, and mortgage revenue bonds to support housing needs. Earlier in this chapter, I mentioned that California issued approximately $2.862 billion in multifamily housing bonds in 2015, which reflects substantial effort toward much-needed affordable rental housing. As shown in Table 28.4, a portion of these bonds, about $132.5 million, was combined with 4 percent federal credits to make the developments possible. Again, these amounts demonstrate considerable state effort toward affordable rental housing in California. In support of homeownership goals, California was less aggressive, with PABs issuing $200 million in MRBs; the state abandoned $1.28 billion in MCC carryover. These decisions are likely related to the high cost of homeownership in the state.

California uses its own resources to complement federal housing programs. Authorized in 1987 by Chapter 1138 of the California statutes, the state tax credit[13] (see Table 28.4) has boosted the effectiveness of the other available subsidies. For example, developers typically combine the state tax credit with bond funds as a financing strategy for affordable rental units. The state tax credits—like the federal LIHTC—are off budget, making them less controversial than direct expenditures. The state also has an HTF, but as of the end of 2016 did not have a dedicated, reliable source of revenue to fund it. Campaigns to adopt a real estate transfer fee for funding the HTF have been unsuccessful in the past. At the same time, California voters have been willing to pass GO bonds to fund affordable housing programs. In 2002, Proposition 46 passed, allowing California to issue $2.1 billion in GO bonds for housing (see Basolo and Scally 2008 for more details). In 2006, California again approved an issue of GO bonds called the "nation's largest housing bond" by the California Department of Housing and Community Development (HCD) (California Department of Housing and Community Development n.d.); this time the amount was $2.85 billion and included language encouraging low-income housing development in urban settings near transit. In 2014, a $600 million GO bond was approved by voters (Proposition 41) aimed at affordable multifamily supportive housing for the homeless and veterans. The 2002, 2006, and 2014 bonds have resulted in sustaining some existing programs and creating new programs aimed at the development or preservation of affordable housing. They supported multifamily rental housing, an infill infrastructure program, farmworker housing (rental and ownership), the Innovative Homeownership Program, supportive housing for homeless youth, a transit-oriented development program, the HTF, and other initiatives. However, in its 2015–2016 annual report, HCD reports that Proposition 46 funds are nearly gone, just a small amount of returned funds are available for reprogramming, and only about $90 million in Proposition 1C funds remain (California Department of Housing and Community Development n.d.).

HCD offers other programs to serve housing needs. A mainstay program for HCD has been its Multifamily Housing Program (MHP). This program provides assistance for the development of affordable rental housing, permanent and transitional housing, and has received general funds ($47.5 million in 2015–2016) and GO bond funding. HCD also administers a housing program, funded by state general funds, serving migrant farmworkers and an emergency program authorized by the state legislature to help relocate and provide rental assistance to households without access to clean water due to extreme drought conditions in parts of California.

California has adopted some important policies to boost awareness of housing needs. For example, California uses the mandated local housing plan to encourage thinking about, and development of, affordable housing by localities.[14] Specifically, city housing staff, elected officials, and citizens (through public meetings on the plan) review housing needs and formulate goals and strategies to address these needs; these activities likely would not occur with the same level of effort and involvement if the state did not require local housing plans. Cities and counties in the state are given an allocation of their fair share of the regional housing need by income level, and they must plan for this allocation. Localities are required to do an inventory of sites to ensure they have adequate and suitable sites (identified at the parcel level) available to absorb the housing allocation. If adequate sites are unavailable, the jurisdiction must rezone to accommodate the shortfall. California's planning requirements for housing are among the most demanding in the nation.

California's efforts to provide affordable housing are consistent with a progressive agenda. The state legislature and the populace have supported efforts to deliver more affordable rental opportunities and homeownership including a variety of mechanisms to fund housing programs and policies to encourage planning for housing at all income levels. The state has also been responsive to special needs populations such as the homeless[15] and unique circumstances such as drought conditions to ensure California residents live in affordable, decent housing. Nevertheless, California does benefit from significant federal housing funding, which offers vital assistance through financing schemes and grants. However, California has also experienced some loss in the area of affordable housing funding. For example, in 2011, the State Legislature began the process to end local redevelopment agencies (RDAs), which, as part of state law, required city and county RDAs to set aside 20 percent of tax increment revenue from redevelopment project areas for low- and moderate-income housing. State redevelopment law generated substantial resources for affordable housing, but shifted tax revenue from other public activities. In a tight budget atmosphere and with questions about the use of some redevelopment revenue,[16] RDAs and their affordable housing streams were dissolved amid local and housing advocates' protests.[17] Overall, despite challenges, California politicians and policymakers have shown an understanding of housing needs and have acted to some extent to address them.

Conclusion

States are in a unique position to promote affordable housing. Situated between the federal government and localities, states offer a range of housing policies, provide substantial funding support for projects, and can counter the tendency for local governments to resist affordable housing development. Much of state funding for affordable housing comes directly from, or is tax assisted by, the federal government (e.g., CDBG grants, tax credits, and tax exempt bonds). While some states remain very dependent on federal support to fund housing initiatives, most states have established their own housing trust fund with a permanent stream of revenue from the state. However, states vary widely in the degree to which they have actually used HTFs to address low-income housing needs.

States have increased their activity in affordable housing financing, programs, and policies over the past 40 years. California is among the leaders in funding low- and moderate-income rental housing and related programs. Furthermore, California is among only a few states that require cities and counties to produce housing plans; it also reviews these plans for compliance with state law. The transition from an Obama to a Trump administration in Washington has made California's leadership in housing planning even more important. Should President Trump's proposed cuts take place, states are likely to play an increasingly larger role in the funding and delivery of affordable housing. Other states should look to California for ways to move forward.

Notes

1 "Affordable housing" is a general term with various interpretations. In this chapter, affordable is defined as housing for households with low- and moderate-incomes in a region. Typically, the region is a county area, and a low- or moderate-income household is defined as one earning up to the county median household income (in some places, moderate income is defined as up to 120 percent of the area median).
2 The amount does vary among the states, however, because states are allowed to carry forward unused balances for a certain period of time.
3 Federal lawmakers, from time to time, question continuing these tax-exempt bonds. This action was discussed by Congress in developing the 2018 tax reform legislation. However, as passed, this legislation maintains the tax-exempt status of these bonds (Ashraf et al. 2017).
4 The tax credits are applied to an investor's tax liability over a ten-year period, so the value of the yield from the credits typically is below the $1 face value to account for the decline in the value over time.
5 The National Bureau of Economic Research identifies the period from December 2007 to June 2009 as the Great Recession (National Bureau of Economic Research n.d.).
6 The Center for Community Change (2016) reports that housing trust funds began in California and Maryland during the 1970s. It may be that this organization and Nenno (1996) use somewhat different definitions of an HTF.
7 Alaska, Mississippi, and Wyoming do not have a state HTF. Some states have more than one housing trust fund, bringing the total above 50 (Center for Community Change 2016).
8 Local housing plans are prepared by cities and counties and generally are a chapter or element of a local comprehensive plan. Local housing plans are separate documents from consolidated plans, which are required of jurisdictions receiving certain federal funds such as Community Development Block Grant or HOME monies. However, in a jurisdiction preparing both types of plans, some overlap of content such as housing needs and strategies likely will occur.

9 The Tax Credit Allocation Committee has the responsibility of distributing the LIHTCs in the state. This Committee is under the State Treasurer's office, and the State Treasurer along with several other high ranking state administrative officials and several political appointees compose the Committee.

10 Housing wage is defined as the hourly wage needed to afford a two-bedroom housing unit at the fair market rent, as determined by the HUD, without spending more than 30 percent of one's income.

11 Diagnosis of HIV is used as a proxy for AIDS as it is the first stage in a progression of the virus.

12 California prepares numerous housing planning-related documents including a Statewide Housing Assessment, which primarily focuses on housing needs in the state. The current draft of the SHA also offers a very general description of some of the federal programs supporting housing (e.g., LIHTC) and some initial recommendations for policymakers; however, it does not explicitly relate to the QAP or Consolidated Plan.

13 The California tax credit program allows the investor to use the credits to reduce state tax liability over a four-year period. California also has an option for investors to have the tax credits taken as personal property (known as certificated tax credits), which is a workaround losing the state tax deduction on federal income taxes. This approach retains more of the value of the tax credit for investors (Mandel 2016).

14 California State Law requires every city and county to prepare a general (or comprehensive) plan. This general plan addresses eight elements: land use, circulation, housing conservation, open space, noise, safety, environmental justice, and air quality. The housing element (or plan) is unique among the required elements because "Provisions in the housing element are more specific and directive than other elements and contain detailed guidance and reviews" (Governor's Office of Planning and Research 2017, 89). Moreover, it is the only element that is reviewed by a state agency, the California Department of Housing and Community Development, for compliance.

15 Passed in 2007, Senate Bill 2 amended General Plan Law to include the following provision: "At least one zone shall be identified to permit emergency shelters without a conditional use permit or other discretionary action" (Government Code Section 65583 as stated in May 7, 2008, policy memo from the Deputy Director of the Division of Housing Policy Development, Department of Housing and Community Development). This provision is referred to as the "by right" emergency shelter provision by planning practitioners.

16 Governor Jerry Brown argued that RDAs (with their use of tax increment financing) took substantial resources away from other public services such as schools; other critics simply characterized some redevelopment activities as corporate welfare (Mishak 2011).

17 RDAs generated substantial funding for affordable housing, and no alternative funding of this magnitude has been identified or suggested. Recent policy efforts to address California's housing needs include 2016 legislation (Senate Bill 1069) to encourage development of accessory units by limiting obstacles such as parking requirements and a bill currently in process to establish a permanent source of funding for the state housing trust fund.

References

Ashraf, Saba, Charles S. Henck, Emilie R. Ninan, and Linda B. Schakel. 2017. "Private Activity Bonds and Stadium Bonds Saved in GOP Tax Reform Bill." *The National Law Review* (December): n.p.

Ballotpedia. n.d.a. "Rhode Island Affordable Housing Bonds, Question 7 (2016)."

Ballotpedia. n.d.b. "California Elections 2016."

Basolo, Victoria. 1999. "The Impacts of Inter-City Competition and Intergovernmental Factors on Local Affordable Housing Programs." *Housing Policy Debate*, 10(3): 659–688.

Basolo, Victoria, and Corianne P. Scally. 2008. "State Innovations in Affordable Housing Policy: Lessons from California and New Jersey." *Housing Policy Debate*, 19(4): 741–774.

Bratt, Rachel G., and Abigail Vladeck. 2014. "Addressing Restrictive Zoning for Affordable Housing: Experiences in Four States." *Housing Policy Debate*, 24(3): 594–636.

Calavita, Nico, Kenneth Grimes, and Alan Mallach. 1997. "Inclusionary Housing in California and New Jersey: A Comparative Analysis." *Housing Policy Debate*, 8(1): 109–142.

California Department of Housing and Community Development. n.d. *California Consolidated Plan 2015–2020*. Sacramento, CA: California Department of Housing and Community Development.

California Tax Credit Allocation Committee. n.d. *Fast Facts 2015 Totals*. Sacramento, CA: California Tax Credit Allocation Committee.

Career Trends. n.d. "Research Cost of Living."

Center for Community Change [Housing Trust Fund Project]. n.d. "What are Housing Trust Funds?"

Center for Community Change [Housing Trust Fund Project]. 2013. "The Status of State Housing Trust Funds . . . from a 2011 Survey with Updates."

Center for Community Change [Housing Trust Fund Project]. 2016. "Opening Doors to All: The 2016 Survey of Housing Trust Funds."

Centers for Disease Control. 2016. *HIV Surveillance Report: Diagnoses of HIV Infection in the United States and Dependent Areas, 2015*. Atlanta, GA: Centers for Disease Control.

Center on Budget and Policy Priorities. 2017. *United States Fact Sheet: Federal Rental Assistance*. Washington, DC: Center on Budget and Policy Priorities.

Choi, Sang Ok, Sang-Seok Bae, Sung-Wook Kwon, and Richard Feiock. 2010. "County Limits: Policy Types and Expenditure Priorities." *The American Review of Public Administration*, 40(1): 29–45.

Council of Development Finance Agencies. 2016. *CDFA Annual Volume Cap Report*. Columbus, OH: Council of Development Finance Agencies.

Cowan, Spencer M. 2006. "Anti-Snob Land Use Laws, Suburban Exclusion, and Housing Opportunity." *Journal of Urban Affairs*, 28(3): 295–313.

Davis, John Emmeus. 2006. "Between Devolution and the Deep Blue Sea: What's a City or State to Do?" In *A Right to Housing: Foundation for a New Social Agenda*, edited by Rachel G. Bratt, Michael E. Stone, and Chester Hartman, 364–398. Philadelphia, PA: Temple University Press.

Ellen, Ingrid Gould, Karen Horn, Yiwen Kuai, Roman Pazuniak, and Michael David Williams. 2015. *Effect of QAP Incentives on the Location of LIHTC Properties*. Washington, DC: U.S. Department of Housing and Urban Development.

Fisher, Lynn M. 2008. *Reviewing Chapter 40B: What Gets Proposed, What Gets Approved, What Gets Appealed, and What Gets Built?* Cambridge, MA: Rappaport Institute for Greater Boston, Harvard Kennedy School.

Fisher, Lynn M., and Nicholas J. Marantz. 2015. "Can State Law Combat Exclusionary Zoning? Evidence from Massachusetts." *Urban Studies*, 52(6): 1071–1089.

Goetz, Edward G. 1993. *Shelter Burden: Local Politics and Progressive Housing Policy*. Philadelphia, PA: Temple University Press.

Goetz, Edward. 1995. "Potential Effects of Federal Policy Devolution on Local Housing Expenditures." *Publius: The Journal of Federalism*, 25(3): 99–116.

Governor's Office of Planning and Research. 2017. *State of California General Plan Guidelines*. Sacramento, CA: Governor's Office of Planning and Research.

Joint Center for Housing Studies of Harvard University. 2015. *The State of the Nation's Housing*. Cambridge, MA: Joint Center for Housing Studies of Harvard University.

Kirp, David L., John P. Dwyer, and Larry A. Rosenthal. 1995. *Our Town: Race, Housing, and the Soul of Suburbia*. New Brunswick, NJ: Rutgers University Press.

Krefetz, Sharon Perlman. 2016. "Unlocking the 'Gates' and Climbing Over the 'Walls': Opening up Exclusive Communities. In *Private Communities and Urban Governance: Theoretical and Comparative Perspectives*, edited by Amnon Lehavi, 239–256. Basel, Switzerland: Springer International Publishing.

Mandel, Richard. 2016. "Certificated State Tax Credit Bill Will Increase Equity Pricing."

Mishak, Michael J. 2011. "California Gov. Jerry Brown Defends Cutting Redevelopment Agencies." *Los Angeles Times*, January 27.

Mueller, Elizabeth J., and Alex Schwartz. 2008. "Reversing the Tide: Will State and Local Governments House the Poor as Federal Direct Subsidies Decline?" *Journal of the American Planning Association*, 74(1): 122–135.

National Alliance to End Homelessness. 2016. *The State of Homelessness in America*. Washington, DC: National Alliance to End Homelessness.

National Bureau of Economic Research. n.d. *U.S. Business Cycle Expansions and Contractions*. Cambridge, MA: National Bureau of Economic Research.

National Council of State Housing Agencies. 2016a. *HFA Directory*. Washington, DC: National Council of State Housing Agencies.

National Council of State Housing Agencies. 2016b. *State HFA Factbook: 2014 NCSHA Annual Survey Results*. Washington, DC: National Council of State Housing Agencies.

National Low-Income Housing Coalition. 2016. *Out of Reach 2016*. Washington, DC: National Low-Income Housing Coalition.

Nenno, Mary K. 1996. *Ending the Stalemate: Moving Housing and Urban Development Into the Mainstream of America's Future*. Lanham, MD: University Press of America.

Nguyen, Mai Thi, Victoria Basolo, and Abhishek Tiwari. 2013. "Opposition to Affordable Housing in the USA: Debate Framing and the Responses of Local Actors." *Housing, Theory, and Society*, 30(2): 107–130.

Novogradac and Company. n.d. "Affordable Housing Resource Center."

Pendall, Rolf. 2007. *From Hurdles to Bridges: Local Land-Use Regulations and the Pursuit of Affordable Rental Housing*. Cambridge, MA: Joint Studies for Housing Studies of Harvard University.

Ramsey-Musolf, Darrel. 2016. "Evaluating California's Housing Element Law, Housing Equity, and Housing Production (1990–2007)." *Housing Policy Debate*, 26(3): 488–516.
Ramsey-Musolf, Darrel. 2017. "State Mandates, Housing Elements, and Low-Income Housing Production." *Journal of Planning Literature*, 32(2): 117–140.
Rosan, Christina. 2014. *Policy Shift: How the U.S. Developed a Hybrid Model of Affordable Housing Provision*. Washington, DC: Wilson Center.
Scally, Corianne P. 2009. "State Housing Finance Agencies Forty Years Later: Major or Minor Players in Affordable Housing?" *Journal of Planning Education and Research*, 29(2): 194–212.
Scally, Corianne P. 2012. "The Past and Future of Housing Policy Innovation: The Case of U.S. State Housing Trust Funds." *Housing Studies*, 27(1): 127–150.
Scally, Corianne Payton, and J. Rosie Tighe. 2015. "Democracy in Action?: NIMBY as Impediment to Equitable Affordable Housing Siting." *Housing Studies*, 30(5): 749–769.
Smith, Kevin B., and Alan Greenblatt. 2017. *Governing States and Localities*. Thousand Oaks, CA: CQ Press.
Technical Assistance Collaborative. 2014. *State Funded Assistance Programs*. Boston, MA: Technical Assistance Collaborative.
U.S. Bureau of the Census. 2016. *American Community Survey, 2015 5-Year Estimates*. Washington, DC: U.S. Bureau of the Census.
U.S. Department of Housing and Urban Development. 2015a. *Low-Income Housing Tax Credits*. Washington, DC: U.S. Department of Housing and Urban Development.
U.S. Department of Housing and Urban Development. 2015b. *CDBG Expenditure Reports, PY02–PY14*. Washington, DC: U.S. Department of Housing and Urban Development.
Washington State Finance Committee. 2002. *Meeting Notes Dated May 14, 2002*. Olympia, WA: Washington State Finance Committee.

29
PREDICTION IS DIFFICULT
The Future of Housing Policy and Housing Studies

Alan Mallach

Introduction

As Niels Bohr, or perhaps Yogi Berra, famously said, making predictions is difficult, particularly about the future. Since no one can know the future, predictive writing generally relies on extrapolation from present trends; where it does not, it is often better characterized as science fiction. A glance at the past few decades, however, makes clear how unreliable extrapolation is as a guide to the future; extrapolation would have enabled few to predict the fall of the Soviet empire, the Great Recession, and the banking crisis that continues to unsettle the European economy, the resurgence of populism and the politics of fear in the United States, or the wave of refugees that has shaken the European political scene. Thus, any predictive project needs to be pursued with humility, recognizing that even the near future, as with the past, is likely to be rife with unexpected shifts and reversals.

Bearing that in mind, in this chapter I will attempt to predict trends over the next ten or so years at three distinct levels. First, to discuss the underlying trends likely to drive future housing policy in developed nations such as the United States, Germany, or Japan, such as population decline, immigration, or economic growth; second, to explore the potential policy responses to those trends; and finally, to speculate on how the field of housing studies is likely to change in response to changes in both housing conditions and policies. Inevitably, my predictions are far more grounded in extrapolation from present trends, rather than speculation about what cannot possibly be known.

Trends Driving Future Housing Policy

The transformation of housing conditions for the great majority of the population of the world's developed nations is one of the great success stories of the twentieth century. During the course of the century, particularly during the decades immediately following World War II, millions of people who in previous decades would have lived in overcrowded, physically inadequate housing came to live in housing meeting modern standards of health and safety. Physical inadequacies, such as the lack of a private bathroom or adequate light and air, as well as severe overcrowding have been dramatically reduced. Despite this manifest achievement, developed nations continue to face growing housing policy challenges, some of long standing, and some only beginning to emerge.

While substandard and overcrowded housing conditions have been sharply reduced, they are no longer the norm; today, instead, they disproportionately affect populations that are to varying degrees marginalized—immigrants, the very poor, and minority communities that continue to be subject to *de facto* discrimination, such as African Americans in the United States or the Roma in Eastern Europe. As Europe's vast stock of social housing built from the 1950s through the 1970s ages, however, problems of housing quality, and the need to rehabilitate or replace this stock, are likely to increase. Moreover, as housing conditions have improved, housing affordability has often declined. More than one of four private market tenants in the European Union (EU) spends 40 percent or more of their disposable income on housing (Eurostat 2015), while nearly half of all tenants in the United States spend 30 percent or more of their gross income on housing (U.S. Bureau of the Census 2015). While homelessness is notoriously difficult to measure, it is a major challenge in both the United States and Europe; point-in-time surveys found that 578,000 people were homeless in the United States on a single night in January 2014 (National Alliance to End Homelessness 2015), while the Canadian Homelessness Research Network estimated that 30,000 Canadians are homeless on a single night (Gaetz et al. 2013).

Wide variations exist, moreover, among those countries generally considered developed nations. While the "severe housing deprivation rate"[1] in the EU as a whole is under 5 percent, it is much higher in the former Socialist countries, particularly Romania, where it is over 20 percent. EU countries, as well as developed nations elsewhere, show extreme variation on many other measures, including cost burden, the homeownership rate, the size of the social housing stock, or the share of young adults living with their parents, which ranges from barely 16 percent in Denmark to over 70 percent in Slovakia (Pittini et al. 2015). The point is that it is often inappropriate to generalize about housing trends across the entire developed world: What may be a crisis in one country may be a matter of little significance in another (see also the chapter authored by W. Dennis Keating).

Larger systemic issues, however, continue to plague developed countries, which are likely to become more severe in coming years. While housing quality as such may be largely adequate, issues of neighborhood condition and the integration of minority and immigrant communities have been a continuing and growing problem. Racial segregation, while slightly declining (Frey 2015), continues to be a major problem in the United States, exacerbated by growing economic segregation, particularly in older urban areas, but increasingly in the suburbs. The problems of integrating the French *banlieues*, despite repeated governmental policy interventions, appear to be as intractable as ever (Cortéséro 2012; Stébé 1999).

As refugees and immigrants continue to press against Europe's borders, pressures that are more likely to increase than decline in coming years, issues of spatial as well as social integration are likely to continue and become more pronounced, to the point of potentially undermining the social fabric of some European communities.[2] This issue cuts in two ways; at the same time as the problems of distressed neighborhoods and communities may be getting worse, gentrification of historically affordable sections of strong market cities like Berlin, Paris, and San Francisco is also on the rise, reflecting the extent to which economic polarization is becoming an endemic condition of many urban areas in the developed world. Gale divides cities into three tiers in terms of the incidence of gentrification, appropriately distinguishing cities like Philadelphia or St. Louis, which "while experiencing significant levels of gentrification, mostly in their cores, still have substantial areas of disinvestment in areas of low market demand," from cities like Detroit or Flint, where isolated, modest signs of gentrification in a few areas are far outweighed by continuing decline of once-vital middle-income and working-class neighborhoods (see the chapter authored by Dennis Gale).

While some nations will be dealing with immigrant integration, others will be grappling with the problem of population decline, which is likely to lead to increasingly difficult challenges of excess housing supply and housing vacancy. Many developed nations are likely to lose population over the coming decades; the hardest hit are likely to be in Eastern Europe, where Bulgaria and Latvia are projected to lose over 40 percent of their peak population by 2050, but they include other countries elsewhere in Europe as well as Japan (United Nations Department of Economic and Social Affairs 2015).[3]

In many of these nations, population loss is taking place simultaneously with growing concentration of the remaining population in major centers, so that secondary cities and regions are experiencing disproportionate population losses, while growing centers are struggling to meet housing demand and manage growth. In Bulgaria, "depopulation [has] pushed 172 towns to the verge of extinction, and completely erased 24 from Bulgaria's map" in 2012 alone (Crowder 2014). Sofia has continued to grow, however, and its share of the nation's population has risen from 12 percent to 17 percent since the 1980s. In Latvia, although the Riga region is more stable than much of the rest of the country, a recent article noted that "a few minutes by tram outside the old town, which is showered with public money, a different reality emerges. Areas such as the Moscow District are crammed with crumbling tenements and empty wooden houses; it could be the set for a ghost town in a low-budget western. The dereliction is leavened only by alcohol and second-hand clothes shops" (Pyzik 2013, n.p.).

Population loss in the United States or the United Kingdom is also a significant issue, but is seen as a regional more than a national problem, affecting both countries' historic industrial regions as well as, in the United States, many rural areas; the German perspective is similar, where despite evidence of shrinkage in some western areas, the issue is still widely seen as one facing the states of the former German Democratic Republic (GDR), rather than a national issue. These changes, and the pressure they generate for greater public investment, however, are taking place at a time when economic growth in most of the developed world is likely to be modest, imposing significant constraints on the availability of resources to address them.

The Course of Future Housing Policy

Housing policy in the developed world is a complicated tangle of laws, policies, and programs, including those designed to promote well-functioning markets, such as mortgage guarantees and secondary markets, supply-side programs to provide social and affordable housing,[4] demand-side programs such as housing allowances, fair housing and anti-discrimination programs, and a variety of programs targeted at particular housing needs, such as those of the homeless population. Within each of these categories, not only do programs and laws vary widely from nation to nation, but what are considered to be appropriate populations to benefit from public investment also varies widely. While the great majority of Austrians, for example, are potentially eligible for that country's social housing and about 80 percent of Singaporean households live in social housing (Kwak 2015), most social housing programs in the United States are limited to families in the lowest 25 percent of the national income distribution. In 2012, 85 percent of all residents in federally subsidized housing in the United States had incomes under $20,000 (Hollar 2014).

The course of postwar housing policy in much of the developed world is generally seen as one of two principal phases: a period from the 1950s through the 1970s characterized by the dominant role of state actors and by large-volume construction of social housing, generally under public sector ownership, and a subsequent period of retrenchment, widely seen as reflecting the rise of a neoliberal agenda, in housing policy as in many other arenas of economic activity. This is, of course, a great oversimplification, as many authors have recognized (Doherty 2004; Dodson 2006). Although neoliberalism has played an important role in framing today's housing policies, as in the sell-off of public housing to tenants in countries like Israel and the United Kingdom, it is typically part of a policy framework that is far from the stereotype of a simplistic reliance on the market as the solution to housing needs. On the contrary, many housing policies that have emerged in recent decades represent thoughtful efforts to balance market and social considerations in light of some of the problems created by the statist policies of the earlier period, including the mix of market construction and social ownership in France (Mallach 2010), as well as inclusionary strategies such as Section 106 agreements in the UK (Monk 2010), and Ireland's Part V (Calavita and Mallach 2010a, 2010b).

Some developed nations, such as the U.S., have expanded demand-side assistance in the form of housing allowances, while in some countries with historically large state housing sectors, such as France and the UK, the creation of social housing continues, but at a more modest rate and largely under the auspices of non-governmental entities (NGOs) rather than public bodies (see also the chapter authored by Rachel Bratt).

In contrast to countries like France and the UK with their highly developed social housing systems, many other countries, particularly many of the former Socialist states in Central and Eastern Europe, have dramatically reduced their social housing stock from their Socialist days and tend to have uneven, leaky social safety nets, particularly with respect to housing. In recent years, many of these formerly Socialist states including Croatia, Lithuania, and the Slovak Republic have adopted new housing programs and policies, but they tend to be small in scale and limited in their impact; that appears to reflect not only these countries' severe fiscal constraints, but also their uncertainty or ambivalence about how best to address the intricacies of the housing sector, not to mention the complex political and social issues involved.

In general, the prognosis for housing policy in most developed nations in the coming decade is not positive. Over the coming decade or more the global environment seems unlikely to offer either the political stability or sustained economic growth on which strong policy development is likely to depend. The reasons for this situation are complex and multi-faceted, but many recent economic, demographic, and migration trends all point to sustained economic and political instability in much of the developed world, as well as most developing nations (Friedman 2016). Within that setting, the magnitude, complexity, and highly contested nature of the emerging housing policy challenges make it even less likely that many nations will be able to respond to them, whether for economic, political, cultural, or other reasons. These circumstances are in sharp contrast to the "*trente années glorieuses*" (30 glorious years) from the late 1940s to the late 1970s, when not only were housing policy challenges, although daunting, more straightforward and more matters of consensus, but the political and economic climate was more supportive of ambitious public goals and policies. In retrospect, however, much of the product of those years, particularly with respect to massive modernist housing estates and the often destructive so-called "renewal" of urban centers and neighborhoods, was far from glorious in its effects.

As suggested in the discussion of the factors driving housing trends, developed nations are likely to see at least seven distinct housing policy challenges over the coming decade or more, listed without any attempt to suggest relative importance:

- Improving housing quality and better integrating minority populations
- Upgrading deteriorating older housing stock, particularly social housing built in the 1970s or earlier
- Stabilizing older urban and suburban neighborhoods in the face of growing economic and demographic challenges
- Reducing the growing cost burden for lower-income households
- Encouraging revitalization of distressed low-income communities, while simultaneously expanding housing opportunities for minorities and lower-income people throughout regions
- Addressing the consequences of shrinking populations, including increased housing vacancies and reduced ability to finance public services
- Accommodating and integrating growing immigrant populations

The last challenge is something of a "wild card," since—in contrast to the first six, where the magnitude of the challenge can be at least approximated—the number of immigrants who will be seeking and gaining entry to the developed world, and Europe in particular, over the next decade is impossible to predict (see also the chapter authored by Alfons Fermin and Frank Wassenberg). Ironically, at least in theory, in the long run large-scale immigration could help ameliorate some other problems, particularly the housing vacancies associated with shrinking urban populations (Coburn 2016). In the short term, however, immigration has become and will continue to be a further destabilizing factor (Traub 2016).

With continuing uncertainty about the European economy, the weakening of the EU, the threat of terrorism, and the rise of xenophobic regimes in countries like Poland and Hungary along with increasing support for similar movements elsewhere, housing policies are likely to suffer. Policies to improve the condition of the Roma minorities in Eastern Europe, however inadequate, have been in no small part a product of both pressure and financial support from the EU (Ram 2003). With both likely to diminish in the future, such policies are likely to suffer. Immigrant integration may not be a major issue in some developed countries, again particularly those of Eastern Europe, because, as has already begun to happen, these nations may bar all but a small number of immigrants from entry.

Again, this is not true for all developed nations. Germany accepted about a million immigrants in 2015 alone; that number is likely to grow over the coming decade, and it is likely that Germany will adapt housing policies and allocate public resources to immigrant integration over the next decade. That notwithstanding, immigrant integration has already been shown to be difficult, and the issues associated with both the speed and scale of the latest immigrant wave are likely to be a destabilizing force in German society for many years.

With adoption of the *Nouveau Programme National de Renouvellement Urbain* in 2014 and a commitment of €5 billion in public funds added to the €17.5 billion previously appropriated to the *Agence Nationale pour la Rénovation Urbaine* (ANRU) in 2003 (*Agence Nationale pour la Rénovation Urbaine* 2015), France appears likely, barring a significant shift in political direction, to continue the large-scale investment in urban regeneration and socioeconomic integration initiated in 2003 with the creation of ANRU. Despite this substantial level of expenditure, most of the investment was directed toward the *banlieues*, although the larger social and political issues involved make the outcomes of these efforts still uncertain. Similarly, New Zealand, relatively less affected by global turmoil, adopted a comprehensive strategy for Maori housing improvement in 2014, which is likely to continue over the decade, although the level of public financial involvement is modest, especially if compared to that of France (Ministry of Business, Innovation and Employment 2014). These are likely to be exceptions; overall the trend is likely to be that of retrenchment or, at best, stabilization of effort, although the recently elected Trudeau administration in Canada has also proposed a significant increase in social housing spending in its 2016 budget (McMahon 2016).

The issue of cost burden is another area where significant new policies appear unlikely, and the problem is likely to become greater rather than being ameliorated over the coming years. A recent report from the Council of Europe Development Bank (CEB) found that "European citizens [. . .] spend more and more on housing, with a disproportionate burden for the people at risk of poverty, while given fiscal consolidation, governments generally are struggling to support the affordable housing sector despite a growing demand. The impact [. . .] seems to be greater now than in previous years, since the crisis and austerity measures taken have a strong time-lag effect" (Athenosy and Revenco 2015, 22). Between 2008 and 2013, the percentage of lower-income households with cost overburden increased in 23 out of 28 EU nations (Fondation Abbé Pierre 2015, 19).

Cost burden is on the rise in the United States as well, where the percentage of renter households spending 30 percent or more of gross income for shelter increased from 37 percent to 48 percent between 2000 and 2013. Housing costs have risen faster than incomes in most developed nations from 2000 to the present. While there are housing allowances in most developed nations, they vary widely on the basis of eligibility and level of benefit; they are more extensive and in some cases provided as a right to income-eligible households in the wealthier countries of Northwestern Europe, less so in the former Socialist countries in the EU, or in the United States and Canada. As a general rule, however, public support for cost burdened families has not risen in proportion to the increase in housing cost burden over the past decade. Even in the countries with the strongest social safety nets, "while the living conditions of their poor households are still preferable to that of other countries, housing inequality in these countries is increasing more rapidly than elsewhere and the holes in the safety net are getting bigger" (Fondation Abbé Pierre 2015, 13). Waiting lists for social housing, and in the United States for housing vouchers as well, are growing far faster than the social housing stock throughout the developed world (Pittini 2012, 8).

Can one expect any significant policy movement toward addressing cost burden over the coming decade? Barring a dramatic increase in economic growth in the developed world, which appears unlikely in the near future, it is hard to imagine that redistributionist national policies will emerge or expand to the extent that would be needed to address an issue that disproportionately affects poor or near-poor households, and moreover, are both highly expensive and—assuming house prices continue to rise faster than incomes—include substantial cost multipliers over time. At best, one can expect a variety of modest enactments to address narrowly targeted populations with clearly defined needs, such as homeless families, elderly households, or people with disabilities. This is true in the United States as well as in Europe. While the conditions of the majority of poor renters in the United States who do not have access to housing vouchers are dire in the extreme (Desmond 2012, 2016), and while the United States certainly has the economic resources to meet their needs, the political climate, particularly since the 2016 presidential election, makes any significant new policy developments in this area or any meaningful steps toward reducing the growing levels of income inequality highly unlikely.

The final areas are those where housing policy overlaps with urban policy writ large, and leads to many difficult challenges. These include not only maintaining the aging urban housing stock, but preserving the vitality of central or inner cities faced simultaneously with destabilization of historically sound neighborhoods and social conflicts exacerbated by gentrification, as well as addressing the consequences of urban shrinkage.

Here, too, for the most part future housing policy will reflect the continuation of policies of recent years. Germany appears committed to a strong urban policy, shown by the cumulative national, state, and local government commitment of over €5 billion to the urban regeneration or reconfiguration programs falling under the *Stadtumbau* (city rebuilding) rubric. It is likely that urban policy will continue to be a priority in Germany and many other Northern European countries over the coming decade, and at least possible, in light of the urgent challenge of immigrant integration, that increasing efforts will be made to link urban regeneration and integration policies.

By contrast, urban policy is less of a priority in most other developed nations. Most post-Socialist countries have pursued few systematic efforts to foster urban vitality beyond the revitalization of downtown areas; not surprisingly, the pace of suburbanization around major Eastern European cities has been rapid, although it has slowed since the onset of the fiscal crisis as well as from the effects of population decline. Moreover, up to now, issues of urban shrinkage—as distinct from general concerns about the population loss associated with migration and demographic trends—have not been consistently addressed as a matter of public policy in these countries (Stryjakiewicz 2013), although they have been a matter of concern at the EU level, which has invested large sums in the revitalization of shrinking cities in post-Socialist countries (Stryjakiewicz et al. 2012).

The same is largely true in Japan, but may be shifting. Although the discussion of low fertility rates, aging, and the depopulation of isolated outlying regions has taken place since the 1970s, it is only in the past few years that attention has begun to be given to the reality of widespread shrinkage as a national problem, particularly with respect to the rapidly growing number of vacant or abandoned houses in secondary cities and postwar suburbs. It is possible, as conditions worsen over the next decade or so, that Japan may embark on policies analogous to the German *Stadtumbau* programs designed to address the particular challenges of shrinking cities, but in light of its continued fiscal constraints, public investment at more than a modest scale appears unlikely.

In sum, I consider the prospects poor for significant policy departures or significant new public investment to address growing housing and neighborhood problems in most developed nations. The likely outcome over the coming decade will be a gradual aggravation of housing problems in many if not most developed nations, including increased housing cost burdens for lower-income households, deterioration of older housing estates, growing social, spatial, and economic disparities within each society, continued and potentially exacerbated isolation of immigration populations, and finally, in some countries, increasing problems associated with shrinking urban populations, housing vacancy, and deterioration of public services.

The Course of Future Housing Studies

Assessing the future of housing studies is, if anything, even more difficult than predicting the future of housing policy, since it is at least easier to identify the central issues around which housing policy will or will not coalesce. The field of housing studies is a large and diverse one, reflected not only in the diversity of multiple journals devoted to housing (such as *Housing Studies*, *Housing Policy Debate*, the *Journal of Housing Economics*, and others), but also in much of what appears in related fields such as urban studies, city planning, sociology, urban economics, and geography. The range of topics covered in a single major journal is vast; recent issues of *Housing Studies* have included contributions on economic studies of prices, studies of population migrations, neighborhood crime, urban regeneration, housing vouchers, post-disaster reconstruction, social housing policy, gentrification, and far more. Recent special issues have addressed the politics of housing, social mix, rental housing, and the future of public housing. Predicting the future behavior of the thousands of scholars from many disciplines and national backgrounds who are engaged in housing studies is akin to predicting the behavior of a particle subject to Brownian motion. Thus, in this section I will discuss those directions that I would like to see the field of housing studies take, as much as those that I expect it will take in the future.

When the focus of a discipline changes, it is likely to do so for one of two reasons. New models or ideas change disciplinary paradigms from within, as when Kahneman and Tversky (1979; Kahneman 2003) brought behavioral economics into the mainstream of economic research or when changing external conditions led researchers to change their focus, as witnessed in the spate of studies on mortgage foreclosure that have appeared since 2007 in the wake of the mortgage crisis in the United States (Frame 2010). Since housing studies is not a single discipline, strictly speaking, but a multidisciplinary field focusing on a cluster of related social, economic, and political issues falling under the housing rubric, change in the field is far more likely to be driven by changes in external conditions. These can be either changes in objective conditions, such as the wave of foreclosures, or policy changes, such as the increased public policy focus on mixed-income housing that emerged in the 1990s in a number of developed nations, which has arguably spawned an entire sub-discipline of scholarly books and articles. Finally, scholars are neither immune to ideological pressures nor to the human tendency to pay particular attention to whatever issues are receiving media attention at any given moment.

In the absence of a significant, visible paradigm shift, however, either in the discipline or in the external conditions that form its subject, disciplines change only gradually and incrementally. Such gradual, incremental change, rather than any dramatic shift in focus, is likely to be true of the housing studies field over the coming decade. Interdisciplinary work integrating spatial, social, and economic perspectives is likely to increase, furthered, perhaps, by the increasing availability of large-scale administrative datasets, including data in such areas as crime incidence, taxation, real estate transactions, and public health, offering new research opportunities. While such datasets have been mined for a number of years by researchers in Sweden and the Netherlands (Musterd and Andersson 2005, among others), they are only beginning to be available in the United States. One also hopes to see more research that links housing and neighborhood issues, including research that looks at neighborhood change in a more integrated fashion than has often been the case, and more creative uses of new data sources to better understand the complex dynamics of neighborhood change.

The field's focus on the housing issues of disadvantaged and marginalized populations, and the relationship between housing issues and social isolation or integration, already strong, may grow. In light of the scale of the refugee crisis that emerged in 2015 and its multiple implications for housing and urban policy, we may see more research on the housing and neighborhood effects of immigration, particularly in Germany, where they are likely to be most pronounced.

More policy-oriented research may emerge, particularly focusing on the challenges of resource allocation under economic constraints in the context of the increasing severity of the challenges represented by growing cost burdens and the aging of the social housing stock. More policy research that looks at critical sub-state actors, particularly state/regional or local governments, would be welcome, in that much current European and Japanese policy research tends to focus on national policy with little sensitivity to significant regional and local variations in how that policy affects actions and behavior, even in societies with highly centralized governmental systems such as Japan.

As populations continue to decline throughout much of the developed world, including its cities, the urban and housing policy issues flowing from shrinkage, including market failure and excess housing supply, are likely to draw increasing attention. In light of the long-term downward population trajectories of many countries, including Japan and many Eastern European counties, the question of whether—and if so, how—cities can thrive while shrinking is likely to be an increasingly important area of investigation. Given the range of issues associated with shrinkage (Dewar and Thomas 2013; Weaver et al. 2016), and the number of different national settings where it will be an issue in coming years, this area may come to be seen as a distinct sub-field within housing studies.

Along similar lines, one hopes to see more cross-national research, which is already being conducted, including the volume on inclusionary housing by Calavita and Mallach (2010a) mentioned earlier, important work by Galster on social mix in housing (Galster 2007), and important work by Varady and his colleagues on public housing restructuring (Kleinhans and Varady 2011; Dekker and Varady 2011). It is, nonetheless, still modest in scope compared to the great volume of studies that deal with a single national context. Given the extent, however, to which developed nations share similar issues, but think about them and address them in very different ways, cross-national research offers valuable opportunities to illuminate salient factors in national policies as well as to make critical distinctions between what dynamics are specific to an issue compared to those that are driven by distinct national political, cultural, and social contexts. In that light, the current international research is seriously imbalanced, with extensive research being published from the United States, United Kingdom, and a few other Northern European countries, and less coming from Eastern and Southern Europe, Russia, and East Asia. While some of this imbalance may have to do with the greater number of scholars pursuing housing studies in the former countries, it also reflects the painful reality that international research in English privileges Anglophone scholars and fails to reflect the scope of national discourses conducted in other languages.

Notes

1 The severe housing deprivation rate is defined as the percentage of the population living in a dwelling that is considered as overcrowded, while also exhibiting at least one of the housing deprivation measures. Housing deprivation is a measure of poor amenities and is calculated by referring to those households with a leaking roof, no bath/shower, and no indoor toilet, or a dwelling considered too dark.

2 Notwithstanding the political furor over immigration and refugee accommodation in the United States, the magnitude of the challenge, relative to the existing population base, is substantially less than in many parts of Europe. For a vivid and distressing picture of the effect of unprecedented migration on Sweden, see Traub (2016).

3 These are the United Nations' medium variant projections and appear to be conservative. The official Japanese population projections, which anticipate a national population of roughly 97 million by 2050, suggest a rate of decline roughly 50 percent greater than that projected by the United Nations.

4 Depending on the circumstances and setting, the terms "social" and "affordable" housing may be treated as generic descriptors, or as in the case of Ireland, terms used to describe two distinct types of housing product.

References

Agence Nationale pour la Rénovation Urbaine. 2015. *Nouveau Programme National de Renouvellement Urban*. Paris: Agence National pour la Rénovation Urbaine.

Athenosy, Lucia, and Viorica Revenco. 2015. *Financial Social and Affordable Housing in Europe: The CEB's Approach*. Paris: Council of Europe Development Bank.

Calavita, Nico, and Alan Mallach, eds. 2010a. *Inclusionary Housing in International Perspective*. Cambridge, MA: Lincoln Institute of Land Policy.

Calavita, Nico, and Alan Mallach. 2010b. "Crisis and Response in Ireland: The Sudden Emergence of Inclusionary Housing." In *Inclusionary Housing in International Perspective*, edited by Nico Calavita and Alan Mallach, 169–202. Cambridge, MA: Lincoln Institute of Land Policy.

Coburn, Jesse. 2016. "Germany is Housing Refugees in Communist Ghost Towns." *Foreign Policy*, January 6.

Cortésèro, Régis, ed. 2012. *La Banlieue Change: Inégalités, Justice Sociale et Action Publique dans les Quartiers Populaires*. Lormont: Editions Le Bord de l'Eau.

Crowder, Nicole. 2014. "The Steep Decline of Bulgaria's Population in its Post-Soviet Era." *The Washington Post*, November 10.

Dekker, Karien, and David P. Varady. 2011. "A Comparison of Dutch and US Public Housing Regeneration Planning: The Similarity Grows?" *Urban Research and Practice*, 4: 123–152.

Desmond, Matthew. 2012. "Eviction and the Reproduction of Urban Poverty." *American Journal of Sociology*, 118: 88–133.

Desmond, Matthew. 2016. *Evicted: Poverty and Profit in the American City*. New York: Crown Publishers.

Dewar, Margaret, and June Manning Thomas, eds. 2013. *The City after Abandonment*. Philadelphia, PA: University of Pennsylvania Press.

Dodson, Jago. 2006. "The 'Roll' of the State: Government, Neoliberalism and Housing Assistance in Four Advanced Economies." *Housing, Theory and Society*, 23: 224–243.

Doherty, Joe. 2004. "European Housing Policies: Bringing the State Back In?" *International Journal of Housing Policy*, 44: 253–260.

Eurostat. 2015. *Housing Statistics*. Brussels: Eurostat.

Fondation Abbé Pierre. 2015. *An Overview of Housing Exclusion in Europe 2015*. Paris: Fondation Abbé Pierre.

Frame, W. Scott. 2010. "Estimating the Effect of Mortgage Foreclosures on Nearby Property Values: A Critical Review of the Literature." *Economic Review: Federal Reserve Bank of Atlanta*, 95: 1–9.

Frey, William. H. 2015. "Census Shows Modest Declines in Black-White Segregation." *The Avenue*, December 8.

Friedman, Thomas L. 2016. "What If." *New York Times*, January 20.

Gaetz, Stephen, Jesse Donaldson, Tim Richter, and Tanya Gulliver-Garcia. 2013. *The State of Homelessness in Canada 2013*. Toronto, ON: Canadian Homelessness Research Network Press.

Galster, George. 2007. "Should Policy Makers Strive for Neighborhood Social Mix? An Analysis of the Western European Evidence Base." *Housing Studies*, 22: 523–545.

Hollar, Michael K. 2014. *Understanding Whom the Low Income Tax Credit Program Serves*. Washington, DC: U.S. Department of Housing and Urban Development.

Kahneman, Daniel. 2003. "Maps of Bounded Rationality: Psychology for Behavioral Economics." *American Economic Review*, 93: 1449–1475.

Kahneman, Daniel, and Amos Tversky. 1979. "Prospect Theory: An Analysis of Decision under Risk." *Econometrica*, 47: 263–291.

Kleinhans, Reinout, and David P. Varady. 2011. "Moving Out and Going Down? A Review of Recent Evidence on Negative Spillover Effects of Housing Restructuring Programmes in the United States and the Netherlands." *International Journal of Housing Policy*, 11: 155–174.

Kwak, Nancy H. 2015. "Myth #8: Public Housing is Only for Poor People." In *Public Housing Myths: Perception, Reality, and Social Policy*, edited by Nicholas Dagen Bloom, Fritz Umbach, and Lawrence J. Vale, 175–188. Ithaca, NY: Cornell University Press.

Mallach, Alan. 2010. "France: Social Inclusion, Fair Share Goals and Inclusionary Housing." In *Inclusionary Housing in International Perspective*, edited by Nico Calavita and Alan Mallach, 203–238. Cambridge, MA: Lincoln Institute of Land Policy.

McMahon, Tamsin. 2016. "Budget Offers $2.3-Billion Boost for Affordable-Housing Measures." *Toronto Globe & Mail*, March 22.

Ministry of Business, Innovation and Employment. 2014. *Improving Housing for Māori and Their Whānau*. Wellington: Ministry of Businesses, Innovation and Employment.

Monk, Sarah. 2010. "England: Affordable Housing through the Planning System: The Role of Section 106." In *Inclusionary Housing in International Perspective*, edited by Nico Calavita and Alan Mallach, 123–168. Cambridge, MA: Lincoln Institute of Land Policy.

Musterd, Sako, and Roger Andersson. 2005. "Housing Mix, Social Mix, and Social Opportunities." *Urban Affairs Review*, 40: 761–790.

National Alliance to End Homelessness. 2015. *The State of Homelessness in the United States*. Washington, DC: National Alliance to End Homelessness.

Pittini, Alice. 2012. *Housing Affordability in the EU: Current Situation and Recent Trends*. Brussels: CECODHAS European Social Housing Observatory.

Pittini, Alice, Laurent Ghekière, Julien Dijol, and Igor Kiss. 2015. *The State of Housing in the EU 2015*. Brussels: Housing Europe.

Pyzik, Agata. 2013. "In Latvia, Riga Is Becoming a Ghost Town." *New Statesman*, August 22.

Ram, Melanie H. 2003. "Democratization through European Integration: The Case of Minority Rights in the Czech Republic and Romania." *Studies in Comparative International Development*, 38: 28–56.

Stébé, Jean-Marc. 1999. *La Crise des Banlieues*. Paris: Presses Universitaires de France.

Stryjakiewicz, Tadeusz. 2013. "The Process of Urban Shrinkage and its Consequences." *Romanian Journal of Regional Science*, 7: 29–40.

Stryjakiewicz, Tadeusz, Przemysław Ciesiółka, and Emilia Jaroszewska. 2012. "Urban Shrinkage and the Post-Socialist Transformation: The Case of Poland." *Built Environment*, 38: 196–213.

Traub, James. 2016. "The Death of the Most Generous Country on Earth." *Foreign Policy*, February 12.

United Nations Department of Economic and Social Affairs. 2015. *World Population Prospects: The 2015 Revision*. New York: United Nations.

U.S. Bureau of the Census. 2015. *American Community Survey 2015*. Washington, DC: U.S. Bureau of the Census.

Weaver, Russell, Sharmishta Bagchi-Sen, Jason Knight, and Amy E. Frazier. 2016. *Shrinking Cities: Understanding Urban Decline in the United States*. New York: Routledge.

INDEX

accessory dwelling unit (ADU) 3, 5, 248, 251–252, 257–258; financing of 265–266; and multigenerational housing 366; NIMBY campaigns against 268; physical accessibility of 261; regulatory barriers on 264–265; social accessibility of 261–262
active aging: and connectivity 232, 246, 248, 251
activities of daily living (ADLs) 248
Addams, Jane 95
Affirmatively Furthering Fair Housing 15–16
affordable housing 3; Affordable Housing and Sustainable Communities (AHSC) program 84; in California 398–402; in Canada 19; commitment to 6; cost overburden in 17, 410; creation of 306–309; and dependence on private sector 208; downside of programmatic goals 378–379; and economic diversity 376–377; European Commission on 5; and exclusionary displacement 76, 332; federal programs for 13; and general obligation bonds 396; and gentrification 297, 386; in high-opportunity neighborhoods 134–135, 136; and homeownership regime 4; as key policy challenge 375–377; and low-income families 195; and mandatory location 15; and middle-income households 379–380; necessity of 376; older adults 249; preservation of 311–312; private-for-profit developers 180, 184; private investment in 215; resistance to 397; and specified affordability levels 378; state policy on 390, 391, 396; *see also* Low-Income Housing Tax Credit (LIHTC); social housing
affordable housing complex (AHC) 77, 78; consensus-building on 85
Affordable Unemployment Program (AUP) 3
aging *see* older persons
aging in community 5
aging-in-place 5, 231–232, *233*, 240, 245–*247*; and multigenerational housing 367; and residential clusters 238
American Dream 106; broadening of concept 267–268; and homeownership 54
American Recovery and Reinvestment Act (ARRA) 2
Americans with Disabilities Act (ADA) 235
area median income (AMI) 80, 81–82, 84, 130; and LIHTC 137
Asians: homeownership rate of 105; and mortgage loans 41; and multigenerational housing 367; and multigenerational living 359
availability: unorthodox approaches to 29–32

Below Market Interest Rate (BMIR) 131; loans 3
Besson Act: and right to housing 351
Blacks/African Americans: concentration in cities 350; and *de facto* discrimination 407; and discriminatory practices 57; financial vulnerability of 45–46; homeownership rate of 54–55, 105; and mortgage loans 40, 41, 42, 43; and multigenerational living 359; and non-conventional mortgages 57; and parental wealth transfers 54, 57–63; racial barriers against 353; re-concentration in low-income neighborhoods 354; unequal returns of homeownership for 48; and subprime mortgage crisis 42–43; and wealth gap 38
Brooke Amendment 178
Brown, Graeme 19
building codes: regulation of 103
Bush, George W. 108

California: and affordable housing 398–402
C&C Development 85, 86, 87
Centre for Equality Rights in Accommodations/Social Rights Advocacy (CERA/SRAC) 19
Chicago Housing Authority (CHA) 135; and public housing 201

Choice Neighborhood Program 130, 179, 209, 326
City Beautiful Movement 95
cohousing 5; financing of 265–266; physical accessibility of 260–261; social accessibility of 261–262; and social capital 263–264
cohousing communities 248–251; origins of 256–257, 258; and popularity in Europe 248, 258; social benefits of 251
Columbus Metropolitan Housing Authority (CMHA) 210
Commission on Severely Distressed Public Housing 209
communal housing 271
community *see* neighborhoods
Community Advantage Panel Study (CAPS) 155, 157
Community Advantage Program (CAP) 46–47, 157
community-based organizations (CBOs) 78
Community Development Block Grant (CDBG) 2, 3, 83, 196; and local government 377; and state government 395, 399
community development corporations (CDCs) 3, 47, 78, 180–183, 196
community housing development organizations (CHDOs) 208
community land trusts 3
Community of Practice 376; and housing plan potential 377, 378; and programmatic goals 378
Community Options NORC 246
Community Reinvestment Act (CRA) (1977) 40; strengthening of 46–47
Consolidated Plan (CP) 398–399
continuing care retirement communities (CCRC) 244
Council of Development Finance Agencies (CDFA) 393
Cox, Michael 357, 362
Cranston-Gonzales National Affordable Housing Act 208
credit: access to 41–44

DeBlasio, Bill 378; and housing initiatives 15
Depot at Santiago 83–85
design: and police approval of social housing design 352; The Magic Mix 31–32; multigenerational housing 272–273, 277–280
Desmond, Matthew 13
discrimination: *de facto* 406–407; and Gautreaux Assisted Housing Programs 143; and housing vouchers as remedy 143–144; and mortgage loans 39–46; and spatial segregation 15
displacement 5–6; definition of 318; direct and indirect 78, 88, 295–296; exclusionary 4, 76, 78; and gentrification 294–296, 318–319; mechanisms of 303; rarity of 332; and subprime financial crisis 16; without relocation 319
Dodd-Frank Wall Street Reform and Consumer Protection Act 2, 46
Donovan, Shaun 3
Dover Kohl: and neighborhood centers 98
droit au logement opposable (DALO) 18
Duany, Plater-Zyberk: and neighborhood centers 98
Durham Central Park Cohousing 250
duty to serve principles 46

Economic Stimulus Act (ESA) of 2008 2
Elder Cottage Housing Opportunity (ECHO) 251–252
Elder Villages 238–239
Elderspirit 257
Emergency Economic Stabilization Act (EESA) 2
Emergency Homeowners Loan Program 3
Enterprise Community Partners 184
Equal Credit Opportunity Act (1974) 40
ethnic populations: spatial concentration of 351
European Commission (EC): and fair competition 222–223; and rental housing 5; and social housing systems 220–226
European Refugee Crisis 23–25
European Union (EU): and social housing 17
eviction: just cause 16, 88, 305; U.S. policy 16
Experimental Housing Allowance (EHAP) 133, 134

Fair Housing Act (FHA) (1968) 40
Fannie Mae *see* Federal National Mortgage Association
Farr, Doug 98
Federal-Aid Highway Act (1956) 349–350
Federal Housing Administration (FHA) 2, 3; establishment of 349; Home Affordable Mortgage Program (HAMP) 2; and mortgage insurance 13; Secure Program 2; Short Refinance 3
federal housing policy 2–3; and affordable housing 15–16; components of 13; decline in support 15; devolution of 207, 213; evolution of 5; expansion of 392; and public housing construction 208; and special populations 398; and tenant-based assistance programs 134
Federal National Mortgage Association ("Fannie Mae") 3; and affordable housing 16–17; and state-sponsored mortgage securitization 118
financial institutions: oversight of 44–46
forced relocation 6, 16, 86–87, 317–318; compensation for 319–320; and counseling 324; and counter-narrative 326; Netherlands 320–321; outcomes of 321–323; *see also* relocation
foreclosure: Black and Hispanic 37–38; prevention of 2–3; protective measures against 120; and subprime mortgage crisis 16

Garden City movement 358
Gautreaux, Dorothy 143
Gautreaux Assisted Housing Programs 143
Gautreaux v. Chicago Housing Authority 143
Gautreaux Consent Decree 143, 144
Gautreaux program 129, 135–136

Geddes, Patrick 95
general obligation bonds (GO) 396, 400
gentrification 4, 5; and affordable housing stock 386; avoidance of 17; classic 287–288; in Europe 290–292; and exclusionary displacement 76–77, 78, 79; and income inequality 292; and middle-class quarter making 331, 335; origins 285–287; modern 288–290; origins of 285–287; prevention of 304–305; residential 302–303, 304; soft 332, 337, 340; state-sponsored 294, 331–333
Germany: and supply-side subsidy system 223–224
ghettoization 17
Gillette, Howard 95
Ginnie Mae *see* Government National Mortgage Association
Global Financial Crisis (GFC) 1; drivers of 119; and homeownership rates 4, 111, 117; in Netherlands 334
Government National Mortgage Association ("Ginnie Mae") 16–17
Great Recession 1; and home-buying market 357; and housing 103; and multigenerational living 359; and racial wealth gap 56

Habitat for Humanity 181
Hardest Hit Fund (HHF) 3
Hartman, Chester 13
Herbert, Gilbert 94
Hispanics/Latinos: and unequal returns of homeownership 48; and financial vulnerability 45–46; and homeownership rate 105; and mortgage loans 40, 41, 42, 43; and multigenerational living 359; and subprime mortgage crisis 42–43; and sustainability of homeownership 46; and wealth gap 38
Home Affordable Foreclosure Alternatives (HAFA) 3
Home Affordable Modification Program (HAMP) 2–3
Home Affordable Refinance Program (HARP) 2–3
Home Forward (Oregon) 210
Home Investment Partnerships Program (HOME) 3, 377, 395, 399
homelessness: in California 398; and Emergency Shelter Grant 395; as global crisis 12; as major challenge 14, 407; and residential displacement 304; and settled accommodations 18–19
Home Mortgage Disclosure Act (1975) 40
homeownership 4; Black/African American 1; characteristics of 57; declining rate of 4, 117–124; and demographics 103, 105, 106–113; and distortion of access 120; as divisive axis 117; and financial returns 45, 48; and mortgage revenue bonds (MRBs) 392–393, 400; in OECD countries 103, 106–113; and parental wealth transfers 54, 55; phases in U.S. 104–105; and polarized access 121–124; promotion of 37; post-World War II 118; and racial and ethnic disparities 47–48; and Right to Buy program 112, 118; social outcomes of 39, 54; strengthening policies on 46; "under water" 16; and wealth gap 37–48; and wealth transfers 57–63
homeownership rate 1; and age 109; decrease in 54–55; in France 348; and freedom of choice 108; and ethnic diversity 109; and legal system 110; and marriage 109; monitoring of 103; policy on 4; post-World War II 104; and privatization of government-owned stock 108, 112; and taxes 110; and urbanization 109
Hope for Homeowners (H4H) program 2
HOPE VI 3, 212, 326; creation of 209; and demolition of public housing 353; and down payment grants 63; and forced relocation 317; and privatization of housing 113; and public housing 130; redevelopment of public housing 179
Housing Act (1937) 190, 191; and upward mobility 70
Housing Act (1949) 133
Housing Act (1959) 193
Housing Act (1961) 193, 194
Housing Act (1965) 194
Housing and Community Development Act (1974) 131, 194, 195
Housing and Economic Recovery Act (HERA) 2
Housing and Urban Development, U.S. Department of: and Elder Cottage Housing Opportunity (ECHO) 251–252; focus on affordability by 133; and public housing 192; and shortage of housing 14; and Small Area Fair Market Rents 69; and transformation of "public" 195
Housing Authority of Snohomish County (Washington) 210
Housing Choice Voucher (HCVP) Program 141; and freedom of choice 354; and location choices 145–146; method of delivery 142–143, 151; and nature of housing choice 149–151; and outcomes 147–149; reform of 129; and suburbanization 353
Housing Choice Vouchers (HCVs) 13, 69, 72, 132, 134, 135, 137, 201; growth of 208; and local government 377; revisions to 4, 13, 5; universal entitlement 13
housing finance agencies (HFAs) 392, 394, 395
Housing First 14
Housing New York: A Five-Borough, Ten-Year Plan 15
Housing Opportunities for Persons with AIDS 395
housing plans: analysis of 380–381; and budgetary support 380; and housing and building codes 388; as management tools 379; and overall supply of housing 380, 382, 387; and subsidized housing 382, 387; summary of local 383–385, 387; usefulness of 377, 378
housing policy: as local priority 387; and role of local government 377; and subsidies 380
housing policy challenges 406–413; economic polarization 407; population decline 407–408; refugees 407, 409–410; severe housing deprivation rate 407

housing satisfaction: drivers of 4–5; homeowners and renters 155–160, 162–163, 165–168; measures of 158–160; regression models on 161–164
Housing Services of Washington County (Oregon), Department of 210
housing stock: maintenance of 376; post-World War II France 347–349
Housing Trust Fund (HTF) 3, 6, 395–396, 401, 402
Howard, Ebenezer 96

I Love Noord 337
inclusionary housing 14; as policy 18
informal housing 12
International Centre on Housing Rights and Evictions 12
International Covenant on Economic, Social, and Cultural Rights (ICESCR) 11, 19

Jacobs, Jane 95
Janus head: 6, 317, 323, 326
Jones, Alan 362

King County Housing Authority (Washington) 210
Koch, Ed 378
Kotlowitz, Alex 144
Krier, Leon 94

La Bodeguita 87
land use 92–93; and public housing 177–178; regulation of 103
land use regulations: and affordable housing 386; and multigenerational housing 369
Lee, Ed 378
Lennar 362, 363; NextGen®Home 361, 362, 363, *364–365*, 366, 367
Lindsay v. Normet 13–14
Living Independently in a Friendly Environment (LIFE) 246
local housing authorities (LHAs) 173
Local Initiatives Support Corporation (LISC) 78, 184
London: and affordable housing 19
Los Angeles, California: as magnet for the homeless 14
Low-Income Housing Tax Credit (LIHTC) 3, 131–132, 134, 137, 208, 209, 212; and affordable housing 80, 81, 85, 173, 308–309, 310, 394–395; and construction of subsidized housing 70; and low-income renters 197, 199; reform of 4, 129, 137; and public housing authorities 215–216; and public-private ventures 191, 192; time limitations of 202

Making Work Pay 2
manufactured housing 3, 251, 252
Maracay Homes 362, 363, 366; Flex Design® 361
Massachusetts: and affordable housing 15
Medicaid 246
Medicare 246
Meritage 363
metropolitan regions in the U.S. 351
mixed-income housing 130; as federal policy 136; focus on 208–209; and HOPE VI 179; and HCV households 137; as social mixing 352; trend toward 215; and urban renewal 317
mixed-use development 86
modular housing 3
Morrison, Taylor 357
Mortgage Credit Certificates (MCCs) 393–394, 400
Mortgage Interest Deduction (MID) 3, 103, 110–111, 112
mortgage loans: and credit scores 41–44; and development of federal mortgage finance system 118; and discrimination 39–46; as end in themselves 119; and loan-to-value (LTV) ratios 111; non-conventional 57; and privatization of risk 44–45; and regulatory oversight 46; and wealth transfer 55, 56
Moving to Opportunity for Fair Housing Demonstration (MTO) 69, 73, 129, 144; establishment of 135–136; impact of 16; and neighborhood conditions 147
Moving to Work (MTW): shift from public housing 200, 201
multifamily housing: and Low-Income Housing Tax Credit (LIHTC) 394; and multifamily housing bonds (MFHBs) 392–393
Multifamily Housing Program (MHP) 401
multigenerational homebuilding 360–369
multigenerational housing 248–249; and accessory dwelling units (ADUs) 366; and aging-in-place 367; attitudes toward 5; definition of 271, 360; demographics of 367; popularity of 357–358; receptiveness to 273–275; and urban sprawl 368; and use of space 275–276, *279*
multigenerational living 357, 358–360
Mumford, Lewis 95

NAACP v. City of Yonkers 143–144
National Affordable Housing Act (1990) 196
National Commission on Severely Distressed Public Housing 178–179
National Council of State Housing Agencies (NCSHA) 395
National Foreclosure Mitigation Counseling (NFMC) 2
National Housing Trust Fund 16, 20
National Law Center for Homelessness and Poverty 12, 14
National Low Income Housing Coalition 16, 179
National Urban Regeneration Program (France) 351
naturally occurring affordable housing (NOAH) 310–311
naturally occurring retirement communities (NORCs) 245–246
naturally occurring retirement communities with supported service programs (NORC-SSP) 238, 245–246

neighborhood centers 92–99; importance of 92, 93, 97, 98; schools 95–97; size consideration 92
Neighborhood Stabilization Program (NSP) 2
neighborhoods 1; and access to high opportunity 6, 134–135, 136; and debate on socially mixed 221; and forced relocation 322–323; racially and ethnically integrated 43; and social mix strategy 353, 354; and social tension 73; and socioeconomic status 73
NeighborWorks 184
Netherlands: and affordable housing 25, 223; and asylum process 27; and broad social housing model 221–222, 225; and forced relocation 320–321; policy agreements on refugees 28–29
New Deal: and discrimination 40, 41; and homeownership programs 104
New Home Company: Life Space 361
New Jersey: and affordable housing 15
New Jersey Supreme Court: *Mt. Laurel I and II* cases 397
New York City Housing Authority (NYCHA) 15, 210; and rent stabilization 15
New Zealand: and right to housing 20
NIMBYism (Not In My Back Yard) 15
Non-Hispanic White: and homeownership rates 54–55, 105; and parental wealth transfers 54, 57–63
nonprofit housing 180–184
nonprofit housing organizations 5
Noorderparkkamer (NPK) 337, 339
Norwood, Aimee 57, 362

Oglethorpe James 95
Older Americans Act (OAA) 246
older persons: connectivity needs of 232, 233; global demographics of 243, 256; and housing quality 231–240; and inappropriate housing arrangements 231; and technology 239–240; transportation issues of 23, 244
Orange Housing Development Corporation (OHDC) 85, 86
Organization for Economic Co-operation and Development (OECD) 103

Pardee Homes 362; GenSmart Suite® 361
Penn, William 95
Perlman, Howard 362, 368
Perry, Clarence 94, 96
Planning the Neighborhood: Standards for Healthful Housing 96
policy: and duty to serve principles 46; "nonprofit-centric" 5
poverty: criticism of policy-based de-concentration efforts 70; de-concentration 4, 68–74, 134; dispersal to suburbs 354; and mixed-income housing 352; LIHTC program and poverty concentration 137; and spatial concentration 351; and structural disadvantage 144
Principal Reduction Alternative (PRA) 3
Privacy Activity Bonds (PABs) 392–393, 400
Pruitt-Igoe development: and faulty assumptions 71; violence in 195
public housing 3; and Australia 20; commitment to 192; definition of 173–174; demolition of 136, 179, 199; as economic driver 177, 190; and gentrification 294; as "housing of last resort" 193; isolation of 17; and low-income families 130, 350; and older persons 193; opposition to 178, 190; privatization of 5, 103, 189–202; and Section 202 program 131; and Section 236 program 131; and Section 818 program 131; and segregation 178; as social housing 5; and Turnkey system 194, 195; upgrading of 179–180; *see also* Federal-Aid Highway Act (1956)
Public Housing Act (1937) 173
Public Housing Assessment System (PHAS) 217
public housing authorities (PHAs) 5, 130; and affordable housing 207; enabling legislation *213*; evolving nature of 216; and new operating environment 209–212; public housing maintenance 208; and reconceptualization 191; and sources of 213–215
public policy: and homeownership rate 103, 107–108, 110–111
public housing 3; and Australia 20; commitment to 192; definition of 173–174; demolition of 136, 179, 199; as economic driver 177, 190; and gentrification 294; as "housing of last resort" 193; isolation of 17; and low-income families 130, 350; and older persons 193; opposition to 178, 190; privatization of 5, 103, 189–202; and Section 202 program 131; and Section 236 program 131; and Section 818 program 131; and segregation 178; as social housing 5; and Turnkey system 194, 195; upgrading of 179–180; *see also* Federal-Aid Highway Act (1956)
public-private partnerships (PPPs) 191, 194

Qualified Allocation Plan (QAP) 399–400
Quality Housing and Work Responsibility Act (QHWRA) 142, 199; and mixed finance 209

Ramirez, Saul 211
redevelopment: and spatial restructuring 79
redevelopment agencies (RDAs) 401–402
redlining: long-term repercussions 47–48; and mortgage loans 41, 42
refugees: Europe 3–4; European response to 17–18; and Netherlands 25, 26–32
relocation: as displacement 318; *see also* forced relocation
rent burden: and gentrification 80–81
rent control 104; and displacement 306, 311; and social housing in Netherlands 324–353
rent habituation: Netherlands 325

Rental Assistance Demonstration (RAD) program 5, 113, 179, 189; and private investment 215, 216; and public housing 201
rental certificates 195; and Housing Choice Vouchers 196
rental housing 3, 4; growth of 120–121; and Housing Choice Voucher Program (HCVP) 141–151; preservation of 309–310; privatization of 118; social 18, 112–113; and spatial exclusion 79; and supply-side programs 130–132; subsidized 129–138
renters 3; federal standard of affordability 13; and freedom of choice 108; and housing satisfaction 161–165; gentrification 294, 295, 302–304; and subprime mortgage crisis 16; and taxation 103
residential mobility programs 16
right to housing 3, 4, 11, 13; European countries 17; explicit 18; fulfillment of 184; as goal 20; implicit 19; international declarations on 11–12; legal 12; in Netherlands 26, 28; opposition to 13; reality of 11–12
Riis, Jacob 96
Roosevelt, Franklin D.: on public housing 177

San Francisco, California: as magnet for the homeless 14
Santa Ana, California 4; and displacement 78
Santa Ana Collaborative for Responsible Development (SACReD) 85
School as a Factor in Neighborhood Development, The (Perry) 96
Scotland in Crisis 19
Second Great Migration 193
Second Mortgage Program (2MP) 3
Section 202 program 181, 193
Section 236 program 194
Section 8 construction program 195; and "lucrative profit potential" 196
Section 8 Housing Choice Voucher 174; Management Assessment Program (SEMAP) 217; New Construction/ Substantial Rehabilitation program 3, 131; Program 132: growth of project-based 208; *see also* Housing Choice Voucher (HCVP) program
segregation: persistence of 43–44
senior citizens *see* older persons
Shea Homes 366
Shelter Scotland 19
Silver Sage Village 249–250, 257
single-family housing: as ideal 358; shift in design 357–358; and suburbanization 349; *see also* American Dream
Small Area Fair Market Rents (Small Area FMRs) 69, 72, 73
social housing 4–5; definition of 173–177; demolition of high rises 351; and Europe 220–226; and European Union 17; focus on 354; and forced relocation 320–321; in France 348–349; limitation in U.S. 408; mandatory 352; and Netherlands 28; and police approval of design 352; pre-World War II France 348; and rent regulation 324–325; redevelopment of 6; role of 5; sale and liberalization of 335
social housing organizations: Europe 207
social integration: importance of housing to 32
social regeneration: and symbolic politics 330–333
social safety net: erosion of 4; and privatization of risk 44–45
Social Security Act (SSA): and discriminatory practices 57
St. Louis plan 95
state and local housing policy 14
state housing policy: and Consolidated Plan (CP) 398–399; and federal programs 390; and housing finance agencies (HFAs) 392; and housing trust fund (HTF) 395–396; and power to mandate 391–392; and Private Activity Bonds (PABs) 392–393; and state housing tax credits 396; and state initiated programs 395; and use of federal programs 400
Stein, Clarence 95
subprime mortgage crisis 1; and "cumulative disadvantage" 42; and displacement 16; emergence of subprime mortgage market 119
suburbanization 6; in post-World War II France 348; in post-World War II U.S. 349–350; in pre-World War II France 347; and single-family housing in U.S. 349, 350; and social housing in France 350
superstar cities 375
Supplemental Nutritional Assistance Program (SNAP) 2
Supreme Court, U.S.: *Hills v. Gautreaux* 143; on right to housing 13–14; *Texas Department of Housing and Community Affairs v. Inclusive Communities Project* 46

taxes: and mortgage interest tax deduction 47, 119; tax relief 2
Taylor Morrison 361–362, 363
tenant rights 332; strengthening of 304–305
Ten Year Capital Plan for Housing (1986) 378
There Are No Children Here (Kotlowitz) 144
Toll Brothers 357; Gen Suite 361
Trailview Development Corporation 257
Troubled Assets Relief Program (TARP) 2
Trudeau, Justin 19
Trudeau, Pierre 19
Truly Disadvantaged, The (Wilson) 144

Uniform Relocation Act 320
United Nations Center for Human Settlements (UNCH) 106
Universal Declaration of Human Rights 11
universal design (UD) 3, 260
Urban Redevelopment Demonstration (URD) 209
urban renewal: and social mixing 351, 352

VDP-buurt 333–337
Veterans Affairs (VA) mortgage program 349
Veterans Home Loan program 104, 105
Village Movement 245

Walker v. HUD 143
Walsh, Martin 378
Ward, Edward 96, 98
wealth gap 4; average incomes 61; Black/African Americans and 4, 37, 38, 39; drivers of 37; income and earnings 56–57; and mortgages 54–56; non-Hispanic White and 4, 37–38; and parental transfers 54, 57–63; widening of 44
Wilson, William Julius 144
Wooden Floor, The 86
Wright, Henry 95

Ymere 333–334, 335, 336, 337

zoning 3; adoption of inclusionary 377; and affordable housing 15; cohousing and ADUs 264–265; and economic discrimination 397; exclusionary 354; inclusionary 308; and restrictive covenants 57; and spatial redevelopment 79; and superstar cities 375

For Product Safety Concerns and Information please contact our EU representative GPSR@taylorandfrancis.com Taylor & Francis Verlag GmbH, Kaufingerstraße 24, 80331 München, Germany

Batch number: 10722200

Printed by Printforce, the Netherlands